MULTIVARIATE ANALYSIS IN VEGETATION RESEARCH

DEDICATION

To the memory of
Ferenc Tuskó
Forester and Educator

MULTIVARIATE ANALYSIS IN VEGETATION RESEARCH

László Orlóci

University of Western Ontario

SECOND EDITION

DR. W. JUNK B.V. – PUBLISHERS – THE HAGUE – BOSTON 1978

ISBN 90 6193 567 9

PREFACE TO THE SECOND EDITION

The favorable reception of the book by users has lead to the decision to produce the second, enlarged edition in the same spirit as the first. The general framework of six chapters, Glossary and Appendix is retained, but with rearrangements of sections to accommodate new materials, including many new references. The references are now collected in a single list in the back with chapters of occurrence indicated.

My wife, Márta, gave invaluable help in the preparation of the manuscript and proofreading. Mmes. Patricia Horn and Stefani Tichbourne typed the manuscript and Miss Evelyn Hamilton assisted with graphs and references. Miss Victoria Jackson helped with the proofs. Mr. Peter Fewster contributed to revisions of the Appendix. I would like to thank them for their help.

L. Orlóci

PREFACE TO THE FIRST EDITION

It seems that vegetation ecology has become increasingly dependent on the use of statistical and other mathematical methods in the solution of its problems. Such methods obviously represent an invaluable tool in the hands of the trained ecologist who can clearly understand their intrinsic limitations.

There is growing interest among ecologists in the methods of multivariate analysis including many conventional and not so conventional procedures. This is not in the least surprising, considering that ecological data are normally multidimensional and that the analysis of such kind of data is most profitably accomplished *via* a multivariate method. Such a method of analysis, and data, are set apart from their univariate counterparts by the important fact that they can provide information about variate interactions, or correlations in general, which the univariate methods and data cannot.

The present book addresses multivariate analysis from the viewpoint of the plant ecologist experimenting with new directions of data analysis. It presents materials which have been used in a graduate plant ecology course given for students with some mathematical and statistical training in their background. The book's objectives are two-fold. On the one hand, it should provide an introduction to concepts and procedures that need to be known about the methods before applications. On the other hand, it should also serve as a source for worked examples and useful computer routines. With these dual objectives in mind the contents are offered for class use in courses with an interest in data analysis, and also, to research workers interested in broadening their experience with data analysis.

The contents are presented in a monographic style, arranged in six chapters, subdivided into sections and paragraphs, a Glossary, Indices, and an Appendix. Chapter 1 discusses steps prior to data collection and analysis. Chapter 2 represents resemblance functions and Chapter 3 treats ordinations. Chapter 4 is concerned with classifications, Chapter 5 describes different identification methods, and Chapter 6 gives the concluding remarks aimed at helping the reader to choose between the different methods. The Appendix contains computer programs. Each chapter is followed by a list of references.

Because of limitations imposed on the scope by various conditions, omissions must be accepted. The most noted of these are the family of open-model factor analysis which R.B. Cattell considers in detail in a review article 'Factor analysis: an introduction to essentials', *Biometrics* 21: 190-215, 405-435 (1965), and the procedures of overlapping clustering which N. Jardine & R. Sibson describe with great eloquence in their book 'Mathematical Taxonomy', London, Wiley (1971). Graph theory clustering represents another, although not a serious omission, considering that the method of single linkage clustering is dealt with in some detail in Chapter 4. Regarding details of the relevant aspects of graph theory clustering the reader may consult G. Estabrook's article 'A mathematical model in graph theory for biological classification', *J. theoret. Biol.* 12: 297-310 (1966) and references therein.

To complete this *Preface* I would like to record my thanks to my colleagues Theodore J. Crovello, Robert C. Jancey, Robert P. McIntosh and James B. Phipps for helpful suggestions after kindly reading the draft, to Miss Catherine Pitt for help with the program listings in the Appendix, and to Miss Joanne Lemon and Mrs. Gail Roney for the typing. I have done much of the preliminary work on this book in the Department of Botany at the University of Hawaii during my sabbatical year. My thanks are due to Noel P. Kefford and Dieter Mueller-Dombois at U.H. for a pleasant and stimulating environment in their department.

L. Orlóci

CONTENTS

Chapter 1

INTRODUCTION

In this chapter we discuss ideas and present facts which will help to set the stage for multivariate analysis. We shall identify possible objectives to be pursued, discuss the advantages of formal methods, as compared to the informal techniques which characterize conventional vegetation ecology, and consider a number of decisions that must be made about variables, sampling design, etc., before the collection and analysis of the data may begin. In the discussions we make the point repeatedly: the methodological decisions at different stages in a vegetation study are interrelated.

1.1 Objectives

There are characteristics of the vegetation which readily appear on first sight to the trained observer. For example, a brief field survey may suffice to discover that the composition of the vegetation changes with altitude in a mountainous area, or to verify that species occur in zones with changing elevation on a flood plain.

Properties of this sort exemplify descriptive characteristics readily apparent without any statistical or other mathematical analysis of precise measurements. But when interests turn to what Greig-Smith (1964) describes as the finer details, the ecologist may have to rely on formal methods of sampling, precise measurements and powerful methods of data analysis to be able to reveal trends and classes of variation in the vegetation. When the ecologist relies on such formal methods of sampling and analysis, the objectives of a vegetation study can be broadened and the details investigated can be refined.

In a vegetation study the objectives tend to be concerned with:

1. Spatial arrangements of individuals or species on the ground or in the vertical profile of the community.
2. Trends and classes of variation in the similarity or dissimilarity relations of species or entire stands.
3. Population processes affecting spatial or temporal patterns in an area.
4. Vegetation response to environmental influences, and the dynamics in such a response.

These properties represent grossly different methodological problems but with one important common aspect: they are not normally amenable to *exact* determination, they have to be *estimated* via a limited sample. The estimation may be approached through different avenues – experimental, observational, deductive (cf. Goodall, 1970a) – depending on scale and objective.

The contents of this book are concerned with methods to reveal information about trends and classes of variation in vegetation data and to establish the connections of this variation to environmental influence. The cases discussed are those in which the available evidence is observational. This means that field surveys rather than controlled experiments supply the data for analysis. With book contents so limited, the analysis of spatial arrangements (pattern) will not be discussed. We refer in this connection to excellent expositions by, e.g., Greig-Smith (1964), Pielou (1969a, 1974), Walker, Noy-Meir, Anderson & Moore (1972), Kershaw (1973) and Fekete & Szőcs (1974). The analysis of population processes will not be discussed either. These are treated in great detail by Bartlett (1960), Watt (1968), Pielou (1969a, 1974), Emlen (1973) and, among others, by the contributors to volumes edited by Patten (1971, 1972, 1975).

With the class of methods to be considered identified, we should make a distinction between two broad families: those which handle continous variation and those specifically designed to analyse discontinuous variation. The former are the so called *ordination methods* and the latter the *clustering methods*. These two families may differ sharply in the formulation of algorithms but not so sharply in the objectives for which they are used in vegetation studies. As may happen in many applications, ordinations achieve classificatory objectives and clustering methods derive scales to serve as ordination co-ordinates. We shall discuss these in the sequel.

What exactly is meant by continuous or discrete variation, ordination, clustering? These are terms which will be explained in the following text. At this point, we refer to meaning established in common ecological usage and to concise definitions in the Glossary.

1.2 Formal methods

Vegetation ecology is still changing very rapidly. The early phytosociological approach, in which the *art* of intuitive classifications on the basis of *preferential* sampling occupies a central position, is no longer predominant. A new approach has emerged. This has lead in recent years to a massive introduction of formal methods in vegetation research, including sampling and experimental design, data analysis, and statistical inference. Experience suggests that vegetation ecology benefits greatly from the use of such methods. Firstly, these methods require the user to be precise and explicit in the formulation of the problem, foreseeing the implications for the type of data to be collected, sampling design used, and method of data analysis and statistical inference chosen. This is in direct contrast with the informal phytosociological methods in which such considerations have only limited potential because they operate on the basis of personal preference and without mathematical foundations. Secondly, the formal methods require from the user uniformity and consistency in the implementation of the operational rules and decisions throughout the entire process of data collection, analysis, and inference. In contrast, the informal methods do not use nor can they offer such a uniformity and consistency. Thirdly, the formal methods, because of their uniformity and consistency, naturally lend themselves to computer processing. In this respect they have certain logistic advantages over the informal methods. But beyond the logistic advantages, they have also qualitative advantages due to the fact that they can promote interdisciplinary cooperation between the ecologists, the programmers and systems men, and other groups of specialists from whom a feedback of useful ideas and methodological advances can be expected.

1.3 Steps prior to data collection

Before the collection and analysis of data can begin the ecologist is expected to have defined the problem and made a number of initial decisions consistent with the information available. It is not impossible, however, that as new information comes to light the problem will appear more clearly and may have to be redefined. The process of defining and redefining the problem is suggestive of a feedback system of the kind discussed by Tukey & Wilk (1966).

We discuss in the following sections the preliminary steps, including:

1. Statement of the problem.
2. Choice of mathematical model.
3. Determination of essential variables.
4. Selection of sampling design and method of data analysis.

Only after completing these steps can the implementation of data collection and data analysis begin, followed by hypothesis testing, ecological interpretations, or even further analyses when required. It should be quite clear to the user that steps 1 to 4 are sequentially related. This implies that, for example, without an explicit statement of the problem no intelligent choice of a mathematical model, variables, sampling design, or method of data analysis can possibly be made. Because of the interdependence of the steps, decisions at each step are expected to limit the user's freedom of choice in the subsequent steps. Dale (1968) and Crovello (1970) discuss similar points in connection with general procedures.

1.4 Statement of the problem

It should be recognized that the success of the analysis depends on how clearly the problem can be defined. In any case, the formulation of the problem should precede data collection, and should preferably involve the statement of simple hypotheses (cf. Goodall, 1975). For example, the statement,

H_0: *Given vegetation types are indistinguishable in terms of soil moisture regime,*

is such that it already indicates the possible mathematical model, variables, sampling design, method of data analysis, etc. Once sufficient data have been obtained, and the analysis performed, H_0 can be statistically tested and accepted or rejected.

Although the specific aspects of a problem may be a matter of local considerations, we can make generalizations about classes of possible problems. We may be concerned with different properties of vegetation *structure* or with specific aspects of *function*. It is quite conceivable, for instance, that the solution of a structure oriented problem will require variables, sampling designs, and methods of data analysis which differ radically from what would be considered appropriate when the problem is concerned with function.

Vegetation structure is variously defined as the manner in which species or other components are arranged on the ground or in the vertical profile. In these terms, structure may mean a species structure, biochemical structure, trophic structure, etc. More relevant to the present discussions, structure may also be defined as an abstract property, such as the resemblance of vegetation stands measured as some objective function of their composition. The principal value of the definition of structure in conjunction with the concept of resemblance is related to the role which such a definition plays in data analysis.

Having defined structure, function may be defined simply as the dynamics in the changing structure. The changing resemblance structure under successional changes of the vegetation can be mentioned as an example.

1.5 The mathematical model

Data analysis operates through mathematical models. These models should be visualized not so much as some geometric construct, which may or may not be a relevant analogy, but rather as a set of statements in a formal algebraic language. These statements specify the steps that have to be taken to obtain a result. It is convenient to categorize mathematical models such as:

1. Deterministic.
2. Deterministic with a random component.
3. Stochastic.

A *deterministic* model assumes that population size is small enough to allow all of its elements to be located and measured; the objectives amount to the best possible *approximation* of given population values. The problem of *estimation*[1] does not arise. Where a deterministic model incorporates a random component for sampling or experimental errors, it is variously described as an estimative, statistical or probabilistic model, implying that the objectives are concerned with an estimation of population values. Estimation becomes necessary because not all elements of the population can be located and measured. Lacking a complete enumeration of the elements, the population values cannot be exactly determined, but rather, they have to be estimated on the basis of a limited random sample. The term *stochastic* is meant to indicate probabilistic models of a specific kind. Common to the different stochastic models are (a) a random process, (b) a probability law which specifies the frequency with which certain events occur, and possibly, (c) a memory device which keeps track of the events as they occur.

[1] The term *estimation* should be clearly distinguished from *approximation*. *Estimation* is a statistical problem, *approximation* is a problem in measurement and computation.

The application of a mathematical model may make little sense, however, if it is not embedded in a broader procedure characterized by strict internal consistency. It will be useful to consider some examples in this connection.

1. Assume that a procedure consists of three major steps:
a. Sampling.
b. Data analysis.
c. Inference.

The ecologist using a procedure of this sort should clearly see that the individual components cannot be considered in isolation (Greig-Smith, 1971a). The sample must provide suitable input for analysis which in turn must produce results that fit the requirements of the chosen model for inference.

2. Consider next a procedure incorporating a deterministic model:
a. Species performance is described within contiguous quadrats which completely cover the survey site.
b. The quadrats are classified into groups.
c. The groups are mapped on the ground as vegetation types.

Clearly, in this particular case the problem of statistical estimation does not arise. The reason is that the survey site is completely covered by quadrats, all of which are described in terms of species performance, and all descriptions are used, in turn, to produce a vegetation map for the locality.

3. Considering that estimation of certain population values is a common objective in vegetation studies, the following set of statements have general relevance:
a. Determination of species performance within randomly sited quadrats.
b. Computation of sample estimates for mean and variance.
c. Testing the hypothesis of differential performance between species.

The objectives are estimation and hypothesis testing. These are consistent with the procedure because random sampling produces suitable data for computation of the different estimates, which then can provide a suitable basis for testing hypotheses about species differences in the population.

4. An example for a procedure incorporating a stochastic model is as follows:
a. Determine birth and death rate in a population and designate these by symbols μ and ν.
b. Assume that birth or death is a random process, population growth is exponential, and birth and death rate are constant.
c. Use function $N_t = i \exp (\mu\text{-}\nu)t$ to obtain an estimate of population size at time t when population size at time zero is i.

Because of the stochastic properties, i.e. population size in time being under the influence of random events, the change in population size has to be regarded as a random variable which can be predicted only within certain limits of accuracy (cf. Pielou, 1969a).

In the present book we shall limit the discussions to deterministic models with or without a random component. In this respect, we define a broader scope for data analysis than is usual in statistical monographs where the model is always assumed to have a random component. In any case, regardless of the mathematical model, we must find a suitable definition of *population* before any vegetation survey can begin. A definition may suggest, for instance, that the population consists of discrete units represented by stands of vegetation delimited as quadrats. Another definition may stipulate that the population units are plant individuals of the same species. Still another may specify life-form affiliation, physiognomy, etc., as criteria for recognizing populations. Other definitions are readily conceived. The important thing about a suitable definition is that it allows the user to identify and locate population units without ambiguity.

So far our definitions have assumed that the population units are concrete objects. For the purpose of data analysis, a population unit may be regarded as an observation (simple or multiple) on one or several characteristics of an object. Counts or other measurements of species within quadrats represent an example in which the quantity of species i in quadrat j is the i,j th population unit X_{ij}. In the case of multiple observations, the jth unit is given by $\mathbf{X}_j$ which signifies a p-valued vector, such as the jth column vector in matrix **X** below, whose elements are simple observations $X_{1j}, \ldots, X_{pj}$. This matrix describes a population in terms of N p-valued vectors. A population whose elements can be represented by such vectors is said to be a p-variate population. A subset n of the N population units constitutes a *sample*.

$$\mathbf{X} = \begin{bmatrix} X_{11} & X_{12} & \ldots & X_{1j} & \ldots & X_{1N} \\ X_{21} & X_{22} & & X_{2j} & & X_{2N} \\ \cdot & \cdot & & \cdot & & \cdot \\ X_{i1} & X_{i2} & & X_{ij} & & X_{iN} \\ \cdot & \cdot & & \cdot & & \cdot \\ X_{p1} & X_{p2} & & X_{pj} & & X_{pN} \end{bmatrix}$$

1.6 Variables[2]

Formal vegetation descriptions apparently began in the mid 19th century. It was not, however, until the 1930s or perhaps even later in the 1950s that well-reasoned criteria started to appear to establish connections between data analysis and choice of vegetation variables. We should refer in this connection to the early work on analysis of species pattern (e.g., Greig-Smith, 1952a, b), stand classification (e.g. Goodall, 1953; Williams & Lambert, 1959) and stand ordination (e.g. Cottam, 1949; Curtis & McIntosh, 1950, 1951). The theoretical aspects and practical implications in selecting vegetation variables are now reasonably well understood (cf. Greig-Smith, 1957, 1964; Lambert & Dale, 1964; Goodall, 1970a; Crovello, 1970). There are, nevertheless, new questions to be answered concerning the ecological informativeness of variables, their performance under different methods of data analysis, the constraints which they incorporate that control variation, etc., to which Smartt, Meacock & Lambert (1974) draw attention.

[2] The terms *variable* and *variate* are used with the same meaning in this book.

The ecologist should choose variables which are directly relevant to the problem, and also, which can be handled by the available facilities in the sampling and data analysis. The variables must be meaningful regarding the formulations in the mathematical model, sensitive to changes in the controlling factors, but at the same time, sufficiently buffered against minor influences that could obscure important trends in their variation. The variables, most commonly used in vegetation surveys, include two major types:

Continuous	*Discrete*
Yield	Density
Basal area	Frequency
Cover[3]	Presence
	Cover[4]

While the quantities of variables would best be determined by actual measurements, all too often they are subject to visual estimation based on the use of various arbitrary scales (cf. Becking, 1957). Certain less often used variables include physiognomy, life-form, phenology (periodicity), leaf morphology, seed dispersal type, etc. (cf. Knight, 1965; Ellenberg & Mueller-Dombois, 1966; Fosberg, 1967; Knight & Loucks, 1969; Webb, Tracey, Williams & Lance, 1970; Goodall, 1970a; Mueller-Dombois & Ellenberg, 1974) which are scored irrespective of species affiliation or affiliation to some higher level taxa.

In most applications the variables describe the performance or other properties of species. There are examples, nevertheless, in which the basic taxonomic unit is not the species but *genus, family* or *order* (e.g. van der Maarel, 1972). One invariant aspect in all applications is that once the basic taxonomic unit is chosen, be it a species, genus, or other, all individuals identified as members of one unit are treated alike. It is quite conceivable that the taxonomic unit would be chosen at a subspecific level, or defined differently such as the biological species (Mayr, 1963; Sokal & Crovello, 1970; Sneath & Sokal, 1973), if the problem of identification were resolved.

We categorize variables according to domain and scale. The domain of *continuous variables* consists of all values on the real number axis. The domain of *discrete variables* is confined to positive integer values or zero. In vegetation work, the domain of a variable is almost always truncated. Tree height (a continuous variable) is a case in point. Its domain consists of an infinite number of values but within some natural range. At the lower end, tree height is just larger than zero. At the upper end it may be somewhere in the vicinity of perhaps 100 m. Trees with zero height do not exist. The same can be said of trees with 500 m height.

The number of plant individuals in a given quadrat exemplifies a realization of a discrete variable. The domain of counts is of course truncated due to the impossibility of placing more than a given number of individuals in a quadrat of finite size. Where the domain of a variable is limited to two values (0/1) it is said to be a *binary variable*. Species presence in quadrats is an example.

[3] Defined as an area or length of line intercept.
[4] Determined from point sampling.

Different variables imply different scales of measure. We commonly distinguish between four major types (Torgerson, 1958; Anderberg, 1973; Zar, 1974):

1. *Nominal scale.* This scale is applied when, for instance, life-form is scored. The algebraic operations meaningfully performed on nominal measurements are identity (=) or non-identity (≠). If, for instance, A is a geophyte then B = A implies that B is also a geophyte.
2. *Ordinal scale.* The Braun-Blanquet cover/abundance scale is probably of this type. (The Mohs hardness scale is definitely so.) If one species scores X = 2 on an ordinal scale in one quadrat and Y = 4 in another, the operation X ≠ Y or Y > X is quite meaningful. However, operations such as Y–X or X/Y are meaningless.
3. *Interval scale.* This scale lacks a natural zero point. The Celsius temperature scale is an example. The operations =, ≠, < and – are all defined for measurements on this scale.
4. *Ratio scale.* An interval scale with a natural zero point is said to be a ratio scale. The commonly used metric scales for length, area, volume and weight are of this kind. The Kelvin temperature scale is also of this kind. The ratio scale possesses all properties that the nominal, ordinal and interval scales possess plus the operation that expresses relative magnitude (/).

Variables that are measured on the same scale have values in identical units. A set of such variables is said to be *homogeneous* for scale.

The ecologist should consider a combination of different criteria in any a priori evaluation of the potential relevance of given variables. The order in which these criteria are presented here is not intended to indicate relative importance:

1. *Commensurability.* If two or more variables measure qualitatively comparable properties in identical units they are said to be commensurable. Commensurability is a critical criterion whenever two or more variables are analysed simultaneously. It should be noted in this connection that because qualitative commensurability implies value judgment by the user, it is not possible to lay down strict rules for selecting variables for simultaneous analysis.

The foregoing definition of commensurability is obviously restrictive and a weaker definition may be preferred. This could stipulate possession of a common unit of measure as a sufficient criterion for commensurability. Vegetation data often describe heterogeneous sets of variables for which the data elements are in different units. Such is the case in applications when species of the different strata in a community are measured in different units. Frequency for herbaceous plants, density for shrubs (or tree species in the shrub layer) and basal area for trees over a specified size are examples of variables that are often found in the same data set. When such is the case, and if it is necessary to establish commensurability in such a mixture, different avenues of approach are open. One possibility is to analyse frequency, density and basal area separately. However, following such an analysis the recombination of the results is problematic. Another possibility is to transform to a common (unitless) scale *via* standardization. This can be accomplished according to $(X_{ij}-\bar{X}_i)/S_i$ in which X_{ij} is the jth measurement on the ith variable, $\bar{X}_i$ is the mean of the ith variable and S_i is the standard deviation. We shall discuss equalizing transformations of this sort in greater detail in the sequel. We should however note at this point that while scale differences can be removed by standardization, such a manipulation of the data does not render qualitatively different variables more comparable.

Commensurability may suffer not only from the lack of a common scale, but also from the excessive size differences even if the unit of measure is the same. Such differences may be diminished by certain monotone transformations (Anderberg, 1973) such as the logarithm, square root, etc., with varying degrees of distortion of information content. Noy-Meir, Walker & Williams (1975), Chardy, Glemarec & Laurec (1967) and Noy-Meir & Whittaker (1977) make specific points about these aspects. It should in any case be remembered that transformations can easily alter the information content in the data, even to the extent that it changes the outcome of the analysis.

2. *Additivity.* Commensurable variables are additive if they are also independent. Independence may mean that they have zero covariances, lack mutual information, or do not influence each other's probability of assuming specific states or quantities. Independence may be determined by statistical analysis of actual data, or in some cases, on the basis of reasoning from basic principles. It is a direct consequence of non-commensurability that variables such as density, basal area, or frequency are not additive (Goodall, 1970a). Because of the non-additivity, the use of composite indices of density, basal area and frequency cannot be recommended.

3. *Dependence on the sampling unit.* Species presence and frequency represent clear examples of variables dependent on the sampling unit. It is absolutely essential, therefore, to specify the number and the size of the sampling units when presenting presence or frequency data. Density, cover,[5] basal area and yield represent examples for the opposite.

4. *Seasonal variation.* Certain variables may undergo substantial variation related to the development and growth of new organs, or the dying off of old organs or entire plants in seasonal cycles. Species cover and species presence are good examples. Seasonal variation may be a serious problem in long-range vegetation surveys.

5. *Relativity.* Most variables can be meaningfully transformed into relative quantities, as percentages or proportions. But not so frequency, cover (from point estimates), or species presence, which are quite meaningless as relative quantities when compared between samples in which the size and number of sampling units are different (Goodall, 1970a).

6. *Distribution properties.* In many methods of data analysis the probability distribution of a variable is assumed and the success with the method is dependent on how closely the assumed distribution approximates the true distribution in the sampled population. This piece of information is, however, hard to come by, as Goodall (1970a) points out, if the distribution is influenced by changing ground pattern in the vegetation or by the size of the sampling units.

Empirical and other formulations were offered for the sampling properties of different variables by Cottam & Curtis (1949), Goodall (1952a), Greig-Smith (1957, 1964), Poissonet, Poissonet, Godron & Long (1973) and Pielou (1974) among others. We shall limit the discussion to probability distributions commonly encountered in

[5] As a continuous variable.

ecological work including the *normal, binomial, Poisson, Poisson-Poisson* and *Poisson-logarithmic.* The *normal* or *Gaussian* distribution often represents a reasonable approximation for the probability distribution of continuous variables such as length, area, volume, or weight. The normal distribution may be used also as the limiting case for the distribution of discrete variables such as density and frequency. Therefore, it is logical to expect that the normal distribution is one with which the ecologist may be faced in many vegetation surveys. The normal distribution has two parameters, the mean μ and the variance σ^2. In terms of these parameters the normal *probability density (co-ordinate) function* can be written such as,

$$f(X) = B \exp - \frac{(X-\mu)^2}{2\sigma^2}, \quad -\infty < X < \infty \tag{1.1}$$

which measures the height of the normal curve at point X. In this formula B represents a scale factor. The *cumulative distribution function,*

$$F(X) = B \int_{-\infty}^{X} \exp - \frac{(Y-\mu)^2}{2\sigma^2} \, dY \tag{1.2}$$

measures the area under the normal curve to the left of point X. When B is defined as $1/(\sigma\sqrt{(2\pi)})$, $F(X)$ is characterized by the following properties:

$F(X) = 0$ if $X = -\infty$,
$F(X) = 1$ if $X = \infty$ and
$F(X_1) \leqslant F(X_2)$ if $X_1 \leqslant X_2$.

Formulae (1.1) and (1.2) describe the univariate case where X represents a single random variable associated with observations $X_1, ..., X_N$. In the multivariate case, X is defined as a p-dimensional random variable,

$$(\mathbf{X}_1 \ldots \mathbf{X}_p)$$

whose elements are undimensional random variables, e.g. yield of each of p-species in a community. The ith such variable is associated with the observational vector,

$$\mathbf{X}_i = (X_{i1} \ldots X_{iN}),$$

with probability density function $f(\mathbf{X}_i)$ and distribution function $F(\mathbf{X}_i)$. The joint density of p independent normal variables (those whose covariances are zero), with the jth observational vector $\mathbf{X}_j = (X_{1j} \ldots X_{pj})'$, is a function of their means $\mu_1, ..., \mu_p$ and variances $\sigma_1^2, ..., \sigma_p^2$,

$$f(\mathbf{X}_j) = B \exp - \tfrac{1}{2} \sum_i \frac{(X_{ij}-\mu_i)^2}{\sigma_i^2}, \quad i = 1, ..., p \tag{1.3}$$

where quantity B accords with

$$B = \frac{1}{\sigma_1 \ldots \sigma_p \, (2\pi)^{p/2}} \, .$$

Formula (1.3) gives the probability density at a point $\mathbf{X}_j = (X_{1j} \ldots X_{pj})'$, whose ith co-ordinate X_{ij} is an observed value on variable i of individual j. For any p correlated normal variables with population mean vector

$$\mu = (\mu_1 \ldots \mu_p)',$$

and population covariance matrix

$$\Sigma = \begin{bmatrix} \sigma^2_{11} & \cdots & \sigma^2_{1p} \\ \cdot & \cdots & \cdot \\ \sigma^2_{p1} & \cdots & \sigma^2_{pp} \end{bmatrix},$$

the p-variate density function can be written as

$$f(\mathbf{X}_j) = \frac{1}{|\Sigma|^{1/2}\,(2\pi)^{p/2}} \exp - \tfrac{1}{2} (\mathbf{X}_j - \mu)'\,\Sigma^{-1}\,(\mathbf{X}_j - \mu) \tag{1.4}$$

where Σ^{-1} implies the inverse of Σ and $|\Sigma|$ the determinant. The distribution function is given by the integral,

$$F(\mathbf{X}) = \int_{-\infty}^{\infty} \cdots \int_{-\infty}^{\infty} f(\mathbf{X})\, d\,\mathbf{X}_1 \ldots d\,\mathbf{X}_p = 1 \tag{1.5}$$

In the case of a random sample of n individuals the population parameters μ_i, σ_i, Σ are replaced by their sample estimates $\bar{X}_i$, S_i and $\mathbf{S}$. It is often questioned whether a multivariate normal distribution can represent an appropriate assumption for vegetation data under actual survey conditions. For p random variables, which reflect the influence of many factors acting on the variables in a completely random manner, the *central limit theorem* suggests that their joint distribution tends toward multivariate normality. This observation should, in a way, encourage a freer hand in the use of multivariate statistical methods under a relatively broad range of circumstances. We note that tests for multivariate normality exist (cf. Reyment, 1971 and reference therein), but these can be computationally very tedious.

The *binomial* distribution is often used as a reference distribution for discrete variables on which the observations can be associated basically with two possible outcomes. As an example we should consider species frequency, determined by inspection of n random sampling units within quadrats, counting the units occupied by a particular species. The resulting distribution has a single parameter p at any given value of n. A term in this distribution,

$$P(k) = b_n(k) = \frac{n!}{k!\,(n-k)!}\, p^k q^{n-k} \tag{1.6}$$

(representing a general term in the binomial expansion $(p+q)^n$), gives the probability of finding k occupied sampling units in n random trials when the probability of finding one occupied unit in a single trial is p and the failure $q = 1 - p$. The absolute mean of the binomial distribution is np and its variance is npq. Because of the limiting property, the normal distribution with parameters np and npq may replace the binomial distribution when the number of random trials n is reasonably large and p is not an extreme value.

Similar considerations apply to data which we obtain on the basis of counting individual plants within sampling units. In this case, the usual assumption is that the unknown probability distribution is *Poisson* with

$$P(k) = \frac{e^{-\lambda}\lambda^k}{k!} \tag{1.7}$$

representing a general term. This $P(k)$ measures the probability of finding k individuals in a random sampling unit when the mean number of individuals per sampling unit is λ. It must of course be assumed that the occurrence of an individual in a unit is the consequence of a perfectly random process. In the case of Poisson variables the normal distribution with mean λ and also variance λ is likely to give a good approximation for the actual distribution. This, of course, does not hold true if λ is small, in which case the Poisson distribution is very asymmetric. The symmetry of a Poisson distribution can be improved by transformation after which it may resemble more closely the normal distribution. Transformations may also help to break up the relationship between the mean and variance. Similar considerations apply when p is not an extreme value in the binomial distribution; there too, symmetry can be improved and the relationship of the mean and variance can be broken up by certain transformations.

Under most survey conditions in natural vegetation the simple Poisson distribution cannot adequately represent count data. Pielou (1969a) considers alternatives to represent the distribution of counts in aggregated populations. Of these the *Poisson-Poisson* is an appropriate distribution to use under those rarely met circumstances when the number of aggregates per sampling unit is a Poisson variable, and the number of individuals per aggregate is a Poisson variable also. If the number of aggregates per sampling unit is known to be a Poisson variable, but the number of individuals per aggregate is logarithmic, a *Poisson-logarithmic* distribution is indicated.

7. *Measurability*. Variables differ in the difficulty with which they can be measured. To simplify matters, surveyors often resort to various approximations rather than to measure a variable directly. Species cover represents a case in point. It may be determined by actual measurement of areal projections, but this is very difficult to accomplish for most species. We know, however, that certain methods, such as the point quadrat method (using fine needles or cross wire) in certain types of vegetation, or some estimative scale, such as the cover/abundance scale of Braun-Blanquet in other types, may yield reasonably accurate data (cf. Goodall, 1952b; Greig-Smith, 1957, 1964; Kubíková & Rejmánek, 1973; and references therein). Binary (0,1) and multistate nominal variables represent other possibilities to overcome the difficulty of measurement and still to retain sufficient information in the data about variation in the vegetation (e.g. Williams & Lambert, 1959; Knight, 1965; Bannister, 1966; Orlóci, 1966; Knight & Loucks, 1969; van der Maarel, 1972; Moravec, 1975).

8. *Ecological informativeness*. While the potential of binary and nominal, multistate variables must not be underrated (cf. Williams & Dale, 1962; Lambert & Dale, 1964; Orlóci, 1966, 1968a; Austin & Greig-Smith, 1968; Norris & Barkham, 1970; Noy-Meir, 1971a), one should not have illusions about the universal usefulness of these variables. The fact of the matter is that binary variables, such as species presence, will have little information content in samples from communities of high floristic homogeneity expected in sites of little ecological variation. Relevant to this point, Smartt, Meacock & Lambert (1974) assert that variables, in which quantitative variation is partially controlled while the floristic component is allowed to be fully manifested, are likely to be most generally informative about plant/habitat relationships. These authors consider as the main discriminating feature in evaluating variables the extent to which the quantitative component of variation masks variation in the floristic component. They also recognize as important criteria the constraints that can be imposed on the variables to limit variation among species within the quadrats or among quadrats within species. We shall consider different methods for constraining variation. As to their informativeness in ecological terms, we may not really know which variables are informative and which are not before we look at hard data. This then brings up the need for experiments, perhaps simulated or real, in

the context of a pilot survey, to learn about the variables beforehand from which we may have to choose in a subsequent survey.

After describing the quadrats based on measurements on species, we may decide (if good reason calls for it) to equalize, weigh or otherwise transform the species data. Equalization adjusts the data to zero mean, unit range, etc., which can effectively standardize or otherwise normalize variation within species or within quadrats. The species weights are determined locally based on either survey data (e.g. Orlóci, 1973a; Orlóci & Mukkattu, 1973; Noy-Meir & Whittaker, 1977) or other criteria dictated in the method of analysis. Through equalization and weighting, species can be prevented from dominating the analysis and commensurability can be established to facilitate logical comparisons.

If viewed from the point of the raw data, equalization and weighting imply the transformation,

$$A_{ij} = W_i X_{ij} \tag{1.8}$$

in which W_i is specific to species i. The different methods define W_i in different ways. Some are described here:

1. $W_i = 1/\bar{X}_i$, where the last symbol is the arithmetic mean of species i. To this mean n values X_{ij}, $j = 1, ..., n$, contribute. With W_i so defined, A_{ij} of formula (1.8) is in scale free units and the $\bar{A}_i$, the mean of the transformed data, has unit value in all species. In the case of species X_1 of Table 1-1, the associated A_{ij} data are given by

$$\mathbf{A}_1 = W_1 \mathbf{X}_1 = (1.97\ 0.09\ 0.39\ 1.14\ 0.13\ 0.22\ 0.09\ 3.93\ 1.35\ 0.70).$$

The mean of the original values is $\bar{X}_1 = 22.9$ and the standard deviation $S_1 = 27.68$. The mean after transformation is $\bar{A}_1 = W_1\bar{X}_1 = 1$ and the standard deviation $W_1 S_1 = 27.68/22.9$. The range is W_1 (max X_1 − min X_1) = 88/22.9, where 88 is the range (the difference of the largest and smallest value) in the original data. Min and max signify minimum and maximum values in the original data.

2. $W_i = 1/S_i$, where S_i is the standard deviation of species i. If the X_{ij} are expressed as deviations from the common mean, the A_{ij} of formula (1.8)

Table 1-1. Sample data. Counts of individuals within quadrats

Species	*Quadrat* 1	2	3	4	5	6	7	8	9	10
X_1	45	2	9	26	3	5	2	90	31	16
X_2	18	92	32	48	73	80	95	13	92	78
X_3	3	40	5	83	68	27	2	17	1	23
X_4	10	61	11	3	32	2	39	2	8	6
X_5	9	53	99	21	49	81	72	6	90	62

will represent measurements in standard, scale free units with mean zero and standard deviation unity. The range is reduced by a factor W_i. Using species X_1 in Table 1-1 as example, the standardized data are given by the values in

$$\mathbf{A}_1 = W_1 \mathbf{X}_1 = (0.80\ -0.80\ -0.50\ 0.11\ -0.72\ -0.65\ -0.76\ 2.42\ 0.29\ -0.25).$$

The mean in $\mathbf{A}_1$ is zero since the X_{ij} in formula (1.8) are expressed as deviations from the common mean. The standard deviation is unity and the range is 88/27.68.

3. $W_i = 1/(\max X_i - \min X_i)$, where max and min signify maximum and minimum values. This method equalizes the range in the different species and transforms the data to scale free units. Such a transformation is preferred if scale differences have to be removed, but the variance is to be only partially affected. The method allows the extreme values to influence the transformation which is a weakness since extreme values depend very much on accidental events. The range in species X_1 of Table 1-1 is 88. The transformed data are given by

$$\mathbf{A}_1 = W_1 \mathbf{X}_1 = (0.51\ 0.02\ 0.10\ 0.30\ 0.03\ 0.06\ 0.02\ 1.02\ 0.35\ 0.18).$$

The range in A_1 is unity. The mean is equal to 22.9/88 and the standard deviation to 27.68/88.

4. $W_i = \sum_h \chi^2_{hi}/n$, where χ^2_{hi} is a chi square measuring association between two species based on the method outlined below. The summation is over all species except i. The number of species in the sample is p and the number of quadrats n. Quantities of this sort were used by Williams, Dale & Macnaughton-Smith (1964; Macnaughton-Smith, Williams, Dale & Mockett, 1964) as weights for variables. The quantity χ^2_{hi}/n is known as the *mean square contingency* of species h and i. This quantity is computed from binary data with states indicating the presence and absence of a species in a given quadrat. The calculations are based on data which we summarize in a 2 x 2 contingency (classification) table:

	Species i		
Species h	1	0	*Row total*
Present 1	a	b	$a + b = R_1$
Absent 0	c	d	$c + d = R_2$
Column total	$a + c = C_1$	$b + d = C_2$	$n = a + b + c + d$

The letters a, b, c and d indicate respectively the number of quadrats with both species present, only species h present, only species i present and both species absent. The grand total is equal to the number of quadrats in the sample. R_1 is the total number of occurrences for species h and $R_2 = n - R_1$

in the sample. C_1 and C_2 are similarly defined. The C and R quantities are the so-called marginal totals in the table. Based on these symbols, the mean square contingency of species h and i is defined by

$$\chi^2_{hi}/n = \frac{(ad - bc)^2}{R_1 R_2 C_1 C_2} \tag{1.9}$$

It can be shown that this is a correlation measure equivalent to the square of the *simple product moment correlation coefficient.* For this we can rewrite formula (1.9) as

$$\chi^2_{hi}/n = \sum_j A_{hj} A_{ij} \,/\, [\sum_j A^2_{hj} \sum_j A^2_{ij}]^{1/2},$$

where $A_{hj} = (X_{hj} - \bar{X}_h)$ and A_{ij} is similarly defined; $\sum_j A_{hj} A_{ij} = a - R_1 C_1/n$, $\sum_j A^2_{hj} = b - R^2_1/n$ and $\sum_j A^2_{ij} = c - C^2_1/n$. If we consider the fact that the species vectors have unit length and a common origin, and if we also consider that $(\chi^2_{hi}/n)^{1/2}$ is a cosine quantity, we can state that:

a. χ^2_{hi}/n is the squared length of the projection of species vector h on species vector i.

b. The χ^2_{hi}/n values are additive.

c. W_i is that component of the total species sum of squares which can be accounted for by variation in the direction of the ith species vector; i.e. the component that can be loaded on species i.

When we so define W_i it gives a weight to species i equal to that species' potential in absorbing variation in the sample. Some classification methods (to be discussed in Chapter 4) select species for which W_i is maximal and use these to subdivide the sample of quadrats into two groups, one with and the other without species i. In such a subdivision the between-quadrats component of the sum of squares is maximized. When W_i is zero, the species have zero correlations, and when equal to $p - 1$, the correlations are perfect, meaning collinearity of the species vectors in the sample.

A numerical example is given below. We have binary data for species X_1, X_2 and X_3,

$$\mathbf{X} = \begin{bmatrix} \mathbf{X}_1 \\ \mathbf{X}_2 \\ \mathbf{X}_3 \end{bmatrix} = \begin{bmatrix} 0 & 1 & 1 & 1 & 1 & 0 & 1 & 0 & 0 & 1 \\ 1 & 0 & 1 & 1 & 1 & 1 & 0 & 1 & 0 & 1 \\ 1 & 0 & 1 & 1 & 1 & 1 & 1 & 1 & 1 & 0 \end{bmatrix} .$$

Zeros and ones indicate respectively species absence and presence in quadrats. Each of the 10 columns represent a different quadrat. The joint frequencies are the values in the body of the 2 x 2 tables:

Species 1	Species 2 1	0	Total
Present 1	4	2	6
Absent 0	3	1	4
Total	7	3	10

Species 1	Species 3 1	0	Total
Present 1	4	2	6
Absent 0	4	0	4
Total	8	2	10

Species 2	Species 3 1	0	Total
Present 1	6	1	7
Absent 0	2	1	3
Total	8	2	10

The mean square contingency of species is calculated according to formula (1.9), yielding for species *1* and *2*,

$$\chi^2_{12}/n = \frac{((4)(1) - (2)(3))^2}{(6)(4)(7)(3)} = 0.008 \quad .$$

Similar computations give χ^2_{13}/n and χ^2_{23}/n, which we give as elements in the mean square contingency matrix,

$$\chi^2/n = \begin{bmatrix} - & 0.008 & 0.167 \\ 0.008 & - & 0.048 \\ 0.167 & 0.048 & - \end{bmatrix} \quad .$$

The individual weights then have the values

$$W_1 = 0.008 + 0.167 = 0.175$$

$$W_2 = 0.056$$

$$W_3 = 0.215 \quad .$$

5. Determination of species weights may be approached in yet different ways, one suggested by Feoli (1973a). Quadrat similarities are determined by Søensen's (1948) index (see Chapter 2 for formula) or by another symmetric measure of similarity. The sample is then divided into two groups. The first, H^+_i includes quadrats with species h present and the sec-

ond, H_i^- includes the remaining quadrats. If $\bar{S}_{H^+}$ is the average similarity of quadrats in H^+ and $\bar{S}_{H^-}$ is the average similarity of quadrats in H^+ with quadrats in H^-, the weight given to species h is

$$W_h = 1 - \frac{\bar{S}_{H^-}}{\bar{S}_{H^+}} .$$

Since $\bar{S}_{H^-}$ can be less than, equal to or larger than $\bar{S}_{H^+}$, W_h is a number less than or equal to 1. The justification of using W_h of this formulation as a weight for species h is that it measures the sharpness of isolation of group H^+ from H^- in terms of total species composition. If a species is given a weight equal to its respective W_h value, it will in fact be weighted according to its diagnostic value for isolating groups. This is consistent with the following facts:

a. With W_h increasing in value to one, the isolation of H^+ and H^- is increased. If $W_h = 1$, H^+ and H^- are completely isolated.

b. With W_h decreasing, the isolation of H^+ and H^- looses sharpness.

It is obvious that species confined to quadrats of unusual species composition will be given high weights. Given two groups of quadrats (A, B), the $\bar{S}_{H^-}$ value is fixed, but $\bar{S}_{H^+}$ depends on whether A or B is the H^+ group. Clearly, W_h, defined above, is not symmetric. If A and B are completely dissimilar, $\bar{S}_{H^-} = 0$ and $W_h = 1$. The minimum value of W_h is achieved when the quadrats in H^+ are completely dissimilar, except for the common possession of species h, and at the same time, H^+ and H^- have identical contents, except for the absence of species i in the quadrats of H^-. Feoli (1973a) gives a numerical example.

6. $W_i = 1 - 1/a_i = R_i^2$, where $a_i = S_{ii}(S^{-1})_{ii}$ in which S_{ii} is the variance of species i and $(S^{-1})_{ii}$ is a value in the ii cell of the inverse of the covariance matrix **S**. R_i^2 is the square of the multiple correlation coefficient of species i and the remaining $p - 1$ species in the sample. Clearly, whereas W_i of method 4 measures the total species variance that can be absorbed by species i, W_i of method 6 measures that proportion of the variance of species i which can be accounted for by the remaining $p - 1$ species. Putting it in another way, W_i of method 4 is a measure of the success of species i to replace the $p - 1$ species in a variance-oriented analysis, and W_i of method 6 is a measure of success of the same $p - 1$ species in replacing species i in a variance-oriented analysis. Whereas W_i of method 4 measures redundancy in other species of the sample conditional on species i, W_i of method 6 measures redundancy in species i conditional on the other species. We shall further elaborate on the concept of redundancy in the sequel. To give an example, we take the data from Table 1-1. The centered data matrix is given by

$$\mathbf{A} = \begin{bmatrix} 22.1 & -20.9 & -13.9 & 3.1 & -19.9 & -17.9 & -20.9 & 67.1 & 8.1 & -6.9 \\ -44.1 & 29.9 & -30.1 & -14.1 & 10.9 & 17.9 & 32.9 & -49.1 & 29.9 & 15.9 \\ -23.9 & 13.1 & -21.9 & 56.1 & 41.1 & 0.1 & -24.9 & -9.9 & -25.9 & -3.9 \\ -7.4 & 43.6 & -6.4 & -14.4 & 14.6 & -15.4 & 21.6 & -15.4 & -9.4 & -11.4 \\ -45.2 & -1.2 & 44.8 & -33.2 & -5.2 & 26.8 & 17.8 & -48.2 & 35.8 & 7.8 \end{bmatrix} .$$

A characteristic element in A accords with $A_{ij} = (X_{ij} - \bar{X}_i)$, where $\bar{X}_i$ is the ith species mean. For example, $A_{11} = (X_{11} - \bar{X}_1) = (45-22.9) = 22.1$. The covariance matrix is defined by the product $(n-1)^{-1}\mathbf{AA}'$. However, we need not divide by $n-1$, since when we compute a_i, $n-1$ will cancel out. We can base the calculations on the sums of squares and cross products matrix,

$$\mathbf{S} = \begin{bmatrix} 6896.9 & -5611.9 & -1470.1 & -2527.6 & -5445.8 \\ -5611.9 & 9022.9 & 594.1 & 2913.6 & 5646.8 \\ -1470.1 & 594.1 & 7462.9 & 581.4 & -2913.8 \\ -2527.6 & 2913.6 & 581.4 & 3576.4 & 686.2 \\ -5445.8 & 5646.8 & -2913.8 & 686.2 & 9881.6 \end{bmatrix} .$$

The elements of $\mathbf{S}$ are computed[6] according to $S_{hi} = \sum_j A_{hi} A_{ij}, j = 1, ..., n$. This gives for species *1* and *2*,

$$\begin{aligned} S_{12} &= (22.1)(-44.1) + (-20.9)(29.9) + ... + (8.1)(29.9) \\ &\quad + (-6.9)(15.9) \\ &= -5611.90 . \end{aligned}$$

The inverse of $\mathbf{S}$ is given by

$$\mathbf{S}^{-1} = \frac{1}{10^4} \begin{bmatrix} \mathit{6.60261} & 0.516488 & 2.53193 & 3.09038 & 3.87558 \\ 0.516488 & \mathit{2.86104} & -0.547348 & -1.60819 & -1.40001 \\ 2.53193 & -0.547348 & \mathit{2.72322} & 1.32853 & 2.41888 \\ 3.09038 & -1.60819 & 1.32853 & \mathit{5.57034} & 2.62704 \\ 3.87558 & -1.40001 & 2.41888 & 2.62704 & \mathit{4.47869} \end{bmatrix} .$$

From these, we obtain the weight of the first species,

$$W_1 = 1 - \frac{1}{(6896.9)(0.000660261)} = 0.780401,$$

and the weights of the remaining species,

$$\begin{aligned} W_2 &= 0.612626 \\ W_3 &= 0.507949 \\ W_4 &= 0.498037 \\ W_5 &= 0.774045 . \end{aligned}$$

This example notwithstanding, once the weight of a species is determined, it would be best to use residuals to determine the weight of the next species. This brings us to the ranking problem discussed later in this chapter. We note further that any method which relies on matrix inversion runs the risk

[6] Numerical results are accurate within computer rounding errors with the number of digits retained.

of not being operational if the matrix is singular. If **S** is in fact singular, the W_i should be determined differently. We shall return to this problem at a point later in the text when we outline ranking procedures.

From the foregoing descriptions several peculiarities emerge: W_i of methods 1, 2 and 3 can be determined for any species i in isolation irrespective of the other species. The main objective of the data transformations in these is to establish commensurability with or without affecting the variance. Two of the methods (4 and 6) define species weights as functions of species redundancy. Both of these use the notion of correlation and thus imply procedures not defined for an isolated species; only for groups of correlated species. Method 5 is different in that it examines quadrat similarities when determining weights for the species. If we decide to use the species weights (W_i) of methods 4 and 6, it is implicit in the decision that the measurement of species importance based on variance is meaningful in the context of the analysis which we intend to perform on the transformed data. Furthermore, when we use the W_i as weights, it is necessarily implied that scaling in the mean, range and standard deviation by a different factor W_i in each species i is exactly as we intended.

A corollary to the weighting problem is the rescaling of variables to obtain more informative descriptions and/or more manageable data sets. To amplify on this point, we shall consider an example. If we describe a quadrat based on, for instance, the mean height of species X, a great deal of the information that resides in variation about the mean value in the quadrat will be lost. Methods are available to efficiently recover the information that resides in variation about the mean. As one solution to a similar problem in taxonomy, McNeill (1974) proposed rescaling in a *character-state frequency procedure.* The method is directly applicable to quantitative variables such as tree height. The steps involve decomposition of each quantitative variable into a primary variable, several secondary variables, and a number of tertiary variables. In the tree height example, these could include one primary variable with score 1 if species X is present in a quadrat; three secondary variables representing the 10th percentile, median and the 90th percentile in the frequency distribution of height for species X within the quadrat; and three tertiary variables, including the frequency of heights not exceeding the 10th percentile, between the 10th and 90th percentiles, and not less than the 90th percentile. If we followed this procedure through, it would yield not one (average height) but 7 different descriptors for height per species per quadrat. We note that one of the three frequency variables could be eliminated because of the linear constraint on their total. Similar procedures are available for binary and nominal variables (Gower, 1971a).

Decisions in the choice of a weighting strategy can be complicated by the variables being logically dependent. Such is often the case in taxonomy (Kendrick, 1965; Dale, 1968; Williams, 1969; McNeill, 1972). Difficulties arise when nominal variables are involved with respect to any one of which an individual may be in more than one state. Williams (1969) discusses this problem and gives examples. The weighting of variables is on the whole a well-worked problem in numerical taxonomy where it generated controversies and a very rich technical literature (e.g. Sneath, 1957a, 1969; Rescigno &

Maccacaro, 1961; Rogers, 1963; Sokal & Sneath, 1963; Kendrick & Proctor, 1964; Farris, 1969b; Jardine & Sibson, 1971; Cormack, 1971; Sneath & Sokal, 1973; Hansell, 1973; Hansell & Chant, 1973; Hansell & Ewing, 1973).

The transformations so far discussed represent only a limited sample from a broad class with a common, basic objective to equalize or to attach weight. Other objectives may be served by other kinds of transformations. Underwood (1969) used a transformation to remove in-line constraints which force a string of data elements to conform with some quantity of which the elements are arbitrarily made to be a fraction. What the transformation does in this case is a sort of unweighting, to undo the influence of constraints in the method based on which we measure species quantity. Consider species frequency as an example. We count the presence (or absence) of species in k sampling units in each of n quadrats to obtain for species i the quadrat frequencies

$$f_{i1}\ , \ldots, f_{in}$$

in absolute terms, or

$$\mathbf{P} = (P_{i1} \ldots P_{in}) = (f_{i1}/k \ldots f_{in}/k)$$

in relative terms. No frequency will be larger than k or, if relative frequency, larger than 1. Now we may assume that $\mathbf{P}$ describes the distribution of a continuous variable X. If we assume further that this distribution is known, we can use the inverse probability (area) integral to determine a linear co-ordinate value X_{ij} corresponding to a given P_{ij} (area under the graph) of the distribution. If $\mathbf{P}$ describes a normal distribution, the polynomial approximation to the inverse probability integral (cf. Abramowitz & Stegum, 1970) is given by

$$X_{ij} = t_{ij} - \frac{C_0 + C_1\, t_{ij} + C_2\, t_{ij}^2}{1 + d_1\, t_{ij} + d_2\, t_{ij}^2 + d_3\, t_{ij}^3}\ ,$$

whose absolute error is expected to be less than 4.5×10^{-4}. The coefficients in this formula have the following values:

$$t_{ij} = (\ln 1/P_{ij}^2)^{1/2}$$

$$C_0 = 2.515517 \qquad d_1 = 1.432788$$

$$C_1 = 0.802853 \qquad d_2 = 0.189269$$

$$C_2 = 0.010328 \qquad d_3 = 0.001308\ .$$

We assume that $0 \leqslant P_i \leqslant 0.5$. When $P_i > 0.5$ we replace it by $1 - P_i$. One may add to X_{ij} a number, say 5, to avoid negative values. Statistical terminology identifies X_{ij} as a *probit* and the inverse normal probability integral transformation as a *probit transformation.*

Should the graph of the assumed distribution be described by a logistic equation, the transformation

$$X_{ij} = P_{ij} + \ln\,[P_{ij}\,/\,(2 - P_{ij})]$$

would yield *logits* which are linear co-ordinates associated with given area segments (P_{ij}) under the logistic curve. The user is at liberty to select a distribution for X and derive X_{ij} values thereafter. We may stress though that the purpose of the probit transformation, and also of the logit transformation, is to remove in-line constraints and thereby to allow variable X to emerge. Explanations and illustrations are found in the statistical literature (e.g. Finney, 1963).

Often, transformations are sought to stabilize the variance by eliminating its dependence on the mean. The independence of mean and variance is an assumption required in statistical tests based on the normal distribution. Bartlett (1936) suggests a square root transformation $(X_{ij} + 1/2)^{1/2}$ to stabilize the variance when X_{ij} is a count from a Poisson population. Kendall & Stuart (1976) recommend $(X_{ij} + 3/8)^{1/2}$ for the same purpose.

In other cases transformations are sought which will bring the observed distribution of X closer to the normal through reducing skewness and kurtosis. Others improve additivity. Still others remove transformation bias. For us, to elaborate on these would be quite outside the scope of the present book. The reader is referred for information to texts on the theory of statistics (e.g. Kendall & Stuart, 1969, 1973, 1976) and on applied statistics (e.g. Sokal & Rohlf, 1969).

1.7 Sampling considerations

Vegetation data are usually burdened by a great deal of sampling variance. Regarding the consequences of this it is rather important to clarify two points. Firstly, it may be futile to attempt to increase the precision of measurement knowing, as Goodall (1970a) observed, that the extra effort is likely to be swamped out by sampling errors over which we have only limited control. Secondly, precision of measurement and lack of bias in the sampling are not synonymous concepts. Although efforts for high precision may be unrewarding, effort should not be spared to eliminate bias. It is in this respect that the choice of sampling method comes into focus, for the sample that we take should be representative.

A. Which method?

The choice of a sampling method must take into account the problem, the variables, intended mathematical model and the method of data analysis. The choice should also reflect economic and other practical considerations. The sampling problem and the methods used are discussed by Sampford (1962) with a view to statistical properties, and by Greig-Smith (1964), Shimwell (1971), Kershaw (1973), Pielou (1974) and Mueller-Dombois & Ellenberg (1974) regarding statistical as well as different ecological aspects.

While in many applications sampling is a one shot enterprise, in others it is repeated over and over again, as needed, with information from a previous sample modifying the sampling method at a subsequent stage, until the desired results are achieved (e.g. Daget, Godron & Guillerm, 1972).

When the goal is estimation of population parameters, or statistical inference in general, *random sampling designs* are indicated. Such sampling designs give every individual of the population an equal chance to get into the sample. This, in turn, assures that the sample is optimally representative. Some specific objectives, such as vegetation mapping for instance, may call for *systematic sampling* in which the sampling units are chosen at regular intervals. Such a sampling is less costly than random sampling of comparable intensity, and also, it provides a more even coverage of the area with sampling units. Vegetation surveys often use *preferential sampling.* In this method stands of vegetation are chosen because they are considered typical, or plant specimens are selected for measurement because they appear representative.

Among the commonly used sampling designs in vegetation surveys *simple random sampling* has special significance. However, since it requires an ordered presentation of all population units, a random sample cannot be drawn if each unit cannot be located and labelled in the population. If the population units are concrete objects, such as for instance individual plants of a species, or cells in a two dimensional grid laid over the survey site, a random sample is drawn on the basis of random numbers. Some sampling methods allow a sample to be constructed from individuals occurring nearest to random sampling points. However, such a sample will almost certainly be non-representative of the respective population. Pielou (1969a) analysed this problem in connection with plant populations and has shown that an individual's chance of being nearest neighbour to a random point is higher in the low density patches of vegetation than in the high density patches. Because of this, the individuals in the low density patches will be over-represented in the sample and the sample will be biased.

Apart from requiring the tedious task of creating an ordered presentation of all population units prior to sampling, simple random sampling suffers also from other weaknesses including a potentially high sampling variance, and in the case of quadrat sampling, notorious underrepresentation of some parts of the survey area. These problems can be diminished based on the use of high sampling intensity.

If simple random sampling is unsatisfactory certain restrictions may be imposed on randomization. In this connection, Greig-Smith (1964) suggests *restricted random sampling* in which compartments are recognized in the vegetation, and each compartment is sampled at random with sampling intensities proportional to compartment size. The essential thing about restricting randomization is that it allows a stratification of the population on the basis of information available about the units prior to the sampling. Stratification reduces heterogeneity, separates compartments for which separate results are required, provides for more even sampling, and increases the precision of the sample estimates within the compartments (Sampford, 1962). If the compartment samples are not pooled but treated separately as if each compartment represented a different population, the random samples, being representative of the compartments from which they are taken, will be well suited for analysis by any statistical method aimed at comparing compartments. Some complications may however arise if we decide to pool the compartment samples. The reason is that the pooled sample cannot be regarded as a representative sample of the broader population which is stratified into compartments (Sampford, 1962). In spite of this, the compartment samples can still be used to obtain estimates for such population parameters as the true weighted mean, the true sampling variance of the weighted means, etc., across the compartments.

Considering the vast heterogeneity which normally exists within extensive areas of vegetation, compartmentalization is a logical first step in most vegetation surveys. The criteria for delimiting compartments may however vary depending on the local circumstances. The criteria may include, for instance, the dominant species, differential species, past stand history (fire, cultivation, etc.), or some environmental conditions having to do with soils, topography, climates, etc. Whatever criteria are used, it must be kept in mind that the compartments cannot be validly compared on the basis of the criteria by which they are delimited since such a comparison would involve a circularity of argument. An example of such an inadmissible argument is the use of climatic compartments to *prove* the existence of natural zones in vegetation, knowing that climatologists like to draw climatic boundaries between groups of stations to coincide with apparent boundaries between vegetation zones. On this basis alone it would be quite risky for the ecologist to use as evidence for the distinctness of vegetation zones their coincidence with mapped climatic boundaries.

Sampling designs using compartments are not without problems in vegetation surveys. This is related to the need to locate boundaries on the ground before the sampling can begin. It is important, therefore, to have the criteria for compartment recognition such that the boundaries can be marked out without too much difficulty.

In *systematic sampling* of finite populations the units must be numbered to create a complete frame for selecting a sample. After numbering all population units the sampling may start with a random unit, after which other units are taken at fixed intervals from the first. A systematic sample which starts with a random unit is said to be *randomly sited.* Random siting has several advantages:

1. It provides accurate estimates for the population mean and, if repeated several times, also for the sampling variance of the mean.
2. It reduces the sampling costs.
3. It allows an even coverage of the area with sampling units.

These advantages, when combined with stratification in the sampling, superbly qualify the randomly sited systematic sample as a choice of method when the objective is mapping or general inventory of plant communities.

Preferential sampling is the commonest method which has traditionally been used in vegetation surveys and not entirely without advantages (cf. Westhoff & van der Maarel, 1973). It is significant that many of the prevalent concepts about the basic aspects of vegetation have their origins in surveys that were based on this method of sampling. In this, 'typical' stands of vegetation are chosen and sampled in 'preferred' sampling sites. Since a sample of this sort is, by definition, non-representative, the conclusions drawn may not reliably reflect the conditions in the sampled vegetation. Should the ecologist decide to base a vegetation survey on preferential sampling, the data will not be suited for statistical interpretation.

Rather than sample, we may choose to *completely enumerate.* In this type of survey we locate and measure all population units. Since all units of the population are measured, nothing needs to be estimated. The objectives reduce to a precise determination of population parameters. To this extent complete enumeration is a logical part of a deterministic analysis.

B. Sampling unit

Selection of sampling units is a task with considerable difficulties apparent from informative discussions by, e.g., Cain (1938), Bourdeau (1953), Bormann (1953), Goodall (1961), Sampford (1962), Greig-Smith (1964), La France (1972), Smartt & Grainger (1974) and Strahler (1977). The choice is especially difficult when size and shape are concerned since size and shape can influence the outcome of a vegetation survey. Goodall (1973a; Noy-Meir, Tadmor & Orshan, 1970) suggests that if the size is smaller than the *minimal area*, the data collected will in part reflect a mosaic structure of the vegetation and not reveal important trends. Goodall (1961) defines the minimal area 'as that of a square with the side equal to the distance at which variance between samples ceases to be a function of their spatial separation'. Here the term *sample* means a sampling unit such as a quadrat. Others define minimal area differently. Numata (1966) relates it to the height of the vegetation and van der Maarel (1970) to vegetation structure. To Moravec (1973) the minimal area is a function of floristic similarity as well as homogeneity in quadrats (see also Trass & Malmer, 1973; Westhoff & van der Maarel, 1973) and to Vasilevich (1973) it is the *coenoquant.* A minimal area may however not exist at any ground scales (Goodall, 1961), in which case the choice of quadrat size would have to be based on other considerations. Attempts have in fact been made to disassociate sampling unit size from minimal area size. Rice (1967), for instance, suggests a size that gives a reasonably normal distribution in the sample data.

Accounts by Greig-Smith (1957, 1964) and Goodall (1961) amplify on these points and suggest solutions. It will be satisfactory here if we consider only general principles:

1. Sampling units must be clearly distinguishable.
2. Inclusion and exclusion rules must be established and must be followed uniformly and consistently throughout the sampling.
3. Preferably, the sampling units should be uniform in size and shape, or nearly so.
4. A sampling unit of rectangular shape is preferred since it can minimize edge-effect and simplify the matter of establishing its boundaries on the ground.

The size and shape problems may be avoided by using plotless methods of sampling. Problems and potentials are discussed by, e.g., Greig-Smith (1964), Yarranton (1966), Goodall (1970a) and Mueller-Dombois & Ellenberg (1974) who also describe methods and give references. These methods are, however, limited in their usefulness because they can handle only certain kinds of variables. Regarding the arrangement of sampling units, they may be placed at random or systematically, along transects or within a rectangular grid. Line or belt transects are indicated when the surveyor can recognize natural trends of changes in the vegetation and environment. Most often, however, the sampling units are arranged in a rectangular grid.

C. *Logistic considerations*

The choice of sampling design will no doubt be influenced by economic considerations. Attempts to economize may lead to simplifications involving the use of low intensity sampling, preferences for sampling designs other than the random, elimination of species from the survey that appear unimportant, and collection of less costly data. Simplifications of this sort are perfectly legitimate, and they are usually effective in reducing sampling costs, but the subsequent loss of information may also be quite substantial. Economy through reduction of sampling intensity, meaning the use of fewer quadrats in a vegetation survey, is an unwise choice if some other ways are open to the ecologist to reduce the costs. For instance, one may choose to reduce the number of species. This is a logical choice considering that the species are usually correlated and the omission of some species may not result in a significant loss of information.

Some ecologists believe that it is best to base a vegetation study on the largest possible number of species. The reasoning behind their belief involves the idea that when large numbers of species are used, the probability of missing one that is really important can be greatly diminished. While this may be true, and under some circumstances the handling of large numbers of species may be a workable proposition, under many other circumstances inclusion of large numbers of species may lead to serious difficulties in the logistics of sampling and subsequent data analysis. It would not be unusual if in the interest of a more economical survey and data analysis one were forced to delete some of the available species. Once it is decided that some of the species must go, then logically, the problem reduces to finding suitable criteria on the basis of which species can be selected for omission.

In some applications, the original variables are replaced by synthetic variables. If the latter are more efficient than the former, fewer may be used to represent the essential information in the sample. Principal components are of this sort (see Chapter 3) which Quadling (1967) has used to reduce the number of group descriptors before an analysis. The same kind of method could be applied in the case of vegetation samples if it seemed important to reduce the number of species. The analysis of synthetic variables in the place of species would of course be not without problems should the user try to interpret the results in ecological terms.

To find solutions to the problem of reducing species number in a sample, solutions that are not burdened by the difficulty of subsequent interpretations, we may try methods based on which we can measure species importance. Once importance is measured, the unimportant species can be identified and deleted. For this we could use results from a multivariate analysis of variance (e.g. McCabe, 1975), cluster analysis (e.g. Jolliffe, 1973), regression and multiple correlation analysis (e.g. Furnival, 1971; Jolliffe, 1972, 1973; McCabe, 1975; and references) or discriminant analysis (e.g. McKay, 1976). We could also rely on methods which order species based on the arithmetic mean, ratio of standard deviation and mean, etc. (e.g. Grigal & Goldstein, 1971; Grigal & Ohmann, 1975), but these do not take into account species correlations.

We shall focus on methods that measure species importance in terms of sums of squares and cross products in this chapter and describe a method which relies on mutual information (Orlóci, 1976a) in the next. A common but not always explicit feature in all these is the assumption that relative importance is a local matter, and if we know the local circumstances we can actually measure species importance.

Many commonly used methods of data analysis manipulate covariances. By specifying the covariance, the determination of species importance is automatically committed to some procedure which incorporates manipulations of the covariance itself, or some covariance related quantity. After statement of the general points, let us consider Orlóci's (1973a) method which derives an ordering of species that reflect their success in absorbing the variance of the companion species. Let N be the number of quadrats in the pilot sample, and let p represent the number of species in a subsample out of the total P available species. Let $\mathbf{D}_p$ represent an $N \times N$ Euclidean distance matrix (see Section 2.4 for definitions) with elements relating the N quadrats in pairs. The distances in $\mathbf{D}_p$ are possibly burdened by a certain amount of distortion due to the fact that they utilize only p of the available P species. The undistorted distances in matrix $\mathbf{D}_P$, computed from the P species, and the distorted distaces in $\mathbf{D}_p$ can be related in some stress function $\sigma(p; N, P)$ at fixed P, N and varying p. This function measures the distortion in the sample distance structure due to deletion of $P-p$ species when computing the distances in $\mathbf{D}_p$. Given the decision that only p species, or less, will be considered in the survey, the determination of the best p-species sample in the course of a limited pilot survey by one method represents a problem in combinatorials:

1. Compute a value of the function $\sigma(p; N,P)$ for each distinct combination of P species in groups of p species. Note that there are $P!/(p!(P-p)!)$ such combinations.
2. The best p-species combination will minimize $\sigma(p; N,P)$.

If p is not specified, two more steps may have to be performed:

3. Repeat finding the best p-species samples at different values of p, and construct a stress curve (Fig. 1-1) showing the minimum values of $\sigma(p; N,P)$ at $p = 1, ..., P-1$.
4. Inspect the stress curve and choose a species sample from that region of the curve which lies in the vicinity of point A where the curve appears rapidly flattening off. This point may have to be determined subjectively.

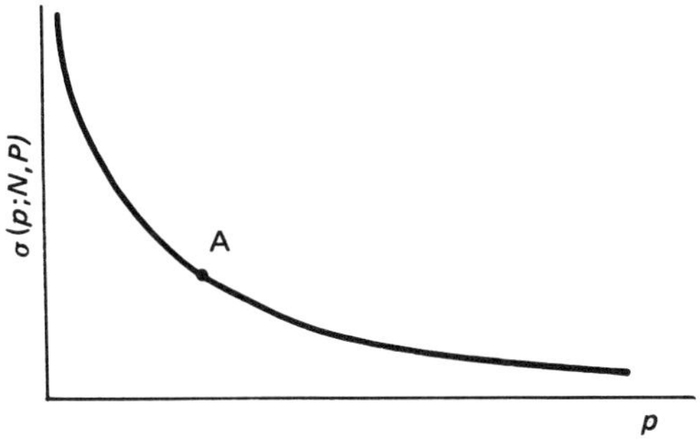

Fig. 1-1. Stress $\sigma(p; N, P)$ in sample structure as a function of species number p. See explanations in the main text.

Even if p is given a priori, a method based on the use of combinatorials may prove to be a computational impossibility because of the extremely high number of combinations to be computed at even relatively small values of P. A computationally simpler method has been described (Orlóci, 1973a), a summary of which is given below:

1. Assume that only p out of a total of P common species are allowed in the analysis.
2. Designate the pilot sample of N quadrats and P common species by $\mathbf{X}$ and an observation by X_{hj}. The subscripts identify the hth species in the jth quadrat.
3. Determine for each species its *independent* share that it can possibly absorb of the total sum of squares and rank the species accordingly.
4. Use the first p species in the survey and in the subsequent data analysis.

If we assume that the data to be collected will be analyzed in some covariance related analysis, it will make sense to rank the species by a sum of squares criterion. To determine the independent components of the sum of squares specific to each species we proceed as follows:

1. Center the data within species (rows of $\mathbf{X}$) to obtain matrix $\mathbf{A}$. The centering is specified by

$$A_{hj} = (X_{hj} - \bar{X}_h)/Q_h, \qquad (1.10)$$

where $\bar{X}_h$ and Q_h represent respectively the mean and a factor of standardization (or other type of adjustment) in species h. Let us define Q_h as 1 at this point.

2. Compute the cross products $\mathbf{S}$ according to $\mathbf{S} = \mathbf{AA}'$ with elements relating species. A characteristic element in $\mathbf{S}$ is given by

$$S_{hi} = \sum_j A_{hj} A_{ij}, \; j = 1, \ldots, N \qquad (1.11)$$

3. Compute dispersion criteria and find their maximum,

$$SS = \max \; [\sum_h S_{h1}^2/S_{11}, \ldots, \sum_h S_{hP}^2/S_{PP}] \; , \; h = 1 \; , \ldots, P \qquad (1.12)$$

The species, labelled m, corresponding to SS, the largest of the P sums, has rank 1.

4. Having determined the rank of species m continue with computing the residual sum of squares and cross products according to

$$S_{hi} := S_{hi} - Y_{hm}\, Y_{im} \text{ for any } h, i = 1, \ldots, P \qquad (1.13)$$

This expression reads 'the entry in the hi cell in matrix **S** is replaced by its own value minus the product of two co-ordinates',

$$Y_{hm} = S_{hm} \,/\, S_{mm}^{1/2} \text{ and } Y_{im} = S_{im} \,/\, S_{mm}^{1/2} \qquad (1.14)$$

5. Compute a new value for SS from the elements of the residual **S**, declare rank 2 on the species which corresponds to the new maximum SS, and then continue with step 4 repeating the steps over and over again until all species are ranked.

The ranking just described orders species according to their independent share of the total sum of squares, i.e. the partitioning of sum of squares is orthogonal. The ratio $SS \,/\, \sum_h S_{hh}, h = 1, \ldots, P$, measures relative importance. It should be stressed that this procedure will produce 'importance values' for species which will represent a meaningful input for further consideration only in those analyses in which S_{hi} itself is a valid parameter. For the best p-species sample we take the first p species whose ranks are highest. When the value of p is not specified, one should use a strategically chosen minimum rank below which no species will be further considered. Ideally, the minimum rank should be such that the species below it will contribute no significant amounts of sum of squares to the total in the sample. But we do not have such a test of significance. The best we can do is to rely on another procedure:

1. Omit the last k species whose relative importance is zero.
2. Compute the stress values $\sigma(p; N,P)$ at fixed N and P and varying p such that $p = P-k, P-k-1, P-k-2, \ldots, 1$.
3. Inspect the stress curve (*e.g.*, Fig. 1.1) and select the cut off point (A) at a value of p where the curve appears rapidly leveling off.

Considering the foregoing ranking method, the following example illustrates the computations. The data are given in Table 1-1. The computer program is RANK in the Appendix. After assuming that standardization is not needed and noting that the presentations will be somewhat repetitive since some of the partial results were already presented in Section 1.6, we proceed as follows:

1. Center data according to formula (1.10) with Q set equal to one to obtain

$$\mathbf{A} = \begin{bmatrix} 22.1 & -20.9 & -13.9 & 3.1 & -19.9 & -17.9 & -20.9 & 67.1 & 8.1 & -6.9 \\ -44.1 & 29.9 & -30.1 & -14.1 & 10.9 & 17.9 & 32.9 & -49.1 & 29.9 & 15.9 \\ -23.9 & 13.1 & -21.9 & 56.1 & 41.1 & 0.1 & -24.9 & -9.9 & -25.9 & -3.9 \\ -7.4 & 43.6 & -6.4 & -14.4 & 14.6 & -15.4 & 21.6 & -15.4 & -9.4 & -11.4 \\ -45.2 & -1.2 & 44.8 & -33.2 & -5.2 & 26.8 & 17.8 & -48.2 & 35.8 & 7.8 \end{bmatrix}.$$

2. Compute the **S** matrix according to formula (1.11). For example,

$$S_{12} = \sum_j A_{1j}A_{2j} = (22.1)(-44.1) + (-20.9)(29.9) + \ldots + (8.1)(29.9) + (-6.9)(15.9) = -5611.90 \,.$$

The complete **S** matrix is given by

$$\mathbf{S} = \begin{bmatrix} 6896.9 & -5611.9 & -1470.1 & -2527.6 & -5445.8 \\ -5611.9 & 9022.9 & 594.1 & 2913.6 & 5646.8 \\ -1470.1 & 594.1 & 7462.9 & 581.4 & -2913.8 \\ -2527.6 & 2913.6 & 581.4 & 3576.4 & 686.2 \\ -5445.8 & 5646.8 & -2913.8 & 686.2 & 9881.6 \end{bmatrix} .$$

3. Determine the sum of squares next according to formula (1.12). For the first species,

$$\sum_h S_{h1}^2 / S_{11} = (6896.9^2 + (-5611.9)^2 + (-1470.1)^2 + (-2527.6)^2 + (-5445.8)^2) \,/\, 6896.9 = 17002.9 \,.$$

After computing values for all species, the criterion sum of squares becomes

$$SS = \max\,(17002.9,\ 17027.2,\ 8982.74,\ 7962.58,\ 17016.5) = 17027.2 \,.$$

This identifies rank *1* for species *2*.
4. Now derive co-ordinates according to formula (1.14). For instance, for species *3* and *4* on vector $m = 2$ these become

$$Y_{32} = S_{32} \,/\, S_{22}^{1/2} = 594.1 \,/\, 9022.9^{1/2} = 6.25441$$

$$Y_{42} = S_{42} \,/\, S_{22}^{1/2} = 2913.6 \,/\, 9022.9^{1/2} = 30.6730 \,.$$

(We use six significant digits in the long-hand results so that we can reproduce the computer figures in the residual **S** matrix.)
5. Compute the elements in the first residual of **S**. For example, for species *3, 4* we have according to formula (1.13),

$$\begin{aligned} S_{34} &:= S_{34} - Y_{32}\,Y_{42} \\ &= 581.4 - (6.25441)(30.6730) \\ &= 389.558 \,. \end{aligned}$$

Similar calculations give the entire matrix of residuals,

$$\begin{bmatrix} 3406.51 & 0.00000 & -1100.59 & -715.452 & -1933.71 \\ 0.00000 & 0.00000 & 0.00000 & 0.00000 & 0.00000 \\ -1100.59 & 0.00000 & 7423.78 & 389.558 & -3285.61 \\ -715.452 & 0.00000 & 389.558 & 2635.56 & -1137.22 \\ -1933.71 & 0.00000 & -3285.61 & -1137.22 & 6347.66 \end{bmatrix} .$$

6. Now calculate the criterion

$$SS = \max(5010.03, 0.00000, 9061.53, 3378.06, 8841.13)$$
$$= 9061.53\ .$$

This identifies species *3* as the second highest ranking species.
7. Compute the elements in the second residual of **S**. This is done similarly as in the case of the first residual to obtain

$$\begin{bmatrix} 3243.35 & 0.00000 & 0.00000 & -657.699 & -2420.80 \\ 0.00000 & 0.00000 & 0.00000 & 0.00000 & 0.00000 \\ 0.00000 & 0.00000 & 0.00000 & 0.00000 & 0.00000 \\ -657.699 & 0.00000 & 0.00000 & 2615.12 & -964.808 \\ -2420.80 & 0.00000 & 0.00000 & -964.808 & 4893.53 \end{bmatrix} .$$

8. From these derive

$$SS = \max(5183.58, 0.00000, 0.00000, 3136.48, 6281.31)$$
$$= 6281.31\ .$$

This indicates rank *3* for species *5*.
9. The third residual of **S** becomes

$$\begin{bmatrix} 2045.79 & 0.00000 & 0.00000 & -1134.98 & 0.00000 \\ 0.00000 & 0.00000 & 0.00000 & 0.00000 & 0.00000 \\ 0.00000 & 0.00000 & 0.00000 & 0.00000 & 0.00000 \\ -1134.98 & 0.00000 & 0.00000 & 2424.90 & 0.00000 \\ 0.00000 & 0.00000 & 0.00000 & 0.00000 & 0.00000 \end{bmatrix} .$$

10. Find the next sum of squares criterion,

$$SS = \max(2675.47, 0.00000, 0.00000, 2956.14, 0.00000)$$
$$= 2956.14\ ,$$

which identifies rank *4* for species *4*.
11. The last residual has a single non-zero element S_{11} = 1514.55 representing the specific sum of squares for species *1*.

We now summarize the results:

Species	*Rank*	*Specific sum of squares*	*Specific sum of squares as a proportion*
1	*5*	1514.55	0.041
2	*1*	17027.20	0.462
3	*2*	9061.53	0.246
4	*4*	2956.14	0.080
5	*3*	6281.31	0.171
	Total	36840.73	1.000

The sum of the specific sum of squares is equal to the sum of diagonal elements in the original S matrix. This indicates that our method partitions the sum of squares additively into components.

To comment further on technique we shall use a spatial analogy. In this we visualize species as vectors, all with a common origin in the centroid. The squared length of the ith vector is S_{ii} and the scalar (inner) product of vector h and i is S_{hi}. Their subtending angle is α_{hi} and $\cos \alpha_{hi} = S_{hi}/(S_{hh} S_{ii})^{1/2}$. As the computations proceed the vectors are projected into spaces of progressively lower dimensions. The projection is always into a hyperplane P_i perpendicular to a given species vector or its residual. Now, if we designate by Q_A the sum of the squared lengths of the species vectors in set A including species i, and by $Q_{A\text{-}i}$ the sum of the squared lengths of the residual vectors in the hyperplane P_i, the sum of squares that can be loaded on the line of vector i is $Q_i = Q_A - Q_{A\text{-}i}$. In one extreme, all vectors are collinear with vector i. They may naturally have different lengths and they may point in the same or opposite direction. If such were the case, our ranking method would show that the total sum of squares of species in A can be absorbed completely by species i. Such a set A is said to be completely redundant with respect to species i. We would conclude just the opposite if the vectors of the companion species were all perpendicular to the vector of species i.

The preceding method thus seeks to measure redundancy of variation in a set A conditional on a given species i which itself is a member of the set. An alternative method is readily conceived which seeks to measure redundancy in species i conditional on the companion species. High redundancy means in this case high success for the companion species in accounting for variation in species i. This definition gives us clues about the method to be used to measure this second type of redundancy. The problem is, in fact, one in multiple regression.

We shall consider the linear model. For this, we designate the sums of squares and cross products matrix of the p species by $\mathbf{S}$ and its inverse by $\mathbf{S}^{-1}$. An element of $\mathbf{S}$ is S_{hi} and an element of $\mathbf{S}^{-1}$ is $(S^{-1})_{hi}$. We seek a partition,

$$S_{hh} = S_{hhs} + S_{hhc}$$

where s signifies a 'specific' component and c a 'common' component. We can define the components in terms of the *multiple correlation* of species h with $p-1$ companion species,

$$\begin{aligned} R_h^2 &= S_{hhc}/S_{hh} \\ &= (S_{hh} - S_{hhs}) / S_{hh} \ . \end{aligned}$$

From this we get

$$S_{hhc} = S_{hh}\, R_h^2 \text{ and}$$

$$S_{hhs} = S_{hh}\, (1 - R_h^2) \ .$$

But since $R_h^2 = 1 - 1/(S_{hh}(S^{-1})_{hh})$,

$$S_{hhs} = 1/(S^{-1})_{hh} \text{ and}$$

$$S_{hhc} = S_{hh} - 1/(S^{-1})_{hh} \,.$$

It is seen that the specific variance S_{hhs} of species h, the variance that cannot be accounted for by correlations with the $p-1$ companion species, is simply defined as the inverse of the hh element in $\mathbf{S}^{-1}$. Such a derivation of values for S in a sample has in fact been suggested by Rohlf (1977). An alternative method, not based on matrix inversion but on a hierarchical decomposition of the sums of squares, has been described by Orlóci (1975a). The two methods give identical results. These are obtained automatically with the help of program SPVAR in the Appendix.

We shall illustrate the analysis on the data in Table 1-1. The sums of squares and cross products matrix **S** and the inverse $\mathbf{S}^{-1}$ are given on page 18. After performing the calculation as outlined, results are obtained which we find in the table below:

Species	*Sum of squares*			*Redundancy*
h	S_{hh}	S_{hhs}	S_{hhc}	R_h^2
1	6896.9	1514.55	5382.35	0.780401
2	9022.9	3495.24	5527.66	0.612626
3	7462.9	3672.12	3790.78	0.507949
4	3576.4	1795.22	1781.18	0.498037
5	9881.6	2232.79	7648.81	0.774045

Inspection reveals that R_h^2 is largest for species *1*. We declare rank *5* on species *1* and continue with forming the first residual of **S** according to formula (1.13). If we subject this residual to matrix inversion and perform the calculations as outlined, we obtain

Species	*Sum of squares**			*Redundancy*
h	S_{hh}	S_{hhs}	S_{hhc}	R_h^2
1	0	0	0	–
2	4456.58	3495.24	961.349	0.215714
3	7149.54	3672.12	3477.42	0.486383
4	2650.08	1795.22	854.855	0.322577
5	5581.59	2232.79	3348.80	0.599972

* First residual

In this case, the value of R_h^2 is maximal for species *5*. To this species we assign rank *4*. Now, we continue computing the second residual of **S** (formula (1.13)), determine its inverse, and do the remaining calculations to get:

Species *h*	*Sum of squares** S_{hh}	S_{hhs}	S_{hhc}	*Redundancy* R_h^2
1	0	0	0	–
2	4191.83	3495.24	696.590	0.166178
3	4175.07	3672.12	502.943	0.120463
4	2342.81	1795.22	547.587	0.233731
5	0	0	0	–

* Second residual

These results indicate rank *3* to be assigned to species *4*. Yet another (3rd) residual is formed and subjected to analysis to yield:

Species *h*	*Sum of squares** S_{hh}	S_{hhs}	S_{hhc}	*Redundancy* R_h^2
1	0	0	0	–
2	3635.01	3495.24	139.774	0.0384521
3	3818.97	3672.12	146.848	0.0384521
4	0	0	0	–
5	0	0	0	–

* Third residual

Since R_h^2 in the last residual is the *squared partial correlation coefficient*, the values for both species are the same. The choice is therefore ambiguous: we are justified to rank either of the two species as first or second. Whereas the ranking in this case is based on the multiple correlation, the specific sum of squares could have been used instead. This may however lead to somewhat different results. The analysis could of course continue as Rohlf (1977) suggests to determine further, potentially informative quantities known as the partial correlations. For any species pair *h, i* this is defined based on the elements in $\mathbf{S}^{-1}$ by

$$R_{hi\cdot} = -(S^{-1})_{hi} / [(S^{-1})_{hh}(S^{-1})_{ii}]^{1/2} .$$

For species *1* and *2* in Table 1-1, $R_{1.1\cdot} = -0.516488/((6.60261)(2.86104))^{1/2} = -0.118834$. We can similarly calculate values for the partial correlation coefficient of other species pairs in the sample. The partial correlation may be positive or negative, but not less than -1 or greater than $+1$. When viewed in the context of residuals, the partial correlation coefficient is the cosine of the subtending angle of the last residuals of two species vectors, representing projections in a subspace, after the covariances with the other $p-2$ species have been removed. To show this, we give the partial correlation coefficient of species *2* and *3*: $R_{23.} = -(-0.547348)/((2.86104)(2.72322))^{1/2} = 0.196092$ or $R_{23.}^2 = 0.0384521$. The latter is exactly the same quantity as those which we gave for species *2* and *3* in the preceding table to show their redundancy.

Several comments are in order regarding the two methods of ranking:

1. We shall refer to the first as the *vector projection method* (Orlóci, 1973a) and to the second as the *regression method* (Orlóci, 1975a; Rohlf, 1977). In the first method the species given the highest rank is the one on which the sum of the squared projections of the other species' vectors is the largest. The ranking is such that species rank decreases with decreasing sum. The method takes into account residuals. In the second method, the species given the lowest rank has the highest multiple correlation with other species. Species rank declines with increasing multiple correlation computed from residuals.
2. The rank order of a species after the first becomes conditional on species whose rank order has already been determined.
3. The similarity of results in the two methods in the examples should not be construed as an indication of what should be expected. As a matter of fact, runs on simulated data confirmed that the two methods tend to produce different results. What should then be the criteria of choice between the two? Whereas the answer to this question is not entirely clear, we can put forward some suggestions. It is obvious that the choice should be such that it would permit ranking species in a pilot sample in a manner compatible with the method of subsequent data analysis. Ranking based on the vector projection method appears to be a natural choice when the data are to be subjected to analysis by a method that itself uses the sums of squares. Some of the linear ordination methods are examples for this which are described in Chapter 3. Several of these extract axes which are uncorrelated and individually maximize the sum of squared vector projections. Here the suitability of the axes to absorb as much as possible of the total sum of squares in the sample is a measure of ordination success. The regression method of ranking thus puts that property into focus which we recognize as the relative uneffectiveness of species in influencing the linear correlation structure in their associated residual sets. When the analysis is directed to reveal such a structure, the species whose redundancy is lowest will have the least amount of information to contribute when compared to other species in the sample.

We shall proceed from here, on the basis of the proposition that the user is interested to perform some analysis on the data and intends to use the distance function

$$d(j, k) = [\sum_h (X_{hj} - X_{hk})^2]^{1/2}, \quad h = 1, ..., P \tag{1.15}$$

or some related quantity, such as the sum of squares, as the classification criterion to arrange N quadrats into discrete groups. Also, suppose that we wish to determine the distortion that would result from reducing the number of species in the analysis from 5 to 4, 3, 2, and finally to 1. Given $N = 10, P = 5$, and $p = 1, 2, 3, 4, 5$, we define a stress function,

$$\sigma(p; 10, 5) = 1 - \rho^2 (\mathbf{D}_p; \mathbf{D}_P),$$

where $\rho^2(\mathbf{D}_p; \mathbf{D}_P)$ is the square of the simple correlation coefficient, used as a mathematical index rather than a statistic, which relates the elements in matrix $\mathbf{D}_P$ defined according to formula (1.15) on the basis of 5 species, and the elements in $\mathbf{D}_p$ defined similarly as $\mathbf{D}_P$ but based on p species. When we perform the computations (see program STRESS in the Appendix), using rank order determined by the vector projection method for the species in Table 1-1, we obtain the following values for stress:

p	*Species in reduced sample*	$\rho(\mathbf{D}_p; \mathbf{D}_P)$	$\sigma(p; 10, 5)$ %
1	2	0.729	46.9
2	2, 3	0.728	47.0
3	2, 3, 5	0.886	21.5
4	2, 3, 5, 4	0.920	15.4
5	2, 3, 5, 4, 1	1.000	0.0 .

In realistic samples of data the curve plotted from the stress values could be examined to decide how many of the P species would be worth further consideration. It is interesting to note that in this example when the two highest ranking species are used, stress slightly increases. This can happen, but the general trend is a reduction in stress with increased number of species. Although the example uses the results of the projection method of ranking for input in stress analysis, any ranking method could be used to order the species prior to stress analysis.

In planning extensive sample surveys problems always arise in deciding what kind of data to be collected. For example, when contemplating count data, it is logical to ask if it is worth the effort to count all individuals within a sampling unit or only up to a predetermined maximum number (Orlóci, 1968a). With k as the maximum count, the surveyor will continue counting individuals within a sampling unit until one of two things happens: either k is reached and the counting stops, or all individuals are counted before reaching the allowed maximum limit k. The question, which can be answered in the course of a pilot survey, can be put this way: *how much information would be lost by not counting individuals beyond the number k?* Finding an appropriate answer to this question has considerable importance in view of the possibility of finding very high numbers of individuals per sampling unit, the counting of which may require a sampling effort far out of proportion with the gain in information. We assume in the following example that we have on hand counts of all individuals in quadrats of a pilot sample, that the quadrats (sampling units) are chosen by random sampling, that the method of analysis to which the data will eventually be subjected is known, and that the species list is reasonably complete. Let us suppose that the objective is a classification of quadrats based on a method in which formula (1.15), or a related function, is a basis for decisions. Let there be p species and N quadrats in the pilot sample. Let $\mathbf{D}$ represent the distance matrix computed from the data of all counts, and $\mathbf{D}_k$ the distance matrix computed from truncated counts, none of which exceeds k. The anticipated loss of information by truncating the counts is proportional to the distortion in $\mathbf{D}_k$ relative to $\mathbf{D}$. This can be measured in terms of the stress function (Orlóci & Mukkattu, 1973),

$$\sigma(\mathbf{D}_k;\mathbf{D}) = 1 - \rho^2(\mathbf{D}_k;\mathbf{D}),$$

where $\rho^2(\mathbf{D}_k;\mathbf{D})$ is the squared correlation coefficient, used here as a mathematical index rather than a statistic. The correlation coefficient relates the elements in $\mathbf{D}_k$ to the elements in $\mathbf{D}$. The stress values at different values of k in the table below, computed in program STRESC in the Appendix, correspond to the data in Table 1-1, which we hereby assume to represent plant counts in given quadrats:

Value of k	$\rho(\mathbf{D}_k;\mathbf{D})$	$\sigma(\mathbf{D}_k;\mathbf{D})$	*Sampling effort* %
10	0.278	92.3	4.3
20	0.489	76.1	8.3
30	0.640	59.0	13.6
40	0.768	41.0	23.1
50	0.858	26.4	30.8
60	0.913	16.6	33.7
70	0.955	8.8	44.1
80	0.983	3.4	60.7
90	0.997	0.6	79.4

If we assume that the sampling effort is directly proportional to the number of individuals counted, it can be seen from the table that stress and sampling effort have an inverse non-linear relationship. In this particular example, for instance, counting up to $k = 50$ means an expenditure of about 31% of the total sampling effort and yet only a 26% distortion. A roughly 100% increase of sampling effort from 31% to about 61% would produce an almost 8-fold reduction in stress.

Presence/absence data represent a special case, equivalent to counting only one. In the case of a sample, such as the one in Table 1-1 in which no absences are registered and in which all variation is associated with the changing quantity of the species, the presence scores would yield totally uninformative data. Very often, however, samples incorporate absences which introduce zeros into the data. In such a case, quantitative data can be partitioned into one component specific to species presence and another representing residuals. Such a partitioning is outlined below following the method of Willams & Dale (1962; Noy-Meir, 1971a):

1. Represent species i by vector $(0, X_i)$ in which zeros indicate absences and the X_i represent quantities. Let M_{Xi} be the mean of the non-zero values. For example, in the observational vector

$$(0, X_i) = (0 \;\; 1 \;\; 0 \;\; 8 \;\; 3 \;\; 7)$$

the mean M_{Xi} is equal to 19/4. The vector $(0, X_i)$ has as many elements as there are quadrats in the sample.

2. Now partition $(0, X_i)$ into two independent components,

$$(0, M_{Xi}) = (0 \;\; 19/4 \;\; 0 \;\; 19/4 \;\; 19/4 \;\; 19/4)$$

for presence, and

$$(0, X_i - M_{Xi}) = (0 \;\; 1\text{-}19/4 \;\; 0 \;\; 8\text{-}19/4 \;\; 3\text{-}19/4 \;\; 7\text{-}19/4) \qquad (1.16)$$

for the residual. The vector product,

$$(0, M_{Xi})\,(0, X_i - M_{Xi})' = 0$$

indicates that the partition is orthogonal.

Ecologists often rely on presence data which greatly simplify relevé protocol, allow an economical use of core space in computer analysis, and also, render the data available for analysis by a number of specialized methods. Anderberg (1973) points out that on a computer with a 60-bit word length, a data matrix of 300 species and 5000 relevés, when given as binary scores, would require only a mere 25000 words core. Should the data be other than binary, 1500000 words of core storage would be required. Opinions divide on the ecological usefulness of presence data. The difference in opinion can however be traced to the objectives that were to be achieved, the type of vegetation surveyed, the method of data analysis used, etc., (cf. Bouxin, 1975a).

If we decide to use presence data we must anticipate an amount of information loss. This loss is proportional to the residual component's (formula (1.16)) contribution to variability in the data. Assuming that the objective is a classification of quadrats based on formula (1.15), the anticipated loss of information is proportional to the stress or distortion in the distance configuration $\mathbf{D}_{PR}$, based on the presence component of the data, relative to the distance configuration $\mathbf{D}$, computed from the unpartitioned data. The function

$$\sigma(\mathbf{D}_{PR}\,;\mathbf{D}) = 1 - \rho^2(\mathbf{D}_{PR}\,;\mathbf{D})$$

quantifies stress as the one-complement of the squared correlation which relates the elements in $\mathbf{D}_{PR}$ to the corresponding elements in $\mathbf{D}$.

A stress analysis of this sort is applicable also to measurement data. In one case it may be meaningful to consider the relative merits of sets of data of different complexity with regard to sampling effort and information content. Tree height (H), basal area (B) and volume (V) represent data that are often collected in surveys of forest communities. In order to determine volume we need both basal area and height – both of them very time consuming for measurement. Now if we want to economize in the survey we may decide to use either height or basal area, but not both. The question is then which of the two would retain more of the total variation associated with volume. A comparison of two stress values may provide a basis for decision:

$$\sigma(\mathbf{D}_B; \mathbf{D}_V) = 1 - \rho^2(\mathbf{D}_B; \mathbf{D}_V)$$

$$\sigma(\mathbf{D}_H; \mathbf{D}_V) = 1 - \rho^2(\mathbf{D}_H; \mathbf{D}_V) .$$

We may choose basal area if $\sigma(\mathbf{D}_B; \mathbf{D}_V)$ is smallest. It is assumed that we have data from a pilot survey for basal area, height, and volume. The $\mathbf{D}_B$, $\mathbf{D}_H$ and $\mathbf{D}_V$ are distance matrices defined depending on the objectives of the subsequent analysis to be performed on the data. Function ρ is defined as before.

In the preceding discussions we focused on methods based on which we can identify unusual species. To identify outlier quadrats, we may use the method of Goodall (1969a). This can give us another means to reduce the number of species through identifying quadrats of unusual species composition and deleting them from the sample. Goodall's (1969a) method involves a general test on the sample or a specific test on a single quadrat. The two tests can be combined into a single procedure. We begin by labelling as unique any species which occurs in only one of the n quadrats. There can be more than one unique species in a sample and also in a quadrat. To proceed, we assume that their number in the sample is r and that their distribution accords with $n_0, n_1, ..., n_r$, where n_i indicates the number of quadrats containing i unique species. The totals are $n_0 + n_1 + ... + n_r = n$ and $n_0 0 + n_1 1 + ... + n_r r = r$. Some n_i will certainly be zero if $r < n$, and some larger than zero if $r \neq 0$.

If the sample of n quadrats were homogeneous for the r unique species, i.e. if the unique species were dispersed at random among the n quadrats, we would expect the observed distribution

$$\mathbf{F} = (f_0 \, f_1 \, ... f_r) = (n_0 0 \;\; n_1 1 \; ... \; n_r r)$$

not to differ significantly from the distribution

$$\mathbf{F}^0 = (f_0^0 \, f_1^0 \, ... f_r^0) = (n_0 \bar{X} \;\; n_1 \bar{X} ... \; n_r \bar{X}),$$

where $\bar{X} = r/n$ is the random expectation for unique species per quadrat and $\sum_i f_i = \sum_i f_i^0$. If the quantity

$$\chi^2 = \sum_i (f_i - f_i^0)^2 / f_i^0$$

$$= \sum_i n_i (i - \bar{X})^2 / \bar{X}, \quad i = 0, ..., r ,$$

is not larger than $\chi^2_{\alpha;n-1}$, which is the α probability point of the chi square distribution with $n-1$ degrees of freedom, then we can declare the sample homogeneous at the chosen probability level α. This means that we would not expect outliers to occur in the sample in the sense of extreme concentrations of unique species in certain quadrats. Goodall's (1969a) example illustrates the case in point. In this $r = 101$, $n = 67$ and

$$\mathbf{F} = (\,(32)(0)\,(13)(1)\,(10)(2)\,(5)(3)\,(1)(4)\,(3)(5)\,(0)(6)\,(0)(7)\,(2)(8)\,(1)(18)\,).$$

A general element in $\mathbf{F}^0$ is $n_i(101/67)$. The test criterion is

$$\chi^2 = (32)(0 - 101/67)^2 / (101/67) + \ldots + (1)(18 - 101/67)^2 / (101/67)$$
$$= 324.22\,.$$

Since $\chi^2 > \chi^2_{0.05;66}$ we declare heterogeneity on the sample at $\alpha = 0.05$ rejection probability. To identify to what extent a given quadrat j is an outlier, we have to compute the probability

$$P = 1 - (1 - P_1)^n\,,$$

in which P_1 is approximated by the quantity

$$\sum_i \{r!/[i!\,(r-i)!\,]\}\,\{(n-1)/n\}^{r-i}\,\{1/n\}^i, \quad i = R, \ldots, r\,.$$

P_1 is the probability that a randomly chosen quadrat q will be equally extreme as the one (j) actually observed having R of the unique species that occurred only in quadrat j. P gives the probability that q will be at least as extreme as j. If we put $r = 101$, $n = 67$ and $R = 18$ and perform the calculations using Goodall's algorithm, we find that $P = 1.11 \times 10^{-12}$. Finding P does not mean that the question of whether or not quadrat j is an outlier has been answered. As a matter of fact many more computational steps may have to be completed. Fortunately, Goodall (1969a) offers a computer program. After determination of P_1, further probabilities $P_2, \ldots, P_k$ are determined, taking sets of species that occur only in quadrat j and one other quadrat, in quadrat j and two other quadrats, and so on. Goodall (1969a) carried the analysis up to $k = 34$ and combined the $P_1, \ldots, P_{34}$ probabilities in the form of

$$\chi^2_j = -\sum_i \ln P_i, \quad i = 1, \ldots, k\,.$$

He determined the value to be 158.21. With $2k = 68$ degrees of freedom the associated probability $P(\chi^2 > 158.21) < 10^{-10}$. At any reasonable rejection probability, quadrat j would have be to regarded as a very extreme outlier. Similar computation could be performed over and over again on all the quadrats until all the outliers were identified, and then, the computations could be performed all over again on groups of two, three, etc., outliers.

Goodall (1969a) points out a peculiarity of his method which could interfere with the identification of outlier quadrats. If there is only one outlier of the same kind it will be detected. However, if there are two of the same kind, to detect them quadrat pairs have to be scrutinized. Similar considerations apply when three and more outliers are of the same kind. The examples provided by Goodall (1969a) indicate that the method can in fact detect outlier quadrats if the data originate in strikingly different communities. But when homogeneous data were analysed that came from the same community outliers were not detected.

A further method suggested for recognizing outliers when the descriptions are quantitative has been described by Wilks (1963). This method uses the ratio

$$R_j = |\mathbf{S}_{-j}| \, / \, |\mathbf{S}|$$

where $|\mathbf{S}|$ and $|\mathbf{S}_{-j}|$ are the determinants of the sums of squares and cross product matrix of p species with and without quadrat j. The quadrat which minimizes R_j is the most likely candidate for an outlier. Minimizing R_j is the same as maximizing

$$D^2(\mathbf{X}_j; \overline{\mathbf{X}}_{-j}) = (\mathbf{X}_j - \overline{\mathbf{X}}_{-j})' \, \mathbf{S}_{-j}^{-1} \, (\mathbf{X}_j - \overline{\mathbf{X}}_{-j})$$

(Rohlf, 1975) which is the square of the so-called generalized distance of a point $\mathbf{X}_j$, representing quadrat j, from the centroid of the sample, i.e. the tip of the sample mean vector $\overline{\mathbf{X}}_{-j}$, excluding quadrat j. The matrix $\mathbf{S}_{-j}^{-1}$ is the inverse of $\mathbf{S}_{-j}$. The generalized distance is discussed in Chaper 2. Here we only give an example using the data in Table 1-2. The sample sums of

Table 1-2. Species abundance estimates in quadrats

Species	*Quadrat* 1	2	3	4	5	*Mean* $\overline{X}$	$\overline{X}^*_{-2}$
1	2	5	2	1	0	2.0	1.25
2	0	1	4	3	1	1.8	2.00
3	3	4	1	0	0	1.6	1.00

* Mean in sample without quadrat *2*. See explanations in the main text.

squares and cross products matrix including quadrat *2* is given by

$$\mathbf{S} = \begin{bmatrix} 14.00 & -2.0 & 12.0 \\ -2.0 & 10.8 & -6.4 \\ 12.0 & -6.4 & 13.2 \end{bmatrix} .$$

The sums of squares and cross products matrix without quadrat *2* is given by

$$\mathbf{S}_{-2} = \begin{bmatrix} 2.75 & 1.00 & 3.00 \\ 1.00 & 10.00 & -4.00 \\ 3.00 & -4.00 & 6.00 \end{bmatrix} .$$

The determinants are $|\mathbf{S}| = 121.6$ and $|\mathbf{S}_{-2}| = 1.0$, and the ratio

$$R_j = |\mathbf{S}_{-j}| / |\mathbf{S}| = 1/121.6 = 0.008\,.$$

We take this to be small. It follows then that quadrat j must be regarded as an extreme outlier. The range of R_j is from 0 to 1 (assuming that $\mathbf{S}$ is not singular). To compute $D^2(\mathbf{X}_j;\overline{\mathbf{X}}_{-j})$ we need the covariance matrix which we derive from the sums of squares and cross products matrix by dividing each element by 3 (one less than the number of data elements per species). This matrix is given by

$$\mathbf{S}_{-j} = \begin{bmatrix} 0.91667 & 0.333333 & 1.00000 \\ 0.333333 & 3.33333 & -1.33333 \\ 1.00000 & -1.33333 & 2.00000 \end{bmatrix}.$$

The inverse is

$$\mathbf{S}_{-j}^{-i} = \begin{bmatrix} 131.962 & -53.9845 & -101.971 \\ -53.9845 & 22.4936 & 41.9880 \\ -101.971 & 41.9880 & 79.4773 \end{bmatrix}.$$

From these we have

$$D^2\,(\mathbf{X}_j;\overline{\mathbf{X}}_{-j}) = 452.124\,.$$

This corresponds to a distance $D(\mathbf{X}_j;\overline{\mathbf{X}}_{-j}) = 21.26$ measured in abundance units. Whether this is large or small is difficult to say. To determine its relative magnitude we may attach to $D(\mathbf{X}_j;\overline{\mathbf{X}}_{-j})$ a probability which measures the change that any quadrat could fall at least as far from the group centroid as quadrat j assuming that the group (quadrats 1, 3, 4, 5 in the example) is a random sample from a multivariate normal population whose mean vector is specified by $\overline{\mathbf{X}}_j$. If this probability is reasonably small, say 0.05, $D(\mathbf{X}_j;\overline{\mathbf{X}}_{-j})$ may be regarded as unusually large. We show how to determine such a probability in Chapter 5.

Rohlf (1975) suggests alternative methods for detecting outliers. In one, he constructs a minimum spanning tree (shortest simply connected graph) and examines the distribution of edge lengths for comformity with the distribution expected when the underlying probability distribution in the data is multivariate normal. In another, he examines the ratio of the edge length to an outlier and the average edge length in the graph. Rohlf (1975) also gives a table of probability points. Regarding spanning trees, the reader is referred to Chapter 4 and the appropriate sections in Sneath & Sokal (1973) for short summary and references.

We shall now turn to a systematic description of resemblance functions of which formula (1.15) represents an example.

Chapter 2

RESEMBLANCE FUNCTIONS

In this chapter we examine the concepts of resemblance and give explicit definitions. We describe and evaluate different functions to measure resemblance and illustrate their computations through numerical examples. It will be seen that these functions are essential in multivariate analysis since they generate input for the different methods.

2.1 General aspects

The term *resemblance* is used here in the colloquial sense. We note that resemblance is a measurable property of objects, or groups of objects, either as a likeness or dissimilarity, as a function of the characteristics which the objects possess. The objects may represent individual species, entire stands of vegetation, or some entities other than these.

There are several general points to be considered before we turn to a detailed discussion of resemblance structure and resemblance functions:

1. The conceptual picture which we form about ecological objects, their role, or their significance, is dependent on our perception of their similarities. This picture about the objects cannot however be more precise than the accuracy with which we can determine similarities. In this respect consistency and objectivity are prime requirements.
2. The formal methods of data analysis often assume that the resemblance of given objects is measured without personal bias. This makes the use of objective functions a basic requirement.
3. Directly related to the previous points, to be consistent, resemblance must be measured in comparable units.
4. Since there are many functions to choose from, the choice between them raises problems. Firstly, there may be restrictions on mathematical admissibility. These require the user to understand the mathematical properties of the resemblance function, and also, the relevant properties of the method of analysis to which the resemblance values are subjected. Secondly, a resemblance function may do more than what the user wants it to do in ways of standardizations or other adjustments which may be definitely to the detriment of the objectives. Thirdly, the different functions are not equally simple for computations.

2.2 Resemblance structure

To the ecologist, resemblance structure often means some kind of a covariance structure which he can readily see in samples of vegetation data. Recognition of such a structure comes naturally to persons whose training is traditionally strong in Euclidean geometry. However, a covariance structure is not the only kind that can be associated with vegetation data. Functions other than the covariance may impose structures which are radically different.

In specific terms, we can define resemblance structure as a set of resemblance values measured between objects based on the use of some objective function. It directly follows from this definition that no single kind of structure may be singularly characteristic for a given sample. In other words, the same sample may be associated with different resemblance structures depending on the resemblance function.

2.3 Sample space

Let us consider the data in Table 1-2. These data represent our sample on which an analysis is to be performed. We shall use symbol p to designate the number of species (p = 3) and symbol n to signify the number of quadrats (n = 5 in the present example). We should note that in more realistic samples there may be many more species and a greater number of quadrats than p or n in the present example.

The raw data in Table 1-2 can be represented in matrix form such as

$$\mathbf{X} = \begin{bmatrix} X_{11} & X_{12} & X_{13} & X_{14} & X_{15} \\ X_{21} & X_{22} & X_{23} & X_{24} & X_{25} \\ X_{31} & X_{32} & X_{33} & X_{34} & X_{35} \end{bmatrix} = \begin{bmatrix} 2 & 5 & 2 & 1 & 0 \\ 0 & 1 & 4 & 3 & 1 \\ 3 & 4 & 1 & 0 & 0 \end{bmatrix}.$$

Each row in $\mathbf{X}$ describes a species and each column holds values which signify performance in one quadrat. A row of $\mathbf{X}$ may be referred to as a *species vector* and a column as a *quadrat vector.* These terms may be interchanged with terms such as *row vector* or *column vector.*

It will be convenient to conceive n quadrats as a multitude of n points in p-dimensional space (H). The relative placements of the points in such a space is in proportion to differences in species composition in the quadrats. The points and the resemblance function $f(j, k)$, which measures the spatial placement of the points, constitute the *sample space.* The function $f(j, k)$ may represent a distance or it can be a measure of similarity. In either case $f(j, k)$ is said to be a *spatial parameter.* Since $f(j, k)$ associates a number with each pair (j, k) of objects, it is quite appropriately referred to as a *pair-function.*

There are several important consequences of the foregoing definition of sample space:

1. The sample space and resemblance structure exist independently from the actual sampling location of the objects or their ground distance.
2. The resemblance structure and the raw data are only indirectly related according to transformations specified in the resemblance function.
3. The information, revealed by a technique of data analysis, is the information which is realized in the sample resemblance structure, and not the information contained in the raw data.
4. When the raw data are given, and a resemblance function specified, the sample space is completely defined.

The following presentations rely on a paper by Orlóci (1972a) for organization and, in some respect, for detail.

2.4 Metrics and metric-related functions

A. Constraints on a metric

Resemblance functions must obey certain simple restrictions. We examine these under the assumption that the function $f(j, k)$ is a measure of distance in H. We further stipulate that $f(j, k)$ is a particular type of distance known as a *metric.* We also note that ecologists have used 'distance' functions that are not of the category of metrics, and that if we define j, k as two quadrats, $f(j, k)$ is an abstract measure, depending on species composition, but not a 'ground' distance.

Functions which ecologists have used to measure distance are conveniently categorized on the basis of the extent to which they satisfy certain conditions known as the *metric space axioms.* We shall examine these on the assumption that a distance is measured between vegetation quadrats:

Axiom 1. If $A = B$ then $d(A, B) = 0$
Axiom 2. If $A \neq B$ then $d(A, B) > 0$
Axiom 3. $d(A, B) = d(B, A)$
Axiom 4. $d(A, B) \leqslant d(A, C) + d(B, C)$

for any A, B or C in H.

Axiom 1 states that quadrats of identical composition ($A = B$) are indistinguishable. *Axiom 2* makes explicit that the distance must be positive, and other than zero, when A and B are non-identicals. *Axiom 3* states the symmetry property, implying that the order in which the quadrats are compared must not influence the resulting distance. *Axiom 4* is known as the *triangle inequality axiom*, according to which the distance between two quadrats A, B cannot be greater than the sum of their individual distances from a third quadrat C.

As a point of interest in method design, attention is drawn to certain properties of metrics:

1. The sum of two metrics is also a metric.
2. Any k-multiple of a metric is itself a metric.
3. The product of two metrics may not be a metric. The same holds true for the square of a metric.
4. Whereas the quantity $d(A, B) + k$ and $d(A, B)/[d(A, B) + k]$ are metrics if $d(A, B)$ is a metric and k is a positive number or zero, the quantities $d(A, B) - k$ and $d(A, B)/[d(A, B) - k]$ may not be metrics.

We find that in certain numerical procedures, e.g., characterization of a hierarchical dendrogram (classification tree, see Chapter 4), metrics are constrained by requiring them to satisfy a more restrictive inequality than in Axiom 4, i.e.

$$d(A, B) \leqslant \max[d(A, C), d(B, C)] \text{ for any } A, B, C \text{ in H.}$$

This is known as the *ultrametric inequality* (Jardine & Sibson, 1968a).

The extent to which the function $f(j, k)$ shares properties in common with the metric $d(A, B)$ is the criterion in our system for categorization. On this basis we can make a sharp distinction between *metric, semimetric* and *non-metric* categories. These categories are quite relevant to us, considering that distances from all of these categories have been used by ecologists in vegetation studies. The user must however realize that distances from the different categories are not equally suited for the definition of resemblance under specific conditions of application.

A distance function which satisfies all four of the metric space axioms is said to be a metric. (The term *metric* may be used either as an adjective or as a noun.) A distance function that satisfies the first three axioms but fails on the fourth is known as a semimetric. Some resemblance functions are metric related in the sense that they can be derived directly from a metric. We shall consider later on in this chapter important metric-related functions, such as the covariance, correlation, etc., after a detailed consideration of metrics.

B. The Euclidean distance, variants of formulae, and related cross product forms

The most familiar expression of Euclidean distance is

$$e(j, k) = [\sum_h (X_{hj} - X_{hk})^2]^{1/2}, \quad h = 1, ..., p \qquad (2.1)$$

In this expression the symbol p indicates the number of species in the sample, and X_{hj} or X_{hk} signifies the quantity or state of species h in quadrat j or k. Formula (2.1) defines a distance between two quadrats as a simple sum of p squared differences. A similar distance measure may be formulated for species, depending on the context of the problem.

Formula (2.1) is a special case of the so-called Minkowski metric (M), i.e. the nth root of the sum of the nth power of absolute differences. When $n = 2$, M becomes the Euclidean distance and when $n = 1$, the absolute value function. A related quantity,

$$d(j, k) = \max (X_{1j} - X_{1k}, ..., X_{pj} - X_{pk})$$

is known as the Chebeychev metric. While the group of Minkowski metrics have demonstrated utility in vegetation studies, the value of the Chebeychev metric in similar applications has yet to be determined.

The computational steps for formula (2.1) are shown on part of the data in Table 1-2:

Species	*Quadrat* 1	2	*Deviation*	*Squared deviation*
X_1	2	5	−3	9
X_2	0	1	−1	1
X_3	3	4	−1	1
			Sum:	11

We take the square root of the sum of squared differences to obtain $e(1,2) = 3.317$.

Distances measured on the basis of formula (2.1) have several peculiarities which may or may not be desired:

1. Requirement of commensurability of variables.
2. No fixed upper bound.
3. Reliance on absolute species quantity.
4. Dependence on species correlations.

Since differences are formed, squared and added together, the requirement of commensurability must be satisfied. This brings to attention important points that are not to be overlooked. Firstly, the same species should be measured in the same units on each occurrence. Secondly, the variables used should measure on an interval or ratio scale, or they should be binary. Strictly speaking, variables of the ordinal or nominal type are unsuited for expressions of difference such as in formula (2.1) and in other formulations which involve similar manipulations. For example, when life-form is coded and X_1, X_2 are data elements, the quantity $X_1 - X_2$ is completely dependent on the coding scheme and for that reason it is quite unsuited to serve as a unique measure of difference between two individuals. Similar considerations would apply if X_1, X_2, ... described order such as the estimated successional status of stands, rank order of species on some arbitrary scale of abundance, etc. Thirdly, the different species for which the differences are summed should be measured on an identical scale. If such were not the case, the distance would not have an interpretable scale.

To recognize that the value of formula (2.1) does not have a fixed upper bound is important because it can render the distances incomparable. For this reason we would normally prefer a relative measure on the basis of which we can decide whether a distance value is small or large. The size of $e(j, k)$ depends on the magnitude of the actual differences in species quantities between the quadrats. Because of the reliance on such differences, an anomalous situation can arise: two quadrats which have no species in common may appear more similar than two other quadrats whose species lists are identical. Three quadrats, with data vectors

$$\mathbf{X}_A = \begin{bmatrix} 0 \\ 1 \\ 1 \end{bmatrix} \qquad \mathbf{X}_B = \begin{bmatrix} 1 \\ 0 \\ 0 \end{bmatrix} \qquad \mathbf{X}_C = \begin{bmatrix} 0 \\ 4 \\ 4 \end{bmatrix},$$

represent an example for such an anomalous situation. In terms of formula (2.1) quadrat A is more similar to B ($e(A, B) = 1.732$) with which it shares none of the species than to C ($e(A, B) = 4.243$) with which it shares an identical species list. The Euclidean distance with such properties is not particularly useful in measuring quadrat resemblances.

The problem can be overcome by computation of relative distances in the following manner:

1. Normalize the quadrat vectors. This is done to set their lengths equal to unity. If, for instance, the vector which describes a quadrat is (2 0 3) with the sum of squared elements equal to 13, the normalized vector is given by

($2/\sqrt{13}$ 0 $3/\sqrt{13}$). When the elements in the normalized vector are squared and summed the sum should be unity if no mistakes were made in the arithmetic.

2. Compute the Euclidean distance based on normalized quadrat vectors. The distance so computed is called a *chord distance* (Orlóci, 1967a) indicating that the length of the chord connecting two points on the surface of a sphere of unit radius has been measured. The chord distance can be obtained directly based on the formula

$$c(j, k) = [2(1 - q_{jk} / (q_{jj} q_{kk})^{1/2})]^{1/2} \tag{2.2}$$

where

$$q_{jk} = \sum_h X_{hj} X_{hk}, \; q_{jj} = \sum_h X^2_{hj} \text{ and } q_{kk} = \sum_h X^2_{hk}, \; h = 1, ..., p.$$

The distance in formula (2.2) is characterized by the following properties:

a. $0 \leqslant c(j, k) \leqslant \sqrt{2}$

b. $c(j, k) = 0$ if $X_{hj}/X_{ij} = X_{hk}/X_{ik}$ for any h or i

c. $c(j, k) = \sqrt{2}$ if $X_{hj} = 0$ implies $X_{hk} > 0$ and $X_{hk} = 0$ implies $X_{hj} > 0$.

These indicate that for the chord distance to be zero the two quadrats need not have the same species in equal quantity; it is sufficient if the species quantities are in the same proportion. If the species quantities are proportional the normalized quadrat vectors will turn out to be the same. The normalized data vectors of quadrats *A, B, C* are

$$\mathbf{X}_A^* = \begin{bmatrix} 0 \\ 0.707 \\ 0.707 \end{bmatrix} \qquad \mathbf{X}_B^* = \begin{bmatrix} 1 \\ 0 \\ 0 \end{bmatrix} \qquad \mathbf{X}_C^* = \begin{bmatrix} 0 \\ 0.707 \\ 0.707 \end{bmatrix}.$$

The quadrat distances are $c(A, B) = c(B, C) = \sqrt{2}$ and $c(A, C) = 0$. The order established in the case of the $e(j, k)$ distances is completely reversed here; *A* and *C* are declared 'identical' and both completely different from *B*.

A further example in the table below utilizes the data in Table 2-1:

Species	*Quadrat* 1	2	$X_{h1}X_{h2}$	X^2_{h1}	X^2_{h2}
X_1	2	5	10	4	25
X_2	0	1	0	0	1
X_3	3	4	12	9	16
			$q_{12} = 22$	$q_{11} = 13$	$q_{22} = 42$

From these we obtain $c(1,2) = (2(1 - 22/((13)(42))^{1/2}))^{1/2} = 0.342$. The program EUCD in the Appendix gives options to compute Euclidean distance based on formula (2.1) or (2.2).

The influence of species correlations on the distance measure leads to reasoning along two rather different lines:

1. Species correlations are the carriers of information about trends in the sample. Linear species correlations associate with linear trends and non-linear correlations with curved trends in the sense of points being placed in a curved line, surface, or some solid of certain complexity in H. Should it be decided to remove the influence of species correlations from a multivariate sample, it would impose a spherical shape on the point configuration with a simultaneous obliteration of existing trends. It is clear then, that when the objectives of an analysis call for trend seeking we shall prefer those distance functions which reflect the influence of species correlations. Formula (2.1) and also formula (2.2) are examples.

2. The influence of species correlations on the distance measure may be completely undesired in the case of analyses in which an individual's class affiliation is identified based on a direct comparison of distances. To clarify this point let us assume that an external quadrat j is compared to quadrat k of vegetation type A and to quadrat m of vegetation type B. The distance $e(j, k)$ or $e(j, m)$ is under the influence of correlations $\mathbf{S}_A$ in type A or $\mathbf{S}_B$ in type B. Distance $e(j, k)$ can logically be compared to $e(j, m)$ only if the two vegetation types are characterized by an identical pattern of species correlations in terms of $\mathbf{S}_A = \mathbf{S}_B$. Otherwise, their comparison lacks a common basis. To incorporate a term for species correlations in the distance measure, and in this way to provide a common ground for comparison, we may use oblique co-ordinate axes with their obliqueness proportional to species correlations, similar to the method of Gengerelli (1963), rely on the use of the Mahalanobis (1936) generalized distance, or utilize other suitable measures, not necessarily metrics, with sufficient provision for species correlations.

An expression for Euclidean distance can be formulated for the case of oblique axes on the basis of Fig. 2-1. In this figure $\cos w_{hi} = S_{hi}$ represents a simple product moment correlation coefficient (used here as a direction cosine rather than a statistic); j and k indicate quadrats with data vectors $(X_{hj}X_{ij})$ and $(X_{hk}X_{ik})$. The correlation coefficient is defined according to

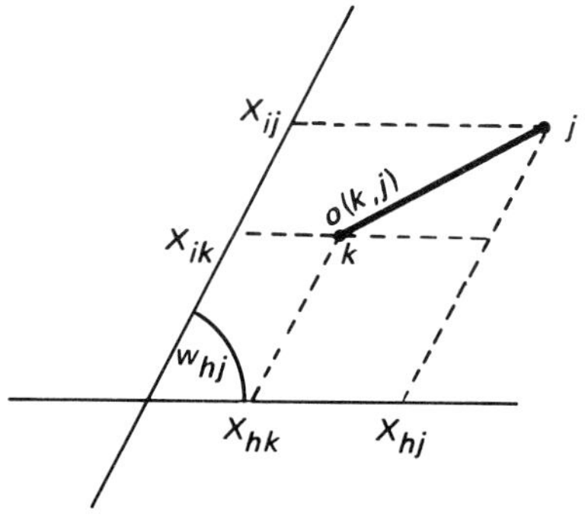

Fig. 2-1. Diagramatic representation of an oblique reference system. See explanations in the main text.

$$S_{hi} = \sum_j (X_{hj} - \bar{X}_h)(X_{ij} - \bar{X}_i) / [\sum_j (X_{hj} - \bar{X}_h)^2 \sum_j (X_{ij} - \bar{X}_i)^2]^{1/2}$$

with summations taken from $j = 1$ to n. Symbols $\bar{X}_h$, $\bar{X}_i$ represent species means. The formula for the squared Euclidean distance of j and k based on oblique co-ordinates then is given by

$$o^2(j, k) = (X_{hj} - X_{hk})^2 + (X_{ij} - X_{ik})^2$$
$$-2(X_{hj} - X_{hk})(X_{ij} - X_{ik}) \cos(180° - w_{hi}).$$

For any number of species p with correlations S_{hi}, $h,i = 1, ..., p$, the distance formula becomes

$$o(j, k) = [\sum_h (X_{hj} - X_{hk})^2 + 2 \sum_h \sum_i (X_{hj} - X_{hk})(X_{ij} - X_{ik}) S_{hi}]^{1/2} \quad (2.3)$$

with summations taken from $h = 1$ to p in the first sum, and $h = 1$ to $p-1$ or $i = h+1$ to p in the second. The first sum represents the squared Euclidean distance of formula (2.1). Added to this sum is a correction term for species correlations. With such a correction we inject new information into the definition of distance. This information is external to the objects (j, k) between which the distance is measured, dependent on the entire sample in which the correlations are measured. The following numerical example illustrates the calculation:

1. Select two quadrat vectors,

$$\mathbf{X}_1 = \begin{bmatrix} 2 \\ 0 \\ 3 \end{bmatrix} \quad \text{and} \quad \mathbf{X}_2 = \begin{bmatrix} 5 \\ 1 \\ 4 \end{bmatrix}.$$

These represent the first and second quadrat in Table 1-2.

2. Compute correlation coefficients for species. These are given as the elements in

$$\mathbf{S} = \begin{bmatrix} S_{11} & S_{12} & S_{13} \\ S_{21} & S_{22} & S_{23} \\ S_{31} & S_{32} & S_{33} \end{bmatrix} = \begin{bmatrix} 1.000 & -0.163 & 0.883 \\ -0.163 & 1.000 & -0.536 \\ 0.883 & -0.536 & 1.000 \end{bmatrix}.$$

3. Compute

$$\begin{aligned} o(1, 2) &= ((2-5)^2 + (0-1)^2 + (3-4)^2 + 2((2-5)(0-1)(-0.163) \\ &+ (2-5)(3-4)(0.883) + (0-1)(3-4)(-0.536)))^{1/2} \\ &= 3.775\,. \end{aligned}$$

Program OBLIK in the Appendix can be used to perform the computations after program CORRS has generated a file for species correlations.

The generalized distance of quadrats j and k is defined by

$$m(j, k) = [(\mathbf{X}_j - \mathbf{X}_k)' \mathbf{S}^{-1} (\mathbf{X}_j - \mathbf{X}_k)]^{1/2} \quad (2.4)$$

In this $\mathbf{X}_j$ and $\mathbf{X}_k$ represent quadrat vectors and $\mathbf{S}^{-1}$ symbolizes the inverse of the species covariance matrix. When the species are uncorrelated, and their variances are equal to unity, formula (2.4) reduces to formula (2.1).

The steps to calculate $m(j, k)$ are illustrated using data from Table 1-2:

1. Determine

$$(\mathbf{X}_1 - \mathbf{X}_2) = \begin{bmatrix} -3 \\ -1 \\ -1 \end{bmatrix}$$

for quadrats *1* and *2*.

2. Compute covariances $\mathbf{S}_{hi}$ according to the formula

$$S_{hi} = \sum_j (X_{hj} - \bar{X}_h)(X_{ij} - \bar{X}_i)/(n-1), \quad j = 1, ..., n,$$

where $\bar{X}_h$ is the mean of species h and $\bar{X}_i$ is the mean of species i. The sample values are given as the elements in

$$\mathbf{S} = \begin{bmatrix} S_{11} & S_{12} & S_{13} \\ S_{21} & S_{22} & S_{23} \\ S_{31} & S_{32} & S_{33} \end{bmatrix} = \begin{bmatrix} 3.5 & -0.5 & 3.0 \\ -0.5 & 2.7 & -1.6 \\ 3.0 & -1.6 & 3.3 \end{bmatrix}.$$

3. Invert $\mathbf{S}$ to obtain $\mathbf{S}^{-1}$. This is easiest to accomplish on a computer. When, however, the covariance matrix is of order three, as in the present example, longhand calculations are not really inhibiting. The first step in finding the inverse of $\mathbf{S}$ is to find the cofactors associated with $\mathbf{S}$. We recall from matrix algebra that the cofactor A_{hi} of S_{hi} is $(-1)^{h+i}$ times the determinant of the submatrix obtained by deleting the hth row and ith column from $\mathbf{S}$. For example the cofactor A_{23} of S_{23} is defined by

$$A_{23} = (-1)^{2+3} \begin{vmatrix} 3.5 & -0.5 \\ 3.0 & -1.6 \end{vmatrix} = (-1)^5 \left((3.5)(-1.6) - (-0.5)(3.0)\right)$$

$$= 4.1 .$$

The cofactors of the remaining elements are similarly determined to obtain the entire matrix

$$\mathbf{A} = \begin{bmatrix} 6.35 & -3.15 & -7.30 \\ -3.15 & 2.55 & 4.10 \\ -7.30 & 4.10 & 9.20 \end{bmatrix}.$$

It is noted that because the upper and lower halves of $\mathbf{S}$ are mirror images of each other it is sufficient to calculate the off-diagonal elements in one half of $\mathbf{A}$. In the next step we calculate the determinant of $\mathbf{S}$,

$$\det \mathbf{S} = \begin{vmatrix} 3.5 & -0.5 & 3.0 \\ -0.5 & 2.7 & -1.6 \\ 3.0 & -1.6 & 3.3 \end{vmatrix} = (3.5)(2.7)(3.3) + (-0.5)(-1.6)(3.0)$$

$$+ (3.0)(-0.5)(-1.6) - (3.0)(2.7)(3.0) - (3.5)(-1.6)(-1.6)$$
$$- (-0.5)(-0.5)(3.3) = 1.9 .$$

We now proceed to finding the inverse by dividing each element in **A** by det **S**,

$$\mathbf{S}^{-1} = \begin{bmatrix} 3.342 & -1.658 & -3.842 \\ -1.658 & 1.342 & 2.158 \\ -3.842 & 2.158 & 4.842 \end{bmatrix} .$$

4. From the foregoing results the generalized distance can be computed according to

$$m^2(1,2) = (-3\ -1\ -1) \begin{bmatrix} 3.342 & -1.658 & -3.842 \\ -1.658 & 1.342 & 2.158 \\ -3.842 & 2.158 & 4.842 \end{bmatrix} \begin{bmatrix} -3 \\ -1 \\ -1 \end{bmatrix} = 7.578$$

or $m(1,2) = 2.753$.

A computer program (GEND) to compute generalized distance is found in the Appendix. Examples of application are numerous in biology (Seal, 1964 and references therein), and also, specifically in ecology (e.g. Hughes & Lindley, 1955; Harberd, 1962; van Groenewoud, 1965; Gittins, 1965). We shall consider some applications in connection with identifications in Chapter 5. It should however be pointed out already at this stage that the generalized distance, as well as other distance functions of those considered so far, are readily generalized to measure distance between groups. To do this, we define $\bar{\mathbf{X}}_A$, $\bar{\mathbf{X}}_B$, $\mathbf{S}_A$ and $\mathbf{S}_B$ as mean vectors and covariance matrices in two groups. The elements in the mean vectors are the group means of species. The elements in the covariance matrices are the group covariances of the same species. We define **S** as the *pooled* covariance matrix with elements

$$S_{hi} = [(n_A - 1)S_{hiA} + (n_B - 1)S_{hiB}] / (n_A + n_B - 2),\ h,i = 1,...,p\ ,$$

in which n_A and n_B indicate the number of quadrats in the two groups. After suitable substitutions in formula (2.1), (2.2), (2.3) or (2.4), using the elements of the mean vectors and, if needed, the pooled covariance matrix, a distance can be measured between the groups. Other measures of groups resemblance are discussed by, e.g., Češka (1966, 1968) and Goodall (1968).

Problems arise in computation of formula (2.4) if **S** is singular. In such a case, we could apply the orthogonal transformation $\mathbf{Y} = \mathbf{B'X}$ of component analysis (see Chapter 3) to derive linearly uncorrelated components to replace the species data. Based on the component scores the formula for generalized distance reduces (see Chapter 5) to

$$m(j, k) = \left[\sum_h (y_{hj} - y_{hk})^2 / \lambda_h \right]^{1/2}, \quad h = 1, ..., t ,$$

where the summation is over t sets. The definition of λ_h is that of the sum of squares in the hth component. As another possibility we could also define $\mathbf{S}^{-1}$ as the generalized inverse of $\mathbf{S}$ (cf. Rao, 1962; Rahman, 1962) and use it with formula (2.4). Finally, $m(j, k)$ could be calculated based on formula (2.1) with X_{hj} defined as a standard score on an orthogonal function (see Chapter 3).

Vegetation data are often presented in binary form, indicating species presence or absence in quadrats. Such data can convey information about the number of species common to different quadrats and the frequency with which the different species occur in the entire sample. We can compute resemblance values for quadrats, or determine the intensity of association between species, from presence/absence data in a 2 x 2 table. An example is given below:

		Quadrat j 1	0	*Total*
Quadrat k	1	a	b	$a + b$
	0	c	d	$c + d$
Total		$a + c$	$b + d$	$p = a + b + c + d$.

In this table symbol a represents the number of common species in the two quadrats, b the number of species which occur in quadrat k but not in quadrat j, c the number of species occurring in quadrat j but not in k, and d the number of species absent from both quadrats. Based on these quantities, the distance of two quadrats is defined as

$$[(b + c)/p]^{1/2} = [1 - (a + d)/p]^{1/2} \tag{2.5}$$

where $(a + d)/p$ is *Sokal's matching coefficient* (Sokal & Michener, 1958). It is noted that division by p, the number of species in the entire sample, in the Euclidean distance $[(b + c)/p]^{1/2}$ may not be a particularly effective operation because apart from scaling down the distances by a factor p, nothing much will happen. A more substantial and possibly undesirable effect results when p is defined as the number of species in the two quadrats, $(a + b + c)$, because such a definition, e.g., *Jaccard's (1901) coefficient, Gower's (1971a) general coefficient*, may alter the scale of measure variably depending on the quadrats compared. This property and its consequences on the measurement of resemblance are discussed in the sequel.

Where formula (2.5) is used, the distance will retain the properties which we considered in connection with formula (2.1). A relative form may be derived by a method other than dividing $(b + c)$ by p. Firstly, we note that the cosine separation of two vectors of binary elements is incorporated in the formula

$$(b + c) = (a + b) + (a + c) - 2\,[(a + b)(a + c)]^{1/2} \cos \alpha_{jk} \ .$$

After simple rearrangement of terms we obtain

$$\cos \alpha_{jk} = a / [(a+b)(a+c)]^{1/2} .$$

This represents the *Ochiai (1957) coefficient* which has been used in a vegetation study by Barkman (1958). A related measure, the squared cosine, was suggested by van der Maarel (1966). From the coefficient we obtain

$$c(j, k) = [2(1 - \cos \alpha_{jk})]^{1/2},$$

which represents the familiar chord distance.

We can use presence scores to define a Euclidean distance in the form of

$$d(j, k) = \{2 [1 - (\chi^2_{jk}/p)^{1/2}]\}^{1/2} \quad (2.6)$$

in which χ^2_{jk}/p is the so-called *mean square contingency function*[1] associated with 2 x 2 tables. It is noted that because $(\chi^2_{jk}/p)^{1/2}$ is the direction cosine of two vectors of binary elements, $d(j, k)$ of formula (2.6) is another expression for the chord distance. The ecological applications of the mean square contingency are numerous (e.g., Williams & Lambert, 1959, 1960, 1961; Dagnelie, 1960; Frey, 1966, 1969a; Ivimey-Cook, Proctor & Rowland, 1975). A relativized chi square function was suggested by Cole (1949, 1957) and Hurlbert (1969), and in the form of

$$\max \chi^2 - \sqrt{\chi^2_{jk}} = \max \chi^2 \, (1 - (\sqrt{\chi^2_{jk}})/\max \chi^2) \quad (2.7)$$

where max signifies a sample maximum of χ^2, by Beals (1965b). The latter defines some multiple of the squared Euclidean distance. Formula (2.6) is nevertheless preferred because it measures relative distance on an unchanging scale.

If we wish to derive a relative chi square we could do it differently. We have the class symbols $A_1, ..., A_r$ in one case and $B_1, ..., B_c$ in another, their joint frequencies are displayed in an $r \times c$ contingency table with grand total $n_{..}$. The possible maximum value of chi square for such a table is $n_{..}$ [min $(r-1, c-1)$]. In a 2 x 2 table, the maximum is $n_{..}$. Since $n_{..}$ is the same for any species pair or quadrat pair in the sample, it follows that if we express chi square as a proportion of the possible maximum value, i.e. if we use the mean square contingency coefficient, it will retain an unchanging scale in all comparisons in the sample. With the maximum so defined, formula (2.7) would qualify as a constant multiple of formula (2.6). But even then the reasons of its introduction would have to be more clearly explained. The fact of the matter is that formula (2.6) does everything that formula (2.7) could do, and significantly, its generic properties are not ambiguous.

Directly derivable from Euclidean distance are certain pair functions known as *scalar (inner) products of vectors* which we compute normally as a

[1] This is more generally defined as $\chi^2/[p(q-1)]$ where the symbol q indicates the smallest dimension of the contingency table and p the table total (cf. Cramér, 1946).

sum of cross products of co-ordinates. The term *scalar product* signifies that this is the product of the lengths of two vectors and the cosine of their subtending angle. The scalar products play an important role in several methods of data analysis as we shall see in Chapter 3 in connection with certain ordinations. The formulations which are most often encountered include the following:

$$S_{hi} = \sum_j A_{hj} A_{ij},\ j = 1, ..., n \tag{2.8}$$

for any two species h and i, and

$$q_{jk} = \sum_h A_{hj} A_{hk},\ h = 1, ..., p \tag{2.9}$$

for any two quadrats j and k. If A_{hj} is defined as

$$A_{hj} = (X_{hj} - \bar{X}_h) / (n - 1)^{1/2},$$

where $\bar{X}_h$ is the sample mean of species h, S_{hi} will represent the *covariance* of species h and i. When A_{hj} is defined according to

$$A_{hj} = (X_{hj} - \bar{X}_h) / [\sum_e (X_{he} - \bar{X}_h)^2]^{1/2},\ e = 1, ..., n,$$

S_{hi} will represent the *product moment correlation coefficient.* The quantity q_{jk} of formula (2.9) is sometimes referred to as the *Q-expression* for the covariance or correlation of quadrats. The Q-expression has special importance as a dual measure replacing formula (2.1) in some methods of ordination which can more conveniently handle scalar products than distances. Scalar products can be computed in program CORRS which we give in the Appendix.

It is noted that whereas the sum of cross products, covariance, or correlation coefficient, are readily recognized as scalar products, some of the more familiar $\rho = a/m$ type coefficients (Orlóci, 1972a, 1973a; Anderson, 1971a), defined for binary data, are not. The Ochiai (1957) coefficient is an example. In this, since m = $[(a+b)(a+c)]^{1/2}$ and a is the sum of cross products, $\rho = \cos\alpha$. Clearly, both a and a/m are vector scalar products. The lengths of the vectors in a are respectively $(a+b)^{1/2}$ and $(a+c)^{1/2}$; in a/m, they have unit length. The vectors, in either case, have common origin in the zero point of the co-ordinate system. Since all such vectors point in the positive direction, the values of a/m range from 0 to 1 inclusive. The cross product a has lower limit at zero and upper limit at p. If quadrats j and k are subjected to analysis, based on presence data, $(a+b)$ and $(a+c)$ will designate respectively the number of species in the two quadrats. Since these quantities define the squared lengths of the quadrat vectors, the data adjustments in a/m are similar to the adjustments implicit in the chord distance. With such adjustments, a/m and its complement $(b+c)/m$ will retain an unchanging scale of measure, uninfluenced by the vectors compared. The coefficient of Russel & Rao (1940; also Crawford, Wishart & Campbell, 1970) in which $m = p = a + b + c + d$ is also a vector scalar product. It measures on a consistent scale, uninfluenced by the quadrats compared. The Jaccard (1901) index in which $m = a + b + c$ will however qualify as a vector scalar product with unchanging scale of measure only in comparisons for which m is constant. But as m can change from pair to pair of the vectors compared, the scale of measure in the Jaccard index, or in its complement $(b+c)/(a+b+c)$, will change also depending on the comparison.

The coefficients so far mentioned differ from the type represented by the product moment correlation coefficient. In this, the common origin of vectors is at the centroid of the system. For this reason the vectors are said to be centered. The values of the correlation coefficient may be positive or negative; the lengths of vectors are normalized. In the case of the covariance the vectors lack such a normalization, and for that reason, the measure lacks a fixed upper or lower bound.

The idea of the scalar product of centered vectors is applicable also to vectors of binary elements. For example, we may write

$$(\chi^2/p)^{1/2} = \frac{(ad - bc)}{[(a+b)\,(c+d)\,(a+c)\,(d+b)]^{1/2}} = \cos\alpha\,,$$

which is a product moment correlation coefficient. The difference

$$A = a - (a+b)\,(a+c)/p$$

is also a scalar product with vector lengths equal to

$$B = [a+b-(a+b)^2/p]^{1/2} \text{ and } C = [a+c-(a+c)^2/p]^{1/2}$$

in which case the product moment correlation coefficient is $A/(BC)$.

A further point of importance is the relationship of the Euclidean distance computed from the A_{hj} type data according to

$$d(j,k) = [\sum_h (A_{hj} - A_{hk})^2]^{1/2}\,,\ h = 1, ..., p \qquad (2.10)$$

and the scalar products determined on the basis of formula (2.9). This relationship is of the form

$$d^2(j,k) = q_{jj} + q_{kk} - 2q_{jk}\,, \qquad (2.11)$$

or in the reverse,

$$q_{jk} = -0.5\,(d^2(j,k) - \overline{d_j^2} - \overline{d_k^2} + \overline{d^2}) \qquad (2.12)$$

where $\overline{d_j^2}$, $\overline{d_k^2}$ and $\overline{d^2}$ are mean squared quadrat distances computed according to

$$\overline{d_j^2} = \sum_k d^2(j,k)\,/\,n,\ k = 1, ..., n$$
$$\overline{d_k^2} = \sum_j d^2(j,k)\,/\,n,\ j = 1, ..., n$$
$$\overline{d^2} = \sum_j \sum_k d^2(j,k)\,/\,n^2,\ j,k = 1, ..., n\,.$$

We shall build on these relationships when we develop an algorithm for component analysis in Chapter 3.

C. *Absolute value function*

This is another formulation for distance defined as a sum of absolute species differences,

$$a(j, k) = \sum_h | X_{hj} - X_{hk} | , \ h = 1, ..., p \qquad (2.13)$$

This distance, similar to the Euclidean distance of formula (2.1), is affected by species correlations. It lacks a fixed upper bound and it suffers from reliance on absolute species differences between quadrats. Because of these, it can give rise to the anomalous situation in which two quadrats with no species in common may appear more similar than two other quadrats which contain an identical set of species. It may seem then that there is little to be gained by the use of formula (2.13) as compared to formula (2.1).

The absolute value function can be made relative (Whittaker, 1952),

$$w(j, k) = \sum_h | X_{hj}/Q_j - X_{hk}/Q_k | , \ h = 1, ..., p \qquad (2.14)$$

where

$$Q_j = \sum_h X_{hj} \text{ and } Q_k = \sum_h X_{hk} ;$$

these are the sums of species quantities in the two quadrats. The values of $w(j, k)$ range from 0 to 2 inclusive. Zero indicates that the two quadrats have the species in similar proportions, i.e. $X_{hj}/X_{ij} = X_{hk}/X_{ik}$ for any species h and i. If the two quadrats have no species in common $w(j, k)$ is equal to 2. In these respects formula (2.14) is similar to formula (2.2). Sokal & Sneath (1963; Sneath & Sokal, 1973) consider the lack of squaring of terms in the formulae an advantage because in this way the differences between the entities compared are not exaggerated. More important, as we shall see later in this chapter, formula (2.14) tends to linearize trends in the data, but in sharp contrast to formula (2.2) it is not Euclidean. Consequently, it has narrower established utility in conventional data analysis.

To illustrate the computation of the relative form of the absolute value function, we use the first two quadrats of Table 1-2 in the example below:

$$Q_1 = 2 + 0 + 3 = 5$$

$$Q_2 = 5 + 1 + 4 = 10$$

$$w(1, 2) = | 2/5 - 5/10 | + | 0 - 1/10 | + | 3/5 - 4/10 | = 0.4 .$$

The distance based on formula (2.13), in which values have not been made relative, is somewhat simpler to compute because it needs no adjustments in the quadrat vectors. For this, we have

$$a(1, 2) = | 2 - 5 | + | 0 - 1 | + | 3 - 4 | = 5 .$$

A certain kind of double adjustment, with conversion from X_{hj} to $X_{hj}/(Q_h Q_j)^{1/2}$, where the Qs represent respectively the hth species total and the jth quadrat total in the sample, has been described by Williams (1952). Such an adjustment, as we shall see in Chapter 3, has a profound effect on the results of certain ordinations.

In some applications, users are concerned with the cost to implement formula (2.1) in hardware. Because of this reason, formula (2.13) has been suggested as a possible approximation to Euclidean distance. The Chebeychev metric has also been suggested for the same purpose. Batchelor (1971) considered these measures and concluded that their weighted sum gives a better approximation than either of the two individually.

D. Geodesic metric

The space in which this metric is a spatial parameter is the surface of an n-dimensional hypersphere. The geodesic metric measures the shorter arc between points. When we write for the geodesic metric,

$$g(j,k) = \text{arc cos} \sum_h X_{hj} X_{hk} / [\sum_h X_{hj}^2 \sum X_{hk}^2]^{1/2}, \quad h = 1, \ldots, p \tag{2.15}$$

the hypersphere has unit radius. The geodesic metric can be derived from the chord distance,

$$g(j,k) = \text{arc cos}\,[1 - c^2(j,k)/2] \tag{2.16}$$

The values of $g(j, k)$, as we defined this distance, range from 0 (when $c(j, k)$ is zero) to $\pi/2$ (when $c(j, k)$ equals $\surd 2$). We can thus see that the geodesic metric properties (cf. Holgate, 1971; Levandowsky & Winter, 1971) which dean. Formulation of the geodesic metric in categorical data is discussed by Edwards & Cavalli-Sforza (1964) and its relationship to different Euclidean measures by Krzanowski (1971).

2.5 Semimetrics and scrambled forms of metrics

If a distance function fails on the triangle inequality property of metrics but not on the others, it is said to be a *semimetric.* A typical case involves subtracting from a metric $d(j, k)$ its observed minimum value, i.e.

$$v(j,k) = d(j,k) - \min\,[d(j,k);\; j < k = 2, \ldots, n]\,.$$

If the minimum is sufficiently large, $v(j, k)$ will fail as a metric. Beals (1960) has used a similar measure on vegetation data.

We shall consider scrambling in connection with the one complement of *Czekanowski's (1909) index* which we write for the distance between two quadrat vectors $(\mathbf{X}_j, \mathbf{X}_k)$ as

$$b(j,k) = 1 - 2 \sum_h \min(X_{hj}, X_{hk})/Q_{jk} = \sum_h |X_{hj} - X_{hk}| / Q_{jk}, \quad h = 1, \ldots, p \tag{2.17}$$

where min signifies the smallest of the pair X_{hj}, X_{hk} and $Q_{jk} = \sum_h (X_{hj} + X_{hk})$, $h = 1, \ldots, p$. When only species presence or absence is recorded in quadrats, formula (2.17) reduces to the one complement of Sørensen's (1948) index, $2a/(2a + b + c)$), given by $(b + c)/(2a + b + c)$. The symbols accord with those of a 2 x 2 contingency table which we already explained. Twice the value of $(b + c)/(2a + b + c)$ has been used by van der Maarel (1969) as a squared distance. Should $2a + b + c$ be a constant quantity in the sample for all quadrat pairs, this distance would qualify as the squared Euclidean metric with range from zero to unity. Another distance measure,

$$E(j, k) = \mu(j \dot{-} k) \, / \, \mu(j + k) \qquad (2.18)$$

where the numerator is a measure of the *symmetric difference* of two sets and the denominator is the set theoretical sum, is known as the *Marczewski & Steinhaus (1958) distance.* In terms of the *a, b, c* symbols of a 2 x 2 table, this distance reduces to

$$E(j, k) = (b + c) \, / \, (a + b + c)$$

which happens to be the one complement of the Jaccard (1901) coefficient.

Levandowsky (1972) used the formulation,

$$E(j, k) = 1 - \sum_h \min(X_{hj}, X_{hk}) \, / \, \sum_h \max(X_{hj}, X_{hk}), \; h = 1, ..., p$$

for the same measure of distance. We note that these variants differ from the one defined in formula (2.17) in so far as these do not take into account the shared quantity (*a* in the case of presence data or the minimum values in quantitative data) with extra weight. This has far reaching consequences on metric properties (cf. Holgate, 1971; Levandowsky & Winter, 1971) which we shall treat in the sequel.

Distance measures similar to formula (2.17) are broadly used (e.g. Odum, 1950; Whittaker, 1952; Bray & Curtis, 1957; Lance & Williams, 1966b), but often in algebraic manipulations for which they are not too well suited. Lance & Williams (1967a) considered it to be to the advantage of data analysis to suggest yet another similar measure for distance,

$$d^k(j, k) = \sum_h |X_{hj} - X_{hk}|^k / (X_{hj} + X_{hk}), \; h = 1, ..., p \qquad (2.19)$$

with k defined as 1. A similar formulation has been used by Switzer (1971) but with k defined as 2. Lance & Williams (1967a; Clifford, Williams & Lance, 1969) regard standardization at the level of the data element an advantage. While it certainly is an effective means for establishing local commensurability between the h species, such a standardization is likely to have an undesired side effect of loss of metric properties by rendering the scale of $d(j, k)$ conditional on the pair (j, k) actually compared. The comparison of such $d(j, k)$ values between pairs of quadrats may therefore be problematic.

It can be seen from formula (2.17) that it is in fact related to the absolute value function (formula (2.13)) from which it differs by the division by Q_{jk}, the sum of species quantities in the two quadrats. Where this quantity is constant for all quadrat pairs in a sample, formula (2.17) will uniformly represent an absolute value function. However, where Q_{jk} itself varies from one quadrat pair to the next, formula (2.17) will not only measure on a new scale with each new value of Q_{jk}, but also, it will no longer qualify as a globally consistent metric (cf. Williams & Dale, 1965; Austin & Orlóci, 1966; Swan, Dix & Wherhahn, 1969; Levandowsky, 1972).

Such loss of a consistent scale of measure is of course not unique to measures of distance derivable from coefficients of the Sørensen type, but also to those that can be derived from the Jaccard index. The latter have however been shown to be metric (Marczewski & Steinhaus, 1958; Levandowsky & Winter, 1971) exemplified by formula (2.18), while the former are semimetric exemplified by formula (2.17). Semimetrics have been used in vegetational ordinations, and often inappropriately because such metrics cannot produce distances from which proper Euclidean triangles can be formed. Consider the expression,

$$Y_k = d(j, k) \cos \alpha_{km} = [d^2(j, k) + d^2(j, m) - d^2(k, m)]/2d(j, m)$$

which Beals (1960) has used to define co-ordinates for quadrats in an ordination. In this formula, symbol Y_k represents the co-ordinate of quadrat k on the first ordination axis through quadrats j and m which represent ordination poles. The distances in the formula are squared quadrat distances. The cosine relation naturally requires that the function generating the quadrat distances be capable of forming ordinary Euclidean triangles. If it cannot then Y_k cannot be uniquely interpreted.

When we carry out the computations and find that $Y_k/d(j, k)$ exceeds unity or is less than -1, we know that the cosine relation does not hold, i.e. the distances fail to produce triangles. This is a clear indication that the distance measure is a semimetric or not a metric at all. This can happen when $b(j, k)$ is substituted for $d(j, k)$ in the formula for Y_k. Let us consider in this connection an example involving three quadrats described by the data vectors,

$$\mathbf{X}_j = \begin{bmatrix} 2 \\ 5 \\ 2 \\ 5 \\ 3 \end{bmatrix} \qquad \mathbf{X}_k = \begin{bmatrix} 3 \\ 5 \\ 2 \\ 4 \\ 3 \end{bmatrix} \qquad \mathbf{X}_m = \begin{bmatrix} 9 \\ 1 \\ 1 \\ 1 \\ 1 \end{bmatrix}.$$

The totals are respectively $Q_j = Q_k = 17$ and $Q_m = 13$. The distance of quadrats j and k is

$$b(j, k) = (|2-3| + |5-5| + |2-2| + |5-4| + |3-3|)/34$$
$$= 0.05882.$$

The distance of quadrats j,m and k,m are $b(\mathrm{j,m}) = 0.6$ and $b(k,m) = 0.53333$. We note the inequality

$$b(j, m) > b(j, k) + b(j, m)$$

is in direct reverse of the inequality stipulated by the fourth metric space axiom. This indicates that in this particular example the distance function has failed in operations in which it has been put to use by its proponents. Where, in spite of this, we force an ordination based on this function, coupled with the formula for Y_k, we end up with a co-ordinate

$$Y_k = (0.05882^2 + 0.6^2 - 0.53333^2)/((2)(0.6)) = 0.06585$$

greater than $b(j, k)$.

One only wonders how did the proponents resolve the problem with $b(j, k)$ when they tried manipulations of Euclidean triangles in the Bray & Curtis ordination, when $b(j, k)$ did not yield distances from which Euclidean triangles could be produced. We suggest that $b(j, k)$ be used with caution as a measure of distance in ordinations, restricting it to such samples where the quantity Q_{jk} remains constant for all quadrat pairs.

Beals' (1960) ecological measure, which can be written as

$$v(j, k) = \max L - L = b(j, k) + e \qquad (2.20)$$

where $\max L$ is the observed maximum of

$$L = 2 \sum_h \min (X_{hj}, X_{hk}) / Q_{jk} \, ,$$

and $e = \max L - 1$, exemplifies further scrambling in the absolute value function. It can be seen that in formula (2.20) a distance $b(j, k)$ is decremented by a constant e. Whenever e is not a zero value, i.e. $\max L < 1$, this measure can fail on all except the third metric space axiom. Because of these reasons formula (2.20) is even *less* suited than formula (2.17) for use in analyses which manipulate Euclidean triangles.

This is not to say, of course, that formulae (2.17) and (2.20) will not have validity as ecological measures because they are not metric functions. What really makes these functions universally undesired is a changing scale which they may incorporate because of the division by Q_{jk}. On the account of this property, it is suggested that the ecologist avoid using formulae (2.17) or (2.20) when Q_{jk} is not a constant quantity for all quadrat pairs in the same sample. No hardship should follow such an action, considering that the roles for which these functions were intended in the first place, i.e. to provide relative distance measures in a Euclidean ordination, can be performed better by other reliable functions of which formula (2.2) represents an example. This measures as a metric, on a consistent scale, never failing to form Euclidean triangles.

Having seen that formulae (2.17) and (2.20) are distances that may fail on one or several of the metric space axioms, we now turn to an example to illustrate that in contrast with Czekanowski's index and Sørensen's coefficient, the Jaccard coefficient can in fact give rise to a metric. We shall use 2 x 2 tables in which relevés are described based on the presence (1)/absence (0) of 324 species:

		Relevé A 1	0	*Total*
Relevé	1	135	70	205
B	0	96	23	119
Total		231	93	324

		Relevé A 1	0	*Total*
Relevé	1	137	24	161
C	0	94	69	163
Total		231	93	324

		Relevé B 1	0	*Total*
Relevé	1	46	115	161
C	0	159	4	163
Total		205	119	324

From these we obtain the values of the complements of the Jaccard coefficient:

$$E(A, B) = (70 + 96) / (135 + 70 + 96) = 0.551495$$

$$E(A, C) = (24 + 94) / (137 + 24 + 94) = 0.462745$$

$$E(B, C) = (115 + 159) / (46 + 115 + 159) = 0.856250 .$$

Evidently, these satisfy the fourth of the metric space axioms. This would be so always with the one complement of the Jaccard coefficient as it has been shown by Marczewski & Steinhaus (1958) and independently by Levandowsky & Winter (1971). Now, we shall consider an example to show what happens with the Sørensen values. The complements are given by

$$E(A, B) = (70 + 96) / (270 + 70 + 96) = 0.380734$$

$$E(A, C) = (24 + 94) / (274 + 24 + 94) = 0.301020$$

$$E(B, C) = (115 + 159) / (92 + 115 + 159) = 0.748634 .$$

Here $E(A, B) + E(A, C) < E(B, C)$ in direct violation of the triangle inequality axiom. This property, while a serious limitation, need not be detrimental to applicability since in many applications metric properties are not assumed. What really makes the Sørensen index and its one complement undesired is that they lack a uniform scale of measure due to the dependence of the quantity $(2a + b + c)$ on the pair of relevés that happen to be compared. Unfortunately, the same limitation plagues the Jaccard index and its one complement.

2.6 Useful non-metric functions

We have in this group different similarity measures which cannot be derived from metrics, and dissimilarity measures which are not metric. We shall make one exception though with Rajski's metric which we include in this section only for convenience of presentation since it is conceptually closely related to other measures of information divergence.

A. A probabilistic similarity index

The resemblance measures which we considered so far define the likeness of objects as a composite of individual bits of resemblance values, each contributed by a different variable. In such a definition the resemblance of a given pair of objects is completely independent from the resemblance of any other pair in the sample. Goodall (1964, 1966a) suggests an index which takes into account comparisons between all pairs of objects when determining the resemblance of any given pair.

Where the data describe species quantity within quadrats, a probability index for quadrats can be calculated based on the steps in the following simplified procedure (see also program PINDEX in the Appendix):

1. Define δ_{jkh} as the dissimilarity of quadrats j and k with respect to species h. If the number of quadrats is n the number of δ_{jkh} values is $n(n-1)/2$ for any species h.
2. Rank the δ values according to size within species h.
3. Define the dissimilarity of quadrats j, k with respect to species h as the proportion $p_{jkh} = P(\delta \leqslant \delta_{jkh})$. This is the proportion of δ values that are smaller or equal to δ_{jkh} within species h; the higher the proportion the more dissimilar j and k are with respect to h. It is important to realize that when p_{jkh} indicates maximum similarity, δ_{jkh} may be a quantity other than zero.
4. Compute p_{jkh} for each of the available r species, and combine the proportions in the product $\pi_{jk} = \Pi p_{jkh}, h = 1, ..., r$. It is obvious that the p_{jkh} values are not zero. We must assume that the p_{jkh} are independent, which we interpret as a lack of correlation between the species, to justify combining the p_{jkh} in a simple product. If linear correlations exist their effect can be removed from the similarity index if the species are replaced by sets of component scores (see Chapter 3) which are uncorrelated.
5. Define an index of similarity in one of two possible ways: (a) Order the individual π_{jk} values, of which there are $n(n-1)/2$ in total, and determine the proportion $P(\pi > \pi_{jk})$; use this proportion as a measure of similarity $S(j, k)$. (b) Compute the quantity,

$$\chi^2_{jk} = -2 \sum_h \ln p_{jkh} = -2 \ln \pi_{jk}, \; h = 1, ..., r \, .$$

This has the χ^2 distribution at $2r$ degrees of freedom on the assumption that the p_{jkh} values represent independent probabilities. A similarity index can be derived as $S(j,k) = 1 - P(\chi^2 \geqslant \chi^2_{jk})$, i.e. the one complement of the probability that a random value χ^2 can exceed the observed value χ^2_{jk}. It may be noted that a small value of χ^2_{jk} associates with a large probability, and low similarity. The reasoning should be clear from the following logical sequence: large $p_{jkh} \rightarrow$ low similarity of quadrats j and k with respect to species $h \rightarrow$ small contribution to $\chi^2_{jk} \rightarrow$ large probability of an equal or greater $\chi^2 \rightarrow$ low similarity of j, k.

Let us consider an example to illustrate the computation of the index for quadrats *1* and *2* in Table 1-1. We define a distance function according to $\delta_{jkh} = | X_{hj} - X_{hk} |$. There are 10(10–1)/2 = 45 such differences to be computed within each species. For example, for quadrats *1* and *2* the first difference is given by $\delta_{121} = | 45-2 | = 43$. The entire vector of δ_{jk1} values is δ_1 = (43 36 19 42 40 43 45 14 29 7 24 1 3 0 88 29 14 17 6 4 7 81 22 7 23 21 24 64 5 10 2 1 87 28 13 3 85 26 11 88 29 14 59 74 15). The number of δ_{jk1} values not exceeding $\delta_{121} = 43$ is 36. This gives p_{121} = 36/45 = 0.8. The remaining four species are similarly treated. The proportions are $p_{122} = 0.911$, $p_{123} = 0.622$, $p_{124} = 0.889$, $p_{125} = 0.622$ (within computer rounding errors). The computed proportions may be combined in $\chi^2_{jk} = -2 \ln ((0.8)(0.911)(0.622)(0.889)(0.622)) = 2.766$. This quantity is compared to the values in the χ^2 table at $2r = 10$ degrees of freedom. From the table we have $P(\chi^2 \geqslant 2.766) \approx 0.986$ and $S(1,2) = 1 - 0.986 = 0.014$ or 1.4%. This indicates a very low similarity between the two quadrats. The alternative method for determining $S(1,2)$ involves ordering the combined probabilities π of which $\pi_{12} = 0.251$. There are 45 such values of which 44 are smaller than 0.251. Thus $P(\pi > 0.251) = S(1,2) \approx 0.022$ or about 4%. This, similar to χ^2, puts a very low value on the similarity of the two quadrats.

The algorithm just described departs from Goodall's (1966a) in that it treats the equalities in $p_{jkh} = P(\delta \leqslant \delta_{jkh})$ differently. The version presented counts all equalities as such, but not so the original in which a nominal equality,

$$[\delta_{jkh} = | X_{jh} - X_{kh} |] = [\delta_{mzh} = | X_{mh} - X_{zh} |] ,$$

does not necessarily imply an equal dissimilarity for the first pair (j, k) as for the second (m, z). The first pair would be regarded more dissimilar than the second, in spite of the equal nominal dissimilarities, if the frequency of data points in the interval with end points X_{jh} and X_{kh} exceeded the frequency of data points in the interval with end points X_{mh} and X_{zh}. No wonder then that the original version tends to produce smaller p_{jkh} values and thus tend to make the quadrat pairs appear more similar than the version we described. The advantage in the present version is computational simplicity, while in the original, the advantage is in giving more weight to uncommon data points. The latter appears to be a desired property in numerical taxonomy (Goodall, 1966b).

Several peculiarities are to be noted. With the given relation signs in $p_{jkh} = P(\delta \leqslant \delta_{jkh})$, the maximum of $S(j, k)$ occurs when the difference δ_{jkh} is consistently the smallest for all h. This maximum is the one complement of the probability of a chi square exceeding $-2 \ln\{1/[n(n-1)/2]\}^r$; the maximum comes close to one when n is very large. The minimum of $S(j, k)$ is zero. This value occurs when δ_{jkh} is consistently the largest for all h. Now, if we change the relation signs so that $p_{jkh} = P(\delta < \delta_{jkh})$, the possible maximum of $S(j, k)$ would be one and its minimum would approach (but never become exactly) zero.

One may be tempted to use $S(j, k)$ as a cosine quantity and convert it into a chord distance $d(j, k) = [2(1 - S(j, k)]^{1/2}$. Strictly speaking, this would not be correct. The reason is simple. If we take the definition of p_{jkh} with relation signs $\leqslant$, for $S(j, k)$ to be zero the subtending angle of the quadrat vectors $\mathbf{X}_j$ and $\mathbf{X}_k$ need not be 90°. A similar difficulty arises if in the definition of p_{jkh} we used the relation sign $<$, in which case, $S(j, k)$ may reach the maximum 1, without the quadrat vectors $\mathbf{X}_k$ and $\mathbf{X}_j$ being collinear. But for the cosine to be unity the subtending angle of the vectors must be zero. With $S(j, k)$ not possessing the cosine property, the distance $d(j, k)$ can in fact fail on all but one (symmetry) of the metric space axioms. Considering the significance of this index, its properties should be further studied, particularly its behaviour under different transformations of the data.

The probabilistic index combines r individual probabilities. Its precision will therefore depend on the degree to which the probabilities are independent. But apart from the difficulty arising from the need to assume that the probabilities are independent, a probabilistic index can be a useful measure of quadrat resemblance. Its principal advantage resides in the flexibility of handling different types of data as Goodall (1966b; Clifford & Goodall 1967) has shown in different publications. The index has the further advantage of being made relative on a zero to one scale which under independence of probabilities is itself a linear probabilistic scale. The definition of resemblance by this index, however, is local to the sample in which the quadrats occur. This means that the same two quadrats may appear differently similar in terms of such an index when embedded in different samples.

B. Calhoun distance

Another non-metric measure has been suggested by Calhoun (Bartels, Bahr, Calhoun & Wied, 1970) to measure topological divergence between individuals. Should we decide to use Calhoun's method, we would associate each quadrat pair (j, k) with a hypervolume and equate the distance $d(j, k)$ with the number of quadrats which fall (as points) inside or on the surface of the hypervolume. In lower dimensions, points would be counted within a cube and on its surface (three dimensions) or within an area and on its boundary line (two dimensions). We shall use the data in the table below to illustrate the calculations:

Quadrat		*1*	*2*	*3*	*4*	*5*	*6*	*7*	*8*	*9*	*10*
Species	X_1	4	7	5	7	5	1	4	8	2	7
	X_2	3	6	4	2	2	7	8	3	2	4

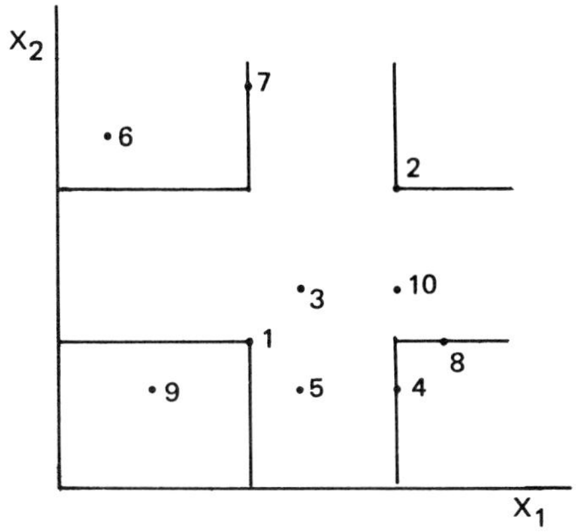

Fig. 2-2. Scatter diagram for the sample data used in the calculation of Calhoun's distance. See details in the main text. X_1 and X_2 identify species, numbers to points identify quadrats.

The quadrats are shown as points in Fig. 2-2. Should we compare quadrat *1* and *2*, the associated area is marked by lines in Fig. 2-2. We inspect each point with respect to location and count those which fall within or onto the boundary line of the area, or tie in value, to obtain the following quantities:
N_a – the number of points in the interior of the hypervolume and its extensions. This is equal to 3 in the example.
N_b – the number of points which lie on the boundary surface, i.e. points which do not lie within the hypervolume but tie in value with quadrat j or k on one or more of the axes. There are three such points in the example.
N_c – the number of points that tie simultaneously with the value of quadrat j and k on one or more axes but do not lie in the interior or on the boundary of the hypervolume. For this to be the case, quadrat j must have an identical value with quadrat k on one or more of the axes. We have no point of this sort in the example.

The Calhoun distance (for which a program HYPD is given in the Appendix) is defined by

$$d(j, k) = 6N_a + 3N_b + 2N_c$$

which for $j = 1$ and $k = 2$ has the value

$$d(1,2) = 9(3) + 3(3) + 2(0)$$
$$= 27.$$

The weights (6,3,2) are arbitrary, and they are offered only as suggestions (Bartels, Bahr, Calhoun & Wied, 1970). Actually, it may be a good idea to omit the term with N_b altogether and possibly N_c too in the formula, since points in these categories do not lie between j and k, and thus, they contribute nothing particularly to promote the appearence of a topological divergence between them. Since the maximum of $d(j, k)$ is $6(n-2)$ when there are n quadrats in the sample, the Calhoun distance may be relativized in the form of $d(j, k)/6(n-2)$.

As a resemblance function, the Calhoun distance deserves attention in vegetation studies. It is a symmetric, scale free measure and it can handle mixed variable sets. It is obviously not a metric (failing the first and fourth metric space axioms) and it is not invariant under rotation of the co-ordinate axes.

C. *Mountford's index*

We can describe a multispecies population on the basis of its frequency distribution,

$$n_1, n_2, n_3, \ldots$$

in which n_i signifies the number of species represented by i individuals when the total number of species is

$$P = \sum_i n_i, \quad i = 1, 2, 3, \ldots$$

and the total number of individuals in all species

$$N = \sum_i in_i, \quad i = 1, 2, 3, \ldots .$$

It has been observed that in some populations the n_i can be approximated by the $\frac{\alpha X^i}{i}$ terms in the algebraic expansion of the logarithm,

$$-\alpha \ln (1 - X) = \sum_i \frac{\alpha X^i}{i}, \quad i = 1, 2, 3, \ldots$$

in which α and X are parameters. This distribution, originally suggested by Fisher, Corbet & Williams (1943), has been used to describe different animal and plant populations (e.g. Pielou, 1969a and references therein). To determine α and X, Fisher's equations (Fisher, Corbet & Williams, 1943),

$$p = \sum_i n_i = -\alpha \ln (1 - X), \quad i = 1, 2, 3, \ldots$$

$$n = \sum_i in_i = \alpha X/(1 - X), \quad i = 1, 2, 3, \ldots$$

have to be solved. In these, p and n represent respectively the observed number of species and the observed number of individuals.

If we consider only presence data, the formula of interest is that which defines p. From this it is seen that α is proportional to the total number of species; it can thus be regarded as a measure of species richness. Williams (Fisher, Corbet & Williams, 1943) has in fact suggested that the parameter α be known as the *index of diversity* on the account of the observation (see Fig. 5 in their paper) that at a given total number of individuals (n) increased values of α correspond to increased numbers of species, and thus, to increased diversity.

The property of α that induced the use of $S = 1/\alpha$ as an index of similarity is its independence from sample size (n) under certain conditions. This property is quite apparent from the graphs presented by Williams (Fig. 5 in Fisher, Corbet & Williams, 1943). Mountford (1962) gives the defining relationship for S as

$$e^{aS} + e^{bS} = 1 + e^{(a+b-j)S}$$

where e is the natural base, a the number of species in quadrat A, b the same in quadrat B and j the number of species held in common between two quadrats. The values of S lie between zero and infinity inclusive. A zero value indicates that A and B have no species in common. But at complete identity of A and B, i.e. $j = a = b$, S is indeterminate. Mountford (1962) suggests the formula

$$\frac{2j}{2ab - (a+b)j}$$

as a reasonably close approximation to S. To illustrate the computation of S on this basis, we use the first and fourth quadrat vectors of Table 1-2:

$$\mathbf{A} = \begin{bmatrix} 2 \\ 0 \\ 3 \end{bmatrix} \qquad \mathbf{B} = \begin{bmatrix} 1 \\ 3 \\ 0 \end{bmatrix} .$$

For **A** and **B**,

$$a = 2$$
$$b = 2$$
$$j = 1 .$$

The approximation for S is

$$S(A, B) = \frac{(2)(1)}{(2)(2)(2) - (2+2)(1)} = 0.5 .$$

Alternatively, one may use Mountford's (1962) chart to obtain $(a + b - j)$ $S(A, B) = X$, and then $S(A, B) = X/(a + b - j)$. The chart is entered with ordinate $y_1 = a/(a + b - j)$ and abscissa $y_2 = b/(a + b - j)$. It is assumed that $b < a$, when not equal. In the example,

$$y_1 = 2/(2+2-1) = 2/3$$

$$y_2 = 2/(2+2-1) = 2/3\ .$$

The X value corresponding to y_1, y_2 is 1.5. The corresponding similarity value is $S(A, B) = 1.5/(2+2-1) = 0.5$.

Mountford (1962) compared $S = 1/\alpha$ with Jaccard's (1901) coefficient and Sørensen's (1948) index, and concluded that whereas S is independent of n, the number of individuals in all species, the other indices are not, even if the sample is taken from a single logarithmic population. If the underlying population is other than logarthmic, S may itself be dependent on n. Mountford's similarity index is a symmetric measure, i.e. $S(A, B) = S(B, A)$, but it is indeterminate for quadrats with identical species lists. We could of course use the quantity,

$$d(A, B) = e^{-S}$$

to measure a relative distance. It appears from random simulation that $d(A, B)$ is a metric. Its minimum, $d(A, B) = 0$, indicates identity and maximum, $d(A, B) = 1$, signifies total dissimilarity of the species lists in A and B. With $d(A, B)$ so defined,

$$R(A, B) = [1 - d^2(A, B)]^{1/2}$$

is a measure of coherence on a 0 to 1 scale. For the data in the example, $d(A, B) = e^{-0.5} = 0.61$ and $R(A, B) = [1 - 0.61^2]^{1/2} = 0.80$.

D. *Measures of information divergence*

It will be seen that when data are condensed into frequency distributions the analysis to measure resemblance can proceed according to methods that rely on information divergence. The logarithmic expressions of Shannon (1948) and Kullback (1959) will be considered. The functions to be considered handle data on nominal, ordinal, interval or even ratio scales of both the discrete and continuous type. But when continuous variables are analysed their underlying distributions need to be known, or the data have to be condensed into a frequency distribution with arbitrarily chosen class intervals. Whatever the data type, the information which lies in order relationships, depending on the magnitude of the data elements, is not utilized.

D 1. Rajski's metric divergence

Before we can define this metric it will be necessary to consider some general concepts about frequency distributions, entropy, and information. Any set of observed values or symbols $\mathbf{X}_h = (X_{h1} \dots X_{hn})$ can be condensed into a frequency distribution $\mathbf{F}_h = (f_{h1} \dots f_{hs_h})$ by simply setting up s_h frequency classes and counting the number of observations falling into each. When the data are categorical, representing counts or frequencies, $\mathbf{X}_h$ itself may be considered as a frequency distribution, so that $\mathbf{F}_h$ is equivalent to $\mathbf{X}_h$.

The distribution $\mathbf{F}_h$ is an observed distribution. There may be a second distribution $\mathbf{F}_h^0$, such that both distributions have identical totals $f_{h.} = f_{h.}^0 = n$, and are identically ordered in the sense that for each element in $\mathbf{F}_h$ there is a corresponding element in $\mathbf{F}_h^0$. Any element in $\mathbf{F}_h$ or $\mathbf{F}_h^0$ can be expressed as a proportion,

$$p_{hj} = f_{hj}/f_{h.} \quad \text{or} \quad p_{hj}^0 = f_{hj}^0/f_{h.}$$

such that

$$\sum_j p_{hj} = \sum_j p_{hj}^0 = 1 \ .$$

Where the distribution $\mathbf{F}_h$ describes the hth species in the sample, the element f_{hj} will indicate the frequency with which the observed values or symbols occur in the jth class. Consider for example the data in Table 2.1.

Table 2-1. Species quantities in quadrats estimated on an arbitrary scale. Data after Pietsch & Muller-Stoll (1968)

Species	*Quadrat*																								
	1	*2*	*3*	*4*	*5*	*6*	*7*	*8*	*9*	*10*	*11*	*12*	*13*	*14*	*15*	*16*	*17*	*18*	*19*	*20*	*21*	*22*	*23*	*24*	*25*
X_1	*1*	*1*	+	+	*1*	+	+	+	.	*r*	+	+	*1*	+	*1*	*4*	*1*	*4*	*3*	*2*	*1*	*1*	*3*	*1*	*3*
X_2	*3*	*4*	*3*	*3*	*1*	*3*	+	*4*	*r*	*2*	+	+	+	*1*	+	*1*	+	.	.	*1*	*1*	*2*	.	*1*	+

X_1 – *Eleocharis ovata*

X_2 – *Carex bohemica*

In this table the entries include numerals, mixed with one letter of the alphabet, a period and a +. A period indicates absence, *r* and + signify species of very low or low quantitative representation; the numerals *1, 2, 3, 4* are cover-abundance estimates for the species which occur in greater quantity. With discrete symbols such as these, it is logical to define ., *r*, +, *1*, *2*, *3*, *4* as the class symbols, and then to obtain the values for the elements in $\mathbf{F}_1$, $\mathbf{F}_2$ by counting occurrences of the class symbols in the data. These counts are given in Table 2-2.

Table 2-2. Distribution of class symbols for species in Table 2.1

Species as frequency distribution	*Class symbol* .	*r*	+	*1*	*2*	*3*	*4*	*Total*
$\mathbf{F}_1$	1	1	8	9	1	3	2	25
$\mathbf{F}_2$	3	1	7	6	2	4	2	25

In this example $\mathbf{F}_1$, $\mathbf{F}_2$ both are 7-valued ($s_1 = s_2 = 7$) and their totals are the same $f_{1.} = f_{2.} = 25$. Furthermore, the elements are identically ordered in both distributions according to the 7 common class symbols. The relative frequencies are obtained based on division by 25. These have been computed and are given in Table 2-3.

Table 2-3. Relative frequencies of class symbols for species in Table 2-1

Species as frequency distribution	*Class symbol* .	*r*	*+*	*1*	*2*	*3*	*4*	*Total*
$\mathbf{P}_1$	0.04	0.04	0.32	0.36	0.04	0.12	0.08	1
$\mathbf{P}_2$	0.12	0.04	0.28	0.24	0.08	0.16	0.08	1

Let us assume that the class frequencies in a distribution $\mathbf{F}_h^0 = (f_{h1}^0 \ldots f_{hs_h}^0)$ have different observed weights specified by the elements in $\mathbf{P}_h = (p_{h1} \ldots p_{hs_h})$. If no such weights are given, the distributions $\mathbf{F}_h$, $\mathbf{F}_h^0$ are taken to be identical. The entropy, or average information, in $\mathbf{F}_h^0$ relative to $\mathbf{F}_h$ is defined as the sum

$$H(\mathbf{P}_h^0) = -\sum_j p_{hj} \ln p_{hj}^0, \; j = 1, \ldots, s_h \tag{2.21}$$

where 'ln' indicates the natural logarithm. The minimum value of $H(\mathbf{P}_h^0)$ corresponds to the condition $\mathbf{P}_h = \mathbf{P}_h^0$,

$$\min H(\mathbf{P}_h^0) = H(\mathbf{P}_h) = -\sum_j p_{hj} \ln p_{hj} = \ln f_{h.} - \frac{1}{f_{h.}} \sum_j f_{hj} \ln f_{hj}, \quad j = 1, \ldots, s_h \tag{2.22}$$

This is Shannon's (1948) entropy function, also known as *entropy of order one* (Rényi, 1961) distinguished from entropies of other higher order,

$$H(\mathbf{P}_h)^k = -\sum_j p_{hj} \ln^k p_{hj}, j = 1, \ldots, s_h$$

for $k > 1$. The ecological applications of the entropy function are varied. We shall consider aspects related to measuring the divergence of groups in classification problems. In other connections, we refer to work by, e.g., Quastler (1956), Margalef (1958), McIntosh (1967a), Juhász-Nagy (1967a), Godron (1968), Pielou (1966a,b,c, 1969a, 1974, and references therein), Daget, Godron & Guillerm (1972), Lausi (1972), Orlóci (1972a,b, 1973a) and Hill (1973a).

Should the frequency distribution of the data be identified, the entropy can be calculated from the distribution parameters:

1. In the Poisson case,

$$H(\mathbf{F}_h) = -\sum_X P(X)_h \ln P(X)_h, \; X = 0, \ldots, \infty \;,$$

where

$$P(X)_h = (e^{-m_h} m_h^X) / X!$$

and

$$m_h = \sum_j X_{hj}/n, j = 1, ..., n.$$

2. In the Binomial case,

$$H(P_h) = - \sum_X P(X)_h \ln P(X)_h, \ X = 0, ..., t$$

where t is the number of random trials per quadrat and

$$P(X)_h = \{t!/[X!\ (t-X)!]\} p_h^X q_h^{t-X}; \quad p_h = \sum_j X_{hj}/tn, j = 1, ..., n,$$

i.e. the probability of one success in one random trial when X_{hj} is the number of successes in t random trials in quadrat j. The number of quadrats in the sample is n. The one complement of p_h is designated by q_h.

3. In most practical applications $H(\mathbf{P}_h)$ is defined in terms of discrete probabilities. However, when the probability distribution of X is continuous and normal, the entropy is defined by

$$H(\mathbf{P}_h) = - \int f(X)_h \ln f(X)_h \, dX = \ln S_h \, (2\pi e)^{1/2}, \quad X = -\infty, ..., \infty \, .$$

In this, $f(X)_h$ is the probability density at point X and S_h is the standard deviation of species h. Since $H(\mathbf{P}_h)$ is proportional to S_h, the entropy is scale dependent.

The entropy, $H(\mathbf{P}_h)$, has a number of properties that are useful to know before attempting to apply any of the formulae to vegetation data:

1. Continuity for values of p in the interval $0 < p < 1$.
2. Symmetry in the sense that the order in which the p_{hj} are considered has no influence.
3. Maximum value at $p_{h1} = p_{h2} = ... = p_{hs_h} = 1/s_h$.
4. Additivity in the sense of $H(\mathbf{P}_h\mathbf{P}_i) = H(\mathbf{P}_h) + H(\mathbf{P}_i)$ where $\mathbf{P}_h\mathbf{P}_i$ indicates the direct product of two sets of independent frequencies.
5. Consistency and approximate normality when X_h, from which $\mathbf{P}_h$ is derived, represents a random sample from a normal population (Basharin, 1959). Based on normality, and neglecting terms of order of magnitude N^{-2} or less, the large sample expectation of H_h is $H_h' - (S_h - 1)/f_{h.}$, where H_h' is the true population value with estimated variance,

$$S(H_h) = \frac{1}{f_{h.}} \left[\sum_j (p_{hj} \ln^2 p_{hj}) - H_h^2\right], \ j = 1, ..., s_h$$

(cf. Miller & Madow, 1954; Khinchin, 1957; Basharin, 1959; Fraser, 1965; Pielou, 1966a). We note that because H_h is a sample value, $S(H_h)$ is only an approximation and the computed value may even be negative in which case it has to be equated to zero.

6. Improved normality with increased sample size.

Considering the data in Table 2-1, we can compute $H(\mathbf{P}_h)$ of formula (2.22) according to the equivalent formula,

$$H(\mathbf{P}_h) = \frac{1}{f_{h.}} \left[f_{h.} \ln f_{h.} - \sum_j f_{hj} \ln f_{hj} \right] \qquad (2.23)$$

which lends itself to easy table look-ups. The corresponding numerical values are,

$$H(\mathbf{P}_1) = \frac{1}{25}(25 \ln 25 - (3 \ln 1 + 8 \ln 8 + 9 \ln 9 + 3 \ln 3 + 2 \ln 2))$$
$$= 1.5752$$

$$H(\mathbf{P}_2) = \frac{1}{25}(25 \ln 25 - (3 \ln 3 + 1 \ln 1 + 7 \ln 7 + 6 \ln 6 + 4 \ln 2$$
$$+ 4 \ln 4)) = 1.7795 \ .$$

In this example we have used data which consist of discrete symbols. It was natural to identify the different symbols as class symbols when we constructed the frequency distributions. A similar procedure could be followed in the case of continuous data, but with modification the frequency classes would have to be defined as non-overlapping intervals. If the data are categorical, the original observations can be regarded as frequencies and used directly in formula (2.23).

Relative frequencies and the average information are sometimes less convenient for handling than their $f_{h.}$ multiples,

$$I(\mathbf{F}_h) = f_{h.}\, H(\mathbf{P}_h) = -\sum_j f_{hj} \ln f_{hj}/f_{h.}$$
$$= f_{h.} \ln f_{h.} - \sum_j f_{hj} \ln f_{hj},\ j = 1, ..., s_h \tag{2.24}$$

The corresponding numerical values for the two species in Table 2-1 are given by

$$I(\mathbf{F}_1) = 25\, H(\mathbf{P}_1) = 39.379$$
$$I(\mathbf{F}_2) = 25\, H(\mathbf{P}_2) = 44.486\ .$$

Where a population distribution $\mathbf{P}_h^0$ is replaced by the observed sample distribution $\mathbf{P}_h$, the information generated can be measured as a divergence,

$$H(\mathbf{P}_h\, ; \mathbf{P}_h^0) = H(\mathbf{P}_h^0) - H(\mathbf{P}_h) = \Sigma\, p_{hj} \ln p_{hj}/p_{hj}^0,\, j = 1, ..., s_h \tag{2.25}$$

Formula (2.25) measures *information of order one* (Rényi, 1961). Its multiple

$$2I(\mathbf{F}_h; \mathbf{F}_h^0) = 2 \sum_j f_{hj} \ln f_{hj}/f_{hj}^0,\ j = 1, ..., s_h \tag{2.26}$$

is the *minimum discrimination information statistic* (Kullback, 1959). It should be noted that formula (2.25) or (2.26) represents a one-way or I-divergence, $\mathbf{P}_h^0 \rightarrow \mathbf{P}_h$ or $\mathbf{F}_h^0 \rightarrow \mathbf{F}_h$, which is not the same as $\mathbf{P}_h^0 \leftarrow \mathbf{P}_h$ or $\mathbf{F}_h^0 \leftarrow \mathbf{F}_h$ where the comparison is reversed. A two-way divergence $\mathbf{P}_h \leftrightharpoons \mathbf{P}_h^0$ or $\mathbf{F}_h \rightleftharpoons \mathbf{F}_h^0$ is a J-divergence. The J-divergence is a symmetric divergence, with respect to $\mathbf{F}_h$ and $\mathbf{F}_i$,

$$J(\mathbf{F}_h;\mathbf{F}_i) = \sum_j (f_{hj} - f_{ij}) \ln f_{hj}/f_{ij},\ j = 1, ..., s_h \tag{2.27}$$

In this, the order in which $\mathbf{F}_h$ and $\mathbf{F}_i$ are compared is immaterial and $s_h = s_i$. $J(\mathbf{F}_h;\mathbf{F}_i)$ is a semimetric (Kullback, 1959).

We shall consider other I-divergences in the sequel. As far as their characterization is concerned it will be sufficient to note at this point that they are not influenced by zero elements in $\mathbf{F}^0$ or $\mathbf{F}$, and the order in which the f/f^0 terms are taken has no influence as long as f and f^0 are not interchanged. (In a J-divergence f_h and f_i can be interchanged.) The I-divergence of formula (2.26) and the J-divergence of formula (2.27) both have sampling distributions comparable to the χ^2 distribution with $s_h - 1$ degrees of freedom under the null hypothesis that $\mathbf{F}_h$ represents a random sample from a population completely specified by $\mathbf{F}_h^0$, or that $\mathbf{F}_h$ and $\mathbf{F}_i$ are random samples from the same population.

Kullback's (1959) results provide a basis for derivation of a sampling distribution for $2I(\mathrm{F};\mathrm{E})$. Consider the relation,

$$1 - e/f \leqslant \ln f/e \leqslant f/e - 1$$

which holds true when $f/e > 0$, with equality when $f = e$. Use as a first approximation to $\ln f/e$ the average of the two limits, i.e.

$$\begin{aligned}\ln f/e &\approx [(f/e - 1) + (1 - e/f)]/2\\ &= (f/e - e/f)/2\\ &= (f^2 - e^2)/2fe\end{aligned}$$

or $2f \ln f/e \approx (f^2 - e^2)/e$. After observing that

$$\sum_i (e_i^2 - f_i e_i)/e_i = 0,\ i = 1, ..., s$$

because $\Sigma f_i = \Sigma e_i$, derive the chi square approximation as the quantity

$$\begin{aligned}2I(\mathrm{F};\mathrm{E}) &= 2\sum_i f_i \ln f_i/e_i\\ &\approx \sum_i (f_i^2 - e_i^2)/e_i + \Sigma\,(2e_i^2 - 2f_i e_i)/e_i\\ &= \sum_i [(f_i^2 - e_i^2) + (2e_i^2 - 2f_i e_i)]/e_i\\ &= \sum_i (f_i^2 - 2f_i e_i + e_i^2)/e_i\\ &= \Sigma\,(f_i - e_i)^2/e_i,\ i = 1, ..., s\end{aligned}$$

– which is the familiar expression for chi square. The approximation improves as the ratio f/e approaches unity.

Under specified circumstances the quantity $\Sigma(f_i - e_i)^2/e_i$, and the $2I(\mathrm{F};\mathrm{E})$ too, can be referred to the chi square distribution with $s - 1$ degrees of freedom. This should not obscure the fact that under most practical situations we use the chi square distribution only as an approximation with these quantities, knowing that their true sampling distribution is likely to be different. Obviously then, if we use the chi square distribution as a reference distribution for $2I(\mathrm{F};\mathrm{E})$ we may have a potentially weak, double approximation on hand – an approximation of the quantity

$$\sum_i (f_i - e_i)^2/e_i$$

by the quantity

$$2 \sum_i f_i \ln f_i/e_i \quad ,$$

and another of the true sampling distribution of $\Sigma(f_i - e_i)^2 / e_i$ under the prevailing local conditions by the hypothetical chi square distribution.

An alternative derivation of the relationship between $2I$ and χ^2 is described by Hamdan & Tsokos (1971). They point out connections to other measures (Linfoot, 1957) and show the advantages in using the form,

$$r_I = [1 - \exp(-2I/f_{..})]^{1/2}$$

as a coefficient of contingency over the Pearson measure (see Kendall & Stuart, 1973),

$$P = [(\chi^2/f_{..})/(1 + \chi^2/f_{..})]^{1/2} \,.$$

The symbol $f_{..}$ signifies table total.

Any two distributions $\mathbf{F}_h$ and $\mathbf{F}_i$, such as the two species in Table 2-2, and the relationship between them can be represented in an $s_h \times s_i$ table (Table 2-4). In such a table the individual frequency distributions appear as the principal marginal distributions, and an element $f_{hj,\,ik}$ in the body of the table indicates the joint frequency of the jth class value (or class symbol) in $\mathbf{F}_h$ and the kth class value (or class symbol) in $\mathbf{F}_i$. In other words, $f_{hj,\,ik}$ indicates the frequency of the joint observation in which one element can be identified with the jth class in $\mathbf{F}_h$, and another with the kth class in $\mathbf{F}_i$. We can define the following information quantities on Table 2-4:

1. Information in the marginal distributions

$$I(\mathbf{F}_h) = f_{h.} \ln f_{h.} - \sum_j f_{hj} \ln f_{hj}, \; j = 1, \ldots, s_h$$
$$I(\mathbf{F}_i) = f_{i.} \ln f_{i.} - \sum_j f_{ij} \ln f_{ij}, \; j = 1, \ldots, s_i \;.$$

Table 2-4. Tabular representation of frequency distributions and the relationships between them

		Class value or class symbol in $\mathbf{F}_i$				*Total in* $\mathbf{F}_h$
		1	2	...	s_i	
Class value or class symbol in $\mathbf{F}_h$	1	$f_{h1,i1}$	$f_{h1,i2}$	...	f_{h1,is_i}	f_{h1}
	2	$f_{h2,i1}$	$f_{h2,i2}$	...	f_{h2,is_i}	f_{h2}
	.	.	.	...	.	.
	s_h	$f_{hs_h,i1}$	$f_{sh_h,i2}$	...	f_{hs_h,is_i}	f_{hs_h}
Total in $\mathbf{F}_i$		f_{i1}	f_{i2}	...	f_{is_i}	$f_{h.} = f_{i.} = f_{h.,i.}$.

2. Joint information

$$I(\mathbf{F}_h, \mathbf{F}_i) = f_{h.,\,i.} \ln f_{h.,\,i.} - \sum_j \sum_k f_{hj,ik} \ln f_{hj,ik},$$

$$j = 1, \ldots, s_h; \; k = 1, \ldots, s_i \tag{2.28}$$

3. Mutual information

$$I(\mathbf{F}_h; \mathbf{F}_i) = I(\mathbf{F}_h) + I(\mathbf{F}_i) - I(\mathbf{F}_h, \mathbf{F}_i)$$

$$= \sum_j \sum_k f_{hj,ik} \ln f_{hj,ik} f_{h.,\,i.} / (f_{hj} f_{ik}),$$

$$j = 1, \ldots, s_h; \; k = 1, \ldots, s_i \tag{2.29}$$

$I(\mathbf{F}_h; \mathbf{F}_i)$ is an I-divergence. The quantities $I(\mathbf{F}_h)$, $I(\mathbf{F}_i)$ or $I(\mathbf{F}_h, \mathbf{F}_i)$ represent multiples of entropy of order one.

From these quantities we obtain the equivocation information,

$$E(\mathbf{F}_h; \mathbf{F}_i) = I(\mathbf{F}_h, \mathbf{F}_i) - I(\mathbf{F}_h; \mathbf{F}_i)$$

$$= -\sum_j \sum_k f_{hj,ik} \ln (f_{hj,ik}/f_{hj})(f_{hj,ik}/f_{ik}),$$

$$j = 1, \ldots, s_h; \; k = 1, \ldots, s_i \tag{2.30}$$

where $f_{hj,ik}/f_{hj}$ and $f_{hj,ik}/f_{ik}$ are conditional probabilities. When $\mathbf{F}_h$ and $\mathbf{F}_i$ represent two species, $f_{hj,ik}/f_{hj}$ specifies the probability for species i to have state k when the state of species h is know to be j. Similarly, $f_{hj,ik}/f_{ik}$ gives the probability for species h to have state j if the state of species i is known to be k. A definition of Rajski's (1961) metric divergence $d(\mathbf{F}_h; \mathbf{F}_i)$ follows directly:

$$d(\mathbf{F}_h; \mathbf{F}_i) = \frac{E(\mathbf{F}_h; \mathbf{F}_i)}{I(\mathbf{F}_h, \mathbf{F}_i)} = 1 - \frac{I(\mathbf{F}_h; \mathbf{F}_i)}{I(\mathbf{F}_h, \mathbf{F}_i)}. \tag{2.31}$$

Rajski's metric is a relative measure of the two-way divergence of $\mathbf{F}_h, \mathbf{F}_i$ with 0 and 1 as limit values. We can regard $d(\mathbf{F}_h; \mathbf{F}_i)$ as the sine of an angle. From this metric a similarity measure can be derived according to

$$r(\mathbf{F}_h; \mathbf{F}_i) = \sqrt{[1 - d^2(\mathbf{F}_h; \mathbf{F}_i)]} \tag{2.32}$$

This is the *coherence coefficient.* Its limits are one and zero. Rajski (1961) gives an equivalent formula in terms of relative frequencies,

$$r(\mathbf{F}_h; \mathbf{F}_i) = \frac{\left[\sum_j \sum_k p_{jk} \ln \dfrac{p_{jk}}{p_j p_k} \sum_j \sum_k p_{jk} \ln \dfrac{p_j p_k}{p_{jk}^3} \right]^{1/2}}{-\sum_j \sum_k p_{jk} \ln p_{jk}},$$

$$j = 1, \ldots, s_h; \; k = 1, \ldots, s_i,$$

where $p_{jk} = f_{hj,ik}/f_{h.,i.}$, $p_j = f_{hj}/f_h$ and $p_k = f_{ik}/f_i$. with symbols similarly defined as in Table 2-4. Rajski's metric has been used in vegetation studies (e.g. Godron, 1968; Orlóci, 1968b, 1969a, 1972a) and in numerical taxonomy (e.g. Estabrook, 1967; Hawksworth, Estabrook & Rogers, 1968) as a measure of dissimilarity.

Turning to an example, we shall calculate $d(\mathbf{F}_1;\mathbf{F}_2)$ and $r(\mathbf{F}_1;\mathbf{F}_2)$ for the two species in Table 2-1. The joint frequencies are given in Table 2-5.

Table 2-5. Joint frequencies for two species in Table 2-1

		Class symbol in $\mathbf{F}_2$							*Total in* $\mathbf{F}_1$
			r	*+*	*1*	*2*	*3*	*4*	
Class symbol in $\mathbf{F}_1$	.		1						1
	r					1			1
	+			3	1		3	1	8
	1			3	3	1	1	1	9
	2				1				1
	3	2		1					3
	4	1			1				2
Total in $\mathbf{F}_2$		3	1	7	6	2	4	2	25

We note in connection with this table that, for instance, the joint frequency of the third class symbol (*+*) in $\mathbf{F}_1$ and the sixth class symbol (*3*) in $\mathbf{F}_2$ is $f_{13,26} = 3$. But this is not the same as $f_{16,23}$ which happens to be 1. Clearly, the sequence in which the class symbols occur, whether (*+, 3*) or (*3, +*) in the example, must be carefully checked when counting their joint frequencies. We also note that the cells in the body of the table that correspond to unrealized combinations are left blank, and the corresponding terms in the formula are arbitrarily set to zero. The information quantities have the following numerical values:

$$I(\mathbf{F}_1) = 39.379$$

$$I(\mathbf{F}_2) = 44.486$$

$$I(\mathbf{F}_1, \mathbf{F}_2) = 25 \ln 25 - 4\,(3 \ln 3) - 2 \ln 2 = 65.902$$

$$I(\mathbf{F}_1; \mathbf{F}_2) = 39.379 + 44.486 - 65.902 = 17.963$$

$$E(\mathbf{F}_1; \mathbf{F}_2) = 65.902 - 17.963 = 47.939$$

$$d(\mathbf{F}_1; \mathbf{F}_2) = 47.939/65.902 = 0.727$$

$$r(\mathbf{F}_1; \mathbf{F}_2) = \sqrt{(1 - 0.727^2)} = 0.686\,.$$

Program INFC in the Appendix automatically computes the different information terms from input data of the kind which were used in the example. We can see that the information quantities so far considered are particularly well-suited to analyse vegetational data which consist of mixed symbols, such as data of cover/abundance estimates. We shall consider the case of count and frequency data later in connection with different information type divergences.

D2. Other measures of information divergence

In this section we look at information theory functions which, similar to Rajski's metric, can measure the divergence of frequency distributions. We begin with a categorization of frequency distributions according to their totals and the ordering of elements:

1. The first category includes distributions whose totals are identical and the elements are identically ordered. Identical ordering implies that for every element in one of the distributions there is a corresponding element in every other. The distributions $\mathbf{F}_1$ and $\mathbf{F}_2$ in Table 2-2 are examples of this category.
2. Distributions of the second category are most frequent in vegetation data. They are characterized by an identical ordering of elements, but different totals. The rows in Table 1-1 represent examples of this.
3. The elements of distributions in the third category are unordered and their totals are also different. Such distributions could arise if we subdivided Table 1-1 into two groups, for instance, A from quadrat *1* to *4* and B from *5* to *10*. After subdivision, the species vector $\mathbf{X}_{1A}$ in group A and its counterpart $\mathbf{X}_{1B}$ in group B present a problem in comparisons, requiring considerations of their totals rather than their elements.

The information functions that may serve as measures of divergence in these categories are basically I- or J-divergences. When a standard $\mathbf{F}^0$ is specified, formula (2.26) is the relevant expression, but when no standard is given formula (2.27) is appropriate. $I(\mathbf{F}_h; \mathbf{F}_i)$ of formula (2.29) represents an example for an I-divergence with observed frequencies $\mathbf{F} = (f_{hj,ik})$ and expectations $\mathbf{F}^0 = (f_{hj}f_{ik}/f_{h.,i.})$. The elements in distribution $\mathbf{F}$ and $\mathbf{F}^0$ are identically ordered and their totals are also identical. We shall consider formulations on the I-divergence information for distributions whose elements are identically ordered but their totals are different.

Let $\mathbf{F}_1 = (f_{11} \ldots f_{p1})$ and $\mathbf{F}_2 = (f_{12} \ldots f_{p2})$ represent two quadrats described on the basis of the frequencies of p species. The frequencies are such that $\sum_h f_{h1} \neq \sum_h f_{h2}$, $h = 1, \ldots, p$; but for every element f_{h1} in $\mathbf{F}_1$ there is a corresponding element f_{h2} in $\mathbf{F}_2$. The criterion which identically orders the frequency in $\mathbf{F}_1$ and $\mathbf{F}_2$ is the common set of species which they completely share. In this particular case we can write an I-divergence as

$$I(\mathbf{F}_1; \mathbf{F}_2) = \sum_h I(\mathbf{F}_h; \mathbf{F}_h^0) = \sum_h \sum_j f_{hj} \ln 2f_{hj}/(f_{h1} + f_{h2}),$$

$$h = 1, \ldots, p; \; j = 1, 2 \qquad (2.33)$$

where p is the number of species. In this formula, the hth observed frequency distribution is defined by $\mathbf{F}_h = (f_{h1} \; f_{h2})$ and the standard by $\mathbf{F}_h^0 = [(f_{h1} + f_{h2})/2 \; (f_{h1} + f_{h2})/2]$. When $\mathbf{F}_1$ and $\mathbf{F}_2$ are allowed to designate quadrats *1* and *2* in Table 1-1, we have

$$\mathbf{F}_1 = (45 \;\; 18 \;\; 3 \;\; 10 \;\; 9) \text{ and } \mathbf{F}_2 = (2 \;\; 92 \;\; 40 \;\; 61 \;\; 53)\,.$$

From these we can derive $\mathbf{F}_h$ or $\mathbf{F}_h^0$ for any species. For instance, for the fourth species, we have

$$\mathbf{F}_4 = [10 \;\; 61] \text{ and } \mathbf{F}_4^0 = \left[\frac{71}{2} \;\; \frac{71}{2}\right].$$

The I-divergence of the two quadrats is the quantity,

$$\begin{aligned} I(\mathbf{F}_1;\mathbf{F}_2) \;=\; & I(\mathit{quadrat\ 1;\ quadrat\ 2}) = \sum_h I(\mathbf{F}_h;\mathbf{F}_h^0) = 45 \ln 45/ \\ & ((45+2)/2) + 2 \ln 2/((45+2)/2) + 18 \ln 18/((18+92)/2) \\ & + 92 \ln 92/((18+92)/2) + 3 \ln 3/((3+40)/2) \\ & + 40 \ln 40/((3+40)/2) + 10 \ln 10/((10+61)/2) \\ & + 61 \ln 61/((10+61)/2) + 9 \ln 9/((9+53)/2) \\ & + 53 \ln 53/((9+53)/2) = 108.102\,. \end{aligned}$$

In this, we were able to use the data from Table 1-1 directly since the data represent counts of individuals. Species frequencies could be similarly used, but not so measurements for which a different I-divergence may have to be defined (see Kullback, 1959).

Formula (2.33) involves a summation of information quantities $I(\mathbf{F}_h;\mathbf{F}_h^0)$, $h = 1, \ldots, p$, each of which is specific to one species. Strictly speaking, such a summation may be objectionable if the species are not independent, i.e. if they have mutual information, since it can bias the measured divergence in the sense that $I(\mathbf{F}_1;\mathbf{F}_2)$ will indicate an I-divergence for the two quadrats greater than their actual divergence. To overcome this problem, we may regard $\mathbf{F}_1$ and $\mathbf{F}_2$ as columns in a $p \times 2$ contingency table and compute an information quantity in terms of

$$I(\mathit{species;\ quadrats}) = \sum_h \sum_j f_{hj} \ln f_{hj} f_{..}/(f_{h.} f_{.j})\,, \quad h = 1, \ldots, p;\; j = 1, 2 \qquad (2.34)$$

This relates two criteria of classification – the first is the species affiliation of an observation and the second its quadrat of occurrence. The frequency symbols in formula (2.34) are as follows:

f_{hj} = quantity of species h in quadrat j

$$f_{..} = \sum_h \sum_j f_{hj}, \quad f_{h.} = \sum_j f_{hj}, \quad f_{.j} = \sum_h f_{hj}\,.$$

Formula (2.34) can be used as a measure of divergence between $\mathbf{F}_1$ and $\mathbf{F}_2$ without the need for assuming that the p species are independent. We may compare formula (2.34) and (2.33) to discover that their difference is

$$\begin{aligned}\Delta(1, 2) &= I(\mathbf{F}_1; \mathbf{F}_2) - I(species;\ quadrats)\\ &= \sum_h \sum_j f_{hj} [\ln 2 f_{hj}/f_{h.} - \ln f_{hj} f_{..} / (f_{h.} f_{.j})]\\ &= \sum_j f_{.j} \ln 2 f_{.j}/f_{..}\,, \quad j = 1, 2 \end{aligned} \qquad (2.35)$$

Because $\sum_j f_{.j}\ \ln\ 2f_{.j}/f_{..}$ is always a positive quantity, or zero, $I(\mathbf{F}_1; \mathbf{F}_2)$ is always larger than *I(species; quadrats)* when they are not equal.

Formulae (2.33), (2.34) and (2.35) each represent a different *I*-divergence information. We have already considered an example for formula (2.33). We can use the same data to illustrate the computations in the case of formulae (2.34) and (2.35). A $p \times 2$ table is constructed first:

Species	*Quadrat* 1	2	*Total*
X_1	45	2	47
X_2	18	92	110
X_3	3	40	43
X_4	10	61	71
X_5	9	53	62
Total	85	248	333

Then we calculate based on formula (2.34),

$$I(species;\ quadrats) = \sum_h \sum_j f_{hj} \ln f_{hj} + f_{..} \ln f_{..} - \sum_h f_{h.} \ln f_{h.}$$

$$- \sum_j f_{.j} \ln f_{.j} = 45 \ln 45 + 2 \ln 2 + 18 \ln 18 + 92 \ln 92$$

$$+ 3 \ln 3 + 40 \ln 40 + 10 \ln 10 + 61 \ln 61 + 9 \ln 9$$

$$+ 53 \ln 53 + 333 \ln 333 - 47 \ln 47 - 110 \ln 110$$

$$- 43 \ln 43 - 71 \ln 71 - 62 \ln 62 - 85 \ln 85 - 248 \ln 248$$

$$= 66.440\ .$$

This is a substantially smaller quantity than $I(\mathbf{F}_1; \mathbf{F}_2)$ based on formula (2.33). The difference according to formula (2.35) is

$$\begin{aligned}\Delta(1, 2) &= I(\mathbf{F}_1; \mathbf{F}_2) - I\ (species;\ quadrats)\\ &= 108.102 - 66.440\\ &= 41.662\ .\end{aligned}$$

This measures a divergence based on the distribution totals. Note that formula (2.34) represents a less restrictive measure of divergence than formula

(2.33), for whereas a zero value of *I(species; quadrats)* of formula (2.34) may imply only proportionality of species quantities in the quadrats, such as $f_{h1}/f_{h2} = f_{i1}/f_{i2}$, a zero value of the $I(\mathbf{F}_1; \mathbf{F}_2)$ of formula (2.33) always implies identity such as $f_{h1} = f_{h2}$ regardless of species.

An *I*-divergence may be defined not only between individual distributions but also among groups of distributions. The table below indicates how the data may be presented:

Species	*Quadrat in Group A* 1	2 ...	n_A	*Group total A*	*Quadrat in Group B* 1	2 ...	n_B	*Group total B*	*Total A + B*
X_1	f_{A11}	f_{A12}	f_{A1n_A}	$f_{A1.}$	f_{B11}	f_{B12}	f_{B1n_B}	$f_{B1.}$	$f_{.1.}$
X_2	f_{A21}	f_{A22}	f_{A2n_A}	$f_{A2.}$	f_{B21}	f_{B22}	f_{B2n_B}	$f_{B2.}$	$f_{.2.}$
.	.	.	.	.	.	.	.	.	.
X_p	f_{Ap1}	f_{Ap2}	f_{Apn_A}	$f_{Ap.}$	f_{Bp1}	f_{Bp2}	f_{Bpn_B}	$f_{Bp.}$	$f_{.p.}$
Total	$f_{A.1}$	$f_{A.2}$	$f_{A.n_A}$	$f_{A..}$	$f_{B.1}$	$f_{B.2}$	$f_{B.n_B}$	$f_{B..}$	$f_{...}$

The *I*-divergence of *A* and *B* based on formula (2.33) can be written as

$$I(A;B) = \sum_k \sum_h \sum_j f_{khj} \ln (n_A + n_B) f_{khj}/f_{.h.} - \sum_h \sum_j f_{Ahj} \ln n_A f_{Ahj}/f_{Ah.}$$
$$- \sum_h \sum_j f_{Bhj} \ln n_{B} f_{Bhj}/f_{Bh.}$$
$$= \sum_h \sum_k f_{kh.} \ln (f_{kh.}/n_k)/[f_{.h.}/(n_A + n_B)],$$
$$k = A, B;\ h = 1, ..., p;\ j = 1, ..., n_A \text{ or } n_B \qquad (2.36)$$

As in formula (2.33), in formula (2.36) also, we assume that the species are independent.

A less restrictive measure can be obtained based on formula (2.34), in terms of

$$I(A;B) = I(species;\ quadrats)_{A+B} - I(species;\ quadrats)_A$$
$$- I(species;\ quadrats)_B = \sum_h \sum_k f_{kh.} \ln [f_{kh.} f_{...}/(f_{k..} f_{.h.})],$$
$$h = 1, ..., p;\ k = A, B \qquad (2.37)$$

In this case, species independence need not be assumed. The least informative measure takes the form

$$\Delta(A;B) = \sum_k f_{k..} \ln (f_{k..}/f_{...})(n_A + n_B)/n_k,\ k = A, B \qquad (2.38)$$

in which the weighted group means are compared.

Let the first four quadrats of Table 1-1 be included in group *A* and the last six in group B. The group totals are given by numbers in the body of the table:

Species	*Total A*	*Total B*	*Total A + B*
X_1	82	147	229
X_2	190	431	621
X_3	131	138	269
X_4	85	89	174
X_5	182	360	542
Total	670	1165	1835

Based on formula (2.36), we have

$$\begin{aligned} I(A;B) &= 82 \ln ((82/4)/(229/10)) + 190 \ln ((190/4)/(621/10)) \\ &+ 131 \ln ((131/4)/(269/10)) + 85 \ln ((85/4)/(174/10)) \\ &+ 182 \ln ((182/4)/(542/10)) + 147 \ln ((147/6)/(229/10)) \\ &+ 431 \ln ((431/6)/(621/10)) + 138 \ln ((138/6)/(269/10)) \\ &+ 89 \ln ((89/6)/(174/10)) + 360 \ln ((360/6)/(542/10)) \\ &= 24.386\,. \end{aligned}$$

We apply formula (2.37) in the expanded form,

$$\begin{aligned} I(A;B) &= \sum_k \sum_h f_{kh.} \ln f_{kh.} + f_{...} \ln f_{...} - \sum_k f_{k..} \ln f_{k..} - \sum_h f_{.h.} \ln f_{.h.} \\ &= 82 \ln 82 + 190 \ln 190 + 131 \ln 131 + 85 \ln 85 + 182 \ln 182 \\ &+ 147 \ln 147 + 431 \ln 431 + 138 \ln 138 + 89 \ln 89 + 360 \ln 360 \\ &+ 1835 \ln 1835 - 670 \ln 670 - 1165 \ln 1165 - 229 \ln 229 \\ &- 621 \ln 621 - 269 \ln 269 - 174 \ln 174 - 542 \ln 542 \\ &= 19.685\,. \end{aligned}$$

After these, we have for formula (2.38),

$$\Delta(A;B) = 24.386 - 19.685 = 4.701\,.$$

Information divergences can be used probabilistically, based on the relationship between $2I$ and χ^2 at specified degrees of freedom, under specified circumstances:

Formula	Degrees of freedom	Null hypothesis tested based on 2I
(2.33)	p	Distributions F_1 and F_2 which represent two quadrats are random samples from the same multispecies population. Species are assumed independent.
(2.34)	$p-1$	An observation's species affiliation is unpredictable from its quadrat of occurrence in the population of which the two quadrats represent a random sample.
(2.35)	1	The parent populations from which F_1 and F_2 are taken at random have identical totals.
(2.36)	p	The parent populations from which groups A and B are taken at random have specified total frequencies ($f_{.1.}$ ... $f_{.p.}$). Species are assumed independent.
(2.37)	$p-1$	An observation's species affiliation is unpredictable from its groups of occurrence in the populations from which the two groups represent random samples.
(2.38)	1	The parent populations from which groups A and B are random samples have identical totals.

Under the null hypothesis, $2I$ is distributed approximately as a χ^2 variate with the specified degrees of freedom. The approximation improves with increased number of observations.

When the elements in the observed frequency distributions are not ordered by a single criterion, they cannot be paired up in any meaningful way between distributions. But still an I-divergence can be defined based on the distribution totals in a manner exemplified by formulae (2.35) and (2.38). These formulae actually represent comparisons between distribution means weighted by the number of observations.

We can formulate an information divergence in the general context of formula (2.36), such as

$$K(A;B) = \sum_h \frac{1}{2} I(\mathbf{F}_{Ah}; \mathbf{F}_{Bh}),\ h = 1, ..., p \ .$$

This measures the divergence of group A from B. $\mathbf{F}_{Ah} = (f_{Ah1} \ldots f_{Ahn_A})$ and $\mathbf{F}_{Bh} = (f_{Bh1} \ldots f_{Bhn_B})$ are distributions representing the hth species in the two groups. The p individual terms of $I(\mathbf{F}_{Ah}; \mathbf{F}_{Bh})$ accord with the definition,

$$I(\mathbf{F}_{Ah}; \mathbf{F}_{Bh}) = \sum_j f_{Ahj} \ln f_{Ahj}/f^0_{hj}$$
$$+ \sum_j f_{Bhj} \ln f_{Bhj}/f^0_{hj} \ ,$$

with summations taken from $j = 1$ to n_A in the first sum and $j = 1$ to n_B in the second. When we define f^0_{hj} as $f_{.h.}/(n_A + n_B)$ for any j, then it is clear that $\frac{1}{2} I(\mathbf{F}_{Ah}; \mathbf{F}_{Bh})$ measures the average divergence of $\mathbf{F}_{Ah}$ and $\mathbf{F}_{Bh}$ from an equidistribution $\mathbf{F}^0_h$ with the mean of the frequencies of species h in $n_A + n_B$ quadrats as elements. In the sense of a geometric analogy, $\frac{1}{2} I(\mathbf{F}_{Ah}; \mathbf{F}_{Bh})$ is the radius of a circle on which $\mathbf{F}_{Ah}$ and $\mathbf{F}_{Bh}$ are opposite points. Apparently, the same measure is described by Sibson (1969; Jardine,

1971; Jardine & Sibson, 1971) who did not establish connections to a broader class of formulations, known as the *minimum discrimination information statistic* (Kullback, 1959) with which $2I(F_{Ah}; F_{Bh})$ is here identified.

Other constructs are encountered in measuring association. The one most widely used is based on presence data summarized in a 2 x 2 contingency table:

Species B	*Species A*		*Total*
	Present	*Absent*	
Present	a	b	$a + b$
Absent	c	d	$c + d$
Total	$a + c$	$b + d$	$n = a + b + c + d$.

Of the n quadrats, a contained both species, d neither, b only species B and c only species A. The association of A and B is measurable as the quantity,

$$2I(A; B) = 2\left(a \ln \frac{a}{\alpha} + b \ln \frac{b}{\beta} + c \ln \frac{c}{\gamma} + d \ln \frac{d}{\delta}\right),$$

where $\alpha = (a + b)\ (a + c)/n$. β, γ and δ are similarly defined. If $a > \alpha$ the association is said to be positive, otherwise zero ($a = \alpha$) or negative ($a < \alpha$). The sampling distribution of $2I(A; B)$ is approximately a chi square distribution with 1 degree of freedom if the null hypothesis 'the n quadrats represent a random sample from a population completely specified by the frequency distribution ($\alpha\ \beta\ \gamma\ \delta$)' is true. This is, of course, a strong assumption implying that upon resampling a particular population over and over again by n random quadrats, in each case the emerging 2 x 2 table would have the same marginal totals. For this to happen is an obvious impossibility, and for that reason, the chi square distribution may not be quite appropriately used. Goodall (1973b) points out in this connection that the nominal probability of a more extreme value of chi square than the observed is expected to exceed the actual but unknown value. It follows from this that the test is likely to be conservatively biased. This means that in such a test, the null hypothesis of no association will be more often accepted than should be without bias. Pielou (1969a, 1974) analysed this problem and considered alternative methods to determine the probability of an observed 2 x 2 table.

2.7 Inequalities and partitions

A. *Information*

It is obvious from the preceding discussions that information content in the union group $A + B$ cannot be less than the sum in the subgroups. This property is stated in the inequality,

$$I_{A+B} \geqslant I_A + I_B \,.$$

Because of this property, information superbly qualifies for hierarchical partitioning into additive components.

We shall consider this topic in some detail. We begin with observing that we can derive schemes for partitioning information that parallel the schemes which conventionally use the sum of squares. Whereas the methods of ranking in Section 1.7 are of this kind in that they partition sum of squares, Orlóci (1976a) described a method which relies on mutual information. With a view to illustrating the method, we give E. van der Maarel's data (Orlóci, 1976a) in the following table:

Species	*Quadrat*
1. *Aster tripolium*	003200033200020322232323035553230330332030025200200
2. *Spartina patens*	00000000000000000000000000000000000000000000099999
3. *Atriplex hastata*	00000000000000000000000000000000000087997000000000
4. *Suaeda maritima*	00002552002020023222322288555789552002050000000000
5. *Salicornia radicans*	00000778980000000000000000000000000500000000000000
6. *Spartina alterniflora*	00000000000000000000000000000000000000009899900000
7. *Salicornia europaea*	00002502020000055537887702333523353000050000000000
8. *Puccinellia maritima*	00000872350000087755003220000230008300050023000000
9. *Spartina maritima*	79898533330000000000000000000000000000000000000000
10. *S. townsendii*	00000000007889953333333335255333555557550000000000

In this table the entries are codes for cover-abundance estimates of species in quadrats taken from salt marsh communities. There are 50 quadrats in the table. The method of ranking to be outlined partitions the mutual information in the table into components specific to the different species. We have already defined mutual information in formula (2.29). This can directly be generalized from two species to any higher number p without difficulty.

We regard the individual numbers in the table as class symbols, of which there are $s = 7$ different kinds. From these, we construct a p-dimensional frequency distribution of p^s cells (many of which may correspond to unrealized symbol combinations in the data). Let $I(1, \ldots, p)$ designate the mutual information in the p-dimensional distribution and let $I(1, \ldots, p{-}1 \mid m)$ designate the mutual information in the $p{-}1$ dimensional distribution after species m has been removed. We assign rank 1 to species m if the divergence

$$\Delta_m = I(1, \ldots, p) - I(1, \ldots, p-1 \mid m)$$

is largest among all species in the sample. This gives us the clue as to the algorithm that determines Δ values for the remaining species:

1. Having assigned a rank to species m, remove row m from the data.
2. Test each species in the residual set until a species z is found such that

$$\Delta_z = I(1, \ldots, p-1 \mid m) - I(1, \ldots, p-2 \mid m, z)$$

is a maximum. The term $I(1, \ldots, p-2 \mid m, z)$ is the residual of the mutual information in the data table with both rows m and z removed. Species z for which Δ_z is largest is given rank 2.

3. Continue as in steps 1 and 2 until all $p - 2$ of the species are ranked. The two species remaining last are given equal values for rank.

The algorithm partitions the total mutual information into p additive components,

$$I(1, ..., p) = \Delta_1 + \Delta_2 + ... + \Delta_{p-1} + \Delta_p \quad .$$

of which the last is always zero (indicating that the mutual information is defined only for two or more species).

A computer program which performs the calculations automatically is given under the name RANKIN in the Appendix. When run with the salt marsh data as input, the following results are produced:

Species *i*	*Number of species in set with species i*	*Mutual information in set with species i* a	*Mutual information in set without species i* b	*Mutual information specific to species i* Δ_i=a-b	*Proportion of total (282.11) accounted for by species i* %	*Rank of species i*
10	10	282.11	210.94	71.17	25.23	1
7	9	210.94	141.93	69.01	24.46	2
4	8	141.93	82.59	59.34	21.03	3
8	7	82.59	43.55	39.04	13.84	4
9	6	43.55	20.36	23.19	8.22	5
1	5	20.36	4.05	16.31	5.78	6
5	4	4.05	1.74	2.31	0.82	7
6	3	1.74	0.56	1.18	0.42	8
3, 2	2	0.56	0.00	0.56	0.20	9

The entries in columns 5, 6 are additive; the sum of the Δ_i is 282.11 and the sum of the percentage values is 100. The values in column 5 measure the mutual information that the individual species share with the others in their respective residual sets. We stress ***residual*** because the Δ_i value assigned to any species i is not the value that it would have should it be considered with the complete set of $p - 1$ species in the sample. With respect to the example, it is seen from the results that the species of low rank are those occurring in few of the quadrats. These species, while unimportant in the context of an analysis that uses mutual information to express species correlation, may be quite important in other respects such as, for instance, their diagnostic value in identifying vegetation types (or other synsystematic categories).

At this point, we should note that an entire class of statistical techniques, that parallel the techniques which partition sums of squares in the analysis of variance, relies on the quantity $2I(\mathrm{F}; \mathrm{F}^0)$. The fundamental reference on this topic is Kullback (1959). Kullback, Kupperman & Ku (1962) present essentials and illustrate applications in a practical context. Examples that use information in a related sense on ecological data are numerous (e.g. Juhász-Nagy, 1967a; Pielou, 1967; Williams, Lance, Webb, Tracey & Dale, 1969; Orlóci, 1971a, 1972b; Stanek & Orlóci, 1973; Staniforth & Cavers, 1976). We shall outline four basic models for partitioning information in this section and consider others in the sequel:

1. The first to be considered is a *one factor model* with different levels of the factor

representing c treatments. The response variates are discrete. The response of the kth replicate to the jth treatment is f_{jk}, and the number of replicates is n_j. The symbols accord with those in the table,

Replicate	*Treatment*				
	1	*2*	...	*c*	
1	f_{11}	f_{21}		f_{c1}	
2	f_{12}	f_{22}		f_{c2}	
.	.	.		.	
n_j	f_{1n_1}	f_{2n_2}		f_{cn_c}	
Total	$f_{1.}$	$f_{2.}$		$f_{c.}$	$f_{..}$.

There are c columns in this table each representing a different treatment with a different number of replicates. Treatment totals are indicated by substituting a period for the replicate subscript, e.g., $f_{j.}$ signifies the total in the jth treatment. When both subscripts are missing, as in $f_{..}$, the symbol then represents the grand total for the table. Since the observations are categorical, i.e. they consist of counts or frequencies, the analysis produces information partitions with components as follows:

$$2I(\bar{f}_{*.};\bar{f}_{..}) = 2 \sum_{j=1}^{c} f_{j.} \ln \frac{f_{j.}/n_j}{f_{..}/n_.} ;$$

this defines the component generated by divergences among treatment means.

$$2I(f_{**};\bar{f}_{*.}) = 2 \sum_{j=1}^{c} \sum_{k=1}^{n_j} f_{jk} \ln \frac{f_{jk}}{f_{j.}/n_j} ;$$

this is the error component associated with variability between replicates within treatments.

$$\begin{aligned} 2I(f_{**};\bar{f}_{..}) &= 2I(\bar{f}_{*.};\bar{f}_{..}) + 2I(f_{**};\bar{f}_{*.}) \\ &= 2 \sum_{j=1}^{c} \sum_{k=1}^{n_j} f_{jk} \ln \frac{f_{jk}}{f_{..}/n_.} ; \end{aligned}$$

this is the total information. The analysis is summarized in:

Source	*Information*	*Degrees of freedom*
Treatment	$2I(\bar{f}_{*.};\bar{f}_{..})$	$c - 1$
Error	$2I(f_{**};\bar{f}_{*.})$	$n_. - c$
Total	$2I(f_{**};\bar{f}_{..})$	$n_. - 1$

Each term in this table can be regarded as an independent chi square variate with the given degrees of freedom under the null hypothesis of zero population divergence between treatments and homogeneity within populations.

2. The second model incorporates *subsampling within treatments* in the one factor model. Each value of the response f_{ijk} is specified by three subscripts – one for treatment, one for sample within a given treatment, and one for replicate within a sample. Sample totals, treatment totals and the grand total are represented respectively by symbols $f_{ij.}$, $f_{i..}$ and $f_{...}$. This model permits estimation of all components of information defined for the one factor model, plus one more, the component for the sampling error. The symbolic data are given in:

Replicate	*Treatment* 1		*Treatment* 2		... *Treatment* c	
	Sample within treatment					
	1 ...	s_1	1 ...	s_2	1 ...	s_c
1	f_{111}	f_{1s_11}	f_{211}	f_{2s_21}	f_{c11}	f_{cs_c1}
2	f_{112}	f_{1s_12}	f_{212}	f_{2s_22}	f_{c12}	f_{cs_c2}
.	.	.	,	.	.	
n_{ij}	$f_{11n_{11}}$	$f_{1s_1n_{1s_1}}$	$f_{21n_{21}}$	$f_{2s_2n_{2s_2}}$	$f_{c1n_{c1}}$	$f_{cs_cn_{cs_c}}$
Sample total	$f_{11.}$	$f_{1s_1.}$	$f_{21.}$	$f_{2s_2.}$	$f_{c1.}$	$f_{cs_c.}$
Treatment total		$n_{1.}$		$n_{2.}$		$n_{c.}$
		$f_{1..}$		$f_{2..}$		$f_{c..}$
Grand total						$f_{...}$ $n_{..}$

Once again, to analyse such data, information partitioning is appropriate. Such an analysis gives the following components:

$$2I(\bar{f}_{*..};\bar{f}_{...}) = 2\sum_{i=1}^{c} f_{i..} \ln \frac{f_{i..}/n_{i.}}{f_{...}/n_{..}} ;$$

this is the information generated by divergence among treatment means.

$$2I(\bar{f}_{**.};\bar{f}_{*..}) = 2\sum_{i=1}^{c}\sum_{j=1}^{s_i} f_{ij.} \ln \frac{f_{ij.}/n_{ij}}{f_{i..}/n_{i.}} ;$$

this is the information associated with sampling error.

$$2I(f_{***};\bar{f}_{**.}) = 2\sum_{i=1}^{c}\sum_{j=1}^{s_i}\sum_{k=1}^{n_{ij}} f_{ijk} \ln \frac{f_{ijk}}{f_{ij.}/n_{ij}} ;$$

this is the information associated with experimental error.

$$2I(f_{***};\bar{f}_{...}) = 2I(\bar{f}_{*..};\bar{f}_{...}) + 2I(\bar{f}_{**.};\bar{f}_{*..}) + 2I(f_{***};\bar{f}_{**.})$$

$$= 2\sum_{i=1}^{c}\sum_{j=1}^{s_i}\sum_{k=1}^{n_{ij}} f_{ijk} \ln \frac{f_{ijk}}{f_{...}/n_{..}} ;$$

this is the total information.

The different components are summarized in:

Source	*Information*	*Degrees of freedom*
Treatment	$2I(\bar{f}_{*..};\bar{f}_{...})$	$c - 1$
Sampling error	$2I(\bar{f}_{**.};\bar{f}_{*..})$	$\sum_{i=1}^{c}(s_i - 1)$
Experimental error	$2I(f_{***};\bar{f}_{**.})$	$\sum_{i=1}^{c}\sum_{j=1}^{s_i}(n_{ij} - 1)$
Total	$2I(f_{***};\bar{f}_{...})$	$n_{..} - 1$

The individual terms in the table can be regarded as independent chi square variates with the given degrees of freedom under the assumption of zero divergence between the treatment populations and homogeneity within them.

3. The third model assumes an $r \times c$ *factorial design.* Each response f_{ijk} is specific to the combination of two factor levels, i and j, and the replicate k. The difference between this and the previous model should be noted. The symbolic data are given in:

Level of factor C	*Replicate*	*Level of factor R* 1	2	...	*r*	*Total*
1	1	f_{111}	f_{211}		f_{r11}	
	2	f_{112}	f_{212}		f_{r12}	
	.	.	.		,	
	n_{i1}	$f_{11n_{11}}$	$f_{21n_{21}}$		$f_{r1n_{r1}}$	
Total		$f_{11.}$	$f_{21.}$		$f_{r1.}$	$f_{.1.}$ $n_{.1}$
.	.	.	.		.	.
c	1	f_{1c1}	f_{2c1}		f_{rc1}	
	2	f_{1c2}	f_{2c2}		f_{rc2}	
	.	.	.		.	
	n_{ic}	$f_{1cn_{1c}}$	$f_{2cn_{2c}}$		$f_{rcn_{rc}}$	
Total		$f_{1c.}$	$f_{2c.}$		$f_{rc.}$	$f_{.c.}$ $n_{.c}$
Grand total		$f_{1..}$ $n_{1.}$	$f_{2..}$ $n_{2.}$		$f_{r..}$ $n_{c.}$	$f_{...}$ $n_{..}$

The model with subsampling within treatments represents a problem in the nested (hierarchical) decomposition of information. The decomposition in the present model is different in that firstly, the joint effect and the experimental error are defined.

Secondly, the joint effect is further decomposed into terms for main effect and interaction. On the assumption that the data are categorical, information partitioning is appropriate:

$$2I(\bar{f}_{*..};\bar{f}_{...}) = 2\sum_{i=1}^{r} f_{i..} \ln \frac{f_{i..}/n_{i.}}{f_{...}/n_{..}} ;$$

this is the information associated with variability in factor R.

$$2I(\bar{f}_{.*.};\bar{f}_{...}) = 2\sum_{j=1}^{c} f_{.j.} \ln \frac{f_{.j.}/n_{.j}}{f_{...}/n_{..}} ;$$

this is the information generated by variability in factor C.

$$2I(R;C) = 2\sum_{i=1}^{r}\sum_{j=1}^{c} f_{ij.} \ln \frac{f_{ij.}f_{...}}{f_{i..}f_{.j.}} \frac{n_{i.}\,n_{.j}}{n_{ij}\,n_{..}} ;$$

this is the interaction of R and C, representing the independence component generated by the departure of the treatment combination means, $\bar{f}_{ij.}$, from their random expectations based on marginal totals and grand total in the contingency table.

$$2I(\bar{f}_{**.};\bar{f}_{...}) = 2I(\bar{f}_{*..};\bar{f}_{...}) + 2I(\bar{f}_{.*.};\bar{f}_{...}) + 2I(R;C)$$

$$= 2\sum_{i=1}^{r}\sum_{j=1}^{c} f_{ij.} \ln \frac{f_{ij.}/n_{ij}}{f_{...}/n_{..}} ;$$

this is the joint information of factors, related to the divergence of treatment combination means from the grand mean.

$$2I(f_{***};\bar{f}_{**.}) = 2\sum_{i=1}^{r}\sum_{j=1}^{c}\sum_{k=1}^{n_{ij}} f_{ijk} \ln \frac{f_{ijk}}{f_{ij.}/n_{ij}} ;$$

this gives the information generated by experimental error.

$$2I(f_{***};\bar{f}_{...}) = 2I(\bar{f}_{**.};\bar{f}_{...}) + 2I(f_{***};\bar{f}_{**.})$$

$$= 2\sum_{i=1}^{r}\sum_{j=1}^{c}\sum_{k=1}^{n_{ij}} f_{ijk} \ln \frac{f_{ijk}}{f_{...}/n_{..}} ;$$

this is the total information associated with the divergence of individual responses from the common mean. The analysis is summarized in:

Source	*Information*	*Degrees of freedom*
Main effect R	$2I(\bar{f}_{*..};\bar{f}_{...})$	$r-1$
Main effect C	$2I(\bar{f}_{.*.};\bar{f}_{...})$	$c-1$
Interaction RC	$2I(R;C)$	$(r-1)\ (c-1)$
Joint effect	$2I(\bar{f}_{**.};\bar{f}_{...})$	$rc-1$
Experimental error	$2I(f_{***};\bar{f}_{**.})$	$\sum_{i=1}^{r}\sum_{j=1}^{c}(n_{ij}-1)$
Total	$2I(f_{***};\bar{f}_{...})$	$n_{..}-1$

The individual terms can be regarded as independent chi square variates with the given degrees of freedom.

4. An $r \times c \times d$ *factorial model* is considered next. The data are arranged in the cells of a three dimensional solid with replicates as a string of numbers within each cell. The contents of a cell in the main body of the solid register response to a treatment combination. Each response is specified by four subscripts as in f_{ijkl}, the first three indicating treatment combination and the last signifying a replicate. A partition of information, analogous to partitioning of sums of squares in the analysis of variance, gives the following components:

$$2I(\bar{f}_{*...};\bar{f}_{....}) = 2\sum_{i=1}^{r} f_{i...} \ln \frac{f_{i...}/n_{i..}}{f_{....}/n_{...}} ;$$

this is the information component associated with the main effect of factor R related to variation among the observed marginal means $\bar{X}_{i...}$, $i = 1, ..., r$.

$$2I(\bar{f}_{.*..};\bar{f}_{....}) = 2\sum_{j=1}^{c} f_{.j..} \ln \frac{f_{.j..}/n_{.j.}}{f_{....}/n_{...}} ;$$

this is the component generated by the effect of factor C.

$$2I(\bar{f}_{..*.};\bar{f}_{....}) = 2\sum_{k=1}^{d} f_{..k.} \ln \frac{f_{..k.}/n_{..k}}{f_{....}/n_{...}} ;$$

this is the component associated with the effect of factor D.

$$2I(R;\, C;\, D) = 2\sum_{i=1}^{r}\sum_{j=1}^{c}\sum_{k=1}^{d} f_{ijk.} \ln \frac{f_{ijk.}f_{....}^2}{f_{i...}f_{.j..}f_{..k.}} \frac{n_{i..}\,n_{.j.}\,n_{..k}}{n_{ijk}\,n_{...}^2} ;$$

this component is specific to the interaction (independence) of three factors.

$$\begin{aligned} 2(\bar{f}_{***.};\bar{f}_{....}) &= 2I(\bar{f}_{*...};\bar{f}_{....}) + 2I(\bar{f}_{.*..};\bar{f}_{....}) \\ &+ 2I(\bar{f}_{..*.};\bar{f}_{....}) + 2I(R;\, C;\, D) \\ &= 2\sum_{i=1}^{r}\sum_{j=1}^{c}\sum_{k=1}^{d} f_{ijk.} \ln \frac{f_{ijk.}/n_{ijk}}{f_{....}/n_{...}} ; \end{aligned}$$

this is the joint information of the three factors.

$$2I(f_{****};\bar{f}_{***.}) = 2\sum_{i=1}^{r}\sum_{j=1}^{c}\sum_{k=1}^{d}\sum_{l=1}^{n_{ijk}} f_{ijkl} \ln \frac{f_{ijkl}}{f_{ijk.}/n_{ijk}} ;$$

this is the error component generated by variability of response to the same treatment combination.

$$2I(f_{****};\bar{f}_{....}) = 2\sum_{i=1}^{r}\sum_{j=1}^{c}\sum_{k=1}^{d}\sum_{l=1}^{n_{ijk}} f_{ijkl} \ln \frac{f_{ijkl}}{f_{....}/n_{...}} ;$$

this is the total information in the data.

The components so far derived are summarized with the associated degrees of freedom in the table,

Source	Information	Degrees of freedom
Main effect R	$2I(\bar{f}_{*...}:\bar{f}_{....})$	$r-1$
Main effect C	$2I(\bar{f}_{.*..};\bar{f}_{....})$	$c-1$
Main effect D	$2I(\bar{f}_{..*.};\bar{f}_{....})$	$d-1$
Interaction RCD	$2I(R;C;D)$	$rcd-r-c-d+2$
Joint effect	$2I(\bar{f}_{***.};\bar{f}_{....})$	$rcd-1$
Experimental error	$2I(f_{****};\bar{f}_{***.})$	$\sum_{i=1}^{r}\sum_{j=1}^{c}\sum_{k=1}^{d}(n_{ijk}-1)$
Total	$2I(f_{****};\bar{f}_{....})$	$n_{...}-1$

Further apportionment of the independence component $2I(R;\ C;\ D)$ can be accomplished in a similar manner as in the case of the sums of squares. The existence of the following additive sequences can be shown:

a. $2I(R;\ C;\ D) = 2I(C;\ D) + 2I(R;\ (C,\ D)),$

where

$$2I(C;\ D) = 2\sum_{j=1}^{c}\sum_{k=1}^{d} f_{.jk.}\ \ln \frac{f_{.jk.}f_{....}\,n_{.j.}\,n_{..k}}{f_{.j..}f_{..k.}\,n_{.jk}\,n_{...}}$$

and

$$2I(R;\ (C,\ D)) = 2\sum_{i=1}^{r}\sum_{j=1}^{c}\sum_{k=1}^{d} f_{ijk.}\ \ln \frac{f_{ijk.}f_{....}}{f_{i...}f_{.jk.}}\ \frac{n_{i..}\,n_{.jk}}{n_{ijk}\,n_{...}}\ .$$

b. $2I(R;\ C;\ D) = 2I(R;\ C) + 2I(D;\ (R,\ C)),$

where

$$2I(R;\ C) = 2\sum_{i=1}^{r}\sum_{j=1}^{c} f_{ij..}\ \ln \frac{f_{ij..}f_{....}}{f_{i...}f_{.j..}}\ \frac{n_{i..}\,n_{.j.}}{n_{ij.}\,n_{...}}$$

and

$$2I(D;\ (R,\ C)) = 2\sum_{i=1}^{r}\sum_{j=1}^{c}\sum_{k=1}^{d} f_{ijk.}\ \ln \frac{f_{ijk.}f_{....}}{f_{..k.}f_{ij..}}\ \frac{n_{..k}\,n_{ij.}}{n_{ijk}\,n_{...}}\ .$$

c. $2I(R;\ C;\ D) = 2I(R;\ D) + 2I(C;\ (R,\ D))$,

where

$$2I(R;\ D) = 2\sum_{i=1}^{r}\sum_{k=1}^{d} f_{i.k.}\ \ln \frac{f_{i.k.}f_{....}}{f_{i...}f_{..k}}\ \frac{n_{i..}\,n_{..k}}{n_{i.k}\,n_{...}}$$

and

$$2I(C;\ (R,\ D)) = 2\sum_{i=1}^{r}\sum_{j=1}^{c}\sum_{k=1}^{d} f_{ijk.}\ \ln \frac{f_{ijk.}f_{....}}{f_{.j..}f_{i.k.}}\ \frac{n_{.j.}\,n_{i.k}}{n_{ijk}\,n_{...}}\ .$$

These sequences are summarized in the following table:

Source	*Information*	*Degrees of freedom*
Interaction RCD	$2I(R;C;D)$	$rcd-r-c-d+2$
Interaction CD	$2I(C;D)$	$(c-1)(d-1)$
Interaction R(CD)	$2I(R;(C,D))$	$(cd-1)(r-1)$
Interaction RC	$2I(R;C)$	$(r-1)(c-1)$
Interaction D(RC)	$2I(D;(R,C))$	$(rc-1)(d-1)$
Interaction RD	$2I(R;D)$	$(r-1)(d-1)$
Interaction C(RD)	$2I(C;(R,D))$	$(rd-1)(c-1)$

We can derive further components for the interaction term $R(CD)$, $D(RC)$ and $C(RD)$:

a. $2I(R;(C,D)) = 2I(R;C) + 2I(RC;CD) = 2I(R,D) + 2I(RD;DC)$.

where

$$2I(RC;CD) = 2\sum_{i=1}^{r}\sum_{j=1}^{c}\sum_{k=1}^{d} f_{ijk.} \ln \frac{f_{ijk.}f_{.j..}}{f_{ij..}f_{.jk.}} \frac{n_{ij.}n_{.jk}}{n_{ijk}n_{.j.}}$$

and

$$2I(RD;DC) = 2\sum_{i=1}^{r}\sum_{j=1}^{c}\sum_{k=1}^{d} f_{ijk.} \ln \frac{f_{ijk.}f_{..k.}}{f_{i.k.}f_{.j.k}} \frac{n_{i.k}n_{.jk}}{n_{ijk}n_{..k}} .$$

b. $2I(D;(R,C)) = 2I(D;R) + 2I(DR;RC) = 2I(D;C) + 2I(DC;CR)$,

where

$$2I(DR;RC) = 2\sum_{i=1}^{r}\sum_{j=1}^{c}\sum_{k=1}^{d} f_{ijk.} \ln \frac{f_{ijk.}f_{i...}}{f_{i.k.}f_{ij..}} \frac{n_{i.k}n_{ij.}}{n_{ijk}n_{...}}$$

and

$2I(DC;CR) = 2I(RC;CD)$ which we already defined.

c. $2I(C;(R,D)) = 2I(C;R) + 2I(CR;RD) = 2I(C;D) + 2I(CD;DR)$,

where

$2I(CR;RD) = 2I(DR;RC)$ and $2I(CD;DR) = 2I(RD;DC)$,

both already defined.

The components derived have associated degrees of freedom as shown in,

Source	*Information*	*Degrees of freedom*
Interaction (RC)(CD)	$2I(RC;CD)$	$c(r-1)(d-1)$
Interaction (RD)(DC)	$2I(RD;DC)$	$d(r-1)(c-1)$
Interaction (DR)(RC)	$2I(DR;RC)$	$r(c-1)(d-1)$

All information terms can be regarded as independent chi square variates with specified degrees of freedom. It should however be noted once again that the relationship between chi square and information is only an approximate one. It therefore is important to compute information from a large sample to improve the approximation.

B. *Sums of Squares*

An inequality can be stated for the sums of squares,

$$Q_{A+B} \geqslant Q_A + Q_B,$$

i.e. the sum of squares in the union group $A + B$ cannot be less than the sum of the sums of squares in the subgroups. This inequality, similarly as the information inequality, plays an important role in classifications requiring comparisons between groups. Some classifications use the difference

$$I_{A+B} - I_A - I_B \text{ or } Q_{A+B} - Q_A - Q_B$$

as a decision criterion for fusions or subdivisions. We shall consider them in Chapter 4.

2.8 Transformations implicit in resemblance functions

Resemblance functions often incorporate transformations or other adjustments which have the same effect as if the raw data were manipulated first and then a similarity or dissimilarity value computed from the manipulated data. The transformations may be so substantial that when implemented the data may retain little resemblance to the original observations. We shall consider a number of important transformations here and refer to Noy-Meir (1973a) and Noy-Meir, Walker & Williams (1975) for others.

1. The *Euclidean distance* is defined by formula (2.1). It incorporates no implicit adjustments of any sort. The resemblance structure which it imposes on the quadrats is the same as the resemblance structure imposed on the quadrats by the raw data elements as rectangular co-ordinates. The points in Fig. 2-3*a* illustrate this based on the data in Table 2-6.

Table 2-6. Quadrats from a simulated two-species community. Entries in the table indicate species abundance in the quadrats

Species	*Quadrat*									
	1	*4*	*3*	*5*	*10*	*2*	*6*	*9*	*7*	*8*
X_1	0.6	5.0	4.0	4.0	0	1.6	1.6	0.1	0.6	0.3
X_2	0.1	1.6	0.6	4.0	0.3	0.3	5.0	0.6	4.0	1.6

2. The *chord length* (formula 2.2)) is different in that it incorporates normalization of quadrat vectors. Since this has the effect of setting the length of each quadrat vector to unity, the net result is one of projecting the sample points onto the surface of some sphere (circle in two dimensions) of unit radius. This is illustrated in Fig. 2-3*b*. It can be seen that it drastically alters the structure of the point configurations as the horseshoe-shaped sequence unfolds into a circular arrangement.

3. The transformation in formula (2.3) brings about an *infolding* of the species axes until their obliqueness becomes proportional to the sample species correlations. Fig. 2-3*c* illustrates the effect of such an infolding of axes on the points. The angle enclosed by axes X_1, X_2 is arc cos S_{12}, where S_{12} is the product moment correlation coefficient. In the example S_{12} = 0.1841 and arc cos S_{12} = 79° 23′ approximately. Note that this transforma-

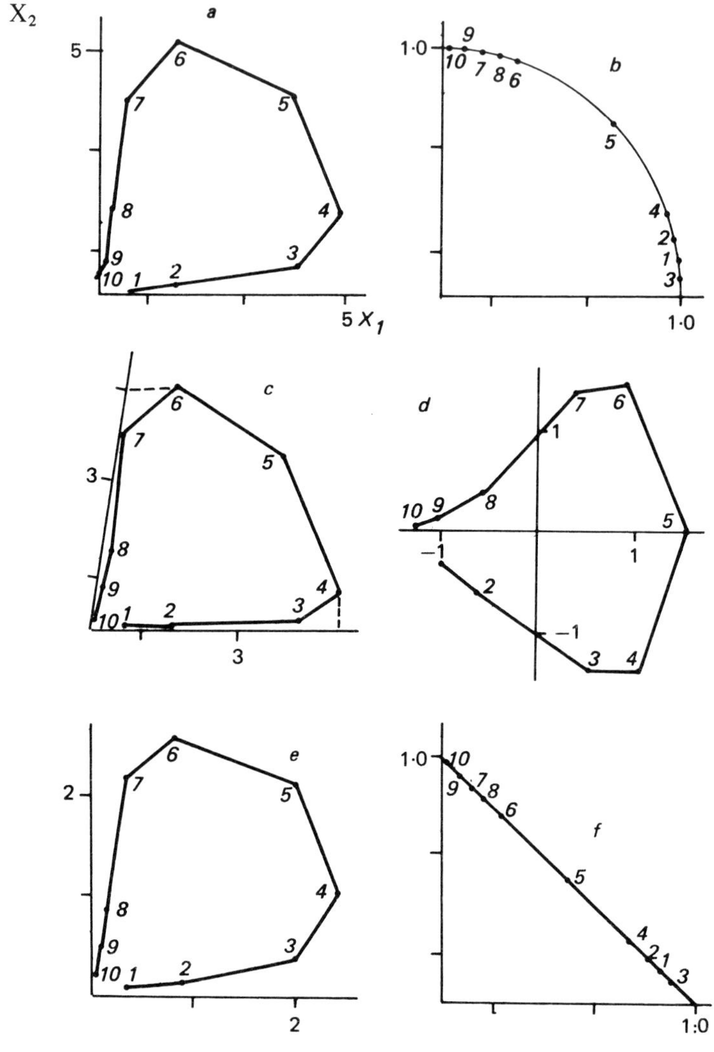

Fig. 2-3. The effect of transformations implicit in resemblance functions on the joint distribution of species. *Legend to figures: (a)* no transformation; *(b)* normalization within quadrat vectors; *(c)* oblique axes; *(d)* standardization within the components; *(e)* standardization within species; *(f)* transformation to unit sum within quadrats. Numbers inside diagrams are quadrat labels. Symbols X_1 and X_2 signify species in Table *2-6.*

tion, if it has an effect, exaggerates the curvature in the point configuration.
4. The *generalized distance* (formula (2.4)) incorporates a double transformation. The axes X_1, X_2 are rigidly rotated to an orthogonal position Y_1, Y_2 (zero covariance), and then, the co-ordinates are standardized on the rotated axes. The rotation has no effect on the shape of the point configuration. The change in Fig. 2-3*d*, as compared to Fig. 2-3*a*, is a consequence of the standardization. This does not seem to reduce the curvature in the point configuration.
5. When S_{hi} or q_{jk} are computed, representing expressions for the simple product moment correlation or its *Q*-equivalent, defined by formulae (2.8) and (2.9), the species vectors are centered and normalized. The effect of these on the resemblance structure is seen from Fig. 2-3*e*. Obviously, here the strong original curvature in the point configuration is at least retained if not intensified.
6. The Whittaker transformation which adjusts the quadrat vectors to unit sum in formula (2.14) has the most drastic effect toward increasing *linearity* (Fig. 2-3*f*). It may be convenient to incorporate this transformation in the Euclidean distance formula to obtain

$$u(j, k) = [\sum_h (X_{hj}/Q_j - X_{hk}/Q_k)^2]^{1/2} \qquad (2.39)$$

where

$$Q_j = \sum_h X_{hj} \text{ and } Q_k = \sum_h X_{hk},\ h = 1, ..., p .$$

A doubly adjusted form of formula (2.39) is described by Briane, Lazare, Roux & Sastre (1974; Chardy, Glemarec & Laurec, 1976) to which they referred as the chi square distance. We refer in this connection to Escofier-Cordier (1969).
7. Whereas the functions which we already examined in this section measure linear distances, the *geodesic metric* of formula (2.15) is noted for non-linearity. It resembles the chord distance of formula (2.2) in so far as it incorporates a similar polar projection of points onto a sphere or a circle. However, the geodesic metric differs from the chord distance in that it measures the distance of points along the shorter arc rather than in the direction of the chord. In this way it can transform a very complex non-linearity in the point configuration into a simpler spherical non-linearity. This, and especially the linearizing transformation in formula (2.14) or (2.39), is considered as a highly desired property in predictive ordinations (cf. Whittaker & Gauch, 1973) to be considered in Chapter 3.
8. The foregoing resemblance functions are consistent in their transformations in the sense that they retain the same scale within the same sample. Formulae (2.17) and (2.20) *do not always* have this property and can change their scale depending on the objects compared. The lack of a uniform scale of measure can lead to a very peculiar situation under which these formulae cannot uniquely describe a given point configuration. This means that as their scale keeps changing the points keep migrating through broad regions of the sample space. The actual migration of the point rep-

resenting quadrat *1* in Table 2-6, under the influence of a changing scale in formula (2.17) due to the changing quantity Q_{1k}, is traced in the scatter of

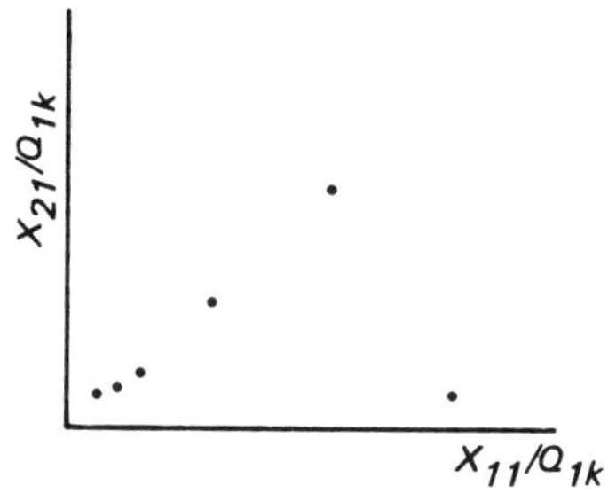

Fig. 2-4. Migration of a point, representing quadrat *1* in Table *2-6* in sample space, under the influence of a changing scale in formula (2.17). Point positions indicate comparisons with second, third, fourth, etc., quadrats.

point positions in Fig. 2-4. Each time Q_{1k} changes its value, quadrat *1* also changes its position to occupy a different point in the diagram. For example, when quadrat *1* is compared to quadrat *2* for which $Q_{12} = Q_1 + Q_2 = 0.7 + 1.9 = 2.6$, quadrat *1* will be represented by a point whose co-ordinates are

$$Y_1 = X_{11}/Q_{12} = 0.6/2.6 = 0.231$$

$$Y_2 = X_{21}/Q_{12} = 0.1/2.6 = 0.039\ .$$

These values accord with the transformations in formula (2.17). When the comparison is with quadrat *3* the point representing quadrat *1* migrates to a new position marked by the new co-ordinates,

$$Y_1 = X_{11}/Q_{13} = 0.6/5.3 = 0.113$$

$$Y_2 = X_{21}/Q_{13} = 0.1/5.3 = 0.019\ .$$

The varying Y_1, Y_2 values as co-ordinates thus trace the positions of quadrat *1* in the sample space in response to the changing sum Q_{1k}. Functions that cause such a migration of points in sample space are thoroughly undesirable in data analysis. Here we refer to the discussion in Section 2.5.

9. We could convert the *probabilistic index* of Section 2.6 into a quadrat distance $[2(1 - S_{jk})]^{1/2}$,but the resemblance structure which this distance imposes on the quadrats may bear little resemblance to the covariance structure which appears from the raw data. The reason is quite simply that this particular index uses the data only to derive probabilities which then impose a completely new structure on the sample. Similar considerations apply also to *Rajski's metric* for which the data are condensed into frequency distributions. These distributions redescribe the objects in a form suitable for com-

putation of a metric information divergence. Other *information divergences* routinely include such a condensation of the data into frequency distributions, with the case of categorical data representing an obvious exception, and impose on the sample a structure which can be radically different from the covariance.

The idea, that monotone transformations of resemblance functions produce functions that themselves are measures of resemblance, underlies much of the effort to describe resemblance functions. We refer in this connection to Goodman (1972) concerning Euclidean functions and to Lerman (1969, 1970) with regard to 2 x 2 tables. Many of the transformations lead to substantial loss of detail. One such case is the incorporation of a signum transformation in the resemblance measure to affect observations selectively or indiscriminately (e.g. Bonner, 1964; Estabrook, 1966; Parker-Rhodes & Jackson, 1969).

2.9 Further remarks on resemblance functions

The preceding treatment of resemblance functions provides a convenient framework of reference within which the ecologist can consider, categorize and evaluate still other resemblance functions which have not been discussed in this monograph. In this connection we refer to articles by, e.g., Cole (1949, 1957), Goodall (1952a, 1973a), Morisita (1959), Dagnelie (1960, 1965a), Greig-Smith (1964), Češka (1966, 1968), Orlóci (1968b, 1972a, 1973a), Green & Rao (1969), Cormack (1971), Brisse & Grandjouan (1971) and Goodman (1972) for further information. Statistical and other mathematical properties are treated by, e.g., Goodman & Kruskal (1954, 1959) and Anderberg (1973). The applications in numerical taxonomy are discussed in detail by Sokal (1961), Sokal & Sneath (1963), Jardine & Sibson (1971) and Sneath & Sokal (1973).

Evidently, to the ecologist, computation of a resemblance value rarely represents a goal in itself. Often such values provide input in different analyses such as classification, ordination, etc. It is reasonable then to require the resemblance functions, to be admissible in ecological work, to satisfy specific criteria:

1. Meaningfulness in mathematical as well as ecological terms.
2. Uniform scale of measure.
3. Desired sampling (probabilistic, statistical) properties.
4. Reasonable computational load.

In any method of data analysis the resemblance function must be mathematically meaningful. But by meeting the conditions of mathematical meaningfulness, the function may not necessarily be desired in ecological terms. Formula (2.1) is an excellent example in this connection; while it represents a mathematically meaningful formulation, it has only limited usefulness in ecological applications because it does not measure on a relative scale. The lack of such a scale can lead to anomalous situations which we have discussed in Section 2.4.

When it comes to choosing a resemblance function for analysis the mathematical properties may remove certain functions from consideration. The changing scale, which we discussed in connection with formulae (2.17) and (2.20), weighs heavily in our decisions. Once a function is identified as one having a variable scale dependent on the objects being compared, it may be completely pointless even to consider ecological meaningfulness or other aspects related to applications. The scale of measure, whether absolute or relative, linear or non-linear, is another important internal criterion for consideration. Classifications of the same set of quadrats, for example, when based on formula (2.1) in which the scale is absolute, may produce drastically different clusters than other classifications based on formula (2.2) in which the scale is relative.

Not all measures of resemblance are suitable for application in a statistical context, and conversely, not all applications require the knowledge of the sampling properties of resemblance measures. It is however well to remember that knowledge of the statistical properties will add to applicability. Although numerous studies were concerned with sampling distributions, only few produced results that are directly applicable to resemblance measures used in vegetation studies (e.g. Goodall, 1967, 1973b). It appears from these that the traditional axiomatic approach of mathematical statistics may not be as promising in providing usable results as the heuristic approach involving local experiments based on random simulations (e.g. Orlóci & Beshir, 1976). We give an example in Chapter 4.

When the sampling distribution of a resemblance function is known, hypotheses can be tested about population values based on samples:

1. If we assume p-variate normality in the population from which the data at hand represent a random sample, we may find $(1/4)m^2(j, k)$ of formula (2.4) as a reasonably good approximation to the Hotteling T^2, and $F = \{(n-p)/[p(n-1)]\}T^2$ to the variance ratio statistic with p and $n-p$ degrees of freedom. $(1/4)m^2(j, k)$ is the generalized distance of $\mathbf{X}_j$ (or $\mathbf{X}_k$) from point $(\mathbf{X}_j + \mathbf{X}_k)/2$. Symbol n indicates the number of quadrats in the sample and p the number of species. When the condition

$$F_{1-\alpha/2;p,n-p} \leqslant F \leqslant F_{\alpha/2;p,n-p}$$

is satisfied, $m(j, k)$ expresses neither greater nor smaller dissimilarity than would be expected should quadrats j and k be samples from a homogeneous, multispecies population. $F_{\alpha/2;p,n\text{-}p}$ is the $\alpha/2$ probability point of the F distribution with p and $n-p$ degrees of freedom, and $F_{1\text{-}\alpha/2;\,p,n\text{-}p}$ is the $1-\alpha/2$ probability point whose nominal value is $1/F_{\alpha/2;n\text{-}p,p}$.

2. Goodall (1967, 1973b) considered the sampling distribution of the matching coefficient $M = (a + d)/s$ (with symbols defined as in formula (2.5) but p is replaced by s). He suggests (1967; Sokal & Sneath, 1963), as a first approximation, the cumulative binomial distribution as a sampling distribution for M. A general term in the binomial distribution is

$$P(X) = \frac{s!}{X!\,(s-X)!}\; p^X\,(1-p)^{s-X}\ ,$$

where s is the number of species in the sample and

$$p = s^{-1} \sum_i p_i = s^{-1} \sum_i [(f_i/n)^2 + (1-f_i/n)^2], i = 1, ..., s .$$

Species i occupies f_i of the n quadrats. The ith term in the sum gives the probability that a random pair of quadrats will agree in possession or lack of species i on the assumption that the occurrence of a species in a quadrat is a completely random event. The cumulative probability associated with M is

$$F(M) = \sum_X P(X),\ X = 0, 1, ..., a + d .$$

The hypothesis that an observed M measures (with α probability) neither greater nor smaller similarity between two quadrats than could be expected based on random assortment of species between quadrats is accepted if the condition,

$$\alpha/2 \leqslant F(M) \leqslant 1 - \alpha/2$$

is satisfied, or in terms of M, if the condition

$$X_{\alpha/2} \leqslant sM \leqslant X_{1-\alpha/2}$$

holds true. $X_{\alpha/2}$ is the value of X for which

$$\sum_Y P(Y) = \alpha/2,\ Y = 0, ..., X .$$

Goodall (1967) expects this test to be conservative, i.e. the actual width of the $1 - \alpha$ interval with end points $X_{\alpha/2}$ and $X_{1-\alpha/2}$ greater than its nominal value.

For a second approximation, Goodall (1967) suggests the standard normal variate,

$$Z = (\text{arc sin}\, M^{1/2} - \text{arc sin}\, p^{1/2}) / \{V(M)/[4p(1-p)]\},$$

in which

$$V(M) = \frac{p(1-p)}{s} - \frac{(1/s) \sum_i (p_i - p)^2}{s},\quad i = 1, ..., s .$$

When the condition

$$-Z_{\alpha/2} \leqslant Z \leqslant Z_{\alpha/2}$$

is satisfied, the observed similarity M of two quadrats is not greater than could be expected if species were assorted randomly between the quadrats. Goodall (1967) sees even greater accuracy achieved in the test if the sampling distribution of M is locally derived in a Monte Carlo experiment. For this he offers a computer program.

3. Goodall (1973b; Frey, 1966; Holgate, 1971) also considered the sampling

distribution of the Sørensen coefficient and Jaccard's index for which he derived random expectations.

4. Users are often tempted to calculate a chi square value for two quadrats, as we have done for χ^2_{jk} in formula (2.6), and then, to regard it as a chi square variate with a single degree of freedom. Their objective is to test the hypothesis that quadrats j and k are not more similar than should be if species occurrence in the quadrats were a random event. A test on such a basis would however be quite unreliable, for the chi square distribution could not, unless by accident, serve as a suitable reference distribution for the χ^2_{jk} values (cf. Juhász-Nagy, 1964; Frey, 1966; Goodall, 1973b).

5. Certain types of scalar products can be associated readily with certain standard types of distributions when the sampling is random in a multivariate normal population. The product moment correlation coefficient is a case in point (formula (2.8)). Tests on this coefficient are described in, e.g., Sokal & Rohlf (1969) among other textbooks which treat theory and application.

6. Whereas χ^2_{jk} is unlikely to have the chi square distribution when it measures association between two quadrats, χ^2_{hi}, with a single degree of freedom, would be expected to qualify as a chi square variate if it measured association between two species in terms of a 2 x 2 contingency table. We may note also that $d^2(j, k)$ of formula (2.19) and $d^2(h, i)$ of the same formula (revised) is expected to behave similarly to χ^2_{jk}/p and χ^2_{hi}/p but with $p - 1$ degrees of freedom (cf. Switzer, 1971) in the large sample case.

7. If the data are of the presence/absence type, the squared Euclidean distance of two quadrats (j, k) is given by

$$d^2\,(j, k) = b + c\ .$$

The b and c are symbols of a 2 x 2 contingency table. Assuming that species presence or absence in quadrats is a simple random event, so that the probability that a species h would occur in quadrat j rather than in quadrat k is $p = 0.5$ and for the reverse $q = 1 - p = 0.5$, Goodall (1969a) suggests to attach probability to $(b + c)$ by multiplying by 2 the cumulative probability in the shorter tail of the binomial distribution,

$$\begin{aligned} P(b+c) &= 2\sum_r \frac{(b+c)!}{r!\,(b+c-r)!}\, p^r q^{(b+c-r)} \\ &= 2\sum_r \frac{(b+c)!}{r!\,(b+c-r)!} \left[\frac{1}{2}\right]^{(r+b+c-r)} \\ &= 2^{-d+1} \sum_r \frac{(b+c)}{r!\,(b+c-r)!}\ , \quad r = 1, \ldots, m\ , \end{aligned}$$

where $d = b + c$ and $m = \min\,(b, c)$. With $P(b + c)$ falling in value below a small, threshold probability, say 0.05, the hypothesis that quadrats j and k are not more similar or dissimilar than could be expected, if presence of species in quadrats were a random event is rejected.

Goodall (1973b) warns that resemblance measures in general, and especially when relating unknown populations rather than fixed individuals, are often context dependent. The circumstances in which they are applied are important to specify. They can be influenced by the size of the sampling units, whether small or large relative to local pattern in the vegetation, by the properties of the variables, whether correlated or not, or by other factors. The effect of quadrat size on association measures represents an example. If determined from small quadrats, association may appear strong, but weak when quadrats of larger size are used. The generalized distance may or may not qualify as a statistic depending on the sampling method which supplies the data. The covariance, while meaningfully applied when the variables are commensurable, may lack interpretability when the variables are not commensurable. It can be expected that more often than not the resemblance measure is under the influence of local factors which reside with causes already mentioned. To account for their potential influence on the resemblance measure must therefore not be overlooked in interpretations.

In ecology, resemblance functions in most applications are subjected to non-statistical uses. This is particularly so in classifications whose goal is cluster recognition. Quite often, however, classificatory analyses try to find the class which represents a most likely parent population for a particular individual. This is clearly a statistical problem requiring the probabilistic use of resemblance functions. There is though one big problem with such uses of resemblance functions quite obvious from the preceding discussions: we rarely know the exact sampling distributions with which they should be associated. The generalized distance is one of the examples which we have considered for the opposite with known sampling distribution under random sampling in a multivariate normal population. But functions which do not qualify as probabilistic measures because of the circumstances of sampling and data collection may still be used as efficient mathematical descriptors of the vegetation. Such a deterministic use of resemblance functions is most common in ecological practice.

The need for heavy computations may rule out the use of some resemblance functions under specific circumstances. Rajski's metric and the generalized distance are two examples of computationally difficult functions. They are not normally contemplated for longhand calculations. In larger samples they cannot even be attempted if a suitable computer program is unavailable.

It is hardly possible to lay down strict guidelines for the selection of resemblance functions. The foregoing discussions should, nevertheless, prove helpful in the search for the right kind of resemblance function. It may be added that the ecologist should select resemblance functions of known properties and *try* to avoid those not sufficiently described or understood. Functions of known probability distributions should be preferred provided that they have an appropriate scale of measure and that the computations are not inhibiting.

Keeping in mind the foregoing points about resemblance functions, we turn in the next chapter to a discussion of ordinations and their uses in vegetation studies.

Chapter 3

ORDINATION

Once we impose on the sample a certain resemblance structure by a suitably chosen resemblance function, our further choice of method may be ordination. In this chapter we consider the different objectives that call for ordination, describe different methods and outline the conditions of their application. We shall see that ordinations, contrary to popular belief, represent no 'general preference' methods for data analysis, but rather, they are techniques specifically designed to serve certain narrow objectives.

3.1 Ordination Objectives

It is apparent from accounts concerning history that modern ordination methodologies evolved under two distinct sources of influence. There were the schemes that mapped vegetation units (quadrats, stands, types, etc.) or species as sample points onto lines or planes in a reference space of environmental variables (e.g. Ramenski, 1938; also Whittaker, 1956, 1967, 1973a; Loucks, 1962; Sobolev & Utekhin, 1973; and references therein). There were also the uni- and multidimensional schemes which used species composition (e.g. Ramenski, 1930; Curtis & McIntosh, 1951) or some linear functions of species composition (e.g. Goodall, 1954a) to achieve a direct ordering of vegetation units.

Although ecologists were practising the art of ordination for many decades, their work has not reduced, but increased diversity in objectives as well as in methods. Goodall (1954a,b) – who coined the term ordination in the spirit of the German *Ordnung* used by authors in the early literature (e.g. Ramenski, 1930; Ehrendorfer, 1954) – defined ordination as the 'arrangement of units in a uni- or multidimensional order'. Ordination so conceived extracts from suitable data an arrangement of vegetation units in series, in contrast with classification which extracts an arrangement in classes.

Whereas Goodall's definition is broad and it allows practically any of the methods that analyse continuous variation to be included, the opposite can be said of Whittaker's (1952, 1956, 1967) definition which restricts the scope of ordinations to predictive techniques. The objective with the prediction is to reveal ecological information either *directly* based on ordination of environmental data or *indirectly* based on vegetation data. But the use of the adjectives *direct* and *indirect* has generated controversy. McIntosh (1958, 1967b) suggested to describe direct ordinations as *gradient analysis* and indirect ordinations as *continuum analysis.* Austin (1968) has used a rather straight forward terminology: *environmental ordination* and *vegetational ordination*, to designate two broad categories of methods.

Numerous proposals to categorize ordinations notwithstanding (e.g.

Whittaker & Gauch, 1973, 1977; Dale, 1975; Noy-Meir, 1977; Noy-Meir & Whittaker, 1977), and in view of the inevitable fact that the designation and redesignation of old methods by new names add nothing to understanding the intrinsic properties, we elect not to be bound by previous suggestions. We shall identify as ordination any method, descriptive or predictive, ecological or taxonomical, etc., if it is suitable to unfold continuous variation into series through arranging objects on axes. The methods in this category can serve different objectives (see following text) as has been recognized from time to time by different authors (e.g. Goodall, 1954b; Greig-Smith, 1964; Lambert & Dale, 1964; Pielou, 1969a; Brisse & Grandjouan, 1971; Kershaw, 1973; Orlóci, 1973a; Dale, 1975). Thus we place ordinations in the family of continuous multivariate techniques (Gittins, 1969) with actual or potential applications transgressing interdisciplinary boundaries.

Ordination methods have been put to many uses in vegetation studies to achieve different goals. Among these:

1. Summarization.
2. Trend seeking and prediction.
3. Multidimensional scaling.
4. Reciprocal ordering.
5. Classification.

Summarization means reduction of dimensionality in the data for the purpose of convenient handling and interpretations. Summarization is regarded efficient when the reduction is as great as possible without loss of information. To elaborate on this point further we should consider a sample of 20 quadrats described on the basis of 100 species. When we conceive the quadrats as a cluster of 20 points in a space of 100 dimensions it is not too difficult to accept the idea that such a representation must be inherently redundant because it requires 100 dimensions to represent a point cluster whose intrinsic dimensionality cannot be more than 19. Summarization should remove the redundant dimensions and produce a parsimonious representation of the points in a space of dimensions equal to the number of intrinsic dimensions in the cluster. When a summarization succeeds in reducing the sample's dimensionality to the number of intrinsic dimensions, it is said to be *efficient.* We recognize, of course, that to force certain relationships to appear in the data, a drastic reduction in dimensionality may be necessary to eliminate random variation which causes background noise (Greig-Smith, 1971a) and which makes the interpretation of results rather difficult.

Trend seeking is concerned with the discovery and simple representation of trends in vegetational variation and their correlation with environmental variables. These are often part of the total task in *predictive ordinations* whose objective is to make inferences about the environmental influence to which the vegetation responds. The observations may represent similarity (proximity) values which are not suited for direct input in certain methods of data analysis. *Multidimensional scaling*, as presented here, refers to the process of extracting metric co-ordinates from such observations. Although such a task can be performed by most methods of ordination, multidimensional scaling as a means of metricizing observations rarely constitutes an objective in isolation from others that can be served by ordinations. In *reciprocal ordering* (correspondence analysis) the direct objective is to arrange species and quadrats into an order that maximizes some measure of correlation. The ordering may use species scores for quadrats or quadrat scores for species, such that the species scores correspond to points of optimal response on an environmental gradient, and the quadrat scores represent co-ordinates on extracted axes identified as the canonical variates of a certain type of scalar product matrix. As a secondary function ordinations may perform *classifications*, but for this they may not be ideally suited.

Any of these objectives calls for suitable methods, the choice of which will depend to a considerable extent on the data structure in the sample.

3.2 Data structure

The ecological circumstances of a general vegetation survey can be characterized by certain common properties:

1. We would expect at least one major environmental gradient where the

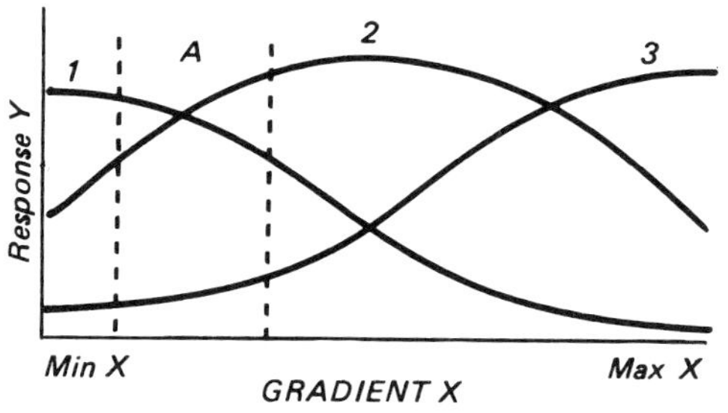

Fig. 3-1. Schematic representation of a hypothetical ecological universe. Curves indicate changes in species response Y along a predominant gradient X. Labels *1*, *2* and *3* signify species. Symbol A designates a limited segment on the gradient where the response is quasi-linear.

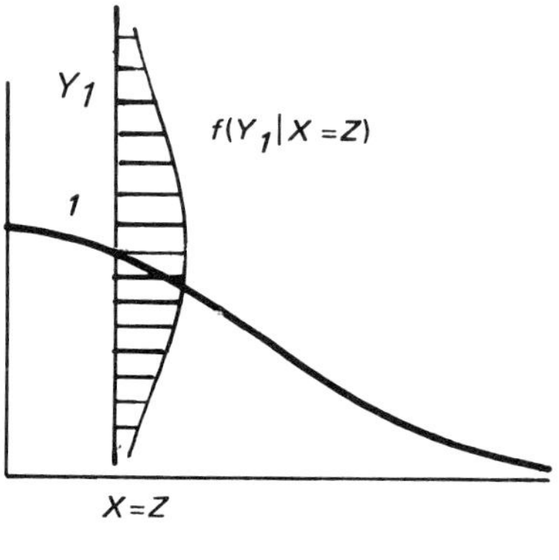

Fig. 3-2. Distribution of the response of species *1* at point $X=Z$ of the gradient. Symbol Y_1 indicates species response and $f(Y_1 | X=Z)$ signifies the density (co-ordinate) function of the probability distribution of response.

survey site is sufficiently large. A gradient is a continuous unidirectional progression of states signifying rising or falling levels of environmental influence. Levels of soil moisture regime from dry to wet represent an example. Any given state may have realizations at different points on the landscape.

2. Species response to changing levels of environmental influence along a gradient is expected to be non-linear if the gradient is sufficiently broad. Such a response is manifested by rising or falling species performance.

3. Performance declines away from the optimal environment resulting in patterns often representable by unimodal curves.

Postulates of properties of the mentioned kind are at the basis of formulations which represent the ecological universe as a set of response curves approximately bell-shaped in form, but often skewed and truncated (e.g. Whittaker, 1956, 1967). Fig. 3-1 illustrates the hypothetical case of three species responding to environmental influence. It is understood that although a given level of influence, say $X = Z$, defines a point on the gradient, such an influence can occur and reoccur at geographically isolated points on the landscape. Furthermore, at each occurrence, it may trigger a similar response of the same species. We can thus expect each point on the gradient, because of the reoccurrence property, to be associated with a p-dimensional (p is the number of species) probability distribution of joint species response. A unidimensional component of this distribution is pictured in Fig. 3-2. It is obvious that it would be poor practice to represent species performance by a mere response curve, without showing in what manner the swarm of points, representing response, envelopes the curve in the sample or in the population.

It is expected that different species will respond differently to environmental influence along a gradient. If the environmental amplitude is wide,

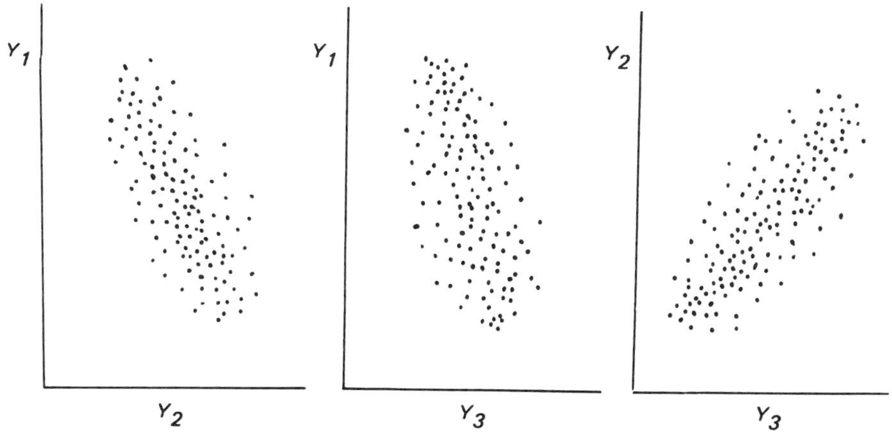

Fig. 3-3. Schematic representation of the distribution of joint response of species in region *A* of the gradient in Fig. *3-1.* The manner in which the points are scattered in the diagram is indicative of linear species correlations and a linear structure in the coordinate data. Symbols Y_1, Y_2, Y_3 identify species.

the response is likely to be non-linear. But if it is sufficiently narrow, the response may be linear (van Groenewoud, 1965). The important point to be made is that the manner in which species respond has an influence on the nature of correlations among the species. When species response is linear (a case of quasi-linearity is pictured in segment A of Fig. 3-1), the species themselves are linearly correlated (pictured in Fig. 3-3). While linear response determines linear species correlations, a non-linear response triggers non-linear correlations in general, except in the trivial case of coinciding or parallel response graphs (curves or curved surfaces), or in the case of response graphs that are mirror images of one another in the direction of the response axis.

Species correlations generate the data structure in the sample. What do we mean by data structure? It is a property easiest to illustrate based on Goodall's (1954a,b) spatial analogy. The space to be used has as many dimensions as the number of species (p). Each spatial point is the tip of a p-valued response vector, potential or realized, associated with a given point on the gradient. Of the N actual realizations there are n corresponding to the quadrats in the sample. When all N points fall into a single cluster, so that only a single density phase of points can be recognized, the data structure is said to be *continuous* (Figs. 3-3 and 3-4). If two or more density phases are present, the data structure is said to be *disjoint* (Fig. 3-5). The actual recognition of disjointness or continuity is of course a difficult matter and requires a statistical decision (Juhàsz-Nagy, 1967b). In any case, the individual clusters may be *linear* or *curved*. In the linear case (Fig. 3-3) the points fill out an ellipse, and in the non-linear case (Fig. 3-4) they fall into a curved area. In higher dimensions they would fill out a hyper-volume, or fall on the surface of some simple or complex surface which could turn and twist through space in any regular or irregular fashion.

It is reasonable to suspect on the basis of Fig. 3-3 and Fig. 3-4 that there may be a connection between the shape of the point cluster and the rela-

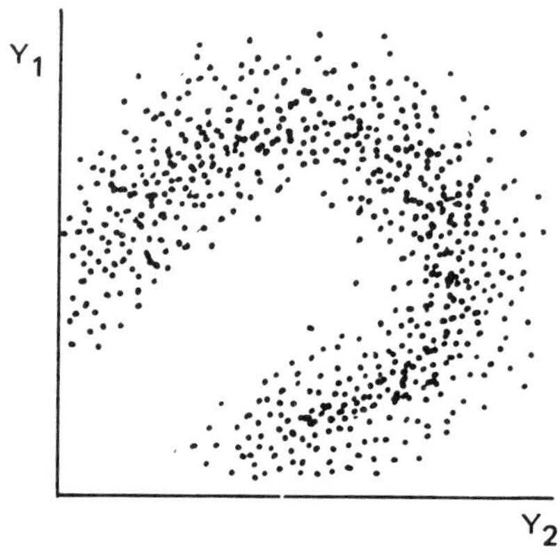

Fig. 3-4. Schematic representation of the joint distribution of non-linear species response. The curved shape of the point cluster is indicative of non-linear species correlations and a non-linear structure in the co-ordinate data.

tionship of the species as axes. We can postulate that linear clusters associate with linear correlations of the species and non-linear clusters with correlations that have in addition a non-linear component. To complicate matters, a non-linear correlation can have a linear component, or linear and non-linear clusters may have the same underlying linear correlation structure, in spite of their differences in point arrangements.

There are no explicit tests to identify exactly the data structure in a vegetation sample, and what is even worse, the data structure revealed can depend on the method used. The types of structure in the sample may have to be inferred from clues found in the course of the survey or from results of different analyses by which we probe the data. Where the survey site shows continuous environmental variation of a wide amplitude, the assumption of non-linear species response may be quite justified. This would almost certainly imply that species correlations are non-linear. Continuity in environmental variation and non-linear species correlations in turn are properties of data structures that are continuous and curved. If the survey site encompasses environmental variation of a lesser amplitude the assumption of linear species response and a linear continuous data structure is a logical inference. Linearity may also be inferred if scatter diagrams of joint species response indicate ellipsoidal clusters of points, or non-linearity if the clusters are curved.

If we sampled in a survey site across broad environmental differences, we would not be surprised to discover that certain levels of environmental influence are rather frequent while others are missing or under-represented. When such is the case, the points in the sample space would be expected to form phases of high and low density. While this would be indicative of a

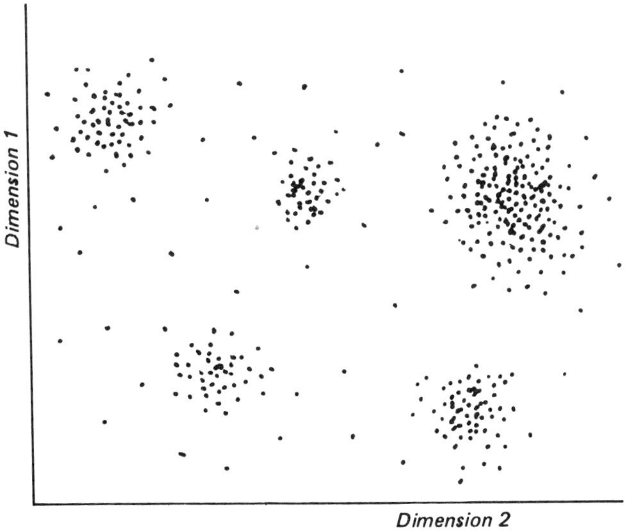

Fig. 3-5. Two-dimensional unfolding of a p-dimensional sample space. Alternating low and high density phases of points are indicative of a disjoint data structure.

disjoint data structure, the lack of density phases, i.e. when all levels of environmental influence were equally frequent in the survey site would be indicative of a continuous data structure. To actually describe density phases in the sample we may rely on different clustering techniques. Scatter diagrams may also offer clues about continuity or discontinuity in the sampled population (e.g. Goodall, 1973a).

The data structure may or may not be described efficiently by certain commonly used resemblance functions depending on complexity. The covariance is an efficient descriptor of the data structure if it is continuous and linear. However, if the data structure is disjoint and particularly if non-linear, the covariance can serve only as an inefficient descriptor. What do we mean by an inefficient descriptor? We mean, if it is the covariance, that it describes the linear component without responding to the non-linear component of species correlation. Goodall's (1973b) example illustrates a case in point. Consider the hypothetical data:

Species	*Cover % in quadrats*						
A	0	3	6	15	30	42	60
B	0	9.5	18	37.5	50	42	0

Although there is an exact mathematical relationship between the two sets of figures, namely $18B = 60A - A^2$, the product moment correlation coefficient (0.169) is totally unindicative of this. This is exactly what we would expect from a linear measure where the data structure is non-linear.

Methods of ordination which use the covariance interpret two clusters of points as identical if their covariance structures are identical, even if they happen to differ in other important respects. This does not mean that such methods of ordination cannot be used to represent non-linear data structures. They actually can, which will be shown, but such representations are likely to prove inefficient because the ordination axes do not respond to non-linear correlations. To be regarded as efficient, an ordination must represent a given data structure in the fewest possible dimensions. In the linear case efficiency is achieved by making the axes linearly independent. In the non-linear case the interpretation of efficiency is somewhat more complicated because it may mean different things depending on the existing nonlinearity. For instance, if there is a single continuous cluster of points, which may turn and twist through space essentially in a single direction, efficiency could mean that the ordination axis turns and twists exactly the same way. In other words, an efficient ordination finds the best fitting path through the cluster and represents it by a suitable curved axis. If, however, the points fall on the surface of a manifold, efficiency would have to be defined differently. It may mean finding a simple curved surface, or a suitable set of curved axes, on which the cluster of points can be unfolded or flattened with the least amount of distortion.

3.3 Ordination as a transfer system

We may view ordinations as transfer systems (Orlóci, 1974b) in which information flows from raw data to ordination co-ordinates. Naturally, we may expect from such transfer systems certain peculiar properties:
1. Different results by different technique of transfer.
2. Information loss under most circumstances.
3. Optimal performance confined to specific objectives.

The ordination techniques, with differences in their handling of the data, are expected to produce different results. Their differences, however, need not lead to radically different ecological interpretations. On the contrary, ordinations are quite robust in that they have a strong tendency toward convergence of the ecological information which they reveal.

Information transfer is controlled by two basically different transformations in most ordinations. The first yields resemblance values from raw data for input in the second which produces the ordination co-ordinates. Whereas no new information can be generated in the process, a great deal of information may be lost when the transformations are inefficient. We shall discuss the consequences of inefficiencies, as distortions, in the sequel after we have presented the different methods.

3.4 Summarization and multidimensional scaling – linear case

When all sample points fall into a single linear cluster, and the objective is to summarize such a structure on axes, thereby reducing it to some simpler form better suited for handling and analysis, *component analysis* is indicated. Component analysis is an efficient summarizer, since the components extracted, unlike the species, are uncorrelated and thus free from the redundancies attributable to linear correlations. Since the components are not redundant, fewer are needed to completely redescribe a set of quadrats than the number of species on which the original descriptions were based.

It appears that the method of component analysis, as we know it today, is a direct outgrowth of Hotelling's (1933) work with correlation matrices. However, the modern applications are rarely associated with the correlation, but with the covariance, or equivalently, with the Euclidean distance. Since the covariance and the correlation are efficient descriptors only of linear data structures, component analysis will be an efficient summarizer when the data structure is linear. A separate aspect of linearity is obvious in the basic model,

$$Y_{ij} = \sum_h b_{hi} X_{hj} = b_{1i} X_{1j} + \ldots + b_{pi} X_{pj} \qquad (3.1)$$

This specifies linear transformations of species quantities (or other variates) $X_{1j}, \ldots, X_{pj}$ which yield the *component score* Y_{ij} for quadrat j. The component score Y_{ij} can be used as a rectangular co-ordinate on the ith component representing an axis. The coefficients $b_{1i}, \ldots, b_{pi}$, relating the species to the ith component, are called *component coefficients.* As direction cosines they satisfy the condition $\mathbf{b}_i'\mathbf{b}_i = b_{1i}^2 + \ldots + b_{pi}^2 = 1$. The symbol $\mathbf{b}_i$

represents the ith column vector in matrix **B** (given below) and $\mathbf{b}_i'$ is its transpose.

Given a set of raw observations, component analysis extracts component coefficients for the species and component scores for the quadrats. If there are p species and t is the number of intrinsic dimensions in the sample, the transformations based on formula (3.1) require $p \times t$ component coefficients,

$$\mathbf{B} = \begin{bmatrix} b_{11} & b_{12} & \cdots & b_{1t} \\ b_{21} & b_{22} & \cdots & b_{2t} \\ \cdot & \cdot & \cdots & \cdot \\ b_{p1} & b_{p2} & \cdots & b_{pt} \end{bmatrix}.$$

We should note that as far as the computations are concerned the column vectors in **B** represent the normalized eigenvectors $\mathbf{b}_1, ..., \mathbf{b}_t$ associated with the eigenvalues $\lambda_1, ..., \lambda_t$ of the covariance (or correlation) matrix **S**. Keeping in mind that the direct computational objectives include the component coefficients and component scores, we shall turn to an example for illustrations while we refer to PCAR in the Appendix as our computer program:

1. The vegetation sample to be analyzed contains two species given as row vectors in the matrix

$$\mathbf{X} = \begin{bmatrix} X_{11} & X_{12} & X_{13} & X_{14} & X_{15} \\ X_{21} & X_{22} & X_{23} & X_{24} & X_{25} \end{bmatrix} = \begin{bmatrix} 2 & 5 & 2 & 1 & 0 \\ 0 & 1 & 4 & 3 & 1 \end{bmatrix}.$$

The columns of **X** represent quadrats.

2. We may decide at this point to proceed on the basis of the correlation coefficient or the covariance. It is relevant to our choice that if the variables were given in widely different units their linear compounds (formula (3.1)) would not be particularly meaningful. Standardization would be required in a manner implicit in the correlation coefficient (see Section 1.6). While such a standardization would no doubt be preferred by some ecologists when the variables are measured in different units, there may be other ecologists who would rather implement scaling by the range. Still others could argue that the unit of measure is an intrinsic, logically inseparable property of the variables, and on this ground, express their objections to standardization by any method. Where, in the ecologist's opinion, standardization is not required, the use of the covariance is indicated. In the present example we opt for the covariance.

3. The covariance matrix for the two species in **X** is given by

$$\mathbf{S} = \begin{bmatrix} s_{11} & s_{12} \\ s_{21} & s_{22} \end{bmatrix} = \begin{bmatrix} 3.5 & -0.5 \\ -0.5 & 2.7 \end{bmatrix}.$$

4. If there are only two species in the sample, longhand calculations are

quite feasible. But should the sample be large we would not proceed without having access to a digital computer. One of the longhand methods is based on solving equations. Firstly, we solve the determinantal equation

$$\begin{vmatrix} 3.5-\lambda & -0.5 \\ -0.5 & 2.7-\lambda \end{vmatrix} = (3.5-\lambda)(2.7-\lambda)-(-0.5)(-0.5)$$
$$= \lambda^2 - 6.2\lambda + 9.2 = 0$$

to obtain two eigenvalues (roots),

$$\lambda_1 = [-b + (b^2 - 4ac)^{1/2}]/2a = (6.2 + (6.2^2 - (4)(9.2))^{1/2})/2$$
$$= (6.2 + 1.28)/2 = 3.74$$
$$\lambda_2 = (6.2 - 1.28)/2 = 2.46\ .$$

Next, we solve the characteristic equation

$$\mathbf{S}\,\mathbf{b}_1 = \lambda_1 \mathbf{b}_1$$

which can be written in the form of

$$\begin{bmatrix} 3.5 & -0.5 \\ -0.5 & 2.7 \end{bmatrix} \begin{bmatrix} b_{11} \\ b_{21} \end{bmatrix} = \lambda_1 \begin{bmatrix} b_{11} \\ b_{21} \end{bmatrix}.$$

This gives rise to two equations,

$$3.5\,b_{11} - 0.5\,b_{21} = 3.74\,b_{11}$$
$$-0.5\,b_{11} + 2.7\,b_{21} = 3.74\,b_{21}$$

which can be reduced to

$$-0.24\,b_{11} - 0.5\,b_{21} = 0$$
$$-0.5\,b_{11} - 1.04\,b_{21} = 0$$

and further to

$$-0.74\,b_{11} - 1.54\,b_{21} = 0\ .$$

Through a rearrangement of terms we obtain,

$$b_{11} = -\frac{1.54}{0.74}\,b_{21}\ .$$

This relation specifies the size of b_{11} relative to b_{21}. Based on this, the normalized eigenvector $\mathbf{b}_1$ corresponding to the largest eigenvalue λ_1 has the solutions,

$$\mathbf{b}_1 = \begin{bmatrix} b_{11} \\ b_{21} \end{bmatrix} = k_1 \begin{bmatrix} -2.081 \\ 1 \end{bmatrix} = 1/(2.081^2+1)^{1/2} \begin{bmatrix} -2.081 \\ 1 \end{bmatrix} = \begin{bmatrix} -0.901 \\ 0.433 \end{bmatrix}.$$

Note that the condition $\mathbf{b}_1'\mathbf{b}_1 = 1$ is satisfied (within rounding errors).

Having determined the first set of component coefficients associated with λ_1, we proceed to determine the component coefficients associated with λ_2. The characteristic equation to be solved is given by

$$\mathbf{S}\,\mathbf{b}_2 = \lambda_2 \mathbf{b}_2 \,.$$

The solution specifies relative sizes,

$$b_{12} = \frac{0.26}{0.54}\, b_{22} \,.$$

the eigenvector $\mathbf{b}_2$ then has the solutions,

$$\mathbf{b}_2 = \begin{bmatrix} b_{12} \\ b_{22} \end{bmatrix} = k_2 \begin{bmatrix} 0.481 \\ 1 \end{bmatrix} = 1/\,(0.481^2+1)^{1/2} \begin{bmatrix} 0.481 \\ 1 \end{bmatrix} = \begin{bmatrix} 0.433 \\ 0.901 \end{bmatrix}.$$

The solution satisfies the condition $\mathbf{b}_2'\mathbf{b}_2 = 1$ (within rounding errors).

5. We summarize the results in a table,

Eigenvalues	
$\lambda_1 = 3.74$	$\lambda_2 = 2.46$
Eigenvectors	
$\mathbf{b}_1$	$\mathbf{b}_2$
−0.901	0.433
0.433	0.901

Note that the sum of the eigenvalues is equal to the sum of species variances $S_{11} + S_{22}$. The efficiency of the components with which they individually account for variation in the sample is

$$E_1 = 100\,\frac{3.74}{6.2} = 60.3\%$$

and

$$E_2 = 100\,\frac{2.46}{6.2} = 39.7\% \,.$$

6. To compute the component scores we may have to rewrite formula (3.1) in terms of

$$Y_{ij} = \mathbf{A}_j'\mathbf{b}_i = \sum_h b_{hi} A_{hj} \qquad (3.2)$$

where $\mathbf{A}_j'$ is a transpose of $\mathbf{A}_j$ whose elements are defined to conform with the definition of **S**. When **S** is the covariance matrix the elements of **A** accord with $A_{hj} = (X_{hj} - \bar{X}_h)/(n-1)^{1/2}$. With such a definition of the elements in **A** the solution for Y_{ij} based on formula (3.2) will differ from formula (3.1) only by a factor of $1/(n-1)^{1/2}$. The translation $X_{hj} - \bar{X}_h$ is not essential, it only shifts the origin of the co-ordinate system. Let us now consider the computations based on formula (3.2):

a. Adjust the data to obtain

$$\mathbf{A} = \begin{bmatrix} A_{11} & A_{12} & A_{13} & A_{14} & A_{15} \\ A_{21} & A_{22} & A_{23} & A_{24} & A_{25} \end{bmatrix} = \begin{bmatrix} 0.0 & 1.5 & 0.0 & -0.5 & -1.0 \\ -0.9 & -0.4 & 1.1 & 0.6 & -0.4 \end{bmatrix}.$$

b. Compute the component scores,

$$Y_{11} = (-0.901)(0.0) + (0.433)(-0.9) = -0.390$$

$$Y_{21} = (0.433)(0.0) + (0.901)(-0.9) = -0.811\,.$$

The component scores for other quadrats are similarly determined. The complete set of results is given in the table,

Component (ordination axis)	*Component score (co-ordinate)* 1	2	3	4	5	*Eigenvalue*
Y_1	−0.390	−1.525	0.476	0.711	0.728	3.74
Y_2	−0.811	0.289	0.991	0.324	−0.793	2.46

The sum of squared component scores is equal on each component to the associated eigenvalue (within rounding errors in the example). The component scores are rectangular ordination co-ordinates and can be plotted accordingly to form a scatter diagram (Fig. 3.6). Such a diagram shows the joint distribution of the components.

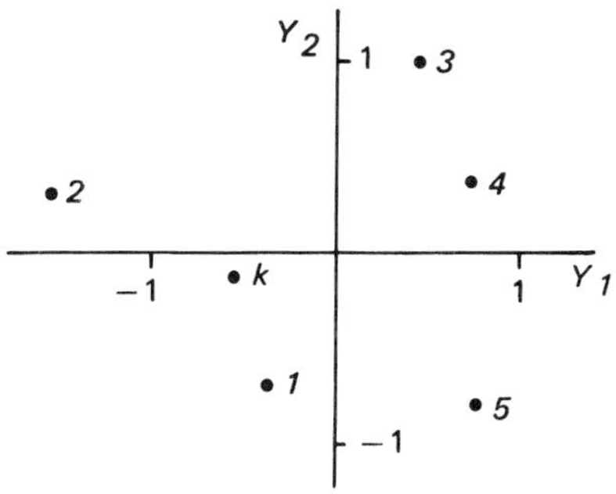

Fig. 3-6. Scatter diagram showing the joint distribution of components. See explanations in the main text.

The algorithm in the foregoing example is going to be referred to as the *R-algorithm*. It has the specific advantage of providing component coefficients before yielding component scores. The component coefficients can facilitate mapping of new individuals in a scatter diagram or computing various correlation quantities. If no objectives call for component coefficients, the component scores may be determined simply in another way. We shall consider two alternatives:

1. *The Q-algorithm*. The descriptions follow Orlóci (1966, 1967b). We begin by defining the scalar product matrix as $\mathbf{S} = \mathbf{A}\ \mathbf{A}'$. In this, $\mathbf{A}$ contains the adjusted or standardized data. We then define the characteristic equation of $\mathbf{A}\ \mathbf{A}'$ as $\mathbf{A}\ \mathbf{A}'\mathbf{b}_i = \lambda_i \mathbf{b}_i$ in which $\mathbf{b}_i$ and λ_i represent respectively the ith eigenvector and the ith eigenvalue of $\mathbf{A}\ \mathbf{A}'$. After pre-multiplying both sides by $\mathbf{A}'$ we obtain $\mathbf{A}'\mathbf{A}\ (\mathbf{A}'\ \mathbf{b}_i) = \lambda_i\ (\mathbf{A}'\ \mathbf{b}_i)$. Here $\mathbf{A}'\mathbf{A}$ is an $n \times n$ matrix $\mathbf{Q}$ with elements q_{jk} defined according to formula (2.9) and $\mathbf{A}'\mathbf{b}_i$ is the ith eigenvector of $\mathbf{A}'\mathbf{A}$; this contains after adjustments the scores on the ith component. Because $\mathbf{A}_j'\mathbf{b}_i$ is the same quantity as Y_{ij} in formula (3.2) the component scores obtained from $\mathbf{Q}$ or $\mathbf{S}$ are one and the same. Based on these relations we can determine component scores for quadrats from matrix $\mathbf{Q}$ as follows:

a. Compute a scalar product matrix according to formula (2.9) to obtain an $n \times n$ matrix $\mathbf{Q}$.

b. Determine the non-zero eigenvalues ($\lambda_1 \ldots \lambda_t$) and associated eigenvectors ($\mathbf{Y}_1 \ldots \mathbf{Y}_t$) of $\mathbf{Q}$.

c. Adjust the elements in the eigenvectors so that any eigenvector $\mathbf{Y}_i$ will satisfy the condition $\mathbf{Y}_i'\mathbf{Y}_i = Y_{1i}^2 + \ldots + Y_{ni}^2 = \lambda_i$. The elements ($Y_{1i} \ldots Y_{ni}$) in $\mathbf{Y}_i$ are the component scores for n quadrats on component i.

2. *The D-algorithm.* We can develop this algorithm most simply through a numerical example. For this we have two quadrats j and k with data vectors

$$\mathbf{X}_j = \begin{bmatrix} 3 \\ 4 \end{bmatrix} \text{ and } \mathbf{X}_k = \begin{bmatrix} 8 \\ 4+\sqrt{11} \end{bmatrix}.$$

Matrix $\mathbf{Q}$ is based on formula (2.9),

$$\mathbf{Q} = \begin{bmatrix} 9 & -9 \\ -9 & 9 \end{bmatrix}$$

and matrix $\mathbf{D}$ is based on formula (2.1),

$$\mathbf{D} = \begin{bmatrix} 0 & 6 \\ 6 & 0 \end{bmatrix}.$$

The elements of $\mathbf{Q}$ and $\mathbf{D}$ are related (formula (2.11)),

$$d^2(j,k) = q_{jj} + q_{kk} - 2q_{jk} = 9 + 9 + 18 = 36$$

or in the reverse

$$q_{jk} = -0.5\, d^2(j, k) + (q_{jj} + q_{kk})/2 = -18 + 9 = -9 \,.$$

It is reasonable to suspect after observing the relationship of **Q** and **D** that when subjected to analysis they will yield related component scores. To verify this we perform on **Q** and **D** the following analysis:

a. Compute

$$\mathbf{D}^* = \begin{bmatrix} 0 & -18 \\ -18 & 0 \end{bmatrix}$$

with an element defined according to $-0.5\, d^2(j, k)$.

b. From the determinantal equation of **D***,

$$\begin{vmatrix} 0-\lambda & -18 \\ -18 & 0-\lambda \end{vmatrix} = \lambda^2 - 324 = 0$$

we find $\lambda = 18$.

c. From the determinantal equation of **Q**

$$\begin{vmatrix} 9-\lambda & -9 \\ -9 & 9-\lambda \end{vmatrix} = \lambda^2 - 18\lambda = 0$$

we have $\lambda_1 = (18 + \sqrt{324})/2 = 18$ and $\lambda_2 = (18 - \sqrt{324})/2 = 0$. Note that in this example **D*** has a single non-zero positive eigenvalue, which happens to be identical to the single non-zero positive eigenvalue of **Q**. The reader can verify on the basis of other examples that irrespective of the order of **Q** and **D*** the non-zero positive eigenvalues are identical (within rounding errors).

d. The characteristic equation of **D***, $0b_{11} - 18b_{21} = 18b_{11}$, indicates that $-b_{21} = b_{11}$. Using the constraint $\mathbf{b}_i'\mathbf{b}_1 = b_{11}^2 + b_{21}^2 = \lambda$ we find

$$\mathbf{b}_1 = \begin{bmatrix} b_{11} \\ b_{21} \end{bmatrix} = \begin{bmatrix} 3 \\ -3 \end{bmatrix} .$$

The characteristic equation of **Q**, $9b_{11} - 9b_{21} = 18b_{11}$, reduces to $-9b_{21} = 9b_{11}$ and therefore $-b_{21} = b_{11}$. Under constraint of $b_{11}^2 + b_{21}^2 = \lambda$ we find

$$\mathbf{b}_1 = \begin{bmatrix} b_{11} \\ b_{21} \end{bmatrix} = \begin{bmatrix} 3 \\ -3 \end{bmatrix} .$$

This is the only non-zero eigenvector of **Q**. We conclude from these that **Q** and **D*** have not only identical eigenvalues but also identical eigenvectors. The reader can compute other examples to find that regardless of the order of **D*** and **Q** the eigenvectors associated with the non-zero positive eigenvalues are identical (within rounding errors). For theory we refer to Gower (1966).

We now summarize the **D**-algorithm (given as program PCAD in the Appendix):

a. Compute matrix **D*** with elements $-0.5\, d^2(j, k)$ where $d^2(j, k)$ is a squared Euclidean distance.

b. Extract all non-zero positive eigenvalues $\lambda_1, ..., \lambda_t$ and associated eigenvectors $\mathbf{b}_1, ..., \mathbf{b}_t$ of **D***.

c. Adjust the eigenvectors $\mathbf{b}_1, ..., \mathbf{b}_t$ to conform to the condition $\mathbf{b}_i'\mathbf{b}_i = \lambda_i$ for any i. The elements in the eigenvectors so adjusted can be used as rectangular ordination co-ordinates. The similarity between the **D**- and **Q**-algorithm in steps *b* and *c* is obvious.

A direct implication of the foregoing results is that when quadrat distances are determined from the same data matrix, from which we would compute **S** or **Q**, we will obtain the same sets of component scores for the quadrats. It should however be noted that whereas the *R*-algorithm produces component coefficients, these are not available in the *Q*- or *D*-algorithm. The computational load normally is heaviest in the *R*-algorithm. One (but not the most efficient) algorithm for finding the eigenvalues and eigenvectors of a square symmetric matrix is described in lucid terms by Greenstadt (1960). His algorithm relies on the *Jacobi (Neumann) iterative method.* An ALGOL program is described by Grench & Thacher (1965).

Component analysis suffers from three major defects that should not be left unmentioned. Firstly, components summarize linear variation. This is a direct consequence of the fact that the method relies on the covariance in measuring the resemblance of objects and that it also relies on the covariance in defining the independence of components. Secondly, the components are scale sensitive (Gower, 1967a). Unless all variables are measured on the same scale, the components will be influenced by the scale of measure that happened to be chosen. For instance, if the *h*th of *p* variables were measured in kg units and rescaled to gr units, the change in scale would be reflected in the components extracted. If a correlation matrix is analysed, i.e. the variates are standardized before computing the covariance, the problem of scale differences will certainly disappear, but at the same time other problems may arise by altering the data structure. Thirdly, related to the second, outlier individuals or outlier clusters of individuals are expected to have improportionately high weight in influencing the orientation of the components (Anderson, 1971b). This certainly can complicate interpretations.

The statistical potential of estimation and hypothesis testing in component analysis (Bartlett, 1950; Morrison, 1967) has not been fully utilized in vegetation studies. The reason is that it requires the assumption of multivariate normality in the data which the users have not been prepared to accept in most ecological applications. It is however suggested that when the data represent measurements of species quantity on a continuous scale, the assumption of multivariate normality may not be too unrealistic. Seal

(1964) believes that if the statistical aspects are not dominant, component analysis is misused. Such an extreme view notwithstanding, the deterministic use of component analysis can be defended (Crovello, 1970). Deterministic applications are in fact wide spread, many with success (e.g. Goodall, 1954a; Orlóci, 1966; Ivimey-Cook & Proctor, 1967; Yarranton, 1967a,b,c; Austin, 1968; Kershaw, 1968, 1973; Fresco, 1969; Gittins, 1969; van der Maarel, 1969; Wilkins & Lewis, 1969; Chandapillai, 1970; Cuanalo & Webster, 1970; James, 1971; Walker & Wherhahn, 1971; Austin, Ashton & Greig-Smith, 1972; Auclair & Cottam, 1973; Jesberger & Sheard, 1973; Kershaw & Rouse, 1973; Feoli & Feoli Chiapella, 1974; Bouxin, 1975a,b; Fabbro, Feoli & Sauli, 1975; Feoli, 1975; Walker, 1975; Phillips, 1978) but others with results not completely pleasing (e.g. Swan, 1970; Austin & Noy-Meir, 1971; Beals, 1973; Wali & Krajina, 1973; Whittaker & Gauch, 1973; Westman, 1975; Noy-Meir, 1977).

Frequent complaints about component analysis raise the point that the interpretation of the components is difficult in ecological terms. This is not surprising when we consider that the components compound linear variation of the species without retaining a one to one correspondence with them. Moreover, since the components are linear functions of species, they may not show a simple relationship to any of the environmental variables to which species respond individually.

A method of summarization that could improve interpretability relies on the use of orthogonal functions. Whereas the notion of orthogonal functions is one of long-standing practical importance in the computation of statistical constants (cf. Rao, 1952), its use in the context of ordinations is rather recent. Ottestad (1975a) proposed ordination based on orthogonal functions as a possible alternative to component analysis in summarizing linear vegetational variation. The proposed method has certain fundamental properties in which it differs from component analysis:

1. Each orthogonal function, as an ordination axis, corresponds to a given species.
2. The manner in which variation is apportioned between the ordination axes is dependent on the sequence in which the species are presented for analysis. This is because the method relies on the specific variance. The ordering of species for input in the analysis could be based on specific variance as in Orlóci's (1975a) method (Section 1.7) if no hypothesis specifies an a priori sequence.

The orthogonal functions in which we seek the coefficients are given by

$$\begin{aligned} Y_{1k} &= A_{1k} \\ Y_{2k} &= A_{2k} - b_{21}A_{1k} \\ Y_{3k} &= A_{3k} - b_{31}A_{1k} - b_{32}A_{2k} \\ &\cdot \quad \cdots \quad \cdot \\ Y_{pk} &= A_{pk} - b_{p1}A_{1k} - \ldots - b_{pp-1}A_{p-1k} \, . \end{aligned}$$

There are p species and p orthogonal functions. The sequence of orthogonal functions $Y_1, Y_2, ..., Y_p$ is the same as the sequence in which the species are presented for analysis. A_{ik} symbolizes the quantity of species i in quadrat k and Y_{ik} signifies the numerical co-ordinate value of the kth quadrat on the ith ordination axis. Thus the elements in

$$\mathbf{Y}_k = \begin{bmatrix} Y_{1k} \\ Y_{2k} \\ \cdot \\ Y_{pk} \end{bmatrix}$$

are the scores of the kth quadrat on p ordination axes, and the elements in

$$\mathbf{Y}_i = (Y_{i1}\ Y_{i2} \dots Y_{in})$$

are the scores of n quadrats on the ith axis. We note that for $i = 1$

$$\mathbf{Y}_1 = (Y_{11}\ Y_{12} \dots Y_{1n}) = (A_{11}\ A_{12} \dots A_{1n})$$

and further, that $S_{hi} = \mathbf{Y}_h \mathbf{Y}'_i = 0$ for any $h \neq i$.

The basic computational problem for ordinations using orthogonal functions amounts to determining numerical values for the partial regression coefficients,

$$\mathbf{B} = \begin{bmatrix} b_{21} & & \\ b_{31} & b_{32} & \\ \cdot & \cdot & \\ b_{p1} & b_{p2} \dots & b_{pp-1} \end{bmatrix} .$$

There are different methods available to accomplish this:

1. In the first method the partial regression coefficients are defined directly in the form of,

$$\mathbf{B}_i = \mathbf{G}_i \mathbf{S}_{i-1}^{-1}$$

where

$$\mathbf{G}_i = (S_{i1} \dots S_{ii-1})$$

and

$$\mathbf{S}_{i-1} = \begin{bmatrix} S_{11} & \dots & S_{1i-1} \\ S_{21} & \dots & S_{2i-1} \\ \cdot & \dots & \cdot \\ S_{i-11} & \dots & S_{i-1i-1} \end{bmatrix} .$$

The elements in $\mathbf{G}_i$ are the covariances of species i with the other $i-1$ species which preceded it in the list. The elements in $\mathbf{S}_{i-1}$ are the covariances of the first $i-1$ species. The drawback with this method is the need for inverting a matrix and for the redimensioning of $\mathbf{S}_{i-1}$ as the computations progress from $i=2$ to p.

2. Another method (Rao, 1952) is simpler. The orthogonal equations are formulated in terms of the Y_i, and not the A_i, so that

$$Y_{1k} = A_{1k}$$
$$Y_{2k} = A_{2k} - a_{21} Y_{1k}$$
$$Y_{3k} = A_{3k} - a_{31} Y_{1k} - a_{32} Y_{2k}$$
$$\cdot \quad \cdot \quad \cdot$$
$$Y_{pk} = A_{pk} - a_{p1} Y_{1k} - \dots - a_{pp-1} Y_{p-1k} \, .$$

The jth coefficient in the ith equation is given by

$$a_{ij} = b_{ij}/S_{Y_j} ,$$

where

$$b_{ij} = S_{ij} - \sum_t a_{jt} b_{it}, \quad t = i-1, \dots, 1$$

such that $j \leqslant i-1$. For any j,

$$S_{Y_j} = S_{jj} - \sum_i a_{ji} b_{ji}, \quad i = 1, \dots, j-1 \, .$$

In these formulae S_{ij} is the covariance of species i and j. S_{Y_j} is the specific variance of species j. We illustrate the calculations in an example that uses the data in Table 1-2. Program SPVAR produces the specific variances,

Species	*Specific variance*
1	3.5
2	2.62857
3	0.206522

The species are thus presented for analysis in the original sequence (1,2,3). The first orthogonal equation,

$$Y_{1j} = A_{1j}$$

requires no determination of coefficients. There is but one coefficient a_{21} in the second equation for which we have

$$b_{21} = S_{21} - a_{11} b_{21}$$
$$= -0.5-0$$
$$= -0.5 \; .$$

We note that $a_{11} = 0$, and further, that

$$a_{21} = b_{21}/S_{Y_1}$$
$$= -0.5/3.5$$
$$= -0.142857$$

and

$$S_{Y_2} = S_{22} - a_{21} b_{21}$$
$$= 2.7-(-0.142857)(-0.5)$$
$$= 2.62857 \; .$$

The latter happens to be the specific variance of the second species. The coefficients in the third orthogonal equation are respectively,

$$b_{31} = S_{31} - a_{12} b_{32} - a_{11} b_{31}$$
$$= 3 - 0$$
$$= 3 \; .$$

We should note that $a_{11} = a_{12} = 0$ and that

$$a_{31} = b_{31}/S_{Y_1}$$
$$= 3/3.5$$
$$= 0.857143$$

$$b_{32} = S_{32} - a_{22} b_{32} - a_{21} b_{31}$$
$$= -1.6-0-(-0.142857)(3)$$
$$= -1.17143$$

$$a_{32} = b_{32}/S_{Y_2}$$
$$= -1.17143/2.62857$$
$$= -0.445653$$

$$S_{Y_3} = S_{33} - a_{31} b_{31} - a_{32} b_{32}$$
$$= 3.3-(0.857143)(3) - (-0.445653)(-1.17143)$$
$$= 0.206520 \; .$$

These give the orthogonal equations,

$$Y_{1k} = A_{1k}$$

$$Y_{2k} = A_{2k} + 0.142857\ Y_{1k} = A_{2k} + 0.142857\ A_{1k}$$

$$Y_{3k} = A_{3k} - 0.857143\ Y_{1k} + 0.445653\ Y_{2k}$$

$$= A_{3k} - 0.857143\ A_{1k} + 0.445653\ A_{2k} + (0.445653)(0.142857)\ A_{1k}$$

$$= A_{3k} - 0.793478\ A_{1k} + 0.445652\ A_{2k}\ .$$

We are now ready to calculate ordination co-ordinates for quadrats. We have for quadrat *1* of Table 1-2, after adjustments $A_{ij} = X_{ij} - \bar{X}_i$, a set of three scores:

$$Y_{11} = 0$$

$$Y_{21} = -1.8 + (0.142857)(0) = -1.8$$

$$Y_{31} = 1.4 - (0.793478)(0) + (0.445653)(-1.8)$$

$$= 0.597825\ .$$

Similar calculations supply the entire set of quadrat scores,

Axis Y_i			*Quadrat*			s_{Y_i}
1	0.00000	3.00000	0.00000	−1.00000	−2.00000	3.5
2	−1.80000	−0.371429	2.20000	−1.05714	−1.08571	2.62857
3	0.597825	−0.33657	0.380435	−0.271739	−0.369565	0.206522

The computation of coefficients, ordination scores and specific variances is automatically performed in program OFORD given in the Appendix. This program rearranges species according to specific variance before commencing the part on the orthogonal functions.

Broadly speaking all ordinations perform *multidimensional scaling* when they derive co-ordinates for quadrats or other entities. In this sense all ordinations are methods for multidimensional scaling (Rohlf, 1970). But there are few examples where scaling by itself would predominate the objectives, for in most vegetation studies the data are in co-ordinate form. When, for example, we describe quadrats based on species abundance, the abundance values are metric co-ordinates. Such co-ordinates can be manipulated in one analysis or another as the need arises. However, the data may not always qualify as metric co-ordinates. It is conceivable that under specific circumstances *proximities* or *resemblances* are observed directly. The point to be made here is that if the data takes the form of resemblances, few techniques could analyse the data without first extracting from the resemblances metric co-ordinates. The transformation which leads from resemblance matrix to co-ordinates may be described as *multidimensional scaling* (cf. Torgerson, 1952). Regarding the choice of method, scaling techniques are preferred which can efficiently represent the observed data structure. Since the type of structure existing in the data is rarely known precisely, or even approximately prior to analysis, it is often necessary to perform multidimensional scaling by different methods before satisfactory results are obtained.

The different methods of multidimensional scaling, which we describe below, have one common assumption: *the data structure is continuous and linear*. This means that the methods cannot respond to non-linear trends and they are not designed to detect discontinuities in a sample. If we proceed under this assumption with the analysis of a matrix **S** of resemblance values to obtain sets of co-ordinates, such that the co-ordinate sets must be efficient summarizers of linear variation, **S** would have to be defined as a covariance matrix, a matrix of correlation coefficients or distances which may be directly converted into covariances, and we could choose for our method component analysis. But **S** may represent a similarity matrix of another kind in which case the choice may fall on Gower's (1966) method. This method has been applied to vegetation data by, e.g., Adam, Birks, Huntley & Prentice (1975). The method, known as *principal axes analysis*, extracts rectangular co-ordinates from a distance matrix $\mathbf{D}^*$ with a characteristic element $S_{jk} - 1$ where S_{jk} is a similarity value in **S**, interpretable as a direction cosine, relating individuals j and k on a zero to one scale. The analysis performed is mathematically similar to the *D-algorithm* of component analysis, from which it differs in the definition of S_{jk}. Principal axes analysis flattens the sample point configuration into Euclidean planes with potential distortions. The distortion can be measured by the stress index $\sigma = 1 - \rho^2$ where ρ is the product moment correlation of the ordination distances and the corresponding elements in $\mathbf{D}^*$ or **S**.

The example to be outlined admittedly depicts a trivial situation in ecology; but it is simple and suitable to make several important points about applications. The data to be used are given in Table 3-1. Such data could originate in the course of any soil survey giving no vegetational details other than the names of vegetation types. If anything is to be concluded from the data about the relationship of vegetation types and environmental variables,

Table 3-1. Data from a soil survey

Soil variable	*Site* 1	*2*	*3*	*4*	*5*	*6*
pH	6.9	8.0	7.8	7.3	6.5	8.1
Calcium kg/ha	3200	6900	5700	6000	2800	6800
Potassium kg/ha	29	266	390	182	48	314
Nitrate N kg/ha	6	21	15	12	8	17
Vegetation type	G	M	M	L	G	M

it is essential to quantify the information which lies in the type affiliation of the sites. When the surveyor identifies vegetation types he imposes upon the sites a certain kind of resemblance structure. This structure is conveniently described in matrix form with zeros and ones as elements, such as in

	1	*2*	*3*	*4*	*5*	*6*	
S =	1	0	0	0	1	0	*1*
	0	1	1	0	0	1	*2*
	0	1	1	0	0	1	*3*
	0	0	0	1	0	0	*4*
	1	0	0	0	1	0	*5*
	0	1	1	0	0	1	*6*

with site labels listed along the margins. A *one* in this matrix indicates sites of the same classification and a *zero* designates sites of different classifications. From matrix **S** we derive matrix **D*** by subtracting 1 from every element, then proceed to extracting the eigenvalues and eigenvectors of **D***. When the eigenvectors are adjusted to make the sum of squared elements in each equal to the corresponding eigenvalue, the vector elements represent rectangular ordination co-ordinates. These are given in Table 3-2. The co-ordinates represent new data that can be put to further analysis.

Table 3-2. Co-ordinates and efficiency values in two methods of multidimensional scaling

Method[1]	*Axis*	*Co-ordinates for site* 1	2	3	4	5	6	λ_i	*Efficiency*[2] %
P.A.A.	Y_1	−0.724	0.678	0.678	−0.236	−0.724	0.678	2.483	66
P.V.O.	Y_1	−0.713	0.624	0.624	−0.445	−0.713	0.624	2.381	65
P.A.A.	Y_2	−0.379	−0.158	−0.158	0.960	−0.379	−0.158	1.284	34
P.V.O.	Y_2	0.463	0.000	0.000	−0.926	0.463	0.000	1.286	35

[1]P.A.A. – Principal axis analysis; P.V.O. – Position vectors ordination (described below)
[2]100 λ_i/Q where Q is the sum of all positive eigenvalues of matrix **D***

The co-ordinates quantify the information in the vegetation type affiliation of the different sites. When these co-ordinates are correlated with environmental variables the correlations may reveal, for instance, the value of the vegetation classification as a predictor of environmental conditions. The simple product moment correlations of Y_1 and Y_2 of P.A.A. in Table 3-2 with the four soil variables in Table 3-1 are given in the table below:

Principal axis	*Soil variable* pH	Ca	K	N
Y_1	0.799	0.587	0.811	0.818
Y_2	−0.588	−0.542	−0.395	−0.315

It can be seen that the correlations are rather high between the derived co-ordinates of sites on the first principal (vegetation) axis and the measured soil chemical properties. These results indicate that the vegetation types, even though a result of subjective determination, are highly indicative of the measured soil conditions.

We shall consider another example in which multidimensional scaling supplies metric co-ordinates when the sample descriptors are mixed quantitative, binary and nominal variables. Why do we extract metric co-ordinates? The reason is that, whereas the sample of mixed data would serve as suitable input in few methods of multivariate analysis, metric co-ordinates are broadly accepted as input in the different methods. The data to be considered are given in Table 3-3. Such type of data presents a unique

Table 3-3. Sample data of mixed variables from a three species community

*Variable**		*Quadrat*	
	1	*2*	*3*
1. Cover of species 1	0.5	0.2	0.2
2. – „ – 2	0.1	0.4	0.6
3. – „ – 3	0.05	0.1	0.1
4. Soil type	BF	P	P
5. Topography	C	CX	CX
6. Fire history	F	NF	F

* Cover values are proportions. BF identifies brown forest soils, P podzols, C indicates concave and CX convex. F signifies presence of charcoal and NF its absence.

problem if we wish to measure quadrat resemblance. The reason is that nominal variables are not directly amenable to ordinary arithmetic manipulations. Solutions to this problem have been proposed by, e.g., Lance & Williams (1966b), Gower (1967a, 1971a) and Anderson (1971a). If we follow Anderson's method, we may define a similarity measure for quadrats j and k in the form of

$$S_{jk} = \frac{1}{p} \sum_h S_{hjk}, \; h = 1, ..., p \; ,$$

where S_{hjk} measures the resemblance of quadrats j and k with respect to variable h, and p is the number of variables. If the variable is quantitative, and has unit range, $S_{hjk} = X_{hj}X_{hk}$ and X_{hj} or X_{hk} is the value of variable h in quadrat j or k. When the variable is nominal or binary,

$$S_{hjk} = \begin{cases} 0 \text{ if } X_{hj} \neq X_{hk} \\ \\ 0.5 \text{ if } X_{hj} = X_{hk}. \end{cases}$$

The value 0.5 is arbitrary. It can be seen that S_{jk} is interpretable as a non-centered cross product and converted to distance,

$$d^2(j, k) = (S_{jj} + S_{kk} - 2S_{jk}) .$$

A weighting factor f_h may be introduced in the definition of S_{jk} if desired. Without weighting, the similarity values for quadrats are given in

$$\mathbf{S} = \begin{bmatrix} S_{11} & S_{12} & S_{13} \\ S_{21} & S_{22} & S_{23} \\ S_{31} & S_{32} & S_{33} \end{bmatrix} = \begin{bmatrix} 0.2938 & 0.0242 & 0.1108 \\ 0.0242 & 0.2850 & 0.2150 \\ 0.1108 & 0.2150 & 0.3183 \end{bmatrix} .$$

The calculation of a characteristic element in **S**, say S_{23}, accords with,

$$\begin{aligned} S_{23} &= ((0.2)(0.2) + (0.4)(0.6) + (0.1)(0.1) + 0.5 + 0.5 + 0)/6 \\ &= 1.29/6 \\ &= 0.2150 . \end{aligned}$$

The elements in **D*** are calculated from the elements in **S**, e.g.,

$$\begin{aligned} &-0.5 d^2(2,3) = -0.5\ (S_{22} + S_{33} - 2S_{23}) \\ &= -0.5\ (0.2850 + 0.3183 - (2)\ (0.2150) \\ &= -0.0867. \end{aligned}$$

The entire matrix is given by

$$\mathbf{D^*} = -0.5 \begin{bmatrix} d^2(1,1) & d^2(1,2) & d^2(1,3) \\ d^2(2,1) & d^2(2,2) & d^2(2,3) \\ d^2(3,1) & d^2(3,2) & d^2(3,3) \end{bmatrix} = \begin{bmatrix} 0.0000 & -0.2652 & -0.1953 \\ -0.2652 & 0.0000 & -0.0867 \\ -0.1953 & -0.0867 & 0.0000 \end{bmatrix} .$$

Should **D*** be subjected to the eigenvalue and vector analysis of program PCAD, the metric quadrat co-ordinates will be determined automatically. After the analysis is performed, it is found that there is a single non-zero eigenvalue of **D*** and the associated eigenvector has elements **Y** = (0.394960 –0.203357 –0.229181). These, as co-ordinates, redescribe the quadrats in such terms that they can be subjected to further analysis in a large number of possible techniques.

We now summarize the main points in multidimensional scaling:

a. Estimates of resemblance or non-metric co-ordinates are the sources for information about the vegetation in the sites.

b. The input consists of resemblance values which are analysed into rectangular co-ordinates which, in turn, may be subjected to further analysis to reveal correlations between vegetation types and environmental variables.

Metric co-ordinates may be extracted under the conditions stipulated in principal axes analysis more simply in other methods. *Position vectors ordi-*

nation (Orlóci, 1966) is one of them. This method is a close approximation to principal axes analysis under certain circumstances to be explained in the sequel. Examples of application include Orlóci (1966), Martin (1969) and Field & Robb (1970) among others. To illustrate the technique we shall re-analyse matrix **S** of the preceeding section with the same objective to derive metric co-ordinates for the sites:

1. Compute on the basis of matrix **S** matrix **D*** with elements $d^2(j, k) = 2(1 - S_{jk})$.
2. Derive matrix **Q** from **D*** according to formula (2.12).
3. Complete position vectors ordination on matrix **Q** as described below.

Program PVO of the Appendix performs the calculations automatically.
In steps 1 and 2 we obtain matrix **D***,

$$\mathbf{D}^* = \begin{bmatrix} 0 & 2 & 2 & 2 & 0 & 2 \\ 2 & 0 & 0 & 2 & 2 & 0 \\ 2 & 0 & 0 & 2 & 2 & 0 \\ 2 & 2 & 2 & 0 & 2 & 2 \\ 0 & 2 & 2 & 2 & 0 & 2 \\ 2 & 0 & 0 & 2 & 2 & 0 \end{bmatrix},$$

the average of the squared distances per row (column),

$$\begin{bmatrix} \overline{d_1^2} \\ \overline{d_2^2} \\ \overline{d_3^2} \\ \overline{d_4^2} \\ \overline{d_5^2} \\ \overline{d_6^2} \end{bmatrix} = \begin{bmatrix} 1.333 \\ 1.000 \\ 1.000 \\ 1.667 \\ 1.333 \\ 1.000 \end{bmatrix},$$

and the **Q** matrix,

$$\mathbf{Q} = \begin{bmatrix} 0.7222 & -0.4444 & -0.4444 & -0.1111 & 0.7222 & -0.4444 \\ -0.4444 & 0.3889 & 0.3889 & -0.2778 & -0.4444 & 0.3889 \\ -0.4444 & 0.3889 & 0.3889 & -0.2778 & -0.4444 & 0.3889 \\ -0.1111 & -0.2778 & -0.2778 & 1.0556 & -0.1111 & -0.2778 \\ 0.7222 & -0.4444 & -0.4444 & -0.1111 & 0.7222 & -0.4444 \\ -0.4444 & 0.3889 & 0.3888 & -0.2778 & -0.4444 & 0.3889 \end{bmatrix}.$$

In position vectors ordination, similarly in principal axes analysis, the point configuration defined by **S** is flattened into Euclidean planes. The sites are defined in terms of position vectors which represent directed lines to the sites (as points) from the sample centroid (center of gravity of all points). The length of the *j*th position vector is $\sqrt{q_{jj}}$ and the quantity q_{jj} is the *jj* value in **Q**. The sum $q_{11} + \ldots + q_{nn}$ measures the total variation in the sample. Position vectors ordination partitions the total variation into components on strategically selected axes. The first axis coincides with the position vector on which the sum of squared projections of all position vectors is maximum. This position vector is identified by the maximum property:

$$\max\,[\sum_j q_{j1}^2/q_{11}, \ldots, \sum_j q_{jn}^2/q_{nn}]\,, j = 1, \ldots, n\,.$$

If the maximum occurs on the mth vector, the first set of co-ordinates are given by:

$$\mathbf{Y}_1 = (q_{m1}/\sqrt{q_{mm}} \ldots q_{mn}/\sqrt{q_{mm}}) = (Y_{11} \ldots Y_{1n})\,.$$

The second set of co-ordinates cannot be obtained directly from the **Q** matrix. A residual matrix **Q*** must be computed first. We obtain the elements in this matrix according to

$$q_{jk}^* = q_{jk} - Y_{1j}Y_{1k}\,.$$

These are subjected to further manipulations to find

$$\max\,[\sum_j q_{j1}^{*2}/q_{11}^*, \ldots, \sum_j q_{jn}^{*2}/q_{nn}^*]\,,\ j = 1, \ldots, n\,.$$

Assuming that the maximum occurs on the zth vector, the second set of co-ordinates are obtained by:

$$\mathbf{Y}_2 = (q_{z1}^*/\sqrt{q_{zz}^*} \ldots q_{zn}^*/\sqrt{q_{zz}^*}) = (Y_{21} \ldots Y_{2n})\,.$$

Any subsequent set of co-ordinates are determined from a subsequent residual of **Q** in the same manner. There will be altogether t sets of co-ordinates where t is equal to or slightly more than the number of independent dimensions in the sample. Considering numerical values, we find the $\sum_j q_{jk}^2/q_{kk}$ quantities and their maximum to be

$$\max\,(2.281\ 2.3810\ 2.3810\ 1.2983\ 2.2821\ 2.3810) = 2.3810\,.$$

The first value in this expression accords with

$$\sum_j q_{j1}^2/q_{11} = (0.7222^2 + (-0.4444)^2 + (-0.4444)^2 + (-0.1111)^2 + 0.7722^2 + (-0.4444)^2)/0.7222 = 2.2821\,.$$

The remaining values are similarly calculated. The maximum indicates that we could select vector *2, 3* or *6* with the same outcome in the next step when we derive the co-ordinates $\mathbf{Y}_1$. These co-ordinates are given in Table 3-2. To obtain for example the fourth co-ordinate in $\mathbf{Y}_1$ we rely on

$$Y_{14} = q_{24}/\sqrt{q_{22}} = -0.2778/0.6236 = -0.4455\,.$$

The first residual **Q*** consists of elements which are obtained by subtracting from an element of **Q** the product of the corresponding co-ordinates in $\mathbf{Y}_1$. For example, the residual of $q_{34} = -0.2778$ is given by

$$q_{34}^* = q_{34} - Y_{13}Y_{14} = -0.2778 - (0.6236)(-0.4455) = 0.0000\,.$$

Another element say q^*_{14} is similarly calculated,

$$q^*_{14} = q_{14} - Y_{11} Y_{14} = -0.1111 - (-0.7217)(-0.4455) = -0.4286 \,.$$

The entire set of residual elements is given in

$$\mathbf{Q}^* = \begin{bmatrix} 0.2143 & 0.0000 & 0.0000 & -0.4286 & 0.2143 & 0.0000 \\ 0.0000 & 0.0000 & 0.0000 & 0.0000 & 0.0000 & 0.0000 \\ 0.0000 & 0.0000 & 0.0000 & 0.0000 & 0.0000 & 0.0000 \\ -0.4286 & 0.0000 & 0.0000 & 0.8572 & -0.4286 & 0.0000 \\ 0.2143 & 0.0000 & 0.0000 & -0.4286 & 0.2143 & 0.0000 \\ 0.0000 & 0.0000 & 0.0000 & 0.0000 & 0.0000 & 0.0000 \end{bmatrix} .$$

The next vector to serve as an ordination axis is chosen to coincide with the value

$$\max(1.2857\ 0.0000\ 0.0000\ 1.2857\ 1.2857\ 0.0000) = 1.2857 \,.$$

This specifies either the first, second, or third vector. We can take either of these and obtain the second set of co-ordinates $\mathbf{Y}_2$ (Table 3-2) through similar calculations as before. For example,

$$Y_{24} = q^*_{14} / \sqrt{q^*_{11}} = -0.4286/0.4629 = -0.9259 \,.$$

The second residual $\mathbf{Q}^{**}$ has no elements other than zero. We conclude that the two axes completely describe variation in the sample.

Based on the results in Table 3-2 we can see a close resemblance between principal axes analysis and position vectors ordination. Experience suggests that these methods tend to produce similar results when the number of sample points is reasonably large and when the points fall into a single cluster with reasonably uniform density (cf. Field & Robb, 1970). However, position vectors ordination is computationally less complex than principal axes analysis which in certain applications may compensate for the somewhat reduced efficiency in providing parsimonious lower dimensional representations. Under such conditions of uniformity, similar results could be expected from the method of Swan, Dix & Wherhahn (1969). Chardy, Glemarec & Laurec (1976) make further points on a related topic.

Principal axes analysis and position vectors ordination are two of many possible methods which can perform multidimensional scaling; but the best known, and probably the least recommendable, of the scaling techniques is an ordination devised by Bray & Curtis (1957). Their method relies on ordination poles to determine the direction of the axes on which vegetation quadrats or other objects are ordered. This simple method attracted many users in the past, and significantly, there were recent attempts to explain its advantages and to restore it to a role which it once played in vegetation ecology (e.g. Gauch & Whittaker, 1972; Kessell & Whittaker, 1976). In its original form, however, the Bray & Curtis method offers no real advantages (cf. Austin & Orlóci, 1966; Anderson, 1971a; Orlóci, 1974a,b) apart from computational simplicity and personal selection of poles. We shall elaborate on certain aspects which will show that the Bray & Curtis ordination is a rather inefficient method for data reduction, and we shall also discuss possible revisions in the mathematics while, at the same time, retaining the option of personal selection of the ordination poles.

At the roots of the problem are the resemblance function which often lacks a consistent scale, axes that are potentially oblique and non-intersecting, and a dubious method which produces scatter diagrams without taking into account the potential obliqueness of the ordination axes. Suggestions to modify the Bray & Curtis (1957) method were not long forthcoming after its introduction. Beals (1960) supplied an analytical solution for co-ordinates; Orlóci (1966, 1974b) modified the distance measure and perpendicularized the axes; Swan, Dix & Wherhahn (1969) revised the choice of ordination poles and van der Maarel (1969) introduced a new maximization to take into account the negative correlation tendency of stands.

A particularly interesting modification has been suggested by Goff & Cottam (1967). The objective of their method is to derive a unidimensional ordering of species. Their source of information about species relationships is a similarity matrix,

$$\mathbf{S} = \begin{bmatrix} S_{11} & \cdots & S_{1p} \\ S_{21} & \cdots & S_{2p} \\ \cdot & \cdots & \cdot \\ S_{p1} & \cdots & S_{pp} \end{bmatrix}$$

with a characteristic element S_{hi} measuring the similarity of species h and species i. A similarity measure, to be admissible, must be symmetric and, preferably, meaningful as a vector scalar (inner) product. If the measure does not qualify as a vector scalar product, the **S** configuration will be forced into a Euclidean space with distortions. That this actually happens will become apparent when we discuss the method's connections with other more general techniques. The algorithm is described below in which the steps are applied recursively:

1. Specify a trial vector

$$\mathbf{X} = (X_1 \ldots X_p)$$

of species scores. Any arbitrary set of p numbers is acceptable for **X**. If the scores are made to approximate levels in a major vegetational gradient in the data, the iterations will converge more rapidly. Goff & Cottam (1967) suggest a trial vector in which a species (L) with lowest similarity with the other species receives a score of 1, another species (U) with highest similarity receives a score of 10, and any h of the remaining species receives a value,

$$X_h = (S_{Lh} + 10S_{Uh}) / (S_{Lh} + S_{Uh}) .$$

2. Calculate the weighted mean score of species h,

$$X_h = \sum_i S_{hi}X_i/Q_{hh}, \quad h,i = 1, ..., p \tag{3.3}$$

where

$$Q_{hh} = \sum_i S_{hi}, \; i = 1, ..., p.$$

3. Adjust the range in the scores to coincide with 1 to 10. Such an adjustment can be achieved based on

$$X_h := (X_h - \text{min}X)\, R + 1 \tag{3.4}$$

where $\text{min}X = \text{min}(X_1, ..., X_p)$ and $R = 9/(\text{max}X - \text{min}X)$.

4. Iterate through steps 2 and 3 until a stable set of scores is obtained.

To illustrate the method, we begin with a similarity matrix,

$$\mathbf{S} = \begin{bmatrix} S_{11} & S_{12} & S_{13} \\ S_{21} & S_{22} & S_{23} \\ S_{31} & S_{32} & S_{33} \end{bmatrix} = \begin{bmatrix} 1.000 & -0.163 & 0.883 \\ -0.163 & 1.000 & -0.536 \\ 0.883 & -0.536 & 1.000 \end{bmatrix}.$$

This happens to be the correlation matrix of species given in Table 1-2. We note that we could have used a different similarity measure or even the quantity $-0.5d^2$ of formula (2.11) which is used in the *D-algorithm* of component analysis. Next, we have to specify a trial vector,

$$\mathbf{X} = (X_1 X_2 X_3) = (1 \;\; 2 \;\; 3)'.$$

A new set of **X** scores is derived according to formula (3.3),

$$\mathbf{X} = (1.93198 \; 0.760797 \; 2.08686)$$

in which the first value is obtained by:

$$\begin{aligned} X_1 &= ((1.000)\,(1) + (-0.163)\,(2) + (0.833)\,(3))/(1.000 - 0.163 + 0.883) \\ &= 3.323/1.720 \\ &= 1.93198. \end{aligned}$$

The remaining values are similarly calculated. We retain a sufficient number of digits in the numbers to minimize rounding errors. The scores have to be adjusted to fall within 1 and 10 inclusive (formula (3.4)),

$$\mathbf{X} = (8.94881 \; 1.00000 \; 10.0000) .$$

The next iteration gives unadjusted scores,

$$\mathbf{X} = (10.2417 \quad -19.3311 \quad 12.8922)\ ,$$

and adjusted scores,

$$\mathbf{X} = (9.25973 \quad 1.00000 \quad 10.0000)\ .$$

The scores stablize in the 4th iteration to unadjusted values,

$$\mathbf{X} = (10.4237 \quad -19.5006 \quad 13.0974)\ ,$$

and adjusted values,

$$\mathbf{X}_i = (9.26183 \quad 1.00000 \quad 10.0000)\ .$$

The subscript i in $\mathbf{X}_i$ is intended to indicate that the set of species scores given is but one of several independent sets $\mathbf{X}_1, ..., \mathbf{X}_p$ that could be extracted from the sample. To show this we rewrite the iteration in terms of matrix symbols such as in

$$\mathbf{QX}_i := \mathbf{SX}_i\ .$$

This formulation tells us that the ith set of improved scores (left $\mathbf{X}_i$) is obtained from the preceding set (right $\mathbf{X}_i$) by averaging. In this, $\mathbf{X}_i$ is a p-valued column vector, $\mathbf{Q}$ a diagonal matrix of the row totals in $\mathbf{S}$, and $\mathbf{S}$ a $p \times p$ symmetric matrix of similarity values. It can be shown that $\xi_i = (\mathbf{QX}_i)$ is an eigenvector of $\mathbf{S}$ satisfying the characteristic equation

$$\mathbf{S}\,(\mathbf{QX}_i) = (\mathbf{QX}_i)\,\lambda_i\ ,$$

where λ_i is an eigenvalue of $\mathbf{S}$. The Goff & Cottam (1967) procedure can thus be restated as an eigenvalue and vector procedure:

1. Compute a similarity (vector inner product) matrix $\mathbf{S}$.
2. Extract the eigenvectors of $\mathbf{S}$ and arrange them in order of decreasing magnitude of the associated eigenvalues. Obtain the ith set of species scores based on

$$\begin{aligned}\mathbf{X}_i &= \mathbf{Q}^{-1}\,\xi_i \\ &= (\xi_{1i}/Q_{11} \ \dots\ \xi_{pi}/Q_{pp}) \\ &= (X_{1i} \ \dots\ X_{pi}) \end{aligned} \tag{3.5}$$

The elements can be adjusted in each $\mathbf{X}_i$ according to formula (3.4).

The numerical example below utilizes $\mathbf{S}$ of the preceding example. To subject $\mathbf{S}$ to analysis based on the Goff & Cottam (1967) method, first we calculate the species totals in $\mathbf{S}$,

$$\mathbf{Q} = \begin{bmatrix} Q_{11} & 0 & 0 \\ 0 & Q_{22} & 0 \\ 0 & 0 & Q_{33} \end{bmatrix} = \begin{bmatrix} 1.72 & 0 & 0 \\ 0 & 0.301 & 0 \\ 0 & 0 & 1.347 \end{bmatrix}$$

and then the eigenvalues and eigenvectors of **S**,

Eigenvector	*Species score* 1	2	3	*Eigenvalue* λ_i
ξ_1	0.602485	−0.417013	0.680523	
X_1	9.26248	1.00000	10.0000	2.11019
ξ_2	0.508400	0.857801	0.0755465	
X_2	1.77153	10.0000	1.00000	0.856188
ξ_3	−0.615258	0.300462	0.728821	
X_3	1.00000	10.0000	6.96569	0.0336177 .

There are several comments to be made to further clarify procedure and its potential. We note that our version of the Goff & Cottam (1967) algorithm can extract not just a single set of species scores but all sets in the data based on the residuals of **S**. For instance, analysis of the first residual $\mathbf{S}^{(1)}$ with elements defined as

$$S_{hi}^{(1)} = S_{hi} - X_{h1} X_{i1}$$

yields the second set of species scores $\mathbf{X}_2$. X_{h1} and X_{i1} are elements in $\mathbf{X}_m$ (formula (3.5)) when $m = 1$. Further sets could be derived from subsequent residuals of **S**. We also note that the amount of linear variance accounted for by $\mathbf{X}_1$, $\mathbf{X}_2$, ... is proportional to $\lambda_1, \lambda_2, \ldots$. In the foregoing example the overwhelming proportion (about 2.1 or 70%) of the total variance ($\lambda_1 + \lambda_2 + \lambda_3 = 3.0$) lies in the first set of scores. We note that if say $h = 1$ and $m = 2$,

$$\begin{aligned} X_{12} &= [\xi_{12}/Q_{11} - \min(\xi_{12}/Q_{11}, \ldots, \xi_{p2}/Q_{pp})] \; \{9/[\max(\xi_{12}/Q_{11}, \ldots, \\ &\quad \xi_{p2}/Q_{pp}) - \min(\xi_{12}/Q_{11}, \ldots, \xi_{p2}/Q_{pp})]\} \; +1 \\ &= (0.508400/1.72 - 0.0755465/1.347) \left[\frac{9}{\dfrac{0.857801}{0.301} - \dfrac{0.0755465}{1.347}} \right] + 1 \\ &= (0.295581 - 0.056085) \left[\frac{9}{2.84984 - 0.056085} \right] +1 \\ &= 1.77153 \, . \end{aligned}$$

The species scores can be used as ordination co-ordinates. However, when a non-centered similarity measure is analysed, the first eigenvector may have to be discarded as it contains no relevant information about the **S** configuration other than its location relative to the co-ordinate origin. When viewed from the point of an averaging procedure, it will be seen that the method is closely akin to reciprocal ordering which we shall discuss in the sequel. After determining the species scores in such a method, we can derive quadrat scores by averaging,

$$Y_{ij} = \sum_h X_{hi} A_{hj} / \sum_h A_{hj}, \; h = 1, ..., p \;,$$

where Y_{ij} is the ith score of quadrat j, X_{hi} the ith score of species h and A_{hj} the quantity of species h in quadrat j. For example, the first score of quadrat *2* in Table 1-2 is given by

$$Y_{12} = ((9.26248)(5) + (1.00000)(1) + (10.0000)(4))/(5 + 1 + 4)$$
$$= 13.0892 \; .$$

Some revisions of the Bray & Curtis (1957) method (e.g. Orlóci, 1966; Bannister, 1968; Swan, Dix & Wehrhahn, 1969; van der Maarel, 1969; La France, 1972) went perhaps too far in a way by not retaining the option for personal selection of ordination poles which provided a handle on the method to those ecologists who have no confidence in automated ordinations. Even when the goal is unidimensional scaling, the method needs some revision:

1. Compute quadrat (stand) distances preferably on the basis of formula (2.2) or formula (2.39).
2. Designate two quadrats A, B as ordination poles.
3. Determine co-ordinates for the remaining quadrats in graphical manipulations of triangles as suggested by Bray & Curtis (1957), or follow Beal's (1960) procedure in which co-ordinates are determined analytically.

If the objective is a multidimensional ordination, the co-ordinates have to be determined by a method other than originally suggested by Bray & Curtis. The methodological problem of correcting for non-perpendicular axes in the scatter diagram construction can be solved by projection of the ordination axes, specified by points *AB* and *CD*, orthogonally onto a given plane *M* which lies parallel to both of the axes. In such a projection the axes appear as intersecting oblique lines which, if required, can be rotated to perpendicular positions analytically. We describe below graphical and analytical solutions for the projection and rotation. In the descriptions we shall use symbols Y_1, Y_2 to designate axes. The numerical results will be derived for the distance values

C =

	A	*B*	*C*	*D*	*j*
A	0	1.20	0.84	0.96	0.93
B		0	1.02	0.63	0.54
C			0	0.90	0.84
D				0	0.12

The elements in **C** define the distances between quadrats representing the ordination poles *A, B, C, D* and another quadrat designated by *j*. A realistic sample may of course contain more than just 5 quadrats; but the procedure would remain essentially the same. We recommend formula (2.2) or (2.39) to measure distance.

The graphical solution, as well as the analytical solution, assumes that the quadrat distances are Euclidean. The steps are described for Fig. 3-7 below:

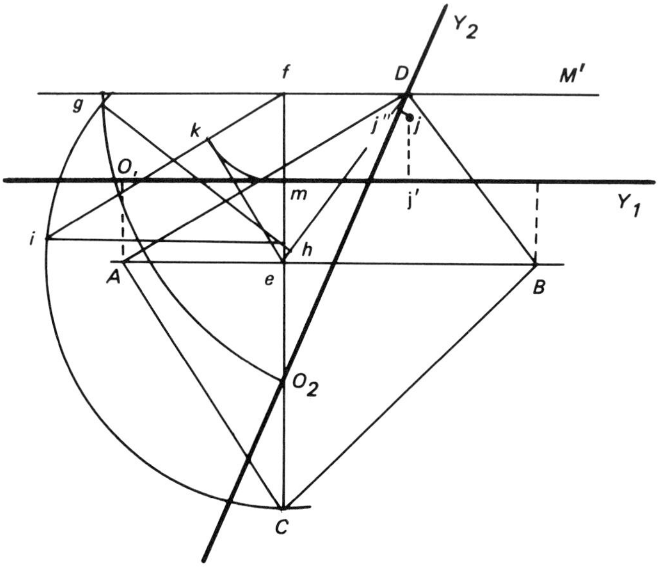

Fig. 3-7. Graphical solution for oblique axes in a modified Bray & Curtis ordination. See description of method in the main text.

1. Draw triangles *ABC* and *ABD* using distances from matrix **C**.
2. Draw line M' through point *D* parallel to line *AB*. M' is the trace of plane *M* on plane *ABD*.
3. Draw line through *C* perpendicular to M' to meet *AB* in *e*, and M' in *f*.
4. Draw arc with center in *D* and radius *c*(*C, D*), the *CD* element in matrix **C**, to meet *Cf* in point O_2. The line Y_2 through *D* and O_2 is the projection image of axis Y_2 on *M*. The origin of Y_2 is in O_2.

At this point we could draw a line through O_2 parallel to *AB* to obtain the Y_1 axis with origin O_1 in O_2. We could then proceed directly to step 10. However, it is possible to determine the exact projection of Y_1 on plane *M* by continuing with step 5:

5. Draw arc through *C* with center in *e* to meet the arc through O_2 in *g*.
6. Draw a line through *g* perpendicular to *eD* to meet *Cf* in *h*. Point *h* is the projection of quadrat *C* on plane *ABD*.
7. Draw *hi* parallel to *AB* to meet *Cg* in *i*. The length of line *hi* is the height of point *C* above plane *ABD*.

8. Draw *ek* perpendicular to *fi* and draw arc through *k* with center in *f* to meet *Cf* in *m*. Point *m* is the projection of *e* in *AB* on *M*.
9. Draw Y_1 through *m* parallel to *AB*. Line Y_1 is the second ordination axis.
10. Transfer Y_1 and Y_2 to a clear sheet of paper and plot points to represent their joint distribution.
Note that Y_1 and Y_2 are oblique lines; their origins are marked by O_1 and O_2. If so desired each axis may be moved parallel to itself until O_1 and O_2 fall in a common point. (This has not been done in Fig. 3-7).

The co-ordinates of quadrats on Y_1 can be determined graphically by actually drawing triangles, as in the original method, or on the basis of formula

$$Y_{1j} = [c^2(A, j) + c^2(A, B) - c^2(B, j)]/2c(A, B) \qquad (3.6a)$$

For quadrat *j* we have

$$Y_{1j} = (0.93^2 + 1.2^2 - 0.54^2) / ((2)(1.2)) = 0.839 .$$

It is important to remember that the co-ordinates on the Y_2 axis are determined by formula

$$Y_{2j} = [c^2(C, j) + c^2(C, D) - c^2(D, j)]/2c(C, D) \qquad (3.6b)$$

and *not* by Beals (1960) second formula which assumes that Y_1 and Y_2 are perpendicular. The Y_{2j} co-ordinate for quadrat *j* is thus obtained,

$$Y_{2j} = (0.84^2 + 0.9^2 - 0.12^2) / ((2)(0.9)) = 0.834 .$$

The point (Y_{1j}, Y_{2j}) is plotted in the scatter diagram according to the following steps:
1. Measure distance Y_{1j} from O_1 on Y_1 and mark the end of this distance by j'.
2. Measure a distance Y_{2j} from O_2 on Y_2 and mark the end of this distance by j''.
3. Draw a line through j' perpendicular to Y_1 and another line through j'' perpendicular to Y_2. These lines meet in *j*, marking the projection of quadrat *j* as a point in the scatter diagram.

Other quadrats would be placed similarly in the scatter diagram. We note at this point that the outlined procedure can be extended to higher dimensions by analytical means; but then the arithmetic becomes somewhat involved and not presented here. We shall concentrate on the analytical solution of rotation. For this purpose, we need to determine first the acute angle α enclosed by the axes, and then consider the problem of transferring from oblique to rectangular co-ordinates. The acute angle between the projections of Y_1 and Y_2 in plane *M* is given in

$$\cos \alpha = \overline{Df} / c(D, C) ,$$

where

$$
\begin{aligned}
\overline{Df} = \; & | \, [c^2(A,D) + c^2(A,B) - c^2(D,B)]/2c(A,B) - [c^2(A,C) + c^2(A,B) \\
- \; & c^2(B,C)]/2c(A,B) \, | \\
= \; & | \, c^2(A,D) + c^2(B,C) - c^2(A,C) - c^2(D,B) \, | /2c(A,B) \, .
\end{aligned}
$$

Using the numerical values in matrix **C**, we have

$$\overline{Df} = | \, 0.96^2 + 1.02^2 - 0.84^2 - 0.63^2 \, | /2.4 = 0.8595/2.4 = 0.358125$$

and

$$\cos\alpha = 0.358125/0.9 = 0.397917 \text{ or } \alpha = 66^\circ \, 33' \, 07'' \, .$$

We transfer from oblique to rectangular co-ordinates according to

$$Y^*_{2j} = Y_{2j} \sin\alpha \pm \Delta_j \, ,$$

where Y_{2j} is a co-ordinate based on formula (3.6b) and Δ_j is a correction term defined by

$$\Delta_j = || \, Y_{1j} - Y_{1C} \, | - Y_{2j} \cos\alpha \, | \, \mathrm{tg}\,(90 - \alpha) \, .$$

Note that we define absolute differences in this expression. Also note that Y_{1C} is the co-ordinate of quadrat C (the origin O_2) on the first axis. We declare Δ_j positive if the condition

$$Y_{2j} \sin\alpha > | \, Y_{1j} - Y_{1C} \, | \, \mathrm{tg}\,\alpha \tag{3.7}$$

is satisfied and negative if it is not. When the right and left sides are equal in formula (3.7) $\Delta_j = 0$. Based on the distances in **C** we have the numerical results,

$$
\begin{aligned}
& Y_{1j} = 0.839 \\
& Y_{1C} = [c^2(A,C) + c^2(A,B) - c^2(B,C)]/2c(A,B) \\
& \qquad = (0.84^2 + 1.2^2 - 1.02^2)/2.4 = 1.105/2.4 = 0.461 \\
& Y_{2j} = 0.834 \\
& \cos\alpha = 0.397917 \\
& \mathrm{tg}\,(90 - \alpha) = 0.433734 \\
& \Delta_j = || \, 0.839 - 0.461 \, | - (0.834)(0.397917) | \, 0.433734 = 0.0204 \\
& \sin\alpha = 0.917421 \\
& \mathrm{tg}\,\alpha = 2.305556 \, .
\end{aligned}
$$

Considering that

$$Y_{2j} \sin \alpha = 0.765 < | Y_{1j} - Y_{1C} | \operatorname{tg} \alpha = 0.872$$

we take Δ_j with negative sign and obtain a rectangular co-ordinate for j,

$$Y_{2j}^* = (0.834)\,(0.917421) - 0.0204 = 0.744\,.$$

The alternative formula,

$$Y_{2j}^* = [Y_{2j}c(C,D) - (Y_{ij} - Y_{1C})(Y_{1D} - Y_{1C})]/[c^2(C,D) - (Y_{1D} - Y_{1C})^2]^{1/2} \qquad (3.8)$$

is due to Beals (1965a). In this $Y_{1D} > Y_{1C}$ is assumed; if not so, the Y_{1D} and Y_{1C} terms have to be interchanged. To illustrate its use, we give

$$Y_{1j} = 0.839$$

$$Y_{2j} = 0.834$$

$$Y_{1C} = 0.461$$

$$Y_{1D} = [c^2(A,D) + c^2(A,B) - c^2(B,D)]/2c(A,B) = 0.819\,.$$

Since $Y_{1D} > Y_{1C}$ formula (3.8) is entered as given to obtain,

$$Y_{2j}^* = ((0.834)\,(0.90) - (0.839 - 0.461)(0.819 - 0.461))/(0.90^2 - (0.819 - 0.461)^2)^{1/2} = 0.745$$

which is the same value as before (within rounding errors). The procedure (BCOAX in the Appendix) puts the construction of scatter diagrams on firm foundations consistent with the original objectives in the Bray & Curtis method. The ordination, however, still remains inefficient as a method for summarization because the axes are likely to be correlated.

3.5 Summarization and multidimensional scaling – non-linear case

It has been recognized early in the ecological applications of multivariate techniques that non-linear species correlations represent a special case that cannot be efficiently handled by linear measures of resemblance (van Groenewoud, 1965). It is now an accepted fact that the linear models of ordination are not suitable for summarizing vegetational data in the presence of non-linear trends to which a linear ordination cannot respond. And yet, in samples which are put to analysis by ordinations the data structure is normally non-linear. What level of efficiency could be expected from a linear ordination when the data structure is non-linear? The answer is that the ordination would certainly tend to be inefficient in direct proportion to the degree and type of non-linearity present. To reflect further on this point, we shall consider applying linear methods to the data in Table 2-6 in which variation has a strong non-linear component. Any given quadrat in this table is represented by a profile which we construct by plotting species abundance in graphs as indicated in Fig. 3-8. Inspection of the profiles, arranged in a certain order, readily suggests the existence of a strong uni-dimensional trend in the data. We subject these profiles to ordinations by different linear methods and give the results in Fig. 3-9. When plotting profiles as points in Fig. 3-9 according to the species (X_1, X_2) axes, the points fall into a strongly horseshoe-shaped configuration (G). The sequence in which the points occur in G is identical to the sequence of the profiles in Fig. 3.8. The horseshoe shape of G indicates that the correlation between species is strongly non-linear. When the same profiles are ordinated by the Bray & Curtis (1957) method, using distance formula (2.1), we obtain axes X, Y which are superimposed as oblique lines on the original diagram. Component analysis of the profiles, with the same distance function, yields axes Y_1, Y_2 which are also superimposed. We can draw general conclusions from these results which, in part, are contrary to the generalizations given by Gauch & Whittaker (1972):

1. The Bray & Curtis method and component analysis both reproduce the original non-linear point configuration G in full dimensions. This is of

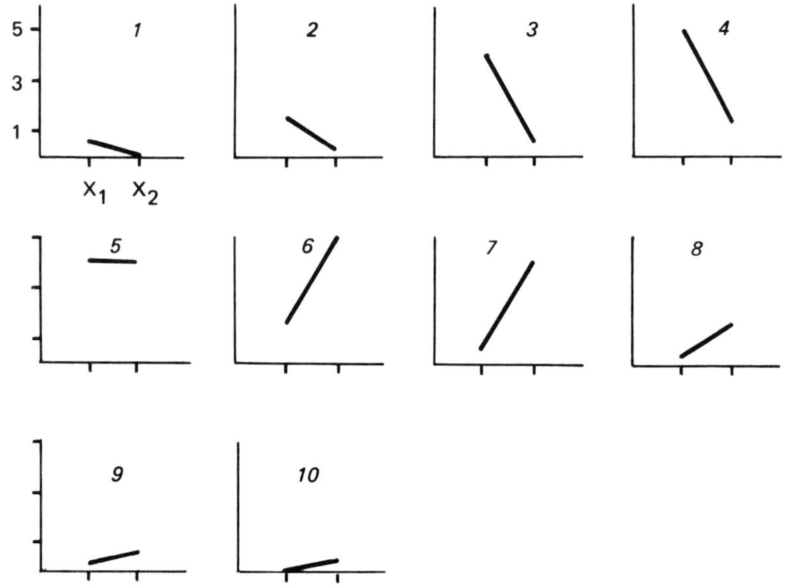

Fig. 3-8. Profiles from quadrat vectors arranged according to labels in Table *2-6*. Vertical scale is species abundance. X_1 and X_2 are species.

course a feat every consistent ordination should be able to achieve without a claim on efficiency in doing so. Linear ordinations may in fact produce one or two dimensional ordination images of the data structure that may be totally unrevealing of an existing non-linear trend. Fig. 3-9 gives an example for this; if the path of *G* were left unmarked the existence of a definite trend from point 1 to 10 could not even be suggested. But this trend is real; it flows from the orientation of the profiles in Fig. 3-8.

2. The Bray & Curtis ordination and component analysis are equally unsuccessful in providing an efficient summarization of the data structure in hands; both methods require two axes to represent an essentially unidimensional trend (*G*) in the data. To a further detriment of the linear methods, none of the axes could individually serve as a reasonably accurate predictor of *G*, for each axis orders the quadrats in a sense completely different from the order of points in *G*.

3. When normalized quadrat vectors are ordinated, a new ordination configuration (*S*) emerges. This configuration differs from the *G* configuration with a substantial reduction in curvature. This is a direct consequence of the normalization which formula (2.2) incorporates. Similar results could be expected from formulae (2.14), (2.15) or (2.39). The *S* configuration is highly suggestive of a unidirectional trend. It can in fact be a basis for

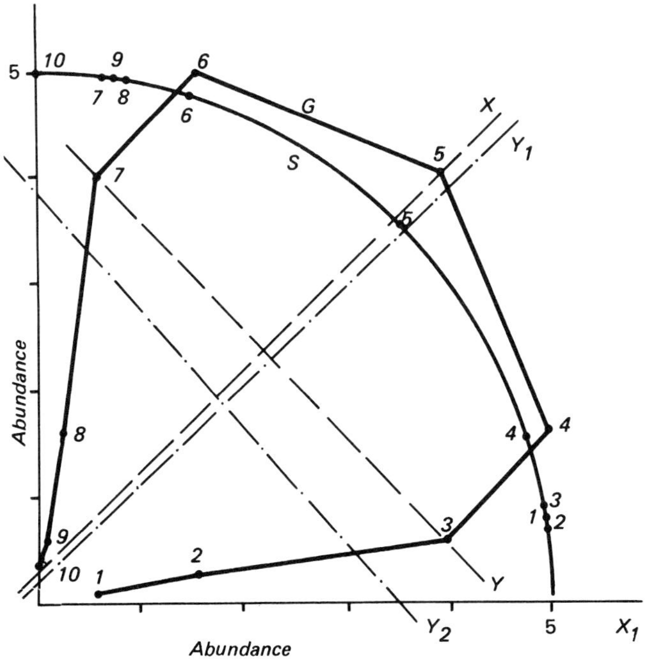

Fig. 3-9. Ordinations by different methods. Legend to symbols: Numbers *1-10* (inside figure) labels for quadrats in Table *2-6;* X_1, X_2 species axes; *X, Y* Bray & Curtis axes; Y_1,Y_2 component axes; *G* point configuration from raw data; *S* point configuration from normalized data. See explanations in the main text.

ordination if quadrat distance is measured on the basis of the geodesic metric (formula (3.15)). In such an ordination the distances from point 1 are co-ordinates on a curved axis (S),

Ordination axis	*Quadrat co-ordinate*									
	1	*2*	*3*	*4*	*5*	*6*	*7*	*8*	*9*	*10*
S	0.000	0.020	0.016	0.145	0.620	1.096	1.257	1.220	1.241	1.406 .

Since S is a segment of a circle in the example, the non-linear co-ordinates could be obtained by simple unfolding of S with the help of the geodesic metric. Even if we consider a more general case, we can similarly define the problem in non-linear ordinations as one of unfolding a curved data structure on one or several ordination axes. The methods that can accomplish this are few in number and rather inefficient. The methods proposed by Kruskal & Carmone (1971; Kruskal, 1964a,b), Sneath (1965), Shepard & Carroll (1966; Shepard, 1962), McDonald (1962, 1967), Sammon (1969) and Phillips (1978) appear to have broader potential.

We have seen already that in ordinations new co-ordinates are extracted on the basis of transformations either directly from a resemblance matrix or indirectly from the raw co-ordinate data. In the Kruskal (1964a,b) method co-ordinates are derived from either of these sources based on a different reasoning. Implicit in the method, similar to the ordinations considered so far, is the assumption that all points in the sample fall into a single cluster with even density. If there are two or more clusters, information will be lost in the course of the analysis since there is no device given which could signal the existence of distinct clusters. There are several versions of the Kruskal method (Kruskal & Carmone, 1971). They all extract metric co-ordinates from the data under the condition that the ordination distances can be derived by monotone transformation from the observed resemblance values. To perform one version one proceeds through the following steps:

1. Specify t the maximum number of ordination axes to be extracted. Each axis will be associated with one set of n co-ordinates for the individuals to be ordinated.
2. Impose on the n individuals a resemblance structure by estimating their pairwise dissimilarities (or similarities). Designate the dissimilarity of individuals j, k by symbol $\delta(j, k)$ and the matrix of all $n \times n$ dissimilarities by Δ.
3. Determine the starting ordination configuration distances. This configuration is obtained by placing n points into a t-dimensional space based on n sets of t arbitrary co-ordinates. Designate the distance of individuals j, k by the symbol $d(j, k)$ and the matrix of all $n \times n$ distances in the ordination configuration by symbol $\mathbf{D}$. At this point an iterative analysis begins.
4. Consider the joint distribution of $\delta(j, k)$ and $d(j, k)$ values as a scatter of n points in the plane. Fit to the scatter the regression line $\hat{d}(j, k) = f[\delta(j, k)]$. The regression performed is a least-squares regression of the ordination dis-

tances $d(j,k)$ on the observed resemblance values $\delta(j,k)$. The regression function to be used is chosen by the user. It may be linear or polynomial, normally not higher than degree 4.

5. After determining the regression coefficients, continue by computing a regression approximation for each $d(j,k)$ corresponding to the $\delta(j,k)$, and designate the matrix of all n x n regression approximations by $\hat{\mathbf{D}}$.

6. Compute a value for the stress (distortion) in $\hat{\mathbf{D}}$ relative to $\mathbf{D}$ based on the Kruskal & Carmone (1971) formula,

$$\sigma(\mathbf{D};\hat{\mathbf{D}}) = \left[\sum_j \sum_k \, [d(j,k) - \hat{d}(j,k)]^2 / \sum_j \sum_k \, [d(j,k) - \bar{d}]^2 \right]^{1/2},$$

$$k > j = 1, ..., n-1$$

where $\bar{d}$ is the mean of the $d(j, k)$ values.

7. After computing stress, all points in the ordination configuration are moved a bit, i.e. their co-ordinates are changed a little, to decrease stress.

8. The procedure starting with a new regression analysis is repeated until finally the value of stress drops below a given threshold limit. The output of the analysis includes t sets of n rectangular co-ordinates which can be plotted jointly in two dimensional scatter diagrams.

Whereas the Kruskal method (with its different variants) has been broadly used for data analysis in different disciplines, there are only few examples of ecological application (e.g. Anderson, 1971b; Austin, 1976a,b; also Noy-Meir & Whittaker, 1977). These, nevertheless, indicate that the method has definite potential in revealing non-linear trends. The Kruskal & Carmone (1971) variant may appeal to the ecologist for several reasons:

1. Extreme flexibility in handling resemblance functions of almost any kind.
2. Reliance on regression analysis the model of which is chosen by the user.
3. Derivation of an ordination configuration which is some pre-selected function of the data.
4. Potential in detecting non-linear trends when the regression function is appropriately chosen.
5. Possibility of using ordination co-ordinates as predictors of the level of environmental influence in the quadrats.

To reflect on the last point further, we shall follow Whittaker (1956) and van Groenewoud (1965) in assuming a unimodal species response along an environmental gradient. To make the assumption more specific, we shall visualize the response curve of species h to be the graph of the function

$$y_{hj} = B\, e^{-X_j^2/2}.$$

In this y_{hj} is the response of species h at a point j of the gradient at which the level of environmental influence is X_j. We assume that $B = 1$ and that the X_j are measured as departures from their common mean in standard units. Symbol e represents the natural base.

When an ordination is maximally predictive of a given gradient, or of given gradients, the co-ordinate X_{ij} on the ith ordination axis will be indicative of the level of environmental influence specific to gradient i in quadrat j. The expected average response at X_{ij} has a magnitude,

$$y_{ij} = e^{-(X_{ij}-a)^2/2} .$$

Since we do not know a, the value of X_{ij} at which the response is maximal, we cannot directly predict the magnitude of the response. We can nevertheless associate the co-ordinate difference $d_{ijk} = X_{ij} - X_{ik}$ with a difference in responses associated with X_{ij} and X_{ik} individually. We can use for this following Gauch (1973a), the integral,

$$d^2(j,k|i) = \int_{-\infty}^{\infty} [e^{-(X-a)^2/2} - e^{-(X+d_{ijk}-a)^2/2}]^2 dX$$

$$= 2(\pi)^{1/2} [1-e^{-d^2_{ijk}/4}] \cdot$$

We can attach meaning to $S_{ijk} = e^{-d^2_{ijk}/4}$ as a similarity measure whose value is unity when $d_{ijk} = 0$, less than unity when $d_{ijk} > 0$, but never smaller than zero. With S_{ijk} so defined, and the constant $\pi^{1/2}$ removed, we can redefine $d(j, k \mid i)$ as

$$d(j, k \mid i) = [2(1-S_{ijk})]^{1/2}$$

which is the chord distance (formula (2.2)) of quadrats j and k conditional on gradient i. Considering t perpendicular ordination axes, the quantity

$$d(j, k) = \frac{1}{t} [\sum_i d^2(j, k \mid i)]^{1/2}, \; i = 1, ..., t$$

measures the total ordination distance of quadrats j and k. Since the ordination distances, $d(j, k)$, are consistent with the Gaussian assumption for species response, the Kruskal method will find the ordination configuration that minimizes the stress function $\sigma(\mathbf{D}; \hat{\mathbf{D}})$, and thus supply quadrat co-ordinates $\mathbf{X}$ for the best fitting t-dimensional Gaussian ordination derivable for the sample. We retain the original definition of Euclidean distance for $\delta(j, k)$ which we compute as a chord distance (formula (2.2)) from the original observations, and define $\hat{d}(j, k) = \delta(j, k)$ when computing stress. We shall comment further on different methods of Gaussian ordination in the sequel.

Kruskal (1964a) gives an example which uses 15 points in the plane. The point configuration is defined by 15(15 − 1)/2 = 105 sample distances which are elements of Δ. From the known Δ, a Δ^* configuration is derived by a monotone distortion of the distances with an added random component. The results of the analysis on Δ^* indicate that the method can recover the Δ configuration with great accuracy. This then suggests that the Kruskal method has a potential to reveal effectively such structural features in the sample which do not readily appear because of random variation in the data. This suggestion has in fact been tested by Austin (1976b) with promising results in vegetation studies, and by Rohlf (1972) with similarly encouraging results in taxonomic applications.

The point cluster in Fig. 3-4 represents a trivial example where a non-linear trend can be readily recognized on first inspection. A curve could be fitted to the points with no great difficulty, the points could be projected on the curve, and the projections could be used as ordination co-ordinates. However, in more realistic situations the problem can get greatly complicated if the intrinsic dimensionality of the point cluster exceeds two or three. In such higher dimensions visual inspection will not help. We need specialized methods for representing non-linear point clusters of such higher complexity. One has been suggested by Sneath (1965). His method seeks out groups of points in a discontinuous sample space, finds the best fitting curve for each based on a gravitational process coupled with single link clustering, and uses the curves as ordination axes. The input data consists of rectangular co-ordinates. Sneath's method can detect branched sequences and rings of points, but not manifolds. The computational difficulty is considerable and the method is sensitive to the choice of parameter values (cf. Noy-Meir, 1974a,b; Dale, 1975).

When the objective of the analysis is to establish a curved co-ordinate system which does for non-linear data structures what component analysis or other linear ordinations do for linear data structures, Shepard & Carroll (1966; Shepard, 1962) suggest *continuity analysis.* To interpret continuity, in this case, we have to consider the relationship of quadrat co-ordinates produced by ordination, and the species scores in the input data. When small changes in the former correspond to small changes in the latter, the ordination is said to be continuous. When continuity is so defined its intensity can be measured.

There are two specific assumptions at the basis of the Shepard & Carroll (1966) method:

1. The data structure is continuous but the single cluster present may turn and twist through space in any possible manner.
2. It is possible to embed the point cluster in a Euclidean space.

Continuity analysis finds t sets of n rectangular co-ordinates for an s-dimensional ordination configuration such that $t \leqslant s \leqslant n - 1$. The constraints are different from those of other ordinations where individuals are placed closer together if they are more similar. In continuity analysis individuals are placed closer or further to one another in such a manner that their placement maximizes the appearance of continuity in the ordination. Shepard & Carroll suggest the function

$$\kappa = \sum_j \sum_k [\delta^{2\alpha}(j, k)/d^{2\beta}(j, k)] \,/\, [\sum_j \sum_k d^2(j, k)]^{\beta/\gamma},$$

$$j = 1, \ldots, n-1;\ k = j+1, \ldots, n$$

as an inverse measure of continuity to be minimized in an iterative procedure. The values of α, β, γ are so chosen that $\gamma = \alpha - \beta$. Values such as

$\alpha = 1$ and $\beta = 2$ were suggested in the original work. Noy-Meir (1974b) found that by emphasizing local continuity, i.e. giving larger values to β, the method is more successful in unfolding complex curvatures in a point configuration. He suggested $\alpha = 4$ and $\beta = 8$ as standards. In the definition of κ $\delta(j, k)$ is an element in Δ, representing a matrix of observed distances between individuals, and $d(j, k)$ is an element in **D** which contains the distances between the points that represent the individuals in ordination space. Index κ is a generalization of the Neumann index for continuity in higher dimensions. The steps in its derivation are discussed by Shepard & Carroll (1966).

In continuity analysis, just as in other methods of ordination, we can conveniently operate on the basis of the assumption that the observed s-dimensional point configuration with resemblance structure Δ can be replaced by a simpler t-dimensional point configuration with resemblance structure **D** without a substantial loss of information. The t-dimensional configuration is derived through an iterative procedure. The main steps in the analysis are outlined below on the assumption that the user has already decided what are the maximum dimensions t within which a simpler structure is sought:

1. Define an arbitrary t-dimensional configuration of n points in terms of t sets of n co-ordinates **X**.
2. Compute a Euclidean distance matrix **D** for the n points from the co-ordinates in **X**.
3. Compute κ.
4. Change the co-ordinates a little so that the value of κ is somewhat reduced.
5. Iterate steps 2 to 4 until the value of κ becomes stationary, or almost so.

The matrix **X** at the stationary stage holds the ordination co-ordinates on the basis of which scatter diagrams may be constructed. Shepard & Carroll (1966) give examples in their paper to illustrate their method. These examples suggest that the method can reveal a simple underlying structure in the data if such a structure actually exists. Since redundancy and non-linearity are a rule rather than exception in vegetation data, no wonder that the method of continuity analysis has generated considerable interest among ecologists. Experience suggests that continuity analysis is capable of recovering non-linear trends in both simulated and natural vegetation data (Noy-Meir, 1974a,b). The method is nevertheless sensitive to discontinuities and performs poorly without dominant trends in the data. The lack of dominant trends, or the appearance of discontinuity, need not of course be inherent in the sampled vegetation. It can be a consequence of the sampling technique which may not provide an even coverage of an existing gradient by sampling units (cf. Kruskal & Carroll, 1966). The choice of parameter values and the heavy computations are other potential sources for difficulties facing applications.

Sammon (1969) described another method of non-linear ordination in the spirit of continuity analysis but without the use of preselected control parameters. The quantity to measure the fit of the ordination configuration is given by

$$E = [1/ \sum_{j,k} \delta(j, k)] \sum_{j,k} [\delta(j,k) - d(j,k)]^2 / d(j,k), \; j<k = 2, ..., n$$

where $\delta(j, k)$ is a Euclidean (or other) measure of distance based on the $p \times n$ observed data **X**, and $d(j, k)$ is the same measure based on the $t \times n$ ordinations co-ordinates **Y**. The elements in **Y** are determined by iteration involving a steepest descent procedure to search for an ordination configuration in t-dimensional space that minimizes E within permissible limits of tolerance. The author claims high efficiency of the method in mapping complex structures.

McDonald (1962, 1967) offers his method of *polynomial ordination* to serve as an extension to component analysis. The method postulates a non-linear model. This is equivalent to postulating that the original variables that describe the species are linear or any non-linear functions of the ordination co-ordinates or their cross products. The method, offered for general use when the data structure is non-linear, is performed in steps:

1. Obtain sets of component scores from analysis of a covariance matrix.
2. Identify the polynomial function that best relates the non-linear structure in the ordination configuration to the components.
3. Rotate the components using an iterative procedure to positions where the orthonormal polynomial fit is maximized.

Since the original variables are non-linear functions of the ordination co-ordinates, McDonald's algorithm could have potential not only as a method of summarization when the data structure is non-linear, but also as a predictive ordination technique. But since the method requires predetermination of the functional form of non-linearity in the ordination configuration, its potential is considerably diminished. This limitation notwithstanding, a restrained optimism seems to prevail among users who considered its application in vegetation studies or elsewhere (e.g. McDonald, 1966; Johnson, 1973; Noy-Meir, 1974b, Dale, 1975). The Phillips (1978) variant of polynomial ordination is more promising, since it uses a curved axis and not just rotated components.

3.6 Predictive Ordinations

A. *Trend seeking*

We can recognize three broadly different approaches for trend seeking. Two of these seek a *direct* representation of trends in species or vegetational variation as a function of certain environmental variables, while the third attempts to predict underlying environmental gradients *indirectly* from trends recognized in the vegetation:

1. *Fitting lines or surfaces by regression analysis.* This may be illustrated by an example in which trends in species response are mapped onto planes in which certain relationships with environmental variables can be clearly observed. Such a map is given in Fig. 3-10, tracing changes in the abundance of the fern *Blechnum spicant* as a function of elevation above sea level and soil moisture regime. There are many relevant sources for examples and

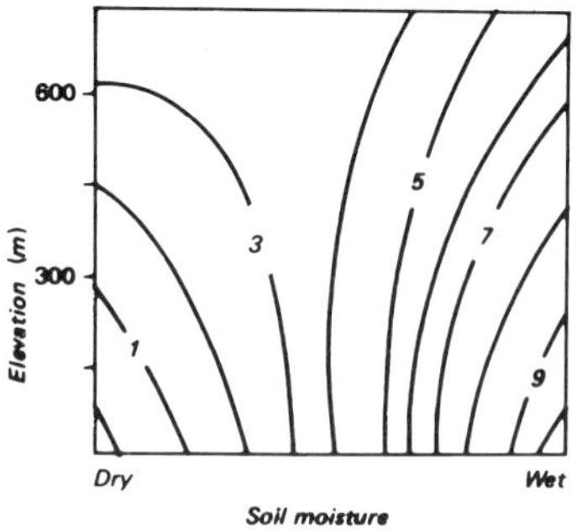

Fig. 3-10. Shown in this diagram are abundance contours for *Blechnum spicant* as a function of elevation and soil-moisture regime. The data originate from sampling in the Coastal Western Hemlock Zone of British Columbia. The contour lines in the diagram represent traces on a quadratic surface whose polynomial expression incorporates an interaction term.

reviews (e.g. Greig-Smith, 1964; Gittins, 1968; Gabriel & Sokal, 1969; Goodall, 1969b; Yarranton, 1969; James, 1970; Austin, 1971, 1972; Forsythe & Loucks, 1972; James & Shugart, 1974; Stout, Deschenes & Ohmann, 1975). Clearly, one cannot question the usefulness of regression analysis. One may even be tempted to suggest that it deserves more attention than it received in vegetation studies (Austin, 1972). It would not however be wise, as Greig-Smith (1971a) points out, if ecologists followed Yarranton's (1967c, 1970) advice and abandoned ordinations in favour of regression analysis as their tool to reveal correlations with environmental variables.

2. *Ordering vegetation stands according to certain environmental criteria.* Whittaker's (1956) direct gradient analysis is one example for such an analysis. In this, the distribution on different vegetation types are mapped within a reference system of moisture regime and elevation. From such maps one may infer about correlations of trends in vegetation composition with environmental changes. Noy-Meir & Whittaker (1977) review other similar approaches to ordering vegetation stands.

3. *Predicting environmental gradients.* In this case the goal is to ordinate sampling units based on vegetational data in such a manner that their order will be maximally predictive of their actual position on a given major environmental gradient. Approaches to such a predictive use of ordination have been described (e.g. Cottam, 1949; Curtis & McIntosh, 1950, 1951; McIntosh, 1957, 1972; Maycock, 1963; Kershaw, 1973) with a common objective to place sampling units, grouped according to leading dominants, on a single axis in such a manner that their order produces unimodal performance curves, at least for the dominant species. Since species performance is made to be a nonlinear function of sampling unit position (co-ordinate) on the ordering axis, it is hoped that the ordination can be predictive of changing levels of environmental influence along an existing environmental gradient. The advantages and limitations of the method are discussed by Cottam, Goff & Whittaker (1973) and Noy-Meir (1974b; see also Gimingham, Pritchard & Cormack, 1966; Buell, Langford, Davidson &

Ohmann, 1966). Précsényi & Szőcs (1969) suggested revisions.

As an interesting attempt to improve the predictive potential of ordinations, we should mention axis rotation (e.g. Dagnelie, 1960; Gittins, 1965). This is done to satisfy predetermined constraints in the hope that in the new position the axes will clearly indicate environmental trends that themselves are representable by the linear axes of the reference system used.

Other possibilities include the use of such other methods which Whittaker (1967) describes as *indirect gradient analysis.* These can be rationalized in the context of a four-stage procedure:

a. Carry out experiments with different resemblance functions and methods of axis construction on the basis of simulated data with a built-in response to a major environmental gradient.

b. Identify the resemblance function and method of axis construction which perform best under the simulated conditions.

c. Use the indicated function and method to analyse an actual vegetation sample.

d. Verify the ecological significance of the ordination by correlation with given environmental variables.

Such a procedure characterized many recent experiments which addressed the problem of finding efficient methods for ordination in vegetation samples of extreme compositional variation. Whereas such variation is likely to signal the existence of at least one underlying environmental gradient in the survey site, it induces involution (horseshoe-shaped or other more complex curvature) in the cluster of sample points if placed in a linear, species-defined reference system. We have already seen that such a non-linear cluster, even if its curvature is unidirectional, cannot be represented by a single linear axis without some degree of distortion. The implications of this are clear: the quadrat (stand) co-ordinates in linear ordinations, such as the component scores, are potentially poor predictors of environmental influence on the vegetation in which the response of species is non-linear.

Designing an efficient ordination in the presence of non-linearity has been explicitly recognized as a central problem in ordination methodology by van Groenewoud (1965), and more recently by, e.g., Swan (1970), Jeglum, Wherhahn & Swan (1971), Noy-Meir & Austin (1970), Austin & Noy-Meir (1971), Austin (1972), Gauch & Whittaker (1972), Gauch (1973a,b), Bachacoun (1973, 1974) and Phillips (1978). Whereas van Groenewoud (1965) foresaw a solution in restricting environmental variation in the sample, Swan (1970) attempted a solution by transformation of zero values and Feoli Chiapella & Feoli (1977) by transformation of the similarity coefficient itself. The logic in restricting environmental variation in the sample is that in this way species response is linearized and ordinability improves.

Determination of the amplitude of environmental variation, at which species response is linear, is problematic (cf. Noy-Meir & Whittaker, 1977) and it may require limiting the sample to such a narrow range of environmental variation that the trace of all major trends will be obliterated in the data. Swan (1970) suggests to improve linear ordinability through replacement of zeros in the data by non-zero values which express degrees of absence. The method of Feoli Chiapella & Feoli (1977) attempts linearization by applying logarithmic transformations on the similarity values for quadrat pairs followed by redescription of quadrats based on new vectors of similarity values. Another obvious possibility is to use variables whose variation is linear, or at least monotone, such as, for instance, the changing stand composition of higher level taxa (e.g. van der Maarel, 1972) along an environmental gradient. It is a well known fact that the transformations implicit in certain resemblance functions can be very helpful to reduce complexity in variation. Austin & Noy-Meir (1971) and Beals (1973), among others, point out possibilities in this respect. We may draw attention to the geodesic metric (formula (2.15)) and the chord distance (formula (2.2)) as two of the functions which will reduce a complex non-linearity in the data structure to a simpler, spherical non-linearity. The relativized absolute value function is especially noted for its linearizing transformation which Gauch & Whittaker (1972) and Gauch (1973b) have used to good advantage in ordination experiments.

In what follows we shall look at methods which assume that non-linearity in the data structure is a consequence of Gaussian species response to environmental changes. The objective of these methods amounts to determining quadrat co-ordinates on a single ordination axis. The co-ordinates must qualify as predictors of the levels of environmental influence in the quadrats. Clearly, differences in co-ordinate values on such an ordination axis should not reflect a mere compositional difference between quadrats, but more than that, a compositional difference as modified by the non-linearity assumption. Gauch (1973a) offers comments on this topic and suggests functional formulations.

In describing the methods, we define y_{ij} as the response of species i in quadrat j to a real but unidentified environmental influence whose level in quadrat j is X_j. We assume further that species response y_i can be described by the function $f(X_j \mid k_i)$ where k_i is a single parameter or a parameter set. Once the functional form of response is assumed, suggestions by Gauch, Chase & Whittaker (1974) can be followed in formulating an iterative algorithm to determine numerical values for $X_1, ..., X_n$ that maximize the fit of $f(X_j \mid k_i)$ to the observed $y_{ij}, i = 1, ..., p; j = 1, ..., n$. The procedure could go like this:

1. Define initial estimates for $X_1, ..., X_n$.
2. Fit $f(X_j \mid k_i)$ to the observed $y_{i1}, ..., y_{in}$ for all i, based on the initial estimates of X_j.
3. Change the X_j values a little to improve the fit; iterate through steps 2, 3 and stop when a desired degree of precision is achieved.

An ordination of this sort is predictive as it is also indirect (Whittaker 1967), but its relevance depends on the accuracy with which the functional form $f(X_j \mid k_i)$ of the response can be guessed. The graph of $f(X_j \mid k_i)$ is often conceived as a rather rounded and continuous curve (e.g. Curtis & McIntosh, 1951; McIntosh, 1967b), more or less bell-shaped (Whittaker 1956, 1967; van Groenewoud, 1965), but often truncated. It has been described as some Gaussian curve with co-ordinate function

$$\hat{y}_{ij} = f(X_j \mid \bar{X}_i, S_i, B_i) = B_i \, e^{-0.5\, t^2} \qquad (3.9)$$

where $\hat{y}_{ij}$ is the fitted species response, X_j the estimated level of environmental influence in quadrat j, $\bar{X}_i$ the level of influence where species i achieves its maximum response, S_i a measure of width of the response curve for species i, B_i a scale factor and e the natural base. The definition of t^2 accords with

$$t = (X_j - \bar{X}_i) / S_i \, .$$

Although the assumption of a Gaussian response is freely made by many users, and it underlies the methods to be described, its validity has been questioned (e.g. Austin 1976a).

If the iterative procedure is followed as outlined, initial values would have to be derived for $X_j, j = 1, ..., n$. Whereas Johnson (1973) obtains these values via a linear ordination, Gauch, Chase & Whittaker (1974) define them as visual or measured estimates of environmental influence presumed to be associated with an existing gradient in the sample, such as changing elevation, moisture regime, etc. To alleviate the problem of heavy computations, a priori sorting of species is desired. Johnson (1973) discards species with three or fewer occurrences. Ihm & van Groenewoud (1975) eliminate those species which occur at environmental extremes. Alternative schemes could be readily conceived (e.g. Orlóci, 1973b, 1976a; Orlóci & Mukkattu, 1973; Gauch, Chase & Whittaker, 1974) to achieve the same objective. However, once the species are selected and the initial values are chosen for $X_j, j = 1, ..., n$, the Gaussian curve (formula (3.9)) can be fitted to the data based on different methods. Gauch & Chase (1974) use a modification of von Hoerner's (1967) technique in which they give an initial value to B_i as the maximum response $[\max(y_{ij}, j = 1, ..., n)]$ and to $\bar{X}_i$ as the estimated level of environmental influence at B_i. The initial value for S_i is chosen to be the weighted average of n solutions,

$$S_{ij} = 0.5 \, [-(X_{ij} - \bar{X}_i)^2 / \ln (y_{ij}/B_i)]^{1/2} , j = 1, ..., n \, .$$

Following the choice of the initial values, $\bar{X}_i$ and B_i are subjected to an iterative analysis in which they are allowed to vary in such a way that the fit of the $\hat{y}_{ij}$ (formula (3.9)) to the observed y_{ij} is improved in steps. The iteration stops if no substantial further improvement occurs. The computations are then performed on the next, and on each of the remaining species. The quantity

$$Q = \sum_{i,j} (y_{ij} - \hat{y}_{ij})^2 , i = 1, ..., p; j = 1, ..., n \qquad (3.10)$$

measures the overall fit of the model to the data. If Q falls below a predetermined threshold value, the computations stop. The $X_1, ..., X_n$ values at this point in the procedure are ordination co-ordinates; these estimate the levels of environmental influence in the quadrats. If Q is considered to be too large, the X_j values are changed and a new surge of iterations begins. The method has been applied by its inventors to vegetation data with some success.

Johnson (1973) describes a similar method which incorporates ordination to determine first approximations for $X_1, ..., X_n$, and regression analysis to estimate the coefficients in

$$\ln \hat{y}_{ij} = a_i X_j^2 + b_i X_j + c_j$$

and to determine the $\hat{y}_{ij}$ values. Alternatively, estimates of the parameters $\bar{X}_i$, S_i and B_i are obtained from

$$X_i = 0.5\, S_i^2 b_i$$

$$S_i = 1/(-a_i)^{1/2}$$

$$\ln B_i = \bar{X}_i / S_i^2 + c_i$$

and the fitted $\hat{y}_{ij}$ values from formula (3.9) corresponding to current values of $X_1, ..., X_n$. The computations are performed recursively for all species.

Johnson (1973) assumes that the distribution of the y_{ij} is normal about the $\hat{y}_{ij}$, with estimated variance

$$Q_i = \sum_j (y_{ij} - \hat{y}_{ij})^2 / (n-1), j = 1, ..., n$$

and then he proceeds to define the likelihood of y_{ij} at X_j as

$$L_{ij} = \frac{1}{(2\pi Q_i)^{1/2}} e^{-0.5(y_{ij} - \hat{y}_{ij})^2 / Q_i} .$$

L_{ij} is at a maximum $(1/(2\pi Q_i)^{1/2})$ when $y_{ij} = \hat{y}_{ij}$. The joint likelihood of the p species $(y_{1j}, ..., y_{pj})$ at X_j is

$$L_{.j} = \Pi\, L_{ij}, \quad i = 1, ..., p \tag{3.11}$$

The analysis selects X_j in such a way that $L_{.j}$ is maximized. The approach to accomplishing this employs an iterative procedure. This involves selection of a new set of values for $X_1, ..., X_n$ and fitting a new set $(i = 1, ..., p)$ of Gaussian curves in each step of iteration. When the $L_{.j}$ can no longer be improved the iterations stop. The essential difference in the methods is that while Gauch minimizes Q of formula (3.10), Johnson maximizes $L_{.j}$ of formula (3.11).

The method of Ihm & van Groenewoud (1975) gives values to $X_1, ..., X_n$ as eigensolutions. The computations are thus relatively less tedious but certain restrictive assumptions are required. These include:

1. Species response is Gaussian.
2. The species variances are all equal to unity.
3. The points $(\bar{X}_1, ..., \bar{X}_p)$ of maximum species response are independently and uniformly distributed in the observed region of the gradient. The method's objective is the estimation of levels of environmental influence $(X_1, ..., X_n)$ in different quadrats. A single predominant gradient is assumed, but extensions to two or more gradients are possible (Ihm & van Groenewoud,

1975). The method includes the transformation

$$\alpha_{hi} = 2 \log \int_{-\infty}^{\infty} f(X|\bar{X}_h,S_h,B_h) f(X|\bar{X}_i,S_i,B_i) dX \tag{3.12}$$

which when $S_1 = ... = S_p = 1$ and $B_i = 1/(2\pi S_i^2)^{1/2}$ reduces to

$$\alpha_{hi} = -\log \pi - 0.5\,(\bar{X}_h - \bar{X}_i)^2 \; .$$

This establishes that the cross product α_{hi} is a linear function of the squared difference of two levels of environmental influence $\bar{X}_h$, at which species h exhibits its maximum response, and $\bar{X}_i$ at which species i has its maximum response.

The problem can be reversed as Ihm & van Groenewoud (1975) showed by reversing the role of X_j and $\bar{X}_i$ so that $\bar{X}_i$ is allowed to vary while X is held constant at X_j. Formula (3.12) can then be rewritten for any two quadrats,

$$\alpha_{jk} = 2 \log \int_{-\infty}^{\infty} f(\bar{X}\,|\,X_j,\, 1,1/(2\pi)^{1/2})\, f(\bar{X}\,|\,X_k,\, 1,1/(2\pi)^{1/2})\, d\bar{X} \; .$$

The quantity,

$$a_{jk} = 2 \log \sum_h y_{hj} y_{hk}, \quad h = 1, ..., p$$

is a sufficiently accurate numerical approximation to the cross product α_{jk} (when assumption 3 holds true). It can be shown that the $X_1, ..., X_n$ values sought are the solutions in the dominant eigenvector of **A** with a characteristic element

$$A_{jk} = a_{jk} - \bar{a}_j - \bar{a}_k + \bar{a} \, ,$$

where $\bar{a}_j$, $\bar{a}_k$ and $\bar{a}$ represent respectively the mean of the jth row (column), kth row (column), and all elements of **A**. In theory, **A** has a single non-zero eigenvalue and associated eigenvector. In sample calculations, however, several non-zero eigenvalues may be expected, of which the largest is chosen. If this is λ, its total share from the trace of **A** is

$$\lambda / \sum_i |\lambda_i| \, , \quad i = 1, ..., t$$

where t is the number of non-zero λ_i (considering also the $\lambda < 0$). We shall use the data in Table 2-6 for illustration. As theory suggests, **A** has a single non-zero eigenvalue $\lambda = 10.2103$. The elements in the associated eigenvector give estimates for the quadrat co-ordinates, are given in the table on page 152.
The rank order of quadrats roughly matches the unidirectional sequence in Fig. 3-5 except for the inversions at the extremes.

In the Phillips (1978) method, $X_1, ..., X_n$ are co-ordinates measured from a common zero origin on a curved axis fitted in least squares regression. This

	Quadrat j									
	1	*2*	*3*	*4*	*5*	*6*	*7*	*8*	*9*	*10*
X_j	0.39	0.37	0.41	0.27	0.03	−0.18	−0.29	−0.26	−0.28	−0.45
Rank order	2	3	1	4	5	6	9	7	8	10

method is more direct than McDonald's which does not use curved axes, but rotated components which maximize the polynomial fit.

The foregoing algorithms are appealing because they produce non-linear predictive ordinations. They are nevertheless limited in potential, as would be any method of ordination so much constrained as these by restrictive assumptions. Unidimensionality is of course another problem, but this is not, at least in theory, an insurmountable difficulty (cr. Johnson, 1973; Ihm & van Groenewoud, 1975). Other methods of gradient analysis focus on the discontinuity property of data structures. The method proposed by van der Maarel, Bridgewater & Fresco (Fresco, 1972) is an example in which similarity and distance measures are used to describe changes in the vegetation in an attempt to recognize boundaries that subdivide a gradient into segments. Some of the predictive methods rely on what McIntosh (1973) described as *plexus diagrams.* In such diagrams, points (species, quadrats, soil profiles, etc.) are linked into a net based on the level of similarity. The species constellations of De Vries, Baretta & Hamming (1954), McIntosh (1957), Agnew (1961) and Looman (1963), the plexus representation by Curtis (1959) for vegetation types and by Hole & Hironaka (1960) for soil types represent examples. When only links that indicate similarities in excess of a threshold limit are shown (e.g.; Hopkins, 1957, Agnew, 1961), the point configuration takes on a distinctly clustered appearance inviting attempts for group recognition. Other examples for plexus diagrams include Prim's (1957) minimum spanning trees, the *k*-link nets of Jardine & Sibson (1971) and the Wagner trees of Farris, Kluge & Eckardt (1970). In a minimum spanning tree, derivable on the basis of single link clustering (cf. Gower & Ross, 1969; Jardine & Sibson, 1971; and Chapter 4), all points are connected; the sum of distances of neighbour points is a possible minimum and there is only one pathway from any given point to any other. The *k*-link net is different in that it allows multiple pathways from point to point. Wagner trees are constructed based on graph theoretical considerations (cf. Farris, Kluge & Eckardt, 1970; also Estabrook, 1966).

B. Reciprocal ordering

Whittaker (1948; see review by Whittaker, 1967) describes a method in which he uses species scores,

$$\mathbf{X} = (X_1 \dots X_p),$$

indicating optima along an environmental gradient, as weights to order stands in unidimensional sequences. His ordering criterion is given by the formula

$$Y_j = \sum_i A_{ij} X_i / A_{.j}, \quad i = 1, ..., p \tag{3.13}$$

in which Y_j represents a stand score, i identifies a species, X_i indicates the ith of p species scores, A_{ij} measures the abundance of species i in stand j, and $A_{.j}$ is the sum of the abundance values for all species in stand j. The execution of the method begins with a set of species scores $X_1, X_2, ..., X_p$ and a set of abundance values (or other measures of species response),

$$\mathbf{A} = \begin{bmatrix} A_{11} & A_{12} & \cdots & A_{1s} \\ A_{21} & A_{22} & \cdots & A_{2s} \\ . & . & \cdots & . \\ A_{p1} & A_{p2} & \cdots & A_{ps} \end{bmatrix}.$$

It is terminated after the first set of stand scores **Y** is determined. In practical applications **A** may represent numerical data of any sort. To illustrate the computations we shall consider the species and quadrats in Table 1-1. Let the vector

$$\mathbf{X} = (X_1\ X_2\ X_3\ X_4\ X_5) = (5\ \ 4\ \ 3\ \ 2\ \ 1)$$

specify species scores on a soil moisture gradient with 5 indicating wet soil and 1 dry soil. From these we calculate the stand scores according to formula (3.13). We have for quadrat *1*

$$Y_1 = ((45)(5) + ... + (9)(1))/85 = 3.94\ ,$$

and for all the quadrats

$$\mathbf{Y} = (3.94\ 2.71\ 1.98\ 3.30\ 2.77\ 2.62\ 2.60\ 4.40\ 2.85\ 2.89)\,.$$

The scores in **Y** define a unique order for the quadrats along the assumed moisture gradient.

While such an ordering of quadrats is simple to accomplish, the interpretation of the results is not without problems. The question of reliability of the species scores as ordering criteria arises. By reliability we mean the consistency of species scores and the data. We can expect that the use of oversized quadrats that lump together different habitats, or the presence of species with broad response curves but with low indicator value, will tend to undermine reliability. Fortunately, we need no information about quadrat homogeneity or species response curves to measure reliability. We need only the species scores **X** and abundance values **A**.

It has been shown by Williams (1952; see also Anderberg, 1973) that the quantity R, given below, is in fact a measure of the reliability of the species scores as ordering criteria:

$$R = \left[\sum_h \sum_i A_{hj} (X_h - \bar{X}) (Y_j - \bar{Y}) / A_{..} \right] / (S_X^2 S_Y^2)^{1/2}$$

$$= \left[\sum_j \left[\sum_h A_{hj} (A_h - \bar{X}) / S_X \right]^2 / (A_{.j} A_{..}) \right]^{1/2}$$

$$= \left[\sum_h \left[\sum_h A_{hj} X_h \right]^2 / (A_{.j} A_{..}) - \bar{X}^2 \right]^{1/2} / S_X ,$$

$$h = 1, ..., p; j = 1, ..., s \qquad (3.14)$$

In this symbols A_{hj}, X_h, Y_j, $A_{.j}$, p and s are defined as before. R is a canonical correlation, $A_{h.}$ the hth species (row) total in **A** and $A_{..}$ the grand total in **A**. The mean species score is

$$\bar{X} = \sum_h A_{h.} X_h / A_{..}$$

and the sum of squares,

$$S_X^2 = \sum_h A_{h.} (X_h - \bar{X})^2 / A_{..} \; .$$

The mean of the quadrat scores is

$$\bar{Y} = \sum_j A_{.j} Y_j / A_{..} \; ,$$

and the variance

$$S_Y^2 = \sum_j A_{.j} (Y_j - \bar{Y})^2 / A_{..} \; .$$

If we replace X_h and Y_h by the standard forms

$$x_h = (X_h - \bar{X}) / S_X$$

$$y_j = (Y_j - \bar{Y}) / S_Y$$

formula (3.14) can be substantially simplified. If R is large, the reliability of the scores is high, if R is low the reliability is also low. Some formulations for quadrat scores do, in fact, incorporate the canonical correlation in the averaging formula (Williams, 1952)

$$y_j = \sum_h A_{hj} x_h / A_{.j} R$$

which, as can be seen from

$$Y_j = y_j R\, S_X + \bar{X} \ ,$$

is related to formula (3.13). Since R is invariant under linear transformations, either y_j or Y_j can be used with complete justification. Due to computational simplicity, Y_j is in fact preferred if standardization is not required.

To illustrate the calculations we turn to the data in Table 1-1. We have the species scores $\mathbf{X} = (5\ \ 4\ \ 3\ \ 2\ \ 1)$ and the species totals,

$$\begin{bmatrix} A_{1.} \\ A_{2.} \\ A_{3.} \\ A_{4.} \\ A_{5.} \end{bmatrix} = \begin{bmatrix} 229 \\ 621 \\ 269 \\ 174 \\ 542 \end{bmatrix}.$$

The quadrat totals are given in $(A_{.1}\ A_{.2}\ A_{.3}\ A_{.4}\ A_{.5}\ A_{.6}\ A_{.7}\ A_{.8}\ A_{.9}\ A_{.10}) = (85\ \ 248\ \ 156\ \ 181\ \ 225\ \ 195\ \ 210\ \ 128\ \ 222\ \ 185)$. The observed abundance values are given in $\mathbf{A}$. We calculate the mean and variance of X,

$$\begin{aligned} \bar{X} &= ((229)(5) + \ldots + (542)(1))/1835 \\ &= 2.90245 \\ S_X^2 &= ((229)(5-2.90245)^2 + \ldots + (542)(1-2.90245)^2)/1835 \\ &= (1.45065)^2 \ , \end{aligned}$$

where 1835 is the grand total $A_{..}$. The standardized score for the first species is $x_1 = (5 - 2.90245)/1.45065 = 1.44594$. Similar calculations give $x_2 = 0.756592$, $x_3 = 0.0672457$, $x_4 = -0.622100$ and $x_5 = -1.31145$. The correlation coefficient is simplest to calculate based on the middle term in formula (3.14). On substituting x_h for $(X_h - \bar{X})/S_X$, we have

$$\begin{aligned} R^2 &= (((45)(1.44594) + \ldots + (9\ )(-1.31145))^2/85 + \ldots \\ &\quad + ((16)\,1.44594) + \ldots + (62)(-1.31145))^2/185)/1835 \\ &= 0.152235 \end{aligned}$$

and $R = 0.390173$.

Considering that R is a sample value, any interpretation of it should take into account the sampling error by which it is likely to be burdened. Williams (1952) asserts that under random sampling in a p-species population with a common expectation for all species scores, i.e. with a population value $E(R)$ of R equal to zero, the sampling distribution of $A_{..}R^2$ is a chi square distribution with $s - 1$ degrees of freedom. Based on the use of this relationship to chi square, we can test the null hypothesis,

H_0: $E(R) = 0$, i.e. the X scores impose a chance order on the quadrats,

in contrast with the alternative hypothesis,

$$H_1 : E(R) > 0 .$$

We have for the example $A_{..}R^2 = (0.152235)(1835) = 279.351$. Since this is a very large value, certainly far in excess of any reasonably chosen probability point of the chi square distribution with 9 degrees of freedom, we reject H_0 and accept H_1. With this decision we commit ourselves to the idea that the Y scores do, in fact, reflect an arrangement of quadrats that is capable of interpretation in terms of soil moisture regime. We must however observe that the order established is rather weak, indicated by the low nominal value of R^2. Unlike in other situations, it would be quite inappropriate here to compare the R value to the theoretical maximum ($R = 1$). The reason for this is that the data may be such that the maximum correlation can never be attained no matter what species scores are used. Fortunately, we can determine the expected local maximum value from the data (matrix **A**) and use this maximum as a reference point to evaluate R.

For exposition of the basics of the method we refer to work by Williams (1952). Determining the maximum correlation is part of a broader analysis which we shall outline. The method, to which we shall refer as *reciprocal ordering*, is not new to ecologists. Hill (1974) gives an historical account and shows connections to other methods (Fisher, 1940). Anderberg (1973) fills in missing steps in derivations and Kendall & Stuart (1973) reflect on underlying statistical properties while also pointing out a weakness in the recommended χ^2 tests. The application to different types of data has been considered by, e.g., Guttman (1959), Benzécri (1969), Escofier-Cordier (1969), Hatheway (1971), Lacoste & Roux (1971), Romane (1972), Hill (1973b), Guinochet (1973), Pemadasa, Greig-Smith & Lovel (1974), Bottliková, Daget, Drdoš, Guillerm, Romane & Ružičková (1976), Bouxin (1976) and Lacoste (1976). A related approach has been pursued with classificatory objectives by Dale & Anderson (1973) to be discussed in Chapter 4.

The name 'reciprocal ordering' signifies that the method achieves a dual purpose, namely an ordering of quadrats based on species scores and an ordering of species based on quadrat scores. Since at least one algorithm (Hill, 1973b) interprets the algebra in averaging terms, the name 'reciprocal averaging' has also been used to describe the method. Hatheway (1971) uses the RQ designation while others refer to it as the 'Analyse Factorielle des Correspondances' or simply just 'correspondence analysis'. We describe the algorithm under the program name RQT in the Appendix.

The generalized method of reciprocal ordering is an outgrowth of work which attempted to supply yet another test of association for contingency tables. From our point of view, however, the relevance of the method lies in its ability to produce sets of metric scores, known as canonical scores,

$$\mathbf{X}_m = (X_{1m} \dots X_{pm}), \quad m = 1, \dots, t$$

for p species, and

$$\mathbf{Y}_m = (Y_{m1} \ldots Y_{ms}), \qquad m = 1, \ldots, t$$

for s quadrats where $t \leqslant \min(p, s)$. Symbol X_{hm} represents the score of species h on the mth canonical variate of species; Y_{mj} signifies the score of quadrat j on the mth canonical variate of quadrats. As we present the method, the canonical variates can be regarded as axes and the scores as co-ordinates on the basis of which scatter diagrams can be constructed or further analyses performed to serve the customary objectives of ordinations.

The classical expositions of reciprocal ordering assume that the data are categorical, presented in summarized form in a contingency table. In such a table the observations are assigned to cells based on two sets of classificatory criteria. Table 1-1 is an example. In this, one set of criteria is the species (row) affiliation of individuals and the other set is the quadrat (column) of occurrence. When we examine such a table, a logical question may be addressed to the strength of association between species and quadrats. This could be measured as a chi square,

$$\chi^2 = \sum_h \sum_j A_{hj}^2 A_{..}/(A_{i.}\, A_{.j}) - 1$$

$$= A_{..}\, R_1^2 + \ldots + A_{..}\, R_t^2 \;.$$

In this, the individual R_m^2 values are canonical correlations defined according to

$$R_m = \frac{\sum_h \sum_j A_{hj}\,(X_{hm} - \bar{X}_m)\,(Y_{mj} - \bar{Y}_m)}{[\sum_h A_{h.}\,(X_{hm} - \bar{X}_m)^2 \sum_j A_{.j}\,(Y_{mj} - \bar{Y}_m)^2\,]^{1/2}}\,,$$

$$h = 1, \ldots, p; j = 1, \ldots, s; m = 1, \ldots, t \qquad (3.15)$$

which may be used individually as measures of association.

We could in principle use any arbitrary scheme to derive the species (**X**) or quadrat (**Y**) scores. The objective of reciprocal ordering, however, requires **X** and **Y** to be such that they maximize χ^2, or equivalently, they maximize the canonical correlations $R_1, \ldots, R_t$. What are the intrinsic properties that characterize reciprocal ordering as an ordination technique? –

1. $\mathbf{X}_m$ and $\mathbf{Y}_m$, $m = 1, \ldots, t$, can be derived from matrix **A** without a need for information from external sources.
2. It is sufficient to determine $\mathbf{X}_m$, and R_m can be calculated according to formula (3.14) which we rewrite here,

$$R_m^2 = \sum_j \left[\sum_h A_{hj}(X_{hm} - \bar{X}_m)/S_m\right]^2/(A_{.j}A_{..})$$

$$= \left[\sum_j(\sum_h A_{hj}X_{hm})^2/(A_{.j}A_{..}) - X_m\right]/S_m^2$$

$$j = 1, ..., s; h = 1, ..., p \qquad (3.16)$$

In these, $\bar{X}_m = \sum_h A_{h.}X_{hm}/A_{..}$ and $S_m^2 = \sum_h A_{h.}(X_{hm} - \bar{X}_m)^2/A_{..}$. The same R_m^2 could be formulated in terms of $\mathbf{Y}_m$. If the $\mathbf{X}_m$ are standardized to zero mean and unit variance, formula (3.16) reduces to

$$R_m^2 = \sum_j \left[\sum_h A_{hj}X_{hm}\right]^2/(A_{.j}A_{..}), j = 1, ..., s; h = 1, ..., p .$$

3. After $\mathbf{X}_m$ is extracted, the corresponding scores in $\mathbf{Y}_m$ are defined by

$$Y_{mj} = \sum_h A_{hj}X_{hm}/(A_{.j}R_m), h = 1, ..., p \qquad (3.17)$$

When $\mathbf{Y}_m$ is extracted first, the scores in $\mathbf{X}_m$ are given by

$$X_{hm} = \sum_j A_{hj}Y_{mj}/(A_{h.}R_m), \; j = 1, ..., s \qquad (3.18)$$

Alternative formulations could be used as we already indicated.

We can rewrite formula (3.17) and (3.18) using matrix symbols,

$$\mathbf{Y} = \mathbf{G}^{-1}\mathbf{A}'\mathbf{X}\mathbf{R}^{-1}$$

$$\mathbf{X} = \mathbf{T}^{-1}\mathbf{A}\mathbf{Y}\mathbf{R}^{-1}$$

and view these as alternating steps in an iterative procedure. **A**, **X** and **Y** are respectively $p \times s$, $p \times t$ and $s \times t$ matrices of abundance data, species scores and quadrat scores. **G** is a diagonal matrix of s quadrat totals and **T** a diagonal matrix of p species totals. $\mathbf{G}^{-1}$ and $\mathbf{T}^{-1}$ are inverses of **G** and **T**. $\mathbf{A}'$ is a transpose of **A**. **R** is a diagonal matrix of t canonical correlations and $\mathbf{R}^{-1}$ is its inverse.

4. With the Hill (1973b) averaging algorithm in mind, we may write for the stable **Y** and **X** scores,

$$\mathbf{X} = \mathbf{T}^{-1}\mathbf{A}(\mathbf{G}^{-1}\mathbf{A}'\mathbf{X})\,\mathbf{R}^{-2} .$$

After substitutions $\mathbf{G}^{-1} = \mathbf{G}^{-1/2}\,\mathbf{G}^{-1/2}$ and $\mathbf{T}^{-1} = \mathbf{T}^{-1/2}\,\mathbf{T}^{-1/2}$, we have

$$\mathbf{X} = \mathbf{T}^{-1/2}\mathbf{T}^{-1/2}\mathbf{A}\mathbf{G}^{-1/2}\mathbf{G}^{-1/2}\mathbf{A}'\mathbf{X}\mathbf{R}^{-2}$$

and

$$(\mathbf{T}^{1/2}\mathbf{X})\mathbf{R}^2 = (\mathbf{T}^{-1/2}\mathbf{A}\mathbf{G}^{-1/2})\,(\mathbf{G}^{-1/2}\,\mathbf{A}'\mathbf{T}^{-1/2})\,(\mathbf{T}^{1/2}\mathbf{X})$$

or

$$\alpha \lambda = \mathbf{DD}'\alpha \qquad (3.19)$$

where $\alpha = \mathbf{T}^{1/2}\mathbf{X}$ and $\lambda = \mathbf{R}^2$. We recognize this as the characteristic equation of the cross product matrix $\mathbf{DD}'$. An element in $\mathbf{DD}'$ is the cross product of two species vectors of $\mathbf{A}$ after the hj element A_{hj} in $\mathbf{A}$ is divided by the product of the square root of the hth species total and the square root of the jth quadrat total. The eigenvalues of $\mathbf{DD}'$ are the squared canonical correlations and its associated eigenvectors are the column vectors in $\alpha = (\mathbf{T}^{1/2}\mathbf{X})$. To obtain the mth score of species h, the hm element (α_{hm}) in α is divided by the square root of the hth species total. Formula (3.19) could of course be restated based on the quadrat scores.

The foregoing discussion give us the clues regarding the algorithm to compute reciprocal ordering:

1. Obtain a $p \times s$ matrix $\mathbf{D}$ from matrix $\mathbf{A}$ with elements

$$d_{hj} = \frac{A_{hj}}{(A_{h.}\ A_{.j})^{1/2}} \ .$$

Experience suggests that λ holds one more non-zero eigenvalue than required. The superfluous value is unity and it can be eliminated if d_{hj} is expressed as a departure from its random expectation. That is, a new matrix $\mathbf{U}$ should be defined with elements

$$U_{hj} = \frac{A_{hj} - A_{h.}\ A_{.j}/A_{..}}{(A_{h.}\ A_{.j})^{1/2}} = \frac{A_{hj}}{(A_{h.}\ A_{.j})^{1/2}} - \frac{(A_{h.}\ A_{.j})^{1/2}}{A_{..}} .$$

2. Compute a $p \times p$ matrix $\mathbf{S} = \mathbf{UU}'$ of species scalar products with a characteristic element given by

$$S_{hi} = \sum_j U_{hj} U_{ij},\ j = 1, ..., s .$$

3. Extract the non-zero eigenvalues $(\lambda_1, ..., \lambda_t)$ and associated eigenvectors $(\alpha_1, ..., \alpha_t)$ of $\mathbf{S}$. It is easily verified that, whereas the mth eigenvalue is in fact the square of the mth canonical correlation, *i.e.* $\lambda_m = R_m^2$, the hth element α_{hm} of the mth eigenvector represents, after adjustments, the score of the mth species on the mth canonical variate. The task of adjusting α_{hm} to obtain a species score X_{hm} is simple enough,

$$X_{hm} = (A_{h.})^{-1/2} \alpha_{hm}, \text{ for any } h \text{ or } m \qquad (3.20a)$$

if no constraint is imposed on the scores other than the requirement that formula (3.16) produce a canonical correlation R_m^2, identical in value to the mth eigenvalue λ_m of $\mathbf{S}$. If the scores are to be also standardized, the transformation becomes,

$$x_{hm} = \frac{(\alpha_{hm} - \bar{\alpha}_m)}{[\sum_h (\alpha_{hm} - \bar{\alpha}_m)^2]^{1/2}} \left[\frac{A_{..}}{A_{h.}}\right]^{1/2} \quad (3.20b)$$

We describe the derivation of formula (3.20). The scores $x_{1m}, \ldots, x_{pm}$ are such that their weighted mean is zero and variance unity:

a. $\sum_h A_{h.}\, x_{hm} / \sum_h A_{h.}$

$= \sum_h A_{h.} \, [(A_{h.})^{-1/2} (\alpha_{hm} - \bar{\alpha}_m)] / A_{..}$

$= \sum_h (A_{h.})^{1/2} (\alpha_{hm} - \bar{\alpha}_m) \,/\, A_{..}$

$= 0, h = 1, \ldots, p$ for any m.

b. $\sum_h A_{h.}\, x_{hm}^2 / A_{..}$

$= \sum_h A_{h.} \, [(A_{h.}^{-1/2} (\alpha_{hm} - \bar{\alpha}_m)/d_m]^2 / A_{..}$

$= \sum_h A_{h.} \, (A_{h.})^{-1} (\alpha_{hm} - \bar{\alpha}_m)^2 / (d_m^2 A_{..})$

$= \sum_h (\alpha_{hm} - \bar{\alpha}_m)^2 / (d_m^2 A_{..})$

$= 1, \quad h = 1, \ldots, p$ for any m.

From a. we have

$$\bar{\alpha}_m = \sum_h (A_{h.})^{1/2} \, \alpha_{hm} / \sum_h (A_{h.})^{1/2},$$

and from b.,

$$d_m = [\sum_h (\alpha_{hm} - \bar{\alpha}_m)^2]^{1/2} / (A_{..})^{1/2}.$$

The required transformation to standard scores then is given by

$$x_{hm} = \frac{(\alpha_{hm} - \bar{\alpha}_m)}{[\sum_h (\alpha_{hm} - \bar{\alpha}_m)^2]^{1/2}} \left[\frac{A_{..}}{A_{h.}}\right]^{1/2} \quad (3.21a)$$

$$= \alpha_{hm}^* \left[\frac{A_{..}}{A_{h.}}\right] \quad (3.21b)$$

in which α_{hm} of formula (3.20a) is replaced by its standardized form. We note that

c. $\sum_h (A_{h.})^{1/2} \alpha_{hm}^* = 0$

d. $\sum_h \alpha_{hm}^{*2} = 1, \; h = 1, \ldots, p$ for any m.

Some computer programs, such as the eigenvalue and eigenvector routine of RQT in the Appendix, will produce α^*_{hm} as solutions. A test for conditions c. and d. can reveal this. If these hold true, the x_{hm} scores can be obtained simply by the transformation in formula (3.21b). If these conditions do not hold true, the x_{hm} scores are derived by formula (3.21a).

The $x_m = (x_{hm} \dots x_{pm})$, $m = 1, \dots, t$, scores are such that

$$S_{mz} = \sum_h A_{h.}\, x_{hm}\, x_{hz} = 0, \quad h = 1, \dots, p \text{ for any } m \neq z\ .$$

The associated quadrat scores are simplest to calculate based on formula (3.17) (with standardized scores y_{mj} and x_{hm} in the place of Y_{mj} and X_{hm}). Formula (3.16) yields the canonical correlations.

An alternative algorithm determines quadrat scores directly from a cross product matrix. It involves the following calculations:

1. Generate a matrix of quadrat cross products $\mathbf{Q} = \mathbf{U'U}$ with elements $q_{jk} = \sum_h U_{hj} U_{hk}$, $h = 1, \dots, p$.
2. The canonical correlations $R_1^2, \dots, R_t^2$ are the eigenvalues of $\mathbf{Q}$ and the unadjusted quadrat scores are obtained from the elements in the associated eigenvectors $\mathbf{V}_1, \dots, \mathbf{V}_t$ according to

$$y_{mj} = A_{.j}^{-1/2}\ V_{mj}\ .$$

If standardized scores are required, the transformation is

$$y_{mj} = \frac{(V_{mj} - \bar{V}_m)}{[\sum_j (V_{mj} - \bar{V}_m)^2]^{1/2}} \left[\frac{A_{..}}{A_{.j}}\right]^{1/2}$$

$$= V^*_{mj} \left[\frac{A_{..}}{A_{.j}}\right]^{1/2}, \quad j = 1, \dots, s\ .$$

We note that

$$\sum_j A_{.j}\, y_{mj} = \sum_j (A_{.j})^{1/2}\, V^*_{mj} = 0$$

$$\sum_j A_{.j}\, y^2_{mj} / A_{..} = \sum_j V^{*2}_{mj} = 1$$

$$\sum_j A_{.j}\, y_{mj}\, y_{zj} = \sum_j V^*_{mj}\, V^*_{zj} = 0, \quad j = 1, \dots, s \text{ for any } m \neq z,$$

Once the quadrat scores are known, the associated species scores are determined according to formula (3.18) and the canonical correlations according to formula (3.16).

Numerical examples are considered below. The data derive from Table 1-1. The program that performs the calculations is RQT (Appendix).

We ran this program with option Z3 = 0; Z4 = 1; Z5 = 0. The adjusted data U_{hj} are given in the matrix,

$$\mathbf{U} = \begin{bmatrix} 0.24651 & -0.121477 & -5.53846E-2 & 1.67591E-2 & -0.110485 \\ -9.14982E-2 & -0.110386 & 0.432377 & 1.46154E-2 & -3.44327E-2 \\[6pt] -4.68582E-2 & 2.05687E-2 & -6.68065E-2 & -3.95330E-2 & -8.41206E-3 \\ 4.02549E-2 & 6.62707E-2 & -0.107534 & 4.54375E-2 & 4.54123E-2 \\[6pt] -6.25645E-2 & 0.014111 & -8.72275E-2 & 0.255903 & 0.142332 \\ -6.92410E-3 & -0.121109 & -9.50661E-3 & -0.129081 & -1.84632E-2 \\[6pt] 1.59525E-2 & 0.180445 & -2.30183E-2 & -7.98067E-2 & 5.39000E-2 \\ -8.95243E-2 & 9.98522E-2 & -6.79272E-2 & -6.64021E-2 & -6.43323E-2 \\[6pt] -7.50388E-2 & -5.52365E-2 & 0.182003 & -0.103641 & -4.99917E-2 \\ 7.19879E-2 & 2.95601E-2 & -0.120759 & 7.04235E-2 & 2.32334E-2 \end{bmatrix} .$$

The **S** matrix is given by

$$\mathbf{S} = \mathbf{UU'} \begin{bmatrix} 0.299985 & -6.84761E-2 & -1.51015E-2 & -5.49618E-2 & -7.99155E-2 \\ -6.84761E-2 & 3.04182E-2 & -1.62509E-2 & 1.15827E-2 & 1.68363E-2 \\ -1.51015E-2 & -1.62509E-2 & 0.129276 & -1.02631E-2 & -5.80477E-2 \\ -5.49618E-2 & 1.15827E-2 & -1.02631E-2 & 7.37661E-2 & -1.12381E-2 \\ -7.99155E-2 & 1.68363E-2 & -5.80477E-2 & -1.12381E-2 & 8.11857E-2 \end{bmatrix} .$$

The solutions in the determinantal equation $|\mathbf{S} - \mathbf{R}^2\mathbf{I}| = 0$ give the canonical correlations,

$$R_1 = 0.592793$$
$$R_2 = 0.404425$$
$$R_3 = 0.280631$$
$$R_4 = 0.144619 \ ,$$

and the solutions in $\mathbf{R}^2(\mathbf{T}^{1/2}\mathbf{X}) = \mathbf{S}(\mathbf{T}^{1/2}\mathbf{X})$ give the eigenvectors,

$$\alpha^{*\prime} = \begin{bmatrix} \alpha^{*\prime}_1 \\ \alpha^{*\prime}_2 \\ \alpha^{*\prime}_3 \\ \alpha^{*\prime}_4 \end{bmatrix} = \begin{bmatrix} 0.915144 & -0.218502 & 3.64655E-2 & -0.180108 & -0.284606 \\ -0.180458 & -6.54975E-2 & 0.866385 & 6.01864E-2 & -0.457055 \\ 6.64177E-2 & 5.74633E-2 & -0.260173 & 0.868200 & -0.413311 \\ 2.72056E-2 & -0.778618 & 0.183748 & 0.339625 & 0.493868 \end{bmatrix} .$$

The species scores $x_{mh} = \alpha^*_{hm}(A_{..}/A_{h.})^{1/2}$ are the elements in

$$x' = \begin{bmatrix} x'_1 \\ x'_2 \\ x'_3 \\ x'_4 \end{bmatrix} = \begin{bmatrix} 2.59054 & -0.375602 & 9.52410E-2 & -0.584892 & -0.523676 \\ -0.510831 & -0.112589 & 2.26283 & 0.195453 & -0.840983 \\ 0.188011 & 9.87786E-2 & -0.679524 & 2.81944 & -0.760493 \\ 7.70121E-2 & -1.33843 & 0.479915 & 1.10292 & 0.908719 \end{bmatrix} .$$

A characteristic element in $\mathbf{x}$, say x_{23} (the score of species *2* on canonical variate *3*), is obtained from α_{23} through the transformation $x_{23} = \alpha^*_{23}\,(A_{..}/A_{2.})^{1/2} = 0.0574633\,(1835/621)^{1/2} = 0.0987786$. Note that $A_{2.}$ is the total for species *2*, and $A_{..}$ is the grand total. The quadrat scores are the elements in

$$\mathbf{y} = \begin{bmatrix} y_1 \\ \\ y_2 \\ \\ y_3 \\ \\ y_4 \\ \\ \end{bmatrix} = \begin{bmatrix} 1.97543 & -0.605376 & -0.502900 & 0.414539 & -0.431461 \\ -0.502718 & -0.729606 & 2.97285 & -4.53156\text{E}{-2} & -0.197281 \\ -0.693500 & 0.463462 & -1.23622 & 2.07723 & 1.19970 \\ -0.230698 & -0.707885 & -0.263207 & -1.09215 & -0.212224 \\ 1.23880 & 1.63748 & -0.978097 & -1.06868 & 0.230041 \\ -1.19631 & 1.07926 & 0.215173 & -0.508063 & -0.677047 \\ 1.72124\text{E}{-3} & 0.324964 & 2.76404 & -6.73696\text{E}{-4} & 0.460377 \\ -0.635438 & -0.579378 & 0.288916 & -0.923847 & -1.09026 \end{bmatrix}.$$

These values correspond to formula (3.17). For example, $y_{19} = \sum_h A_{h9}x_{h1}/(A_{.9}R_1) = ((31)(2.59054) + (92)(-0.375602) + (1)(0.0952410) + (8)(-0.584892) + (90)(-0.523676))/((222)(0.592793)) = -0.0453144$ is the score of quadrat *9* on canonical variate *1*. (The discrepancy in computer calculations and the long-hand results in the lower decimal places is a consequence of rounding errors.)

How to construct a scatter diagram of the scores? This is simple to accomplish by using the scores as rectangular co-ordinates. Considering the first two canonical variates, the point image of species *1* is defined by the co-ordinates $(x_{11}\;\; x_{12}) = (2.59054\;\; -0.510831)$. Similarly, the co-ordinate pair which defines the point image of quadrat *1* in the plane of the first and second canonical variate is $(y_{11}\;\; y_{21}) = (1.97543\;\; -0.693500)$.

It is not to be left unmentioned that any point $(x_{h1}\; \ldots\; x_{ht})$ can be conceived as a group centroid for which group size is $A_{h.}$. Similarly, $(y_{1j} .. y_{tj})$ is a centroid and the associated group size is $A_{.j}$. The observation that every species h associates with a group of size $A_{h.}$ leads us to the idea that group sizes may act as weights exaggerating the influence of the common or rare species on the results. When the common species are to be weighed more heavily, the x_{hm} $(m = 1, \ldots, t;\, h = 1, \ldots, p)$ scores are replaced by the $\alpha^*_{hm} = x_{hm}(A_{h.}/A_{..})^{1/2}$ scores, and new quadrat scores are derived via $y_{mj} = \sum_h A_{hj}\alpha^*_{hm}/(R_m A_{.j})$. The numerical values for α^*_{hm} from the data in Table 1-1 have already been given. The associated y_{mj} values are listed in

$$\mathbf{y} = \begin{bmatrix} y_1 \\ \\ y_2 \\ \\ y_3 \\ \\ y_4 \\ \\ \end{bmatrix} = \begin{bmatrix} 0.654832 & -0.291703 & -0.310683 & 9.14783\text{E}{-2} & -0.228183 \\ -0.305665 & -0.372493 & 1.02895 & -0.14249 & -0.185001 \\ -0.297069 & 7.69326\text{E}{-2} & -0.697011 & 0.746665 & 0.363993 \\ -0.249177 & -0.416949 & -9.63201\text{E}{-2} & -0.572575 & -0.21446 \\ 0.343966 & 0.374549 & -0.690565 & -0.456435 & -9.13426\text{E}{-2} \\ -0.618338 & 0.155652 & 4.33790\text{E}{-2} & -0.371861 & -0.401706 \\ -0.357822 & -0.483372 & 1.27997 & -0.382989 & -0.282584 \\ -0.585433 & -0.814716 & -4.90161\text{E}{-2} & -0.730114 & -0.87511 \end{bmatrix}.$$

Similar considerations may lead to scores $b_{hm} = \alpha^*_{hm}A_{..}/A_{h.}$ in which the rare species receive higher weights. Such a scheme of weighting produces species scores,

$$\mathbf{b}' = \begin{bmatrix} b'_1 \\ b'_2 \\ b'_3 \\ b'_4 \end{bmatrix} = \begin{bmatrix} 7.33314 & -0.645654 & 0.248752 & -1.89941 & -0.963565 \\ -1.44603 & -0.193539 & 5.91010 & 0.634724 & -1.54741 \\ 0.532212 & 0.169799 & -1.77479 & 9.15601 & -1.39931 \\ 0.218001 & -2.30075 & 1.25345 & 3.58168 & 1.67205 \end{bmatrix}.$$

and associated quadrat scores,

$$\mathbf{y} = \begin{bmatrix} y_1 \\ \\ y_2 \\ \\ y_3 \\ \\ y_4 \\ \\ \end{bmatrix} = \begin{bmatrix} 5.78417 & -1.37211 & -0.75377 & 1.43886 & -0.871311 \\ -0.779604 & -1.52327 & 8.51685 & 0.53493 & 1.41604\text{E}-2 \\ -1.69898 & 1.71901 & -2.15356 & 5.64283 & 3.60356 \\ 0.162161 & -1.13174 & -0.776608 & -2.12639 & 7.44271\text{E}-2 \\ 4.2194 & 6.17917 & -0.832964 & -2.50496 & 1.86457 \\ -2.31541 & 4.58118 & 0.831031 & -0.358657 & -0.98006 \\ 1.87285 & 4.07094 & 6.18498 & 1.72396 & 3.51816 \\ -0.231468 & 1.46345 & 1.52420 & -0.763724 & -0.821694 \end{bmatrix}.$$

Either of these sets of scores preserve the maximum correlation property (formula (3.16)).

Among the many alternative weighting systems which could be conceived, we mention two which give weights not to the individual species as in the foregoing schemes, but to the canonical variates themselves. We observe that of the total chi square for the $p \times s$ contingency table, given by

$$\begin{aligned} \chi^2 &= \sum_h \sum_j \frac{A_{hj}^2 \, A_{..}}{A_{h.} \, A_{.j}} - 1 \\ &= A_{..} \, R_1^2 + \ldots + A_{..} \, R_t^2 \end{aligned} \qquad (3.22)$$

with summations taken from $h = 1$ to p and $j = 1$ to s, the mth canonical variate accounts for $A_{..}R_m^2$. The individual contributions to the total are proportional to R_m^2. This gives us the idea that the R_m^2 could be used to give weight to the mth canonical variate in subsequent manipulations of the scores. Another scheme of weighting has been used by Hill (1973b). He chooses to adjust the range of the canonical variate. Adjustment to unit range and zero origin follows:

$$X_{hm} := (X_{hm} - \min \mathbf{X}_m)/(\max \mathbf{X}_m - \min \mathbf{X}_m) \qquad (3.23)$$

where $X_{hm} = (A_{h.})^{-1/2} \, \alpha_{hm}$; min and max indicate minimum and maximum value. Symbol := indicates assignment. It reads 'X_{hm} becomes its original value minus ...'. To adjust the range to be equal to the squared canonical correlation, the X_{hm} of formula (3.23) has to be multiplied by R_m^2.

Hill (1973b) describes another algorithm to determine species and quadrat scores in longhand calculations. The following provide a brief description of steps:
1. Select a trial vector $\mathbf{X}_1 = (X_{11} \ldots X_{p1})$ in which the elements are arbitrary numbers; these serve as first approximations to species scores.
2. Compute an initial approximation for quadrat scores based on formula (3.13). Designate these scores by $\mathbf{Y}_1 = (Y_{11} \ldots Y_{1s})$.
3. Use the scores in $\mathbf{Y}_1$ to compute a new approximation for the scores in $\mathbf{X}_1$,

$$X_{h1} = \sum_j A_{hj} Y_{1j} / A_{h.}\ ,\ j = 1, \ldots, s; h = 1, \ldots, p \qquad (3.24)$$

4. Adjust the species scores,

$$X_{h1} := [X_{h1} - \min(X_{11} \ldots X_{p1})] \,/\, \text{range}\,(X_{11} \ldots X_{p1}) \qquad (3.25)$$

to put the scores on the 0-1 range.
5. Continue with step 3; iterate until the scores stabilize. At this stage, $\mathbf{X}_1$ and $\mathbf{Y}_1$ hold the species scores and quadrat scores on the first canonical variate. The range of the stable scores in $\mathbf{X}_1$, before adjustment to 0-1 range (formula (3.25)) is equal to the first canonical correlation R_1^2. Subsequent sets ($\mathbf{X}_2$, $\mathbf{X}_3$, ... and $\mathbf{Y}_2$, $\mathbf{Y}_3$ ...) can be extracted based on a similar method from residuals. The iterations are illustrated in the example below, using the data (**A**) in Table 1-1. Let $\mathbf{X} = (5\ \ 4\ \ 3\ \ 2\ \ 1)$ be the trial vector. The row totals of **A** are given by

$$\begin{bmatrix} A_{1.} \\ A_{2.} \\ A_{3.} \\ A_{4.} \\ A_{5.} \end{bmatrix} = \begin{bmatrix} 229 \\ 621 \\ 269 \\ 174 \\ 542 \end{bmatrix}$$

and the column totals by

$$(A_{.1}\ A_{.2}\ A_{.3}\ A_{.4}\ A_{.5}\ A_{.6}\ A_{.7}\ A_{.8}\ A_{.9}\ A_{.10}) =$$
$$(84\ \ 248\ \ 156\ \ 181\ \ 225\ \ 195\ \ 210\ \ 128\ \ 222\ \ 185)\,.$$

The first approximations for quadrat scores (formula (3.13)) are the elements in

$$\mathbf{Y}_1 = (3.94118\ \ 2.71371\ \ 1.98077\ \ 3.30387\ \ 2.77333\ \ 2.62051$$
$$2.60000\ \ 4.39844\ \ 2.84685\ \ 2.89189)$$

and the second approximations for species scores (formula (3.24)) are the elements in

$$\mathbf{X}_1 = (3.68346\ \ 2.81212\ \ 3.02294\ \ 2.76413\ \ 2.66058)\,.$$

These are adjusted according to formula (3.25) to give

$$\mathbf{X}_1 = (1.00000\ \ 0.148148\ \ 0.354256\ \ 0.101236\ \ 0.00000)\,.$$

The iterations continue. After two more passes we have

$$\mathbf{X}_1 = (0.47747\ \ 0.148921\ \ 0.210412\ \ 0.12779\ \ 0.129166$$

with a range of 0.34968 and adjusted values,

$$\mathbf{X}_1 = (1.00000\ \ 0.0604287\ \ 0.236278\ \ 0.00000\ \ 0.00393478)\,.$$

These lead to new approximations for quadrat scores,

$$\mathbf{Y}_1 = (0.550964\ \ 0.0694318\ \ 0.0891580\ \ 0.268477\ \ 0.105204\ \ 0.0847821\ \ 0.0404599\ \ 0.740827\ \ 0.167342\ \ 0.142658)\,.$$

If these were longhand results, the iterations would no doubt be terminated at this point. However, it takes another 13 or so passes to have the scores stabilize to 5 or 6 significant digits,

$$\mathbf{X}_1 = (0.470870\ \ 0.142628\ \ 0.194733\ \ 0.119467\ \ 0.126241) \qquad (3.26a)$$

with range 0.351403. After adjustment, X_1 becomes

$$\mathbf{X}_1 = (1.00000\ \ 0.0659090\ \ 0.214187\ \ 0.00000\ \ 0.0192773) \qquad (3.26b)$$

and the associated quadrat scores,

$$\mathbf{Y}_1 = (0.55297\ \ 0.0711807\ \ 0.0903107\ \ 0.261580\ \ 0.103647\ \ 0.0903448\ \ 0.0479890\ \ 0.739169\ \ 0.175733\ \ 0.147364) \qquad (3.27)$$

$\mathbf{X}_1$ of formula (3.26b) holds the species scores on the first canonical axis adjusted to unit range and zero origin. $\mathbf{Y}_1$ of formula (3.27) holds the quadrat scores. The scores in $\mathbf{X}_1$ of formula (3.26a) come out adjusted to a range of R_1^2. These scores, and also the remaining sets of $\mathbf{X}_2$, $\mathbf{Y}_2$, $\mathbf{X}_3$, $\mathbf{Y}_3$ and $\mathbf{X}_4$, $\mathbf{Y}_4$ are computed automatically from the eigenvectors of S in program RQT, option Z3 = 1. This option, however, produces the $\mathbf{X}_m$ scores of formula (3.26b) and not (3.26a).

Since there are t sets of scores available in the complete suite of eigenvectors of the $\mathbf{UU}'$ matrix, the question of how many sets to use has practical significance. Theory suggests that if the parent population from which any pair $\mathbf{X}_m$, $\mathbf{Y}_m$ are taken represents a bivariate normal population, then the second, third, etc., canonical correlations are different powers of the first. This means that if the sequence $R_1, R_2, \ldots$ of the canonical correlations signifies an order of decreasing magnitude, then $R_2 = R_1^2, R_3 = R_1^3$, etc., could be expected. Any departure from these values would be attributed to sampling errors (cf. Kendall & Stuart, 1973). We are fully justified then to take the position that sets of scores other than $\mathbf{X}_1$ and $\mathbf{Y}_1$ are in fact redundant. It would not make sense, however, to throw away scores without an examination of their ecological meaning. This could involve subjecting the sets to analyses which are designed to reveal information about their significance as predictors of environmental conditions.

How to monitor how much of the total chi square in the $p \times s$ contingency table has been accounted for by the canonical variates? The quantity

$$E_m\% = 100\, A_{..}\, R_m^2 / \chi^2$$

may help in this respect. In this, $\chi^2 = \sum_m A_{..} R_m^2$, $m = 1, \ldots, t$. Williams (1952) suggests that $A_{..}R_m^2$ is a chi square variate with $p + s - 2m - 1$ degrees of freedom. This suggestion has however been questioned (Kendall & Stuart, 1973, p. 595 and reference therein).

We note that the correlation is maximal for the first set of species scores and quadrat scores, but correlations decline with the subsequent sets. How should we interpret these correlations? Before we attempt answering this question let us try to picture two extreme situations. In the first, all species within the same quadrat have the same optima, different from species of other quadrats. What this means is that species within the same quadrat score identically on the same environmental gradient. In the other extreme, the species are completely heterogeneous, each with a different score within a quadrat. If we attempt ordering quadrats based on the species scores we would have no difficulty in the first case. The quadrats would fall nicely into order, identically to the species which they incorporate. We would have considerably more difficulty in ordering quadrats in the second case, where the relationship of a quadrat and the species within it are not clearly defined. In these terms then, we may regard the one-complement of the squared canonical correlation $1 - R_i^2$ as an indication of the conceptual difficulty with which the quadrats can be ordered based on the species scores.

At this point, we can revisit the problem of interpreting R of formula (3.14) in the event when the species scores (**X**) are derived from external sources. When we correlated the species scores with the quadrat scores in the example, we obtained $R^2 = 0.152235$. Now, it would not be appropriate to regard variable R as one with a domain of 0 to 1. The reason being that no single set of species and quadrat scores, derived from the data of Table 1-1, can have squared correlation in excess of the largest eigenvalue of the **S** matrix. The value of this is $R_1^2 = 0.351404$. If we regard R^2 and R_1^2 as contributions to the total χ^2, then the ratio.

$$100\, R^2 / R_1^2 = 15.2235/0.351404 = 43.3\%$$

expresses the success of the species scores in X to fit the data in **A**, relative to the best linear fit that can possibly be achieved by a single **X** set.

Is a discrepancy such as $100 - 43.3 = 56.7\%$ large or small? We cannot say, unless a test criterion is found, on the basis of which we can measure the probability that a discrepancy as large as the observed $R_1 - R$ could have arisen by chance in the sample when the expectation of $R_1 - R$ is in fact zero. In our attempt to find a test statistic, we may follow Williams (1952), regard formula (3.13) as a regression equation, and regard $R_1^2 - R^2$ as the square of a multiple correlation coefficient. It then seems justified to treat $R_1^2 - R^2$ as a quantity proportional to a sum of squares with $s-p$ degrees of freedom ($s > p$) and to regard $1 - (R_1^2 - R^2)$ as a quantity proportional to the associated residual sum of squares with $A_{..} - 1 - (s-p)$ degrees of freedom. The test statistic is

$$F = \frac{(R_1^2 - R^2)/(s-p)}{(1 - R_1^2 + R^2)/[A_{..} - 1 - (s-p)]}$$

$$= \frac{(0.351404 - 0.152235)/(10 - 5)}{(1.152235 - 0.351404) \,/\, (1835 - 1 - 10 + 5)}$$

$$= \frac{0.0398338}{0.000437852}$$

$$= 90.9755 \ .$$

Under the null hypothesis that $R_l^2 - R^2$ has zero expectation with 5 and 1829 degrees of freedom the probability of a discrepancy being at least as large as the observed is practically infinitesimal. Thus, we conclude that indeed the species scores

$$\mathbf{X} = (5 \;\; 4 \;\; 3 \;\; 2 \;\; 1)$$

give a very poor fit to the data when compared to the best fitting scores derived in reciprocal ordering.

It is obvious from descriptions of reciprocal ordering and component analysis that they have certain similarities, and also, important differences. Both can be explained in the context of eigenvalue and eigenvector analysis. They are different though in the data adjustments and in the adjustments of elements in the eigenvectors. Furthermore, as originally proposed, reciprocal ordering assumes categorical data. Only then is the expectation of data elements derivable from marginal totals.

Since reciprocal ordering draws entirely on vegetational data, any attempt to interpret the species or quadrat scores in environmental terms will necessarily involve secondary analyses to establish correlations with the suspected environmental variables. In defining such correlations we may fall back on the simple technique of displaying environmental data on scatter diagrams (Greig-Smith, 1964) and deduce from the resulting pattern information about environmental gradients. With a similar purpose in mind, we may apply other techniques such as canonical correlation analysis (Pielou, 1969a) or use trend surface analysis (e.g. Neal & Kershaw, 1973b).

C. Factor analysis

The objective we are concerned with when we contemplate reducing dimensionality of the sample could be estimation of a hypothesized $m < p$-dimensional common covariance structure in the sampled population. We may wish to undertake such an analysis to enhance the identification of specific environmental factors to which vegetation responds by its covariance structure (Dagnelie, 1960, 1973).

When the objective is estimation of a common underlying covariance structure, the appropriate method of ordination is *factor analysis.* Although the biological usefulness of the different methods has been established (e.g. Sokal, 1958; Sokal & Daly, 1961; Sokal, Daly & Rohlf, 1961; Rohlf & Sokal, 1962; Sinha & Lee, 1970), some in connection with vegetation studies (e.g. Dagnelie, 1971, 1973; Szőcs, 1971, 1972; Hinneri, 1972; Pignatti & Pignatti, 1975; Pakarinen, 1976), there are two serious obstacles to wide-spread applications. The first is the reliance on covariance, which limits the range of effective applications to linear data structures. The second is related to the first in that it requires the assumption of multivariate normality. The latter is critical, considering that factor analysis is a statistical method.

Although the concepts at the basis of factor analysis are rather straightforward in their generality, simplicity is often obscured by the use of an awkward terminology, a product of specialized applications. For instance, variables are referred to as 'tests' and the factors that supposedly influence the covariance structure receive the name of 'fundamental abilities'. Such a terminology is obviously alien to users not dealing with human populations, and outright inappropriate in vegetation studies.

The objective in factor analysis has been defined as the explicit elicitation of the ecological conditions that influence species covariance. This is a respectable objective for an ordination. The trouble is that factor analysis, being a linear ordination relying on the covariance, does not square with this objective, except when the data structure is in fact linear and continuous. The immediate computational problem amounts to a reduction of dimensionality from p species axes to m ordination axes. This is to be achieved through elimination of all variation specific to the individual species. In this respect, factor analysis differs from component analysis, in that it happens to be a variance-oriented method of data analysis. There are many methods to achieve a reduction of dimensionality in the factor analysis context about which heavy volumes have been written. And yet, one is left with the impression that the enormous volume of published work on this topic may far outweigh any theoretical or practical importance that a method so much burdened by restrictive assumption, as factor analysis, could possibly have. Dale (1975) offers an informative review and Armstrong (1967) sobering comments.

The linear model,

$$X_h = \sum_i \alpha_{hi} y_i + e_h, \ i = 1, ..., m \ ,$$

tells us that any observed response X_h is a compound of m terms of influence, attributable to m common factors $y_1, ..., y_m$, and a term e_h representing response to a specific factor. The α_{hi} are factor loadings (analogous quantities as the component coefficients). The computational problem is two-fold, involving determination of the $\alpha_{1i}, ..., \alpha_{pi}$, $i = 1, ..., m$, coefficients of species as well as the $e_1, ..., e_p$ values, and then the $z_{i1}, ..., z_{in}$, $i = 1, ..., m$, factor scores. The factor scores, serving as ordination co-ordinates, are not obtained through simple linear transformations as in component analysis but estimated in other ways under specified constraints (cf. Lawley & Maxwell, 1963).

The algebra to determine $\boldsymbol{\alpha}$ and $\mathbf{E}$ is simply stated by Gower (1977a). $\boldsymbol{\alpha}$ is a $p \times m$ matrix with an element α_{hi} in the h, j cell, and $\mathbf{E}$ is a $p \times p$ diagonal matrix with an element e_h in the hth principal diagonal position. The algebra attempts to express the $p \times p$ covariance matrix $\mathbf{S}$ as a sum of two matrices,

$$\mathbf{S} = \alpha\,\alpha' + \mathbf{E}.$$

Whether it succeeds or fails depends on m – the dimensionality of the reduced space. The solution is obtained by iteration. This involves the following:

1. Define the m columns of α_1 as the first m eigenvectors of **S** so adjusted that $\alpha'_{h1}\,\alpha_{h1} = \lambda_h$, where λ_h is the ith eigenvalue of **S**.
2. Replace the diagonal elements of **S** by the diagonal elements of $\alpha_1\,\alpha'_1$ to obtain a new matrix $\mathbf{S}_1$.
3. Extract the eigenvectors which, after adjustment, give the new matrix α_2.
4. Determine a new matrix $\mathbf{S}_2$ by replacing the diagonal elements in $\mathbf{S}_1$ by the diagonal elements in $\alpha_2\,\alpha'_2$.

The procedure is repeated through $\mathbf{S}_2, \mathbf{S}_3, ..., \mathbf{S}^*$ until the diagonal elements stabilize at the specified level of precision. At this stage, the elements in α^* are estimates of the factor loadings in α and $e^*_{hh} = S_{hh} - \alpha^{*\prime}_h\,\alpha^*_h$ is an estimate of the specific variance e_h. We note that conventional factor analysis discards the specifics and focuses on the factor loadings. It is seen from the procedure that the covariance structure of the data is destroyed in factor analysis. This is not so in component analysis in which the covariance structure is preserved.

The assumption that the underlying data structure is linear and continuous is required in factor analysis to justify the use of the covariance. This puts a similar limitation on factor analysis as on component analysis. There are other constraints to be considered. One is the choice of a value for m, the number of common factors, before commencing with the analysis. This is essentially an educated guess and as such it may be far out, so that several trials may be required before the postulated m-factor model will give a reasonably good fit to the data. Yet another limitation is the statistical nature of factor analysis. This is manifested by the probabilistic aspects in that they require the assumption that the sampled population is multivariate normal. The application of factor analysis in vegetation studies has apparently been limited to the least tedious method with some success (e.g. Dagnelie, 1960, 1965a,b). The value of the more complex procedures, based on maximum likelihood solution and various types of rotations, have yet to be assessed.

3.7 Display and evaluation of ordination results

Graphs are widely used to display ordination results and to establish their relationships to vegetational or environmental variation. Such displays commonly involve:

1. Constructing joint distributions for sets of ordination scores as scatter diagrams. Such diagrams are revealing about latent properties in data structures.
2. Mapping species performance in scatter diagrams. Such maps can assist the user to identify the relationship of species and ordination axes.
3. Mapping environmental variables in scatter diagrams. Maps of this sort help to identify environmental influences affecting vegetational variation.

4. Mapping successional or other changes in structure or composition in scatter diagrams. This kind of map can be helpful in studying vegetation dynamics (e.g. van der Maarel, 1969; Goff & Zedler, 1972; James & Shugart, 1974) or other processes in time (e.g. Allen & Skagen, 1973) and in space (e.g. Walker, Noy-Meir, Anderson & Moore, 1972).

Whereas certain types of two dimensional graphs are commonly employed in displays of ordination results (e.g. Goodall, 1954a; Bray & Curtis, 1957; Beals, 1965a,b; Greig-Smith, Austin & Whitmore, 1967; Gittins, 1968, 1969; Orlóci, 1968c; Kershaw, 1968; Chandapillai, 1970; Sokal & Rohlf, 1970; Allen, 1971; Austin, Ashton & Greig-Smith, 1972), advantages deriving from stereoscopic diagrams have largely been overlooked. Fraser & Kováts (1966) describe the technique and provide useful examples. Rohl (1965, 1968; Moss, 1966, 1967) explores their use in numerical taxonomy.

We summarize the technique of construction and give an example. To produce left (L) and right (R) stereograms, our first task is to concentrate as much as possible of the total variation in the sample on three variables. This is a crucial step since stereograms display a three dimensional image of the sample covariance structure. We may use principal components for this purpose. Let $\mathbf{Y}_1$, $\mathbf{Y}_2$, $\mathbf{Y}_3$ hold the component scores for n quadrats associated with the three largest eigenvalues of the covariance matrix $\mathbf{S}$. For convenience we adjust the scores so that

$$Y_{ij} := \frac{a_i(Y_{ij} - \min(Y_{i1} \ldots Y_{in}))}{\max(Y_{i1} \ldots Y_{in}) - \min(Y_{i1} \ldots Y_{in})} \qquad (3.28)$$

In this formula, min and max signify minimum and maximum values and Y_{ij} is the score of quadrat j on component i. The a_i are arbitrary dimensions specifying the new width of range within which the adjusted Y_{ij} will fall,

$$0 \leqslant Y_{1j} \leqslant a_1$$
$$0 \leqslant Y_{2j} \leqslant a_2$$
$$0 \leqslant Y_{3j} \leqslant a_3 .$$

The values of a_1, a_2, a_3 are so chosen that

$$a_1 = fR_1/k$$
$$a_2 = fR_2/k$$
$$a_3 = fR_3/k \qquad (3.29)$$

It is convenient to set f equal to 3.3 and to put $k = \max(R_1 \; R_2 \; R_3)$. The Rs are defined as ranges: $R_i = \max(Y_{i1} \ldots Y_{in}) - \min(Y_{i1} \ldots Y_{in})$ in the unadjusted component scores for $i = 1, 2, 3$.

Our next task is to choose two points P_L and P_R outside the point cluster in the direction of the third component. The co-ordinates

$(Y_{1L}\ Y_{2L}\ Y_{3L})$ define P_L and the co-ordinates $(Y_{1R}\ Y_{2R}\ Y_{3R})$ define P_R. When we view the cluster from P_L with only the left eye, the point P_j with co-ordinates $(Y_{1j}\ Y_{2j}\ Y_{3j})$ appears as a point P_{jR} with co-ordinates $(Y_{1jR}\ Y_{2jR})$ on the (Y_1, Y_2) plane. Similarly, when we view the cluster from point P_R with only the right eye, P_j will appear as point P_{jL} with co-ordinates $(Y_{1jL}\ Y_{2jL})$ on the (Y_1, Y_2) plane. We thus distinguish between left and right point images, and thus, between left and right stereograms. By co-ordinate geometry, we have

$$Y_{1jL} = \frac{Y_{3L}Y_{1j} - Y_{1L}Y_{3j}}{Y_{3L} - Y_{3j}}$$

$$Y_{2jL} = \frac{Y_{3L}Y_{2j} - Y_{2L}Y_{3j}}{Y_{3L} - Y_{3j}}$$

$$Y_{1jR} = \frac{Y_{3L}Y_{1j} - Y_{1R}Y_{3j}}{Y_{3L} - Y_{3j}}$$

$$Y_{2jR} = \frac{Y_{3L}Y_{2j} - Y_{2R}Y_{3j}}{Y_{3L} - Y_{3j}} \qquad (3.30)$$

These co-ordinates map any point P_j $(j = 1, ..., n)$ into the left or right stereogram. When viewed with the help of a stereoscope, the left and right images fuse into a distinct three-dimensional picture.

The example below uses the salt marsh data which Orlóci (1976b) subjected to cluster analysis. The component scores were extracted for 50 relevés based on 27 species, and are given in the Appendix as input for program STEREO. When we transfer to zero origin and adjust the ranges to $a_1 = 2.92557$, $a_2 = 3.3$, $a_3 = 2.80692$ (formulae (3.29)), we obtain the adjusted scores by repeated application of formula (3.28). The adjusted scores are also given in the Appendix with program STEREO. An entry in the adjusted scores, say Y_{19}, is derived as

$$Y_{19} := \frac{a_1(Y_{19} - \min(Y_{11} \ldots Y_{150}))}{\max(Y_{11} \ldots Y_{150}) - \min(Y_{11} \ldots Y_{150})}$$

$$= \frac{(2.92557)(-0.0895978 - (-1.23175))}{0.799955 - (-1.23175}$$

$$= 1.64465 .$$

It is advantageous for stereo viewing to choose P_L and P_R so that

$$Y_{1L} = Y_{2L} = Y_{2R} = (0.39)(3.3) = 1.287 \text{ cm}$$

$$Y_{1R} = (0.64)(3.3) = 2.112 \text{ cm}$$

$$Y_{3L} = Y_{3R} = (3)(3.3) = 9.9 \text{ cm}.$$

By application of formula (3.30), we obtain the four sets of co-ordinates which we also give in the Appendix with program STEREO. To determine a value, say Y_{19L}, we use the first of formulae (3.30),

$$Y_{19L} = \frac{(9.9)(1.64465) - (1.287)(1.14072)}{9.9 - 1.14072}$$

$$= 1.69123 .$$

Note that since $Y_{2L} = Y_{2R}$ we have $Y_{29L} = Y_{29R}$. The left and right sets of co-ordinates produce the left and right stereograms (Fig. 3-11) for viewing under a lense stereoscope. We, of course, realize that by altering the parameters, any model size or viewing point position may be assumed. We note further that the outside frame about the stereograms enhances the appearance of an enclosed space. The groups of relevés recognized on the basis of Orlóci's (1976b) algorithm TRGRPS (see description in Chapter 4) clearly appear as spatially isolated clusters of points.

Graphical displays of ordination results may reveal aspects that require further analysis, more specifically:

1. To correlate ordination axes with species.
2. To place new quadrats (as points) in a scatter diagram.
3. To correlate ordination axes with environmental variables.
4. To reveal trends and classes of variation.

Where the R-algorithm produces component scores, the product moment correlation of components and species may be worth examination since it can reveal information not apparent otherwise from the data. The correlation r_{hi} of the hth species and the ith component is defined by

$$r_{hi} = b_{hi}(\lambda_i / S_{hh})^{1/2} \qquad (3.31)$$

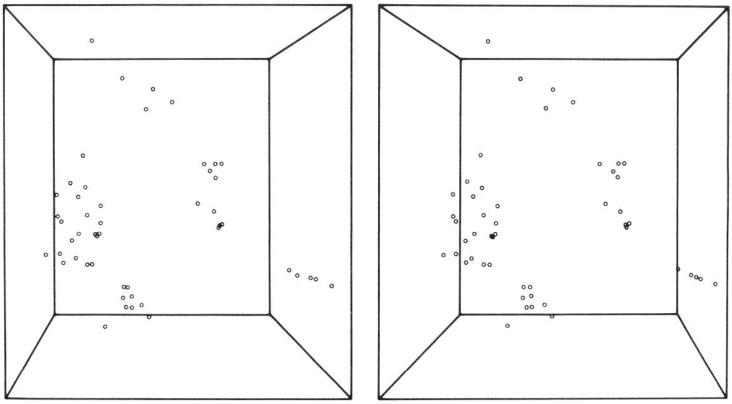

Fig. 3-11. Left and right stereograms of Orlóci's (1976b) salt marsh clusters for stereo-viewing. See the description of method in the main text.

In this, b_{hi} is the ith component coefficient of species h, λ_i the ith eigenvalue of the covariance matrix, and S_{hh} the variance of species h.

One may be tempted to use b_{hi} directly as a measure of correlation. This would be an acceptable proposition if it were to serve as a unique expression of relatedness, considering that b_{hi} is a direction cosine. However, b_{hi} cannot be regarded as a product moment correlation coefficient, except in the case when S_{hh} and λ_i are unity or equal. We may give as a reason for this the peculiar fact that the position of a given component is fixed in space, as is the position of a given species (axis), not by their mutual correlation, but rather by the covariances collectively of all the species in the sample. Accordingly, r_{hi} may be smaller, equal to or larger than b_{hi}.

Considering the species of the example for component analysis in Section 3.4, their correlations with the first and second components are given below:

Species h	*Component* i	b_{hi}	λ_i	S_{hh}	r_{hi}
1	1	−0.901	3.74	3.5	−0.931
2	1	0.433	3.74	2.7	0.510
1	2	0.433	2.46	3.5	0.363
2	2	0.901	2.46	2.7	0.860 .

Since formula (3.31) is operational only if the b_{hi} are available, it can be used only with the species in the sample which produced the b_{hi} coefficients. Whether the b_{hi} are given or not, the product moment correlation coefficient is always defined,

$$r_{hi} = \sum_j (X_{hj}-\bar{X}_h)(Y_{ij}-\bar{Y}_i)/[\sum_j(X_{hj}-\bar{X}_h)^2\sum_j(Y_{ij}-\bar{Y}_i)^2]^{1/2},\ j=1,\ldots,n \tag{3.32}$$

In this, X_{hj} is a value of species h in quadrat j (adjusted if the ordination itself subjected the data to non-linear adjustments), and Y_{ij} is the score of quadrat j on the ith ordination axis. $\bar{X}_h$ and $\bar{Y}_i$ are mean values. In component analysis, $\bar{Y}_i = 0$ and $\sum_j Y_{ij}^2 = \lambda_i$.

To place new quadrats in a scatter diagram is yet another possible objective of post-ordination analyses. If the new quadrat is specified by

$$\mathbf{A}_k = \begin{bmatrix} A_{1k} \\ \cdot \\ A_{pk} \end{bmatrix},$$

and if the A_{hk} are values in quadrat k of the same p species as in component analysis, the component scores that can map $\mathbf{A}_k$ into a scatter diagram are obtained in formula (3.2). We shall consider an example in which the quadrat to be mapped in Fig. 3-6 is described by the data vector,

$$\mathbf{X}_k = \begin{bmatrix} X_{1k} \\ X_{2k} \end{bmatrix} = \begin{bmatrix} 3 \\ 1 \end{bmatrix} .$$

We require the values X_{1k} and X_{2k} to be measured on the same species in the same units as those which produced the scatter diagram. Accordingly, we have for $\mathbf{A}_k$.

$$\mathbf{A}_k = \begin{bmatrix} (3-2)/(5-1)^{1/2} \\ (1-1.8)/(5-1)^{1/2} \end{bmatrix} = \begin{bmatrix} 0.5 \\ -0.4 \end{bmatrix} .$$

The ordination co-ordinates sought are then calculated,

$$Y_{1k} = (-0.901)(0.5) + (0.433)(-0.4) = -0.624$$

$$Y_{2k} = (0.433)(0.5) + (0.901)(-0.4) = -0.144 .$$

Correlations may be sought of the ordination axes with several environmental variables individually or simultaneously. A simple and often entirely sufficient method involves plotting measurements of species performance or the levels of environmental variables on scatter diagrams (e.g. McIntosh & Hurley, 1964; Gittins, 1965; Hulett, Coupland & Dix, 1966; van der Maarel, 1967; Koterba & Habeck, 1971; Arno & Habeck, 1972; Onyekwelu, 1972; Rochow, 1972; Allen & Koonce, 1973). The pattern of variation of values indicate trends in correlations. When the trends are to be quantified, regression lines or surfaces can be fitted to the measurements as functions of the ordination scores. The regression approach is explored by Neal & Kershaw (1973a,b).

The product moment correlation coefficient (formula (3.32)) is a measure that could reveal ecological information. However, when there are several environmental variables the appropriate approach is multivariate and a possible method is *canonical analysis* (e.g. Harberd, 1962; Austin, 1968; Pielou, 1969a; Sinha, Yaciuk & Muir, 1973). We begin the description of this method with identification of the t variables $Y_1, Y_2, ..., Y_t$ of the first set as ordination axes, and the p descriptors $X_1, X_2, ..., X_p$ of the second set as environmental variables. Let us assume that $t \leqslant p$. If $t > p$ we would designate the environmental variables as the first group. The ith set of ordination co-ordinates consists of

$$\mathbf{Y}_i = (Y_{i1} \; ... \; Y_{in})$$

and the ith set of environmental measurements is given by

$$\mathbf{X}_i = (X_{i1} \; ... \; X_{in}) .$$

In canonical analysis, we seek new uncorrelated variables,

$$V_1, V_2, \ldots, V_s$$

in the Y set, and new uncorrelated variables

$$U_1, U_2, \ldots, U_s$$

in the X set, such that the linear correlation of V_1 and U_1 is the greatest among all pairs of the V and U variables, the correlation of V_2 and U_2 is the greatest among all pairs of the V and U variables uncorrelated with V_1 and U_1, and so forth. V_i and U_i can be regarded as ordination axes on which the quadrat co-ordinates (canonical scores) are defined by the transformations,

$$V_{ij} = a_{1i}Y_{1j} + \ldots + a_{ti}Y_{tj} \tag{3.33a}$$

$$U_{ij} = b_{1i}X_{1j} + \ldots + b_{pi}X_{pj} \tag{3.33b}$$

$i = 1, \ldots, s$; $j = 1, \ldots, n$. Here, V_{ij} or U_{ij} is the score of quadrat j on the ith canonical variable in the X or Y set, a_{ih} is a canonical coefficient relating variate i in the Y set and canonical variate h of the same set; b_{ih} is similarly defined for the X set.

The objective of canonical analysis is to determine s canonical correlations,

$$C_1 = r(V_1, U_1), \ldots, C_s = r(V_s, U_s),$$

and canonical coefficients,

$$a_{1i}, \ldots, a_{ti}$$

$$b_{1i}, \ldots, b_{pi},$$

$i = 1, \ldots, s$. It is noted that the ith greatest canonical correlation is the square root of the ith largest eigenvalue of the matrix

$$\mathbf{G} = \mathbf{S}_{YY}^{-1}\, \mathbf{S}_{YX}\, \mathbf{S}_{XX}^{-1}\, \mathbf{S}'_{YX}\,.$$

Each term on the right represents a covariance matrix of the variates within or among the sets indicated; e.g., $\mathbf{S}_{XX}$ is the covariance matrix of variates in set X and $\mathbf{S}_{YX}$ between the variates of sets Y and X. $\mathbf{G}$ is a $t \times t$ matrix not necessarily symmetric; $\mathbf{S}_{YY}$, $\mathbf{S}_{YX}$ and $\mathbf{S}_{XX}$ are respectively $t \times t$, $t \times p$ and $p \times p$ matrices. $\mathbf{S}_{YY}^{-1}$ is the inverse of $\mathbf{S}_{YY}$, $\mathbf{S}_{XX}^{-1}$ of $\mathbf{S}_{XX}$, and $\mathbf{S}'_{YX}$ is a $p \times t$ matrix (the transpose of $\mathbf{S}_{YX}$).

We use the co-ordinates designated P.V.O. in Table 3-2 as the Y set and the last three soil variables of Table 3-1 as the X set in the example below. The $\mathbf{S}_{YY}$, $\mathbf{S}_{YX}$ and $\mathbf{S}_{XX}$ matrices are the partitions in $\mathbf{S}$,

$$\mathbf{S} = \left[\begin{array}{c|c} \mathbf{S}_{YY} & \mathbf{S}_{YX} \\ \hline \mathbf{S}'_{YX} & \mathbf{S}_{XX} \end{array}\right]$$

$$= \left[\begin{array}{cc|ccc} 0.476578 & -4.96336\text{E}-2 & 1030.47 & 93.8368 & 3.54737 \\ -4.96336\text{E}-2 & 0.257243 & -555.600 & -26.5762 & -0.926000 \\ \hline 1030.47 & -555.600 & 3.21867\text{E}+6 & 216207. & 9233.33 \\ 93.8368 & -26.5762 & 216207. & 21196.2 & 666.833 \\ 3.54737 & -0.926000 & 9233.33 & 666.833 & 31.7667 \end{array}\right].$$

After inverting and transposing the appropriate matrices, and completing the arithmetic, we get,

$$\mathbf{G} = \mathbf{S}_{YY}^{-1}\,\mathbf{S}_{YX}\,\mathbf{S}_{XX}^{-1}\,\mathbf{S}'_{YX}$$

$$= \begin{bmatrix} 0.940539 & -0.112479 \\ -0.160843 & 0.694147 \end{bmatrix}.$$

The determinantal equation of $\mathbf{G}$ is given by

$$\begin{vmatrix} G_{11}-\lambda & G_{12} \\ G_{21} & G_{22}-\lambda \end{vmatrix} = \lambda^2 - 1.63469\lambda + 0.634781$$

with roots

$$\lambda_1 = \frac{1.63469 + 0.364811}{2} = 0.999751$$

and

$$\lambda_2 = \frac{1.63469 - 0.364811}{2} = 0.634939\ .$$

The canonical correlations are $C_1 = \lambda_1^{1/2} = 0.999875$ for V_1 and U_1 and $C_2 = \lambda_2^{1/2} = 0.796831$ for V_2 and U_2. It is obvious that V_1 of the ordination scores practically coincides with U_1 of the soil chemical variables; their correlation accounts for $\lambda_1/(\lambda_1 + \ldots + \lambda_s) = 0.999751/1.63469 = 0.61$ or 61% of the total canonical variation in the sample. Our interpretation of the correlations is, in lack of an appropriate test, deterministic.

The canonical coefficients in the first sets are the solutions in a system of homogeneous equations,

$$(\mathbf{S}_{YX}\,\mathbf{S}_{XX}^{-1}\,\mathbf{S}'_{YX} - \lambda_1\,\mathbf{S}_{YY})\,a_1 = 0 \qquad (3.34)$$

After substituting numerical values, and performing operations, we obtain,

$$-0.0202353\ a_{11} \ -0.0384369\ a_{21} = 0$$
$$-0.0384369\ a_{11} \ -0.0730319\ a_{21} = 0 \quad .$$

The solutions are in the proportion of $a_{11}/a_{12} = -1.899857$ or $a_{11} = -1.899857$ and $a_{12} = 1$. Normalization of the coefficients (this is optional) gives $a_{11} = -0.884904$ and $a_{21} = 0.465774$. The canonical scores in the first set are calculated according to formula (3.33a). For instance, site *5* in Table 3-2 has a canonical score,

$$\begin{aligned} V_{15} &= a_{11}\,Y_{15} + a_{21}\,Y_{25} \\ &= (-0.884904)\,(-0.713) + (0.465774)\,(0.463) \\ &= 0.846587 \quad . \end{aligned}$$

Canonical scores are similarly calculated for the remaining sites:

Canonical variate	*Site* 1	2	3	4	5	6
V_1	0.846590	−0.552180	−0.552180	−0.0375244	0.846590	−0.552180 .

Having determined the first set of canonical coefficients ($\mathbf{a}_1$) and canonical scores ($\mathbf{V}_1$), one more set of each ($\mathbf{a}_2$, $\mathbf{V}_2$) could be extracted. For this we would use λ_2 in formula (3.34) after which we would proceed exactly as before.

As far as the soil variables (Table 3-1) are concerned, the canonical coefficients are the solutions in the system of homogeneous equation,

$$(\mathbf{S}'_{YX}\,\mathbf{S}^{-1}_{YY}\,\mathbf{S}_{YX} - \lambda_1\,\mathbf{S}_{XX}) = 0 \tag{3.35}$$

In the actual case, represented by

$$\begin{aligned} -192538\ b_{11} + \ 16626.8\ b_{21} - \ 571.031\ b_{31} &= 0 \\ 16626.8\ b_{11} - \ 1594.62\ b_{21} + \ 68.9000\ b_{31} &= 0 \\ -571.031\ b_{11} + \ 68.9000\ b_{21} - \ 4.12549\ b_{31} &= 0 \ , \end{aligned}$$

the determinant of the coefficients is not zero and thus no solution other than the trivial zero solution exists. This has implications and raises problems. Not being able to determine values for the coefficients b_{11}, b_{21}, b_{31}, we have to decide how to proceed. We have the canonical correlations and we could terminate the analysis. Alternatively, we may remove one of the soil variables from the set, possibly the one that accounts for the least amount of covariance with the others, and then, we may reanalyse the variables in the reduced set.

When we subject the soil variables of Table 3-1 to ranking by program RANK, we get the following results:

Soil variable	*Rank*	*Proportion of sum of squares*
Calcium	3	1.4
Potassium	1	69.8
Nitrate N	2	28.8

These show that calcium is a prominent candidate for omission. Without it, the covariance matrices becomes,

$$\mathrm{S} = \left[\begin{array}{c|c} \mathrm{S}_{YY} & \mathrm{S}_{YX} \\ \hline \mathrm{S}'_{XY} & \mathrm{S}_{XX} \end{array}\right]$$

$$= \left[\begin{array}{cc|cc} 0.476578 & -4.96336\mathrm{E}{-2} & 93.8368 & 3.54737 \\ -4.96336\mathrm{E}{-2} & 0.257243 & -26.5762 & -0.92600 \\ \hline 93.8368 & -26.5762 & 21196.2 & 666.833 \\ 3.54737 & -0.926000 & 666.833 & 31.7667 \end{array}\right].$$

After inverting and transposing the matrices, and performing the required arithmetic, we get

$$\mathrm{G} = \mathrm{S}_{YY}^{-1}\,\mathrm{S}_{YX}\,\mathrm{S}_{XX}^{-1}\,\mathrm{S}'_{YX}$$

$$= \begin{bmatrix} G_{11} & G_{12} \\ G_{21} & G_{22} \end{bmatrix}$$

$$= \begin{bmatrix} 0.909222 & -0.248483 \\ -0.301224 & 0.0845040 \end{bmatrix}.$$

The determinantal equation of **G** is given by

$$\begin{vmatrix} G_{11}-\lambda & G_{12} \\ G_{12} & G_{22}-\lambda \end{vmatrix} = \lambda^2 - 0.993726\,\lambda + 0.00198385 \quad .$$

The two roots are

$$\lambda_1 = \frac{0.993726 + 0.989725}{2} = 0.991726$$

and $\lambda_2 = 0.002000$. The canonical correlations are respectively $C_1 = \lambda_1^{1/2} = 0.995854$ and $C_2 = \lambda_1^{1/2} = 0.0447259$.

We can write for the canonical coefficients in set X (formula (3.35)),

$$-1424.52\ b_{11} + 74.2514\ b_{21} = 0$$

$$74.2514\ b_{11} - 3.87056\ b_{21} = 0\ .$$

The solutions are in the proportion $b_{11}/b_{21} = 70.3808/-1350.27$ or $b_{11} = -0.0521231$ and $b_{21} = 1.00000$. The normalized values are then $b_{11} = -0.0520525$ and $b_{22} = 0.998644$.

After the canonical coefficients are determined, we can apply formula (3.33b) to calculate canonical scores for sites. For instance, site *4* in Table 3-1 has its canonical score given by

$$U_{14} = b_{11}\, X_{14} + b_{21}\, X_{24} = (-0.0520525)\,(182) + (0.998644)\,(12) = 2.51018\ .$$

The remaining sites have their scores,

Canonical variate	*Site* 1	2	3	4	5	6
U_1	4.48234	7.12557	−5.32080	2.51018	5.49064	0.632478 .

We could extract one more set of canonical coefficients ($\mathbf{b}_2$) and canonical scores ($\mathbf{U}_2$) from the X set if we used λ_2 in formula (3.35).

The elements in $\mathbf{V}_1$ and $\mathbf{U}_1$ or $\mathbf{V}_2$ and $\mathbf{U}_2$ could be used as ordination co-ordinates for sites. $\mathbf{V}_1$ could be plotted with $\mathbf{V}_2$ as perpendicular lines, and points could be placed in the ($\mathbf{V}_1$, $\mathbf{V}_2$) plane to construct a scatter diagram for sites. We could also construct a scatter diagram for sites to show the joint distributions of $\mathbf{U}_1$ and $\mathbf{U}_2$, $\mathbf{V}_1$ and $\mathbf{U}_1$ or $\mathbf{V}_2$ and $\mathbf{U}_2$. Canonical analysis has been used in this sense by, e.g., Austin (1968), Pielou (1969a), but in a different sense by Seal (1964). Based on Seal's method, clusters of sites were ordinated by Grigal & Ohman (1975).

It is widely believed that rotation of the ordination axes may improve ecological interpretability (cf. Loucks, 1962; Greig-Smith, 1964; Ferrari & Mol, 1967; Bryant, Crandall-Stotler & Stotler, 1972; Hinneri, 1972; Noy-Meir, 1971b). Rotation was considered advantageous by Dagnelie (1960, 1965b) in factor analysis, by Gittins (1965) in the Bray & Curtis method, and also, by Noy-Meir (1973b) in general. The criteria of rotation varies of course depending on method of ordination and author.

If we consider that all information available in an ordination lies in the resemblance matrix **R**, the question of just how efficient the ordination is may lead one to try to correlate the resemblance structure **D** in the ordination with the resemblance structure defined by **R**. Methods to do this are described by, e.g., Swan & Dix (1966), Austin & Orlóci (1966) and Orlóci (1973a). By correlating resemblance values, a more informative evaluation of ordination efficiency is possible than simply by measuring the total variation accounted for by the ordination axes. The reason is that a comparison of **R** and **D** takes into account the spatial placements of points relative to one another, and not just the dispersion of points in specific directions about some centroid.

In predictive ordinations, efficiency is measured based on how informative the ordination configuration is in revealing the states in the factor or factors which are objects of prediction. Two alternatives are readily conceived. In the first case, actual measurements **X** are available on selected environmental variables. The ordination co-ordinates **Y**, as well as the measurements **X** are assumed to be random samples from given populations. The predictive value of **Y** for **X** is then determined via some correlation or regression analysis. Alternatively, predictive efficiency may be tested experimentally (e.g. Whittaker & Gauch, 1977). We discussed related topics in Section 3.6.

3.8 Sources of distortion

When information is transferred from one form into another with potential losses along the way, such as in ordinations, it is logical to ask the question: *what can possibly justify an analysis which wastes information?* The answer is that through reduction of information important clues may come to light about trends and classes of vegetational variation, and their environmental controls, which could not otherwise be detected in the data. A given ordination's success will of course depend on the efficiency of the transformations which it incorporates. In this respect, the properties of the distance function and the method of co-ordinate extraction are uppermost in importance. We can recognize several types of distortions (Orlóci, 1974b), related to transformations:

1. *Type A distortion.* When a vegetation sample. characterized by specific non-linear trends, is ordinated on the basis of a linear or unfitting non-linear model, a certain amount of distortion can be anticipated. The problem with inappropriate models is that the ordination axes do not respond to the existing non-linearities in the sample, and therefore, existing trends in the data may be left unrecognized. A practical question arises at this point with regard to two often used ordination techniques. *Can the information associated with non-linear trends in a sample be represented or will it be lost in such linear ordinations as component analysis or the* Bray & Curtis *method?* We have already answered this question indirectly in our discussion of Fig. 3-9. The following can be observed further in this connection:

a. Non-linear correlations between species normally yield curved point configurations in sample space.

b. Different resemblance measures define different configurations.

c. Neither component analysis nor the Bray & Curtis method necessarily diminishes non-linearity.

After observing these, we can make the following generalization: any abstract spatial configuration of points, linear or non-linear, which can be completely described by a distance matrix, can be embedded in the geometric space of component analysis, or of the Bray & Curtis ordination, without scrambling the original point distances if the distance function is Euclidean and if the ordination space has sufficient dimensions. It is implied here that the information associated with any trend under the specified constraints can be retained by these ordinations given sufficient dimensions, but at the same time, the linear constraints will render the ordinations potentially inefficient, both as parsimonious summarizes of variation and as predictors of environmental influence.

2. *Type B distortion.* When a t-dimensional point configuration is projected into an ordination space of less than t dimensions, an amount of information will be lost due to distortions by projection. Construction of two-dimensional scatter diagrams is a typical example for this in which normally a great deal of *Type B* distortion must be anticipated. Feoli (1977) associates this type of distortion with the resolving power of ordinations.

3. *Type C distortion.* This is intrinsic in the distance function's own distorting effect. We give as an example the metric,

$$s\,(j,\,k) = \{\,2\,[1 - (\chi^2_{jk}/p)^{1/2}\,]\}^{\,1/2},$$

where χ^2_{jk}/p measures mean square contingency on two quadrats. The distortion, in this case, is a consequence of the *signum* transformation which $s(j,\,k)$ incorporates. After such a transformation only that portion of the information is retained in the data which is associated with species presence and absence.

4. *Type D distortion.* This occurs when the resemblance function is incompatible with the ordination algebra. For example, the different versions of the Bray & Curtis (1957) ordination include operations with triangles which assume that the resemblance function is Euclidean. Now, when a non-Euclidean distance is exposed to such operations, such as in Beals (1960) version, the original point configuration is forced into an ordination space where it cannot fit. The result is an amount of *Type D* distortion.

5. *Type E distortion.* This occurs when unsuitable ordination co-ordinates are used to predict levels of environmental influence along a given gradient. Unsuitability is a consequence of the fact that the method of ordination used did not produce co-ordinates that are related to species performance, in the manner as species response is related to changing environmental influence along the gradient whose existence is hypothesized. Component scores represent an example for inefficient predictors, and the co-ordinates of Gaussian ordination (Johnson, 1973; Gauch & Chase, 1974; Austin, 1976b) for efficient predictors assuming that species response is Gaussian.

It is common practice in ecology to evaluate an ordination's success on the basis of results which it happened to produce in a local application. While we realize that a system of evaluation that relies on measurement of success under very specific survey or experimental circumstances (e.g. Whittaker & Gauch, 1973), and with very specific objectives in mind, can be useful, we must also realize that, in the extreme, the results of evaluation may be so much context dependent that they will have relevance nowhere except in the analysis which produced them. Context dependence is exactly the reason why it is so difficult to generalize the results of recent experiments with ordinations which were derived on the basis of assumptions that are very restrictive. Some even suggest the exact functional form of species response (e.g. Whittaker & Gauch, 1973). The danger here is that if the response is not as assumed the results will have no relevance for the data analysed.

Not infrequently, the evaluation criteria stress the 'ecological meaningfulness' of the results, notwithstanding the fact that there may be such weaknesses in the method which will render the search for ecological meaningfulness completely pointless. Furthermore, some ecological meaning can be interpreted in most ordinations, especially when no concrete objectives are set a priori for the ordination to accomplish. Therefore, it is important that an evaluation minimize the need for subjective value judgements, and use objective criteria instead. A reliable system of evaluation should emphasize axiomatic and other measurable properties in providing reliable answers to two pertinent questions:

1. *Is ordination an appropriate method for data analysis under the circumstances?*
2. *If it is, which kind of resemblance function should be chosen for ordination in association with which method of axis construction?*

Regarding these questions we must clearly define the objectives. Furthermore, we must consider the various sources of potential distortion. Informative discussions are offered on the general topic of choice by, e.g., Noy-Meir (1970, 1971b, 1973a, 1977) and Austin (1976a,b, 1977), and from a similar point of view with regard to predictive objectives by Whittaker & Gauch (1973, 1977). The *Type A* distortion should weigh heavily in any choice of resemblance function and method of axis construction. Unfortunately, a satisfactory solution to the problems springing from such a distortion is not a simple matter, and it may have to be attempted by some indirect means. We may, for instance, restrict the sampling to stands which are characterized by an overall linearity of species correlations. It has been suggested that the linearity condition is likely to be satisfied if sampling units are selected from a relatively narrow segment of a predominant environmental gradient (cf. van Groenewoud, 1965; Austin & Noy-Meir, 1971; Gauch & Whittaker, 1972; Noy-Meir, 1974a,b; Bouxin, 1975a,b; Phillips, 1977). When evidence indicates that species correlations are non-linear, and the non-linearity is suspected to be strong, a linear ordination alone is inappropriate. If perfomed, it should be followed up by other analyses that can reveal the intrinsic non-linear trends in the data. Methods for non-linear ordination, which have potential in reducing the *Type A* distortion, were presented earlier. However, it is relevant to note that non-linear ordinations which assume a specific kind of non-linearity in the sample (e.g. Gaussian species response) will certainly have no more generality for ecological applications than the linear models. Such non-linear models are preferred which require no assumption about the type of the non-linearity.

We refer also to results of comparative studies performed by Jeglum, Wehrhahn & Swan (1971), Whittaker & Gauch (1973) and Gauch (1973a,b). These indicate that there is a relationship between increasing contrast of species composition between stands and the loss of power of the component scores in predicting environmental influence. This would of course be a problem with any linear ordination, or with non-linear ordinations which do not fit the non-linearity in the data. Under assumption of Gaussian species response to a single environmental factor, at a contrast of 4 or so half changes, some authors advise against using linear ordinations. A half change is the length of a gradient segment over which the similarity of species composition in two stands is reduced by a factor of 2. At 4 half changes, the similarity appraoches zero.

While most conventional methods, such as component analysis, may be of little help to control the *Type A* distortion, the Bray & Curtis method comes out worst in controlling the *Type B* distortion. The reason is that the ordination axes, produced by this method, and the principal directions of variation in the sample lack any predictable relationship. The *Type B* distortion is effectively controlled in linear data structures when the analysis is based on an eigenvalue and vector method, such as component analysis and principal axes analysis. When the data structure is non-linear an efficient ordination cannot be achieved by these methods.

The *Type C* distortion is entirely preventable by choice of an appropriate resemblance function. The *Type D* distortion is also preventable, provided that the distance function and the ordination geometry are compatible. Compatibility should be understood in such a sense that the function qualifies as a valid spatial parameter in the sample space of the ordination model. In component analysis, and also in the Bray & Curtis ordination, especially poor choices are the semimetrics, surpassed in lack of appropriateness only by non-metric distance. In these ordinations, Euclidean measures are preferred: $e(j, k)$ or $a(j, k)$ when the influence of absolute species quantity is to be retained, and $c(j, k)$ or $u(j, k)$ when the distance measure is to be put in a relative form.

The *Type E* distortion is just as difficult to control as the *Type A* distortion. It requires specialized ordination methods that can yield co-ordinates which are so related to species response as species response is to changing environmental influence along a gradient. So far, mainly the Gaussian case has been studied (e.g. Johnson, 1973; Gauch, Chase & Whittaker, 1974). Other cases characterized by response curves other than the Gaussian (cf. Noy-Meir, 1977; Phillips, 1978, and references therein) are just being isolated and solutions found.

The sequence in which we discussed the potential sources of information loss in ordinations is not intended to indicate their relative importance. Their importance should be weighed by the user with a view to achieving specific objectives. We may wish, for instance, to summarize vegetational variation on a new set of variables, or equivalently, transform the data into new forms which qualify as co-ordinates on axes, in order to achieve a parsimonious representation of linear or non-linear data structures. The definition of efficiency is straightforward in the linear data structures where it means summarization on axes with zero covariances. But it is not so straightforward in the non-linear case where it may mean different things depending on the type of non-linearity present, as we have seen in our earlier discussions. The objectives may require derivation of co-ordinates for individuals for which only resemblance values are available. The co-ordinates are obtained based on some method of ordination. Those methods should be preferred which can efficiently represent a given data structure. Existing vegetation trends may have to be revealed and correlated with changes in the environment. Since vegetation response is most likely non-linear, methods that can efficiently represent non-linear data structures have special appeal.

If the objective is summarization, it is logical to consider the success of the analysis in terms of the total variation accounted for by the ordination axes. This is an *intrinsic* condition in the data and in the method of analysis. Similarly, when the goal is multidimensional scaling, the logical condition for evaluating an ordination's success is the degree to which it succeeded in representing the existing data structure in the fewest possible dimensions. In neither case would it be necessary to inject *external* criteria in the evaluations, or to try to define 'meaningfulness' in the sense of what the results may mean in ecological terms. Ecological meaningfulness is of course a necessary criterion in evaluating an ordination's success when the objectives call for revealing ecological information.

Among the different methods, component analysis stands out as the single most efficient technique to summarize continuous linear data structures. While this function is performed by component analysis extremely well, this method is not very helpful when the data structure is non-linear. The method of principal axes analysis is mathematically similar to component analysis, and it has similar limitations, except that it is somewhat more open regarding the admissible resemblance functions. Reciprocal ordering, similar to component analysis and principal axes analysis, incorporates a linear definition of the data structure in the sample. This means a serious constraint when it comes to actual applications. The principal justification at the basis of the Bray & Curtis method is its relative computational simplicity and the option for personal choice of ordination poles. This method, however, is a poor substitute for component analysis when the objective is summarization. Nevertheless, it may still render a useful service for those who like to preselect poles to define ordination axes. While position vectors ordination is potentially more efficient than the Bray & Curtis method in summarizing linear data structures, and computationally simpler than component analysis, it is not expected to be any more helpful when the data structure is non-linear. The Kruskal method, the Shepard & Carroll method, Sneath's method, Gaussian ordination, etc., represent a special class of ordinations. These methods do have the advantage of being capable of handling data structures which are not linear.

It should be assumed that the user will choose a method after careful considerations of its suitability for the purpose at hand. No single method should be regarded as a 'general preference' method or used indiscriminately in connection with different objectives. Ordinations should be treated as highly specialized methods which can provide optimal performance with respect to a few, specific goals under particular circumstances.

Chapter 4

CLASSIFICATION

The process which produces new classes constitutes a classification. This should be clearly distinguished from identification which finds the class, among a number of established classes, that represents the most likely parent population for a new individual. In this chapter, we shall consider classifications – the concepts, objectives, methods and conditions of application.

4.1 Informal approaches

Classification is a broadly used tool in vegetation studies, but it often lacks a formal statement of the procedure. The dominant phytosociological schools have traditionally made themselves distinct by the approach in which they pursued informal classifications:
1. *The zonal approach.* Broad regional units are recognized in the vegetation, closely corresponding to climatic and soil zones. This approach reached an early climax with Dokuchaev's (1899) work on Natural Zones.
2. *The dynamic or successional approach.* The main concern in this lies in the recognition of developmental stages in the vegetation, and on that basis, establishment of classification systems. This approach culminated with the work of Clements (1916, and other publications).
3. *The type or association approach.* In this, vegetation units are recognized on the basis of the most abundant species in the undergrowth (Cajander, 1909), dominant species (Tansley, 1939), or diagnostic species (Braun-Blanquet, 1928, 1951).
4. *The holistic or ecosystem approach.* In this, Krajina (1933, 1960) asserts, the ecologist assumes that discrete ecosystems exist and that via the ecosystem units a joint classification of vegetation and environment is possible.

The general aspects of the informal methodologies have been put into perspective by Whittaker (1962), and the more specific aspects by Westhoff & van der Maarel (1973), Whittaker (1973b), Frey (1973), Mueller-Dombois & Ellenberg (1974) and van der Maarel (1975). It is quite evident that the methods can, in fact, detect natural classes (types, associations, etc.) with reliable consistency where vegetational variation is obviously discontinuous.

4.2 Formal clustering methods

It is common to the different fields of science that as their methodologies evolve the changes are cumulative in that the different historical approaches tend to coexist. This is particularly true in phytosociology where the contrasting methodologies are numerous with few ever becoming completely obsolete. While this may seem extraordinary, it is a normal state of affairs, readily explained by the excessive diversity of scientific interests and objectives, and by the varying standards of scientific acceptability of a product among the practitioners of the Art. The trend, in fact, points to further divergence of approach, formal and informal, rather than convergence as it has been hoped (cf. Krajina, 1960, 1961).

Our account of classification methodologies will be devoted to the so-called formal procedures, noted for their suitability to classify sets of data which Hill, Bunce & Shaw (1975) characterized as being visually heterogeneous, too complex and too voluminous to be effectively handled by other methods. The formal clustering methods could of course be mere restatements of procedures that were used informally in previous applications. There are good potentials for such formalized procedures, particularly in the automation of table rearrangements for which algorithms were offered by manv authors (e.g. Beninghoff & Southworth, 1964; Moore & O'Sullivan, 1970, 1973, Lieth & Moore, 1971; Češka & Roemer, 1971; Spatz, 1972; Janssen, 1972; Stockinger & Holzner, 1972; Spatz & Siegmund, 1973; Holzner & Stockinger, 1973; Dale & Quadraccia, 1973), and matrix rearrangements to concentrate high stand or species similarities close to the principal diagonal positions (e.g. Motyka, 1947). Since the discussion of these is beyond the scope of the present book, we refer the reader to references already mentioned and to reviews by Westhoff & van der Maarel (1973), McIntosh (1973) and Mueller-Dombois & Ellenberg (1974) for comments and still further references.

We address in the present chapter the classificatory problem of how to establish classes of objects in a set which has not yet been subjected to analysis with specific classificatory objectives in mind. The problem here is quite different from the problem of identification (to be discussed in Chapter 5) which amounts to locating the class of best fit among given reference classes for each new, as yet unidentified object.

What is usually meant under cluster analysis is a procedure which assigns objects to groups in such a way that each group reveals some distinct characteristic of the sample (cf. Blackith & Reyment, 1971). Clustering methods historically evolved along parallel lines, frequently involving the discoveries of the same method by independent workers. Reviews of methods are offered by Sokal & Sneath (1963), Crovello (1970), Sneath & Sokal (1973) and Clifford & Stephenson (1975) in numerical taxonomy, by Blackith & Reyment (1971) with focus on morphometrics, and by Jardine & Sibson (1971) regarding axiomatic properties as related to taxonomic applications. Cormack (1971) gives an informative overview of clustering methods in general. Relevance to vegetation analysis is shown by Greig-Smith (1964), Lambert & Dale (1964), Kershaw (1964, 1973), Williams & Dale (1965), Gounot (1969), Pielou (1969a,b), Goodall (1970b) and Shimwell (1971). Goodall (1970b, 1973a) reports on the state of the Art, while others (e.g. van der Maarel, 1969, 1970, 1971; Moore, Fitzsimons, Lamb & White, 1970; Ivimey-Cook, 1972; Feoli, 1973b; Kortekaas & van der Maarel, 1973; Whittaker, 1973b; Werger, 1973; Westhoff & van der Maarel, 1973; Bouxin,

1975a; Coetzee & Werger, 1975; Dale & Quadraccia, 1973; Dale & Webb, 1975; Feoli Chiapella & Feoli, 1976; Kortekaas, van der Maarel & Beeftink, 1976; Orlóci, 1976b; Pakarinen, 1976) show connections to the informal methods.

We shall describe different clustering methods which manipulate resemblance matrices to delimit clusters of points in sample space. We shall illustrate the methods by numerical examples. But first, some general topics will be discussed.

4.3 The medium

When one conceives the vegetation as a multispecies population, one implies an aggregate of natural units. What are these units? If they exist, they certainly lack a strong natural basis for identification. At best, they are recognizable only with ambiguity, since they are stands of complex composition whose limits are not easily traced. A logical extension of the notion that the vegetation is a population of individual stands is the idea that the stands may form distinct groups which may be recognized as *types, associations,* etc.

One may of course conceive the vegetation differently, not as a multispecies population, but as a complex of individual species populations representable by a series of smooth response graphs (cf. Curtis & McIntosh, 1951) with maxima ordered according to environmental influence.

When a medium can induce such diametrically opposing conceptualizations about itself, it is no wonder that there should be contrasting views about its basic properties. To hold such obviously opposing views need not however imply inconsistencies in theory or in practice (Goodall, 1963, 1973a) since the stipulated properties may in fact be appropriate descriptors of the vegetation individually or jointly under certain circumstances.

The assumption that the vegetation is an aggregate of recognizable units is directly in line with the views advocated by many phytosociologists, most notably by those in the dominant European schools. They assert that not only are there distinct units, but the units tend to form natural types. The type, in this case, is an abstraction comparable to a population (Westhoff & van der Maarel, 1973). While this viewpoint has been challenged repeatedly in the past (cf. McIntosh, 1958, 1967b, and references therein), this did not deter many phytosociologists from continued erection of classificatory systems based on natural types. Whereas the assumption that natural types exist is central, although not necessary, in the traditional classificatory approach, the assumption of vegetation continuum, or the denial that the vegetation is an aggregate of natural types, is just as fundamental in the early approach, advocated by the American ordinationists, concerned with the discovery and quantification of continuous trends in vegetational variation.

It is quite relevant to the prevailing arguments, whether they promote one or another of the classical views, that in the main they have relied on arbitrary definitions for natural types, or as a matter of fact, for vegetational continua (cf. Scott, 1974; Dale, 1975). It can be suggested that any attempt to prove or disprove the existence of natural types, as discrete populations, has yet to resolve many difficulties:

1. *The difficulty of scale.* At which scale should one look for distinguishable types? Should it be a scale at which only gross vegetational differences can be detected, or should it be a much more refined scale?
2. *The difficulty of measurement.* How accurately should one measure the vegetation before one can be satisfied that the conclusions drawn from the measurements can be accepted with confidence?
3. *The difficulty of continuously changing systems.* Species, communities, or even environments, are not static. The time factor is important. When should the observations take place? What is the expected time span of an observation as a valid fact?
4. *The difficulty of complexity.* The vegetation and the environment have the potential of great complexity. How much of the total complexity must then be taken into account before any statement about the existence or non-existence of natural types can be firmly supported?
5. *The difficulty of population size.* A plant community may have an extent far beyond the possibility of complete enumeration. One may have to be satisfied with statistical estimation (rather than exact determination) of the population parameters. How much error in the estimates is permitted before rendering the results unacceptable in drawing conclusions about the existence of natural types?
6. *The difficulty for the 'intelligent ignoramus'.* One's conclusions about the existence of natural types are affected by experience and personal bias. Which person's views, therefore, should be accepted with confidence in the absence of hard, objective evidence?

In our presentations the validity of the classical views will not be argued, but rather, it will be assumed that a classification has to be done in order to achieve certain utilitarian goals. The principal question thus facing us at this point is not whether a classification can be done but *how* should it be done so that it can best serve the objectives.

4.4 Why classify?

It is quite appropriate to ask this question for two specific reasons. Firstly, classifications represent one of the many possible ways in which vegetational data may be analysed. Secondly, the user must be satisfied that classifications can do something uniquely well under the circumstances which could not be done so well by any other method. What then are the objectives that may call for a classification? These commonly fall into three categories:

1. Problem solving (prediction, hypothesis testing, etc.).
2. Problem recognition (hypothesis generation).
3. Data reduction, inventory, etc.

Different users emphasize different points. To Good (1965), planning or just to have fun, are equally respectable classification objectives. To others derivation of a general purpose framework (Goodall, 1973a), file organization (Borko, Blankenship & Burket, 1968), or even the discovery and description of unsuspected clusters (Fleiss & Zubin, 1969) are sufficient reasons to justify a classification. To Jardine (1970), classifications suggest 'fruitful hypotheses' that may not be suggested by other analyses and to Sokal (1974), they serve description of group structure and group relationships. To Williams & Lance (1965), a main objective is prediction. They can see reasons to perform a classification as an exploratory analysis, even if precise objectives are not specified. But to Rubin (1967) a well-reasoned statement of the objective is important because the user's success of finding the best classification depends on it.

Regarding objectives in the first category, inductive generalizations may be intended to describe the parent population of a sample (cf. Macnaughton-Smith, 1965; Anderberg, 1973). Alternatively, correlations may be sought between the vegetation and the environment to facilitate the use of vegetation classes as predictors of certain environmental conditions (Orlóci, 1972b). We may consider, as an example, an allocation of n units to k classes based on vegetation data, and then, an allocation of the same units to m classes based on soil data. The extent to which certain vegetation classes tend to hold units in common with certain soil classes is an indication of the predictive potential of the vegetation classes. While classifications can be used to perform a predictive function reasonably well (Goodall, 1966c,d), they are generally not suited to serve as a basis for testing hypotheses about the existence of natural types (McIntosh, 1967b). As a matter of fact, just because a sample of vegetation units can be partitioned into groups by some method of classification, one should not conclude that the vegetation consists of discrete types (Pielou, 1969a). Furthermore, groups produced in a classification need not have, as Anderberg (1973) asserts, 'inherent validity or claim to truth'. They are in fact subject to interpretation and may be discarded if found irrelevant.

Predictions based on classification are not without pitfalls. For example, a complete circularity of reasoning would be involved if we hypothesized that the groups of stands, recognized in a classification based on their species content, were samples from discrete types, and if we went about testing this hypothesis knowing that the different stands were assigned to the groups in such a manner that the groups must come out distinct in terms of vegetation. However, it is perfectly legitimate to incorporate a probabilistic decision function in an agglomerative algorithm, such as Goodall's (1970a,b, 1973a), and not to allow fusions to occur for which the null hypothesis, that all of its members are samples from a homogeneous parent population, is untenable. If such an hypothesis were in fact accepted, it would necessarily imply that the different groups identified by the algorithm in the same sample derive from different parent populations. It would also be perfectly legitimate to go about testing the hypothesis of indistinguishability between groups based on vegetation data but excluding the data set from which the groups were actually formed. Furthermore, hypotheses could be tested on the indistinguishability of the groups in terms of environmental variables. The justification for such a test is that the criteria which form the groups, and the criteria on the basis of which they are tested, are different.

Williams & Dale (1965) assert that classifications can serve as hypothesis generating systems. Others believe that there are potential dangers in hypothesis generation in classification. Those who believe this reason that since the groups are products of partitions, when no natural groups exist, a classification will force the data 'into a strait jacket which restricts the domain of possible hypotheses and suggests that some will be generated by the fact of dissection rather than by the data' (Cormack, 1971). The problem of restricting the domain of hypotheses is of course not unique to classifications. The implications in formulating and testing unrealistic hypotheses, having no roots in the data, except in the clustering algorithm, are that they can be time consuming, expensive and misleading. We must however stress the point that vegetation classifications are known to have the potential to display group structures in which the positions of individuals or groups are conducive to new ideas and discovery of new facts about the vegetation.

Among the many uses to which classifications are put, data reduction is by far the most prominent. But unlike in ordinations where data reduction is by strategically chosen axes, here data reduction is by allocations to groups on the basis of suitably chosen criteria. Such groups may serve as a basis for inventory, mapping, organization of future research, and many other utilitarian objectives. In this sense, a classification can provide concise descriptions of the complexity in the data which may lead to more efficient communication about relevant class properties. New information may be revealed in the process, which in turn can illuminate such class properties which would not otherwise become apparent (cf. Greig-Smith, 1964; Jones, 1970; Goodall, 1973a), and which could point the way to identify and to rectify cases where things are not what they appeared to be (Davidson, 1967).

Irrespective of the objectives, whether they call for a special-purpose classification or for the establishment of a general classificatory scheme, the classification itself may be optimal or suboptimal. But optimality can mean different things. It may imply that a certain function, called for by theory, is in fact optimized in the classificatory algorithm (Cormack, 1971; Goodall, 1973a). It may also imply that the classification reveals some useful information, sought by the user, better than could be revealed by other methods of analysis. But, if the definition of optimality relies more and more on judgement of usefulness and practical value, or if it becomes less and less axiomatic, the evaluation of optimality can become context dependent with proportionate losses of generality.

4.5 Classification or ordination

This question, to some, will pose a philosophical dilemma. Yet, to others, it will amount to nothing more than a call for making a pragmatic decision. It should not be surprising then that ecologists of different cultural backgrounds, and at different moments of history, differ in their preference for one or another of the methods.

The strongest proponents of the classificatory approach look upon classification not as a technique, but rather as a general strategy. In their frame of reference classification is so fundamental that, to them, without classification there can be no science of vegetation (e.g. Krajina, 1961; Daubenmire, 1966). In direct contrast stand the ecologists who assume a continuum. To them 'vegetation changes continuously and is not differentiated, except arbitrarily, into sociological entities' (McIntosh, 1967b; also Vogle, 1966; Cottam & McIntosh, 1966). To some holders of the continuum concept of vegetation, classifications have no strong appeal; to them ordination represents the preferred strategy.

Greig-Smith (1964) asserts that ordination and classification as approaches 'are, in theory, quite distinct' but then he adds that 'in practice the divergence is not so great as may appear'. Greig-Smith (1971a) rejects the idea that ordination is necessarily more satisfactory when variation is continuous, and classification is more appropriate when it is discrete. McIntosh (1967b) and Whittaker (1972) make a similar point when they claim that to many ecologists ordination and classification are not necessarily incompatible techniques. To Goodall (1963), the idea of ordination is not inconsistent with that of classification; to Whittaker (1956, 1972), Greig-Smith, Austin & Whitmore (1967), Rhodes, Malo, Campbell & Carmer (1971) and Jesberger & Sheard (1973) classification and ordination are mutually complementary; to Major (1961), the difference between them is just a question of degree; to Kershaw (1973), they are equally valuable, and to Anderson (1965) neither is intrinsically more correct than the other. Noy-Meir & Whittaker (1977) assert, that the question 'ordination or classification' has lost much of its sharpness, and they go on record saying that both ordination and classification can be useful to analyse the same set of data. As a matter of fact, classification and ordination in some problems are so complementary that an entire analysis can be built about their combined application (e.g. Dale & Anderson, 1973).

In line with these views we may add that ordination and classification should not be regarded as some preferential strategies, associated with rigid assumptions, but rather, as techniques which among the many different techniques of data analysis can help the user to accomplish certain objectives. It seems that there is no need for the ecologist to commit the analysis to one or another of the classical hypotheses prior to a vegetation study. On the contrary, an analysis may be best approached without an a priori commitment to one or another of such broad hypotheses.

The problems of choice between ordination or classification is often discussed with a view to what they may do to a given set of data already in the ecologist's possession. This can give a false impression that ecologists ask no questions prior to collecting the data. Appropriately, the choice of method should be the consequence of the questions asked, and so should be the choice of data. Accordingly, the best an ecologist can do is to try to state clearly the objectives to be achieved, and then, to select an appropriate method for sampling, and a suitable method for data analysis. When the objectives are clearly defined, and the methodological implications understood, the choice between ordination and classification should not be too difficult.

4.6 Model for classification

The spatial analogy used in ordinations is directly relevant in classifications of certain kinds. Different accounts treating classifications have used such an analogy (e.g. Goodall, 1963, 1973a). When the data structure is continuous (see Section 3-2) classes can be produced only by dissections. If, however, natural clusters exist, represented by distinct density phases of points in space, such as the density modes of Forgy (1963, 1965), it is tempting to use them as classification categories. The shape of the point cluster is important. Some methods will be able to recognize clusters if spherical in shape, but not if ellipsoidal or curved.

There are three peculiar aspects that may complicate cluster recognition. Firstly, dissections or natural clusters may both result from the same method of cluster analysis. This is likely to happen when the number of classes to be recognized is prespecified and the number of the natural clusters is less than the number of classes sought. Secondly, the appearance of density phases in sample space may be influenced by sampling (Goodall, 1973a). Where the sampling is preferential, the appearance of density phases is likely to be intensified. Thirdly, the data structure may be so complex that the sample space will exhibit a nested pattern of clusters. When this happens, natural clusters could be recognized at different scales. The ecological circumstances under which these types of data structures arise were considered in Section 3.2.

The spatial model which we just described is meaningful in some clustering methods, but irrelevant in others in which density is not a valid concept. In spaces of this sort, clustering may rely on the probability of belonging to a group, the information generated by being fused to a group, etc. Clearly then, the basic classificatory model need not imply spaces of the metric kind; they may be associated with other abstractions which use non-metric spatial parameters.

Classificatory procedures normally include manipulations of similarity or dissimilarity matrices from which the classification parameters are derived. Whereas some of the methods fall back on the raw data after each clustering pass (e.g. Williams & Lambert, 1959), others do not, since they utilize information that can be derived directly or indirectly from the resemblance matrix and its residuals. Agglomerative procedures of the latter type are said to give a combinatorial solution (Williams, 1971). The choice of resemblance function still is critical in either case (Lance & Williams, 1965), since the results of clustering depend upon it.

Procedures incorporating a combinatorial solution are at a computational advantage over those which have to fall back on the raw data. For such algorithms, Lance & Williams (1966a, 1967b; also Rohlf, 1970) have shown that a common underlying transformation exists through which the resemblance of two groups can be determined at any clustering stage. This transformation can be represented in symbolic terms by

$$\delta(j,k)=\alpha_h\delta(j,h)+\alpha_i\delta(j,i)+\beta\delta(h,i)+\gamma\,|\delta(j,h)-\delta(j,i)| \quad (4.1)$$

Wishart (1969a) gives a similar formulation (see also Pritchard & Anderson, 1971; Cormack, 1971; Sneath & Sokal, 1973). Symbol $\delta(j, k)$ indicates the dissimilarity of group j and k where k is the fusion group $(h + i)$. The $\delta(j, k)$

may be replaced by similarities (cf. Cormack, 1971). The exact definition of δ, α_h, α_i, β and γ depend on the classificatory strategy (Lance & Williams, 1967b; Wishart, 1969a; Cormack, 1971). We shall consider this point further in the sequel.

4.7 On concepts and terminology

When should a partition of a sample into groups be called a *classification?* Williams & Dale (1965) answer that question by laying down three axioms. The first requires that individuals in any one group share one or more common characteristics. The second states that group size is not a relevant characteristic, and that allocations must be open-ended in the sense that the classification will in no way limit the number of individuals per group. The third restricts classifications to such subdivisions which produce non-overlapping groups. Classification is however often defined in a broader context to include partitions which conform to the first two axioms, but not the third. These accomplish *clumping*, meaning the formation of overlapping groups (cf. Jardine & Sibson, 1971). When a completely continuous sample is classified, the resulting groups represent *dissections* (Kendall, 1966; Cormack, 1971).

A clustering procedure performs *indirect or transposed-clustering* (Lambert & Dale, 1964; derived-structuring by Williams & Dale, 1965) when it derives a classification of individuals based on an analysis of the variables which describe the individuals. The alternative is *direct* or *self-clustering* based on an analysis of the measured resemblances of the individuals themselves. A further distinction is made between *intrinsic* or *extrinsic* clustering (Williams, 1971). In both, it is assumed that the groups are defined on the basis of properties of the individuals internal to the groups. External variables are not considered. However, whereas in intrinsic clustering the intrinsic pattern of the group is of principal interest to the user, and the predictive value of the classification with regard to the state of external variables may not be important at all, the intrinsic pattern takes second place in importance to predictive value in extrinsic classifications (cf. Macnaughton-Smith, 1965).

Structuring produces groups. *What constitutes a group?* This can be answered simply by saying that any subset of objects that appears distinct either by internal homogeneity or external difference constitutes a group. Cattell & Coulter (1966) recognized two types of groups on this basis and described them as a *homostat* and *segregate*. Williams (1971) finds the value of such a distinction doubtful in solving clustering problems, since users do not normally think of a group as such unless it is both a homostat and a segregate.

To Williams (1971) a given classification procedure implies two things: a *mathematical model* and a *strategy* which controls the model's implementation. But the model and strategy are not necessarily independent: the choice of the model usually dictates strategy or the choice of the strategy dictates the model.

Classifications are implemented under different interpretations of the *sampling status* of individuals. Williams & Dale (1965; also Williams & Lance, 1965) make a sharp distinction between *probabilistic* and *deterministic* (non-probabilistic) strategies of implementation. A typical case in point, calling for deterministic implementations, is when a collection consists of quadrats selected in preferential sampling. Another is a collection of quadrats supplied by complete emuneration. Williams & Lance (1965) believe that in deterministically implemented classifications the notions *true* or *false* and *probable* or *improbable* do not apply. The idea of statistical estimation does not apply either, since the collection itself is the statistical universe. Only *utility* or *profitability* can serve as criteria for judging value and significance. Probabilistic implementations dictate that the collection be regarded as a part of some statistical universe, i.e. a representative sample therefrom.

The mode of implementation has far reaching consequences, since it entails addressing the problem of decision in viewing the objective as one in *estimation* or in *description.* However, even if the implementation is probabilistic, the decision rules for fusions or subdivisions may be deterministic or probabilistic. Probabilistic decision rules are often preferred (e.g. Goodall, 1966a,b,c,d,e, 1970,a,b, 1973a, and references therein) since such rules allow the user not only to locate candidates for optimal fusions or subdivisions, but also, to evaluate the likelihood that the individuals or groups in fusion are in fact members of a common statistical population.

Goodall (1973a) finds it advantageous to distinguish between groups defined by *extension* and groups defined by *intension.* Extension groups are completely described by listing their contents, for an extension group always represents a closed category. Intension groups are open. They are described on the basis of properties, other than content, such as internal variation, spatial orientation, size etc. Goodall (1973b) sees no useful purpose in defining primary biological groups by extension, since the membership in such groups normally represents but a limited sample from an uncountably large parent population. Membership in a group can thus be regarded as a consequence of chance.

We shall follow tradition in distinguishing between classifications of two different kinds based on the overlap, or lack of it, in the groups which they produce:

1. *Overlapping groups.* In this, an individual is allowed to be a member in two or more groups simultaneously. The strategy that allows this to happen is variously described as *clumping* or *non-exclusive clustering* (Williams, 1971). In the present book, we do not deal with clumping techniques but we refer to work that does (e.g. Needham & Jones, 1964; Jackson, 1969; Jardine & Sibson, 1971; Yarranton, Beasleigh, Morrison & Shafi, 1972; Shafi & Yarranton, 1973). We recognize, of course, that a classification which forms non-overlapping groups of individuals may perform clumping in an indirect sense with regard to properties associated with the individuals. For example, when quadrats of vegetation are allocated to non-overlapping groups, the groups will almost certainly overlap for the species which occur in the quadrats. In an analogous sense, in a classification of species with non-overlapping groups, the groups may overlap in terms of the quadrats in which the species occur.

2. *Non-overlapping groups.* These do not allow an individual to be a member in more than one group. Williams (1971) recognizes a dichotomy in this strategy. The groups may be arranged in a hierarchy. The results of *hierarchical clustering* can be presented in a *dendrogram* (classification tree). The groups need not however be arranged hierarchically, but they may be recognized at a single level and presented without showing hierarchical relationships.

In non-hierarchical classifications, groups may be obtained *serially* or *simultaneously*. A serial strategy is one in which a group is formed and removed from the analysis before the formation of another begins. All groups are formed simultaneously in a simultaneous strategy. Williams, Lambert & Lance (1966) suggest that the objective to perform a hierarchical classification is to show the actual pathway of fusions. Others use hierarchical systems merely for cluster recognition, without an interest in the avenues through which the clusters evolve in the subdivisive or agglomerative process (Goodall 1973a).

Clustering may be *subdivisive*, when clusters are formed by subdivisions, or *agglomerative*, when clusters are formed by fusions. Opinions divide on the relative merits of the subdivisive and agglomerative methods. For instance, to Macnaughton-Smith (1965), divisive clustering appeals since the statistical errors are less troublesome than in agglomerative clustering which starts at the 'bottom of the page'. To Cormack (1971), Macnaughton-Smith's reasoning is unjustified, since in neither clustering is a common error structure assumed. Williams & Dale (1965) advocate the use of subdivisive methods on similar grounds as Macnaughton-Smith (1965).

It is true, of course, that subdivisive clustering methods have a potential advantage in that they use the total information in the data at the outset (cf. Noy-Meir, 1973b), and for that reason, unlike in agglomerative clustering, contents in the lower level groups have no influence on the higher level groups. A formal classification that recognizes the advantages of clustering by subdivisions, based on parameters computed over the entire group, is Goodall's (1953). Others in this group were proposed by, e.g., Williams & Lance (1958, 1965), Edwards & Cavalli-Sforza (1965), Noy-Meir (1973b) and Hill, Bunce & Shaw (1975).

It is also true that agglomerative clustering suffers from different weaknesses (cf. Williams & Dale, 1965; Williams, 1971). One is computational, in that the clustering requires the classificatory process to begin at the level of the individual, even if the user were interested in groups at higher levels. Another weakness is the possibility of misclassifications, due to the fact that fusions begin at the level where the possibility of committing errors is greatest.

Advantages and strengths notwithstanding, Gower (1967b) finds the subdivisive methods undesirable in general, because they can irreparably dismember existing groups too early in the clustering process. But quite significantly to the arguments, Goodall (1973a) finds the distinction between agglomerative and subdivisive methods not a firm one in view of the possibility that the same methods can incorporate clustering by both kinds. However, to Williams, Lance, Dale & Clifford (1971), this distinction is important because of the properties with which they are associated, e.g. whereas agglomerative clustering may not be monothetically implemented, subdivisive clustering is superbly suited for such implementation.

Classifications may be categorized according to the characters on the basis of which the partitions are recognized. Sneath (1962), for example, distinguishes *monothetic* and *polythetic* classifications which, respectively, partition a sample on the basis of the presence of one or more common characters or the amount of shared character states or values. In ecology, the term monothetic has been given the meaning of a classification in which groups are recognized based on the *presence* or *absence* of a given species. In typical, monothetic clustering, a group of quadrats is split into two groups such that the first contains and the second lacks a particular species. The clustering method of Williams & Lambert (1959) is of this kind. When fusions or subdivisions are conditional on the discriminating power of two or more species at any step in the clustering process, the algorithm is said to be polythetic. Orlóci's (1967a) clustering method is of this kind. Goodall (1973a) illustrates the meaning of these terms by a spatial analogy. Accordingly, in the monothetic case, when quadrats are classified, each dividing hyperplane is perpendicular to one species axis. In the polythetic case, the dividing hypersurface is likely to be oblique to several axes and may be curved. While monothetic fusions or subdivisions are considerably influenced by chance events, having to do with the presence of a particular species in a quadrat, polythetic fusions or subdivisions are likely to be less affected by chance events (Goodall, 1973a).

Biologists often erect requirements that a good classification must satisfy (e.g. Silvestri & Hill, 1964; Williams, 1967; Sneath & Sokal, 1973). Several of these are desired from an ecological point of view. The first is *stability*. This means that the classification should not be overly influenced by addition of new variables. The second is *robustness*. This means that small changes in the data should produce only small changes in the classification. These properties are well investigated in taxonomy (e.g. Moss, 1968; Crovello, 1968a,b; Sneath & Sokal, 1973). The third is *predictivity*. This follows from stability, for when a classification is stable, it is likely to forecast the presence of properties or property states of individuals in the population that were not involved in the classification. The fourth is *objectivity*. This will be satisfied when the method is such that based on it different workers can produce an identical classification from the same sample of individuals.

The immediate objective in classification is cluster recognition. *What may constitute a cluster?* In giving one answer to this question, we recall the spatial analogy discussed in the preceding section. In this, clusters are pictured as density phases of points in sample space. Recognition of clusters thus entails a search for density phases. It is however difficult to say whether the groups produced in a cluster analysis are in fact distinct density phases or only dissections of a more or less homogeneous sample.

The criteria of cluster recognition differ widely. Some methods stress internal homogeneity, such as the lack of variate (species) correlations or association (e.g. Goodall, 1953; Williams & Lambert, 1959; Gibson, 1959; Lazarsfeld & Henry, 1968). Others rely on high similarity among individuals inside, and low similarity with individuals outside the cluster (e.g. Sneath, 1961; Wirth, Estabrook & Rogers, 1966; Carlson, 1970; Jancey, 1974); on the minimum sum of squares (e.g. Ward, 1963; Ward & Hook, 1963; Edwards & Cavalli-Sforza, 1965; Orlóci, 1967a; Wishart, 1969a); or on nearest neighbour distances (e.g. Sneath, 1957b; Carmichael, George & Julius, 1968; Orlóci, 1976b). In many cases, unfortunately, the recognition of clusters is method dependent.

4.8 Classification of clustering methods

Clustering methods are extremely numerous and difficult to categorize. Yet, in the interest of an orderly presentation, some categorization is desired. For this, we could use a scheme to distinguish between subdivisive and agglomerative, intrinsic and extrinsic, or probabilistic and non-probabilistic categories. We could also use the hierarchical or non-hierarchical characteristic, the discrete or overlapping criterion, and the polythetic or monothetic property to establish categories. We could also group methods according to purpose following Forgy (1965), or we could rely on axiomatic properties which Jardine & Sibson (1971) have used.

Many other criteria have been suggested as a basis to classify clustering methods. Most of these would however lead to rather trivial groupings in the context of our objectives. In what order then do we present the methods? We shall begin with techniques (A) that endeavour to isolate clusters based on a matrix of resemblance measures that relate the objects to be classified or groups of objects directly to one another. Thus we have the *direct or Q-clustering* methods. The resemblance measure may be a sum of squares, such as in minimum variance clustering, a centroid distance such as in average linkage clustering, the direct distance or similarity of pairs of individual objects, such as in single linkage clustering, or some other measure. We shall see that the Q-clustering methods can be implemented hierarchically or otherwise, depending on the algorithm chosen by the user.

We shall establish yet another broad category (B) to include methods which form clusters of objects (quadrats) in which the association of the attributes (species) are minimized. Since the decision whether to subdivide a group or to leave it intact depends on examination of the association of the attributes, we refer to the methods in this category as the *indirect or R-clustering* methods. These implement clustering successively or simultaneously.

We shall also consider a third group (C) of methods which perform clustering as a product of ordination or other gradient analysis. We refer to this group as *gradient clustering.* The fourth group (D) includes methods which we implement in a statistical context to perform estimation or prediction. The methods in this group stand out, because of the statistical device which they incorporate, and which never seem to be incorporated by the methods in the previous groups. The last group (E) is admittedly very arbitrary in that it includes those variants of clustering which can enable the processing of large samples by compromising optimality. We note that with regard to specific aspects of the algorithm, there may be similarities between the methods, even if they are placed in different groups.

The categorization of methods as presented should not claim global relevance irrespective of applications, but it should serve as a guide for those in the author's field interested in a systematic review of clustering techniques.

4.9 Selected techniques

Without attempting to provide a broad review of classifications, which Cormack (1971) has done in a long article, we shall focus attention on methods that can satisfy many of the classificatory needs in vegetation surveys, and consider others only in brief:

A. Q-clustering methods

a. Average linkage clustering

The method of average linkage clustering was first proposed by Sokal & Michener (1958). This method performs a classification which is polythetic, agglomerative, hierarchical. Individuals or groups are fused when their average similarity is greatest. The formula for average similarity will be given later. We illustrate the method on the data in Table 1-2. The objective is a hierarchical classification of quadrats:

1. As the first step, we calculate quadrat similarities. This has been done for the sample and the results are given in the matrix,

$$\mathbf{S} = \begin{bmatrix} s_{11} & s_{12} & s_{13} & s_{14} & s_{15} \\ s_{21} & s_{22} & s_{23} & s_{24} & s_{25} \\ s_{31} & s_{32} & s_{33} & s_{34} & s_{35} \\ s_{41} & s_{42} & s_{43} & s_{44} & s_{45} \\ s_{51} & s_{52} & s_{53} & s_{54} & s_{55} \end{bmatrix} = \begin{bmatrix} 1.00 & 0.94 & 0.42 & 0.18 & 0.00 \\ 0.94 & 1.00 & 0.61 & 0.39 & 0.15 \\ 0.42 & 0.61 & 1.00 & 0.97 & 0.87 \\ 0.18 & 0.39 & 0.97 & 1.00 & 0.95 \\ 0.00 & 0.15 & 0.87 & 0.95 & 1.00 \end{bmatrix}.$$

The elements in **S** represent cosine separations of quadrat vectors according to the formula,

$$s_{jk} = \sum_h X_{hj} X_{hk} \Big/ \left[\sum_h X_{hj}^2 \sum_h X_{hk}^2 \right]^{1/2}, \quad h = 1, \ldots, p \tag{4.2}$$

There is no compelling reason for choosing this particular formula other than convenience. Other symmetric similarity measures could have been chosen if called for by some specific objective. The Sørensen (1948) index is an example which Feoli & Bressan (1972) used on data from bentonic communities.

2. The first operation on **S** involves finding the highest similarity value (excluding self comparisons indicated by unities in the diagonal cells). After

inspecting **S** we find that $S_{34} = 0.97$ is the highest similarity. This indicates fusion between quadrats *3, 4.* In anticipation of work on large matrices, it is advised to keep a fusion register for quadrat labels and similarities.

3. We compute a reduced similarity matrix next. In this, columns (rows) *3* and *4* are replaced by a new third column (row) of average similarities; the fourth column (row) is wiped out. If we designate as *(3+4)* the new third group, formed by the fusion of quadrats *3* and *4*, the average similarity between any quadrat *j* and the third group is given by the Gower (1967b) formula,

$$S_{j(3+4)} = [N_3/(N_3 + N_4)]\,S_{j3} + [N_4/(N_3 + N_4)]S_{j4} + [N_3N_4/(N_3 + N_4)^2\,](1 - S_{34}) = S^*_{j3} \qquad (4.3)$$

The S symbols represent quadrat similarities, the N symbols signify group sizes, and $S_{j(3+4)}$ or S^*_{j3} represents the average similarity of quadrat j and the new third group $(3+4)$. For quadrat *1* this quantity becomes,

$$\begin{aligned} S^*_{13} &= (1/2)S_{13} + (1/2)S_{14} + (1/4)(1 - S_{34}) \\ &= (1/2)(0.42) + (1/2)(0.18) + (1/4)(1 - 0.97) = 0.31 \quad . \end{aligned}$$

Should we put

$$\alpha_h = \frac{N_h}{N_h+N_i}$$

$$\alpha_i = \frac{N_i}{N_h+N_i}$$

$$\beta = -\frac{N_h\,N_i}{(N_h+N_i)^2}$$

$$\gamma = 0 \ ,$$

formula (4.1) will yield the one complement of S_{jk} in formula (4.3),

$$\delta(j,k) = \alpha_h\delta(j,h) + \alpha_i\delta(j,i) + \beta\delta(h,i) \qquad (4.4)$$

where k is a label for the union group $(h + i)$. This transformation identifies our version of average linkage clustering as one of centroid sorting (Lance & Williams, 1967b; Ducker, Williams & Lance, 1965; Hall, 1970). Where S_{jk} qualifies as a cosine, $\delta(j, k)$ measures one half of the squared Euclidean distance of centroids j and k. Should we put $\beta = \gamma = 0$ and retain α_h and α_i as above, the transformation

$$\delta(j,k) = \alpha_h\delta(j,h) + \alpha_i\delta(j,i)$$

would specify the so-called *group average variant* of average linkage clustering (cf. Sneath & Sokal, 1973). We could also put $\beta = \gamma = 0$ and $\alpha_h = \alpha_i = 1/2$, or alternatively, $\beta = -1/4$, $\gamma = 0$ and $\alpha_h = \alpha_i = 1/2$, to specify yet other variants of average linkage clustering known respectively as *weighted group average clustering* and *weighted centroid clustering*. For remarks regarding this nomenclature we refer to Gower (1967b) and references therein.

The remaining values S^*_{23} and S^*_{53} are similarly calculated to obtain a new matrix **S***:

$$\mathbf{S}^* = \begin{bmatrix} 1 & \mathit{0.94} & 0.31 & - & 0.00 \\ 0.94 & 1 & 0.51 & - & 0.15 \\ 0.31 & 0.51 & 1 & - & 0.92 \\ - & - & - & - & - \\ 0.00 & 0.15 & 0.92 & - & 1 \end{bmatrix}$$

$$\mathbf{N} = (1 \quad 1 \quad 2 \quad 0 \quad 1 \,) \;.$$

The elements in vector **N** indicate group sizes after the first fusion. We inspect the reduced matrix and find that the highest similarity is $S^*_{12} = 0.94$; thus we fuse quadrats *1* and *2* in a new first group (*1* + *2*). The quadrat labels are entered in the register, and also the average similarity. Following this, we replace columns (rows) *1* and *2* by a new first column (row) of average similarities and wipe out the second column (row) in **S***.

4. The average similarity of any group *j* and the new first group (*1* + *2*) is defined in symbolic terms by

$$S^*_{j(1+2)} = [N_1/(N_1+N_2)]\,S^*_{j1} + [N_2/(N_1+N_2)]\,S^*_{j2} + [N_1N_2/(N_1+N_2)^2]\,(1-S^*_{12}) = S^{**}_{j1} \quad .$$

The S^* symbols in this indicate similarity values in matrix **S***. For group *3* the average similarity with the new first group is

$$S^{**}_{31} = (1/2)(0.31) + (1/2)(0.51) + (1/4)(1-0.94) = 0.43$$

and for group *5*, $S^{**}_{51} = 0.09$. The entire **S**** matrix becomes:

$$\mathbf{S}^{**} = \begin{bmatrix} 1 & - & 0.43 & - & 0.09 \\ - & - & - & - & - \\ 0.43 & - & 1 & - & \mathit{0.92} \\ - & - & - & - & - \\ 0.09 & - & 0.92 & - & - \end{bmatrix}$$

$$\mathbf{N} = (2 \quad 0 \quad 2 \quad 0 \quad 1 \,) \,.$$

Vector **N** contains group sizes after the second fusion. In **S**** the maximum similarity is $S^{**}_{35} = 0.92$; thus we fuse groups *3* with *5* to obtain the new third group.

5. The average similarity of group *1* with the new third group (*3* + *5*) is

$$S_{13}^{***} = [N_3/(N_3+N_5)]\,S_{13}^{**} + [N_5/(N_3+N_5)]\,S_{15}^{**}$$
$$+[N_3N_5/(N_3+N_5)^2]\,(1-S_{35}^{**})$$
$$= (2/3)(0.43) + (1/3)(0.09) + (2/9)(1-0.92) = 0.33\ .$$

The complete **S***** matrix thus consists of four elements,

$$\mathbf{S}^{***} = \begin{bmatrix} 1 & - & 0.33 & - & - \\ - & - & - & - & - \\ 0.33 & - & 1 & - & - \\ - & - & - & - & - \\ - & - & - & - & - \end{bmatrix}$$

$$\mathbf{N} = (2 \quad 0 \quad 3 \quad 0 \quad 0\ )\ .$$

Vector **N** indicates the sizes of the groups after the third fusion. The fourth and last fusion is between groups *1* and *3* at an average similarity of 0.33.

We now summarize the steps by indicating changes in group contents and similarities. This information is found in the fusion register, which we update after each fusion. Initially the register contains,

Group	*Quadrats in group*	*Fusion similarity*
1	*1*	
2	*2*	
3	*3*	
4	*4*	
5	*5*	

.

After the first fusion the register indicates the changes:

1	*1*	
2	*2*	
3	*3,4*	0.97
4	–	
5	*5*	

.

After the second fusion:

1	*1,2*	0,94
2	–	
3	*3,4*	0.97
4	–	
5	*5*	

.

After the third fusion:

1	*1,2*	0.94
2	–	
3	*3,4,5*	0.92
4	–	
5	–	

After the final fusion:

1	*1,2,3,4,5*	0.33
2	–	
3	–	
4	–	
5	–	

The clustering results are summarized in a dendrogram in Fig. 4.1. Carmichael & Sneath (1969) and Sneath & Sokal (1973) should be consulted for alternative methods of displaying the results of cluster analysis. The program ALC in the Appendix automatically performs the computations.

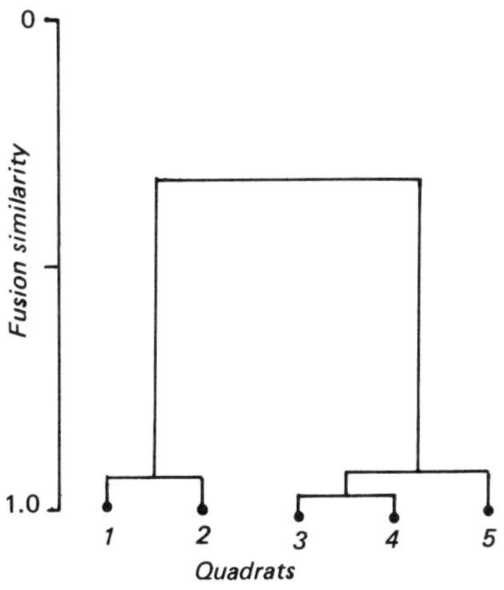

Fig. 4-1. Dendrogram for quadrats in Table *1-2.* The clustering algorithm is average linkage. The method is described in the main text.

The reader should notice that the Gower expression (formula (4.3)) for averaging similarities reduces to

$$S_{j(A+B)} = 0.5(S_{jA} + S_{jB}) + 0.25(1 - S_{AB}) \tag{4.5}$$

for group j and the union group $(A + B)$ when group sizes are not considered. Whether to use formula (4.3) or (4.5) in the averaging is a choice to be made by the user. Weighted and unweighted variants have been described (cf. Sneath & Sokal, 1973).

Average linkage clustering with its many variants (cf. Sneath & Sokal, 1973) is regarded as a good, general purpose method (Sokal & Sneath, 1963; Gower, 1967b; Kaesler & Cairns, 1972) with utility in numerical taxonomy firmly established (e.g. Hodson, Sneath & Doran, 1968; Lee, 1968; Sneath & Sokal, 1973, and references therein). There are few examples of ecological application (e.g. West, 1966; Feoli & Feoli Chiapella, 1974; Poldini & Feoli, 1976), and some not entirely successful (e.g. Cuanalo de la C. & Webster, 1970).

b. Minimum variance clustering

The most prominent method in this family forms groups in which the sum of squares is minimized. Whereas in average linkage clustering the parameter is a centroid distance, in *sum of squares clustering* it is a squared centroid distance weighted by group size. What is the relationship between centroid distance and the sum of squares? We answer this in the course of the following discussions:

Let $\mathbf{X}_m$ represent a $p \times N_m$ matrix of data for p species in N_m quadrats. Let X_{hjm} designate an individual data element in X_m. The subscripts h, j, m identify respectively species, quadrat, and group. The mean of species h in group m is $\bar{X}_{hm}$. In terms of these symbols, we can define the sum of squares as the quantity,

$$Q_m = \sum_j \sum_h (X_{hjm} - \bar{X}_{hm})^2 = \sum_j d^2(j, \bar{X}_m),$$
$$j = 1, ..., N_m ; h = 1, ..., p \ .$$

This gives the sum of squared distances of quadrats from their common group centroid specified by the co-ordinates

$$\bar{\mathbf{X}}_m = \begin{bmatrix} \bar{X}_{1m} \\ \cdot \\ \bar{X}_{pm} \end{bmatrix} .$$

Vector $\bar{\mathbf{X}}_m$ is a *group mean vector.*

The quantity Q_m can be computed directly from quadrat distances,

$$Q_m = (1/N_m) \sum_j \sum_k d^2(j, k; m), \ j = 1, ..., N_m - 1, \ \ k = j + 1, ..., N_m$$

where $d^2(j, k; m)$ is the squared Euclidean distance of quadrat j and k in group m.

It is not too difficult to see that the two formulae for Q_m are equivalent. Consider a group consisting of 2 points A, B. The distance of A and B can be decomposed into additive quantities,

$$d^2(A,B) = d^2(A,\bar{X})+d^2(B,\bar{X})- 2d(A,\bar{X})\, d(B,\bar{X}) \cos \alpha$$

But when $\alpha = 180°$ and $d(A, \bar{X}) = d(B, \bar{X})$, we have

$$d^2(A,B) = 2\left[d^2(A,\bar{X})+d^2(B,\bar{X})\right],$$

and the within group sum of squares,

$$Q = d^2(A,\bar{X}) + d^2(B,\bar{X}) = (1/2)\, d^2(A,B).$$

This directly generalizes to any N points in a multidimensional cluster,

$$\begin{aligned} Q &= \sum_j d^2(j,\bar{X}) \\ &= \frac{1}{N} \sum_{k>j} d^2(k,j), \quad j=1, \ldots, N-1 \quad . \end{aligned}$$

As an example, consider three points A, B, C, in the line with co-ordinates,

$$\begin{aligned} X_A &= 2 \\ X_B &= 5 \\ X_C &= 7\,. \end{aligned}$$

For these $\bar{X} = (2 + 5 + 7)/3 = 4.67$. The individual squared deviations are

$$\begin{aligned} d^2(A,\bar{X}) &= (2-4.67)^2 = 7.1289 \\ d^2(B,\bar{X}) &= (5-4.76)^2 = 0.1089 \\ d^2(C,\bar{X}) &= (7-4.67)^2 = 5.4289 \quad . \end{aligned}$$

The paired squared distances are given by

$$\begin{aligned} d^2(A,B) &= (2-5)^2=9 \\ d^2(A,C) &= (2-7)^2=25 \\ d^2(B.C) &= (5-7)^2=4 \;, \end{aligned}$$

and the within group sum of squared deviations by

$$\begin{aligned} Q &= 7.1289+0.1089+5.4289 \\ &= \frac{1}{3}(9 + 25 + 4) \\ &= 12.6667. \end{aligned}$$

Consider two groups A, B and their fusion $(A + B)$. We can define the sum of squares within $(A + B)$ as the quantity,

$$Q_{A+B} = [1/(N_A + N_B)] \sum_j \sum_k d^2(j, k; A + B),$$

$$j = 1, ..., N_A + N_B - 1; \quad k = j + 1, ..., N_A + N_B \quad .$$

All j, k are members in group $(A + B)$. The symbols N_A, N_B and $d^2(j, k; A + B)$ indicate group sizes in A and B and the squared distance of quadrats j and k within the fusion group $(A + B)$. In clustering we minimize the quantity

$$\begin{aligned} Q_{AB} &= Q_{A+B} - Q_A - Q_B = [N_A N_B/(N_A + N_B)] \sum_h (\bar{X}_{hA} - \bar{X}_{hB})^2 \\ &= [N_A N_B/(N_A + N_B)]\, d^2(A, B), \; h = 1, ..., p \end{aligned} \tag{4.6}$$

where $d^2(A, B)$ is the squared Euclidean distance of the group centroids (tips of group mean vectors). The quantity Q_{AB} has been used as a clustering criterion in vegetational classification (e.g. Orlóci, 1967a, 1973c; Austin, Ashton & Greig-Smith, 1972; Allen & Koonce, 1973; Jesberger & Sheard, 1973; Maka, 1973; Stanek & Orlóci, 1973; Fekete & Szőcs, 1974; Adam, Birks, Huntley & Prentice, 1975; Ayyad & El-Ghonemy, 1976; Pakarinen, 1976) and elsewhere (e.g. Ward, 1970; Ward & Hook, 1963; Edwards & Cavalli-Sforza, 1965; Wishart, 1969a; Phipps, 1972; Lubke & Phipps, 1973; McNeill, 1975).

The strategy in sum of squares clustering is readily interpreted in terms of the equality,

$$d^2(j,k) = \delta(j,k) \quad ,$$

where $d^2(j, k)$ is the squared centroid distance of group j and k and $\delta(j, k)$ accords with formula (4.4). To show this, we follow Lance & Williams (1967b). We designate the groups that fuse in k as h, i. Noting that terms of the squared Euclidean distance are additive, we write for the centroid of the fusion group in one variable,

$$\bar{X}_k = (N_h \bar{X}_h + N_i \bar{X}_i)/(N_h + N_i)$$

and for the squared centroid distance of group j and k,

$$d^2(j,k) = [\bar{X}_j - (N_h \bar{X}_h + N_h \bar{X}_i)/(N_h + N_i)]^2 \quad .$$

This expands into

$$d^2(j,k) = \frac{N_h}{N_h + N_i} d^2(j,h) + \frac{N_i}{N_h + N_i} d^2(j,i) - \frac{N_h N_i}{(N_h + N_i)^2} d^2(h,i)$$

which is exactly the same expression as formula (4.4) with substitution of $d^2(j, k)$ for $\delta(j, k)$, and similarly $d^2(j, h)$, $d^2(j, i)$ and $d^2(h, i)$ for the other δ terms. It thus is seen that the algorithms of sum of squares clustering and centroid sorting (or single linkage clustering in general) are related, even though they can give different results. We must however assume that the distance is Euclidean since the decision is based on

$$Q_{jk} = \frac{N_i N_k}{N_j + N_k} d^2(j,k),$$

and this must be meaningful as a sum of squares.

With the foregoing definition of group distances and sum of squares in mind, we can perform a cluster analysis in the following manner:
1. The quadrats to be clustered are those listed in Table 1-2. We assume that a priori considerations call for the use of the chord distance (formula (2.2)) to measure quadrat distances. Having chosen this definition of distance, we can obtain the quadrat distances from the **S** matrix in the preceding example according to

$$d^2(j,k) = 2(1 - S_{jk})$$

with corresponding numerical values,

$$\mathbf{D}^2 = \begin{bmatrix} 0.00 & 0.12 & 1.16 & 1.64 & 2.00 \\ 0.12 & 0.00 & 0.78 & 1.22 & 1.70 \\ 1.16 & 0.78 & 0.00 & \mathit{0.06} & 0.26 \\ 1.64 & 1.22 & 0.06 & 0.00 & 0.10 \\ 2.00 & 1.70 & 0.26 & 0.10 & 0.00 \end{bmatrix} .$$

In the initial step, when the groups contain only a single quadrat each, the clustering criterion (formula (4.6)) reduces to

$$Q_{AB} = (1/2)\, d^2(A,B)$$

which is half of the squared Euclidean distance. The minimum value occurs between quadrats *3,4*. The labels *3,4* are entered in a register and also the value $Q_{34} = Q_{3+4} = (1/2)(0.06) = (1/2)(0.06) = 0.03$ for the new third group (*3+4*). Note that $Q_{34} = Q_{3+4}$ holds true only because groups *3* and *4* represent a single quadrat each. The same condition would be true also if the groups contained two or more quadrats, provided that all quadrats within a group were identical.
2. Having completed the first fusion we now calculate values for criterion Q_{j3}, involving comparisons between any quadrat j and the new third group (*3 + 4*). Taking quadrat *1* first, we have

$$\begin{aligned} Q_{13} &= Q_{1+3} - Q_1 - Q_3 \\ &= [1/(N_1 + N_3)]\,[d^2(1,3) + d^2(1,4) + d^2(3,4)] - 0 - (1/N_3)d^2(3,4) \\ &= (1/3)(1.16 + 1.64 + 0.06)(1/2)(0.06) = 0.92 \quad . \end{aligned}$$

Another way of calculating Q_{13} would be from the co-ordinates (Table 1-2),

$$Q_{13} = [N_1 N_3/(N_1 + N_3)]\,[2\,(1 - \sum_h X_{h1}\bar{X}_{h3} / \left[\sum_h X_{h1}^2 \sum_h \bar{X}_{h3}^2\right]^{1/2})]$$

$$= (2/3)(2(1 - ((2)(1.5) + (0)(3.5) + (3)(0.05)) /$$
$$= ((4 + 0 + 9)(2.25 + 12.25 + 0.25))^{1/2}))$$
$$= (2/3)(2(1 - 4.5/((13)(14.75))^{1/2})) = 0.90 .$$

There are two things to be noted here. One is the discrepancy in the value of Q_{13} calculated by the two methods; this is a consequence of rounding errors. Another is that the computation is much more tedious if it has to rely on co-ordinates.

We determine the Q_{j3} values for all quadrats and place the values in the third column (row) of matrix **Q**. The values, not in the third column (row), accord with formula $(1/2)d^2(j, k)$. The entire matrix is given by

$$\mathbf{Q} = \begin{bmatrix} 0.00 & 0.06 & 0.92 & - & 1.00 \\ 0.06 & 0.00 & 0.66 & - & 0.85 \\ 0.92 & 0.66 & 0.00 & - & 0.11 \\ - & - & - & - & - \\ 1.00 & 0.95 & 0.11 & - & 0.00 \end{bmatrix}$$

$$\mathbf{N} = (1 \quad 1 \quad 2 \quad 0 \quad 1) .$$

The elements in vector **N** indicate group sizes after the first fusion. After inspection of the sum of squares we find that the next fusion is between quadrats *1* and *2* for which $Q_{1+2} = 0.06$ represents a minimum.

3. The labels *1*, *2* and $Q_{1+2} = 0.06$ are marked down and a new **Q** matrix is constructed by computing values Q_{j1} for the first new group $(1 + 2)$ and any other group j. For the third group $(3, 4)$ we have

$$Q_{31} = Q_{3+1} - Q_3 - Q_1 = [1/(N_1 + N_3)][d^2(1,2) + d^2(1,3)$$
$$+ d^2(1,4) + d^2(2,3) + d^2(2,4) + d^2(3,4)] - (1/N_3)d^2(3,4)$$
$$- (1/N_1)d^2(1,2) = (1/4)(0.12 + 1.16 + 1.64 + 0.78 + 1.22 + 0.06)$$
$$- (1/2)(0.06) - (1/2)(0.12) = 1.16 .$$

For the last group (including only quadrat 5),

$$Q_{51} = Q_{5+1} - Q_5 - Q_1 = [1/(N_1 + N_5)][d^2(1,2) + d^2(1,5)$$
$$+ d^2(2,5)] - 0 - (1/N_1)d^2(1,2)$$
$$= (1/3)(0.12 + 2.0 + 1.7) - (1/2)(0.12) = 1.21 .$$

It is obvious from the transformation $d^2(j, k) = \delta(j, k)$ that another possibility exists in defining Q values without recourse to the elements of the $\mathbf{D}^2$ matrix. Let us designate Q_{31} as $Q_{3(1+2)}$ to clearly indicate that we have in mind group *3* established in an earlier step, and the new first group $(1 + 2)$.

which just formed in the fusion of groups *1* and *2*. We can determine the squared centroid distance of *3* and (*1* + *2*),

$$d^2(3,(1+2)) = \frac{N_1}{N_1+N_2} Q_{31} \frac{N_3+N_1}{N_3N_1}$$

$$+ \frac{N_2}{N_1+N_2} Q_{32} \frac{N_3+N_2}{N_3N_2}$$

$$- \frac{N_1N_2}{(N_1+N_2)^2} Q_{12} \frac{N_1+N_2}{N_1N_2}$$

$$= \frac{(1)(0.92)(3)}{(2)(2)}$$

$$+ \frac{(1)(0.66)(3)}{(2)(2)}$$

$$- \frac{(1)(0.06)(2)}{(4)(1)}$$

$$= 1.16 \quad .$$

The Q_{31} value is the same as the $d^2(3, (1+2))$ value in this case, since the new $N_3N_1/(N_3 + N_1)$ is unity (see vector **N** below).

The next **Q** matrix is given by

$$\mathbf{Q} = \begin{bmatrix} 0.00 & - & 1.16 & - & 1.21 \\ - & - & - & - & - \\ 1.16 & - & 0.00 & - & 0.11 \\ - & - & - & - & - \\ 1.21 & - & 0.11 & - & 0.00 \end{bmatrix}$$

$$\mathbf{N} = (\,2 \quad 0 \quad 2 \quad 0 \quad 1\,) \;.$$

Vector **N** indicates group sizes after the second fusion. The minimum value in the last matrix **Q** is $Q_{35} = 0.11$ indicating the next fusion between the third and fifth group. The labels *3* and *5* are entered in the register, and also $Q_{3+5} = (1/3)(0.06 + 0.26 + 0.1) = 0.14$, the within group sum of squares in the new third group (*3* + *5*).

4. There are only two groups remaining, group *1* including quadrats *1, 2* and group *3* including quadrats *3, 4, 5*. For these, the within group sum of squares is

$$\begin{aligned} Q_{1+3} &= [1/(N_1 + N_3)][d^2(1,2) + d^2(1,3) + d^2(1,4) + d^2(1,5) \\ &\quad + d^2(2,3) + d^2(2,4) + d^2(2,5) + d^2(3,4) + d^2(3,5) + d^2(4,5)] \\ &= (1/5)(0.12+1.16+1.64+2.00+0.78+1.22+1.70+0.06+0.26 \\ &\quad +0.10) \\ &= 1.81 \;. \end{aligned}$$

The corresponding between groups sum of squares is $Q_{31} = 1.81 - 0.06 - 0.14 = 1.61$.

The information about changing group contents and sum of squares is contained in the register. Initially the register indicates,

Group	*Quadrats in group*	*Within group sum of squares*
1	*1*	0
2	*2*	0
3	*3*	0
4	*4*	0
5	*5*	0

After the first fusion:

1	*1*	0
2	*2*	0
3	*3,4*	0.03
4	–	–
5	*5*	0

After the second fusion:

1	*1,2*	0.06
2	–	–
3	*3,4*	0.03
4	–	–
5	–	–

After the third fusion:

1	*1,2*	0.06
2	–	–
3	*3,4,5*	0.14
4	–	–
5	–	–

Finally:

1	*1,2,3,4,5*	1.81
2	–	–
3	–	–
4	–	–
5	–	–

From these results a dendrogram is constructed in Fig. 4-2. The computations are automatically performed in program SSA (Appendix).

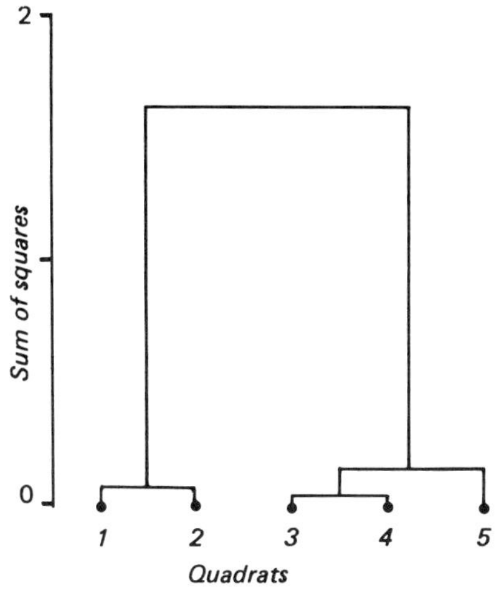

Fig. 4-2. Dendrogram for quadrats in Table *1-2.* The method is sum of squares. See explanations in the main text.

It should be noted in connection with formula (4.6) that the Q_{AB} is influenced by group size. When it is not important to consider the size of the groups, a decision which the user has to make, the clustering procedure reduces to *centroid clustering* in which two groups are fused when the distance between their centroids is the smallest. We have already considered the relationship of centroid clustering and sum of squares clustering in the preceding text.

Sum of squares clustering is a polythetic, hierarchical, agglomerative technique. It forms groups by fusing individuals or groups. We have seen that average linkage clustering and sum of squares clustering have several properties in common. On the basis of the examples given, we may even expect them to produce similar dendrograms. These methods differ, nevertheless, from one another in several important respects:

1. Whereas sum of squares clustering is limited to a Euclidean distance function, practically any similarity measure is admissible in average linkage clustering.
2. Sum of squares clustering is a special purpose method of classification, appropriate when the analysis is committed to a covariance structure. Average linkage clustering is more general with little restrictions on application.

Sum of squares clustering has a close ally in the subdivisive method of Edwards & Cavalli-Sforza (1965; Frey & Võhandu, 1967). This method has the desired property of being capable of identifying the best possible subdivision of a group into two. But for this, $2^{N-1}-1$ splits have to be performed and inspected. When group size N is large this task is hopeless for computation. It has been estimated that clustering in a modest sample of 30 or 40 individuals by this method could require thousands of years of computer work on the fastest electronic computer to complete.

Macnaughton-Smith, Williams, Dale & Mockett (1964) proposed *dissimilarity analysis* which is somewhat related to the method of Edwards & Cavalli-Sforza (1965). Dissimilarity analysis is however suitable for clustering in large samples, since it does not require inspection of all possible splits of every group into two at all hierarchical levels. The algorithm is simple and quick. It begins with identifying quadrat (A) which is most dissimilar to the remaining quadrats in the sample. To this, other quadrats are added one-by-one, each being more similar to A than to those not added to A. The new group and the residual group are then reanalysed in the same manner. The process is continued until either an arbitrary limit of homogeneity is reached or the hierarchy is completed. The algorithm could be made probabilistic with some difficulty by incorporating Goodall's (1966e, 1968) deviant index to identify outliers, or his dissimilarity index to test for fusions.

We should mention yet another clustering method which is similar to those already described in that it forms groups in which heterogeneity is minimized. This method, suggested by Hall (1967a,b), defines heterogeneity as a function of weighted species differences. The weights represent average presence within the groups. Abrupt increases in heterogeneity are given special significance in recognizing isolated groups.

Testing significance in sum of squares clustering, in the manner of an analysis of variance, as Edwards & Cavalli-Sforza (1965) have done, is open to question. The reason is that the sum of squares, used in the test, is also used to identify optimal fusions or subdivisions. The test thus relies on comparison of quantities that are either maximum or minimum values and not averages. This is clearly not compatible with the independence assumptions in the analysis of variance.

A more appropriate method, derived in random simulation, was suggested by Goodall (1973a). It performs a test on the hypothesis that in fusion of any two (A, B) of g groups to form group (A, B), the increase in the variance of the fusion group does not exceed what could be expected if A and B were random samples from two normal populations with equal variance. The null hypothesis is rejected at α probability, and A is not fused with B, if the ratio

$$F = \frac{Q_{AB}}{Q/(N-g)}$$

exceeds $F_{t;1,N-g}$, the t probability point of the F distribution with 1 and $N-g$ degrees of freedom. In these expressions, Q_{AB} is the increment in the sum of squares due to fusion of A and B, $Q = Q_1 + \ldots + Q_g$, the total within group sum of squares in the g groups; $N = N_1 + \ldots + N_g$, the total number of quadrats in the sample; and

$$t = 1 - (1-\alpha)^{1/[g(g-1)/2]} .$$

Should $F < F_{t;1,N-g}$ hold true, A and B would be fused.

The dependence of the fusion criterion in sum of squares clustering on group size has an interesting consequence. To show this, consider Q_{Aj} and Q_{Bj}, both defined on the basis of formula (4.6). Assume that A and B represent groups of N_A and N_B quadrats respectively, while j designates a single quadrat not in A or B. Now allocate j to either group A or group B depending on the size of Q_{Aj} or Q_{Bj}. Fuse j with group A if the condition

$$N_A/(N_A + 1)\, d^2(A,j) < N_B/(N_B + 1)\, d^2(B,j)$$

is true, and with B when the reverse is true. From this we can deduce that,
1. When $N_A = N_B$ the condition for allocation reduces to $d(A, j) < d(B, j)$, the criterion for centroid clustering.
2. When $d(A, j) = d(B, j)$ the allocation of j to group A or B will depend solely on the sizes of $N_A/(N_A + 1)$ and $N_B/(N_B + 1)$.

The latter implies that the allocations are biased in favour of the smallest group, so that quadrats tend to be added onto the small groups, even if they are located closer in distance to the centroid of a large group. Looking at this from the point of view of the affinity of j to A, it is seen that as group size keeps increasing in A, so does the affinity of j to A keep declining. This phenomenon has been compared to an expanding space in which group A recedes further and further from point j, with each increment in group size, making it that much more difficult for j to join group A. But this need not be detrimental to clustering. It may in fact be beneficial, since it can reduce the chaining effect, thereby preventing the large groups from increasing their size more and more rapidly as their size actually increases.

Lance & Williams (1967b; Williams, Clifford & Lance, 1971; Williams, 1971) describe different types of dependence on group size in fusions. In their terminology, sum of squares clustering is *space dilating*, meaning that as the group grows larger its dissimilarity with other groups increases as if it were receding from them in space. Average linkage clustering is *space indifferent* in these terms, but not so single link clustering (described below) which is *space contracting*. The latter means that the larger the group gets, its similarity to other groups will increase. Yet another method, described by Lance & Williams (1967b; see also Stephenson & Williams, 1971) under *flexible clustering*, is capable of different degrees of space distortion depending on the choice of the coefficients in formula (4.1). Experiments with flexible clustering indicate that it can produce strong chaining (β set to nearly 1) as in the space contracting methods, almost no chaining (β set to -1) as in the space dilating methods, and anything in between on demand. Lance & Williams (1967b) suggest $\beta = -0.25$ for general use with flexible clustering. The constraints on the useful domain of coefficients are given by

$$\alpha_h + \alpha_i + \beta = 1$$
$$\alpha_h = \alpha_i$$
$$\beta < 1$$
$$\gamma = 0 \quad .$$

Flexibility, as the name suggests, is a unique feature, and with it comes the advantage that renders clustering eminently suited for controlled experiments with data sets to search for solutions that maximally satisfy a stated objective. But one may find a great deal of arbitrariness in manipulating parameter values to force a solution. Sneath & Sokal (1973) actually prefer not to manipulate parameters, but to set a priori criteria and stick with the results.

c. Single linkage clustering and other similarity methods

In similarity clustering, in the most restrictive case, groups are sought, not unlike the *natural taxa* of Sneath (1961), or the *perfect groups* of Rubin (1967), in which any individual is more like any other in the group than any other outside the group. The definition of a group need not however be so restrictive. Alternatives are readily conceived:
1. Define a group in which every individual is less similar to every outsider than a fixed similarity value.
2. Permit deviation from a perfect group by allowing individuals to have higher similarity to α% of the individuals outside the group than the chosen similarity value which describes a current state of homogeneity within the group.
3. Define a natural group conditional on average similarity.

While the third has been used by Vasilevich (1969a,b), the first appears to be the case in single linkage clustering, and the second in Carlson's (1970; Dabinett & Wellman, 1973; Bradfield & Orlóci, 1975) method.

Most clustering algorithms find clusters in a sample irrespective of the existence, or non-existence, of actual discontinuities. While these algorithms can be very useful when the goal is simply a dissection, they will be of little help when we try to discover discontinuities, or equivalently, recognize natural groups in a sample. We shall begin the description of methods with one, known as *clustering by neighbourhoods.* We can formulate an algorithm for it about the definition that one group of objects is discontinuous with another group if none of its neighbourhoods, with given radius r, overlap with any of the neighbourhoods with the same radius in the other groups. The definitions and concepts are explained in the sequel. The algorithm is given as program TRGRPS (Appendix). When it is hypothesized that discontinuities subdivide the sample into a given number of groups, and the hypothesis is presented to the program, it performs the necessary tests automatically and, if groups are found, it produces group descriptors.

In the description of rationale we shall follow the original presentation (Orlóci, 1976b). We assume that there are N relevés in the sample, labelled A, B, C, etc. We also assume that in the associated sample space the relevés are represented by points in the line (Fig. 4-3). The relative placements of points to one another are important. The line segments connecting the points, with lengths proportional to differences in the species composition of relevés, represent relevé distances. If the sample space is p-dimensional, each dimension corresponds to a different species. It is noted that for measuring distance, any symmetric measure of dissimilarity is admissible.

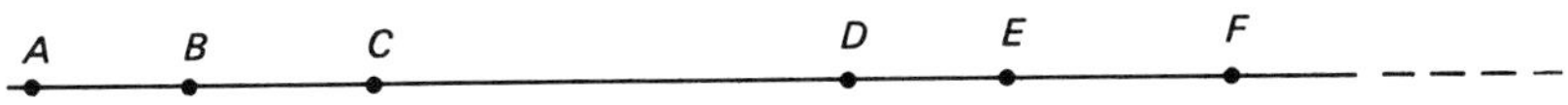

Fig. 4-3. Sample points in the line illustrating neighbourhoods as described in the main text.

The point nearest to another point is said to be that point's *nearest neighbour.* In Fig. 4-3, relevé A is nearest neighbour to B, E to F, etc. Each point may or may not have other points in its *neighbourhood* within a given radius (r) for which the neighbourhood is defined. We can recognize at $r = d(B, C)$, where $d(B, C)$ is the distance of points B from C, the following neighbourhoods:

$\{AB\}$ for point A,

$\{ABC\}$ for point B,

$\{BC\}$ for point C,

$\{DE\}$ for points D and E,

$\{F\}$ for point F .

We note that $\{AB\}$ and $\{BC\}$ are *overlapping* neighbourhoods, i.e. relevé B is common to both, but $\{ABC\}$, $\{DE\}$ and $\{F\}$ are *non-overlapping.* Neighbourhoods that are non-overlapping are said to be *discontinuous* or *disjoint.*

Given $r = 0$ as the initial state of the neighbourhood radius in a clustering procedure, as many discontinuous neighbourhoods can be recognized in the sample as there are non-identical relevés. In Fig. 4-3, these include,

$$\{A\}, \quad \{B\}, \quad \{C\}, \quad \{D\}, \quad \{E\}, \quad \{F\}.$$

Let us increment r next by a constant small quantity, say Δ, step-by-step until the condition $d(A, B) \leqslant r < d(B, C)$ is satisfied, assuming that $d(A, B)$ represents the smallest distance other than zero in the sample and $d(B, C)$ the second smallest distance. At such a radius,

$$\{AB\}, \quad \{C\}, \quad \{DE\}, \quad \{F\},$$

represent discontinuous neighbourhoods. If r is further incremented so that it will satisfy the condition $d(B, C) \leqslant r < d(E, F)$, noting that $d(E, F)$ is the third smallest distance in the sample other than zero, the discontinuous neighbourhoods will have contents,

$$\{ABC\}, \quad \{DE\}, \quad \{F\}.$$

Further increments in r, until $d(E, F) \leqslant r < d(C, D)$ is satisfied, yield discontinuous neighbourhoods,

$$\{ABC\}, \quad \{DEF\},$$

and finally when $r \geqslant d(C, D)$,

$$\{ABCDEF\} .$$

We observe that all neighbourhoods at a given r under the given constraints satisfy our original definition of a discontinuous group: not one of the neighbourhoods in any one group overlaps with another neighbourhood in the others.

It is evident from what we have said that our procedure of clustering by neighbourhoods uses single link fusions, similar to Sneath's (1957b; see also Carmichael, George & Julius, 1968) method. About this method of clustering much theoretical work has revolved in numerical taxonomy (e.g. Jardine & Sibson, 1968a,b, 1971a,b). Our method is also related to the graph theory approach of Wirth, Estabrook & Rogers (1966).

TRGRPS clusters by neighbourhoods and incorporates other devices which enable the user to test hypotheses, in a non-statistical way, about the existence of discontinuous groups of relevés. The flow pattern of operations and decisions is as follows:

a. Read data, specify a hypothesis for the number of groups, compute distances between relevés. Designate the number of groups by R.

b. Specify an initial value of neighbourhood radius $r = C$ and the constant increment $\Delta = D$.

c. Cluster the relevés and count the number of groups. Designate this number by k.

d. If $k = R$ then accept the hypothesis. If $k < R$ then modify the hypothesis to less than R groups and continue with step b. If $k > R$ then increment the neighbourhood radius and continue at c.

We must realize that based on sampling considerations, tests of the kind as in this algorithm should be preferably probabilistic. However, the algorithm as it is presented does not incorporate probabilistic considerations.

TRGRPS evolved through modifications of the original Jancey (1974) algorithm which is abstracted under the name TRGRUP in the first edition of this book. The original descriptions stressed a nearest neighbour property of the groups, although the idea of clustering by neighbourhoods was incorporated in the program TRGRUP. The present treatment corrects this anomaly by recognizing that although the condition that every member in a discontinuous group must have its nearest neighbour within the group is necessary, it is by no means sufficient, because even if this condition is satisfied neighbourhoods can still overlap between the groups. The groups with contents

$$\{AB\} , \quad \{C\} , \quad \{DEF\}$$

represent typical examples for the case in point. In these, the nearest neighbour property is satisfied, since every member in each group has its nearest neighbour within its own group, yet two of the groups, $\{AB\}$ and $\{C\}$, appear continuous because the neighbourhoods of relevés A, B and C overlap at the implied radius $d(E, F)$.

TRGRPS differs from TRGRUP by the specific constraint that none of the relevés are allowed to remain unclustered when the hypothesis is true. This means that for the hypothesis of R groups to be accepted in TRGRPS, the sample must divide into exactly R groups, and not just R noda, plus a number of points that float unclustered among the groups as in TRGRUP. Algorithms TRGRPS and TRGRUP yield identical results when the group structure is strong in the sample.

The principal value of clustering by neighbourhoods appears to be the definition of a discrete group, and two other things which the algorithms do that normally are not done as well by other clustering techniques:

1. Natural groups are found if they exist, and if no groups can be found at the specified constraints (R, C, D), groups will not be formed.
2. The method can find groups at a specified minimum group size S if a solution is possible under the specified constraints R, C, D.

Group recognition is allowed at different levels, and hierarchical connections can be established between groups on the basis of fusions observed in repeated runs, on the same data, under different hypotheses for the number of groups. The method is little restricted regarding admissible resemblance functions. Since the clustering is basically single linkage, the method is sensitive to aberrant individuals which may link the otherwise discrete groups. This problem is difficult to overcome, and the user may have to resort to various arbitrary steps to locate and eliminate the undesired individuals. Methods for this were described by Goodall (1969a) to locate unusual species combinations and by Wishart (1969b) to locate intermediates. We have already considered Goodall's method in Chapter 1. The Wishart method involves selection of a threshold resemblance value r, to serve as the radius of a sphere about a point P_j, and a density (count) limit L for points occurring within the sphere. If the density count is less than L the point P_j is eliminated. In this way, the low density phases are eliminated and the high-density clusters are rendered more distinct. Forgy's (1963) gravitation simulation present yet another possibility along similar lines.

TRGRPS and TRGRUP have considerable potential in vegetation studies as can be seen from published examples (Jancey, 1974; Orlóci, 1976b). The data which Jancey (1974) used in his example are reproduced in Fig. 4.4. This figure incorporates several dense clusters of points, one of which is completely enveloped by a high-density ring. The computations by either of the two algorithms (see the Appendix) reveal all the groups, including the high-density ring, as separate entities. The parameters and results in the analysis based on TRGRPS are summarized below:

Total number of points	85
Number of attributes	2
Minimum group size (S)	4
Number of groups sought (R)	4
Initial value of linkage parameter (C)	4.1
Increment (D)	2.45

Group *1*: High-density ring of 58 points
Group *2*: Cluster of 8 points

Group *3*: Cluster of 15 points
Group *4*: Cluster of 4 points

Trials performed on other data sets, simulated and real, indicate the versatility of the algorithms and suggest possible applications when the objectives call for recognition of vegetation types or classes of habitats as a prelude to vegetation mapping, habitat mapping, or inventory in the course of a general survey (Orlóci, 1976b).

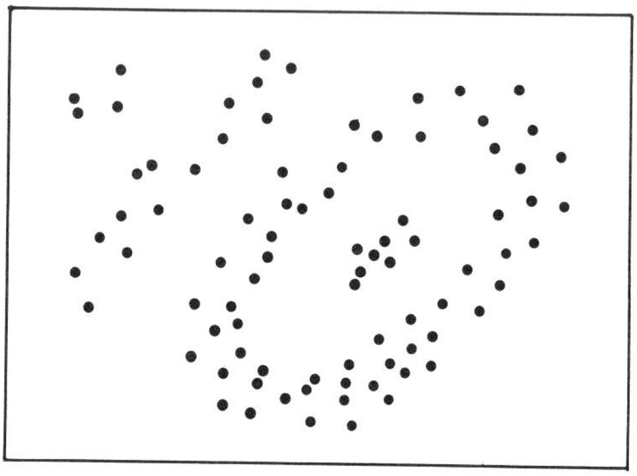

Fig. 4-4. Scatter diagram from point co-ordinates listed under program TRGRPS as RAWD4 in the Appendix. The analysis is described in the main text. Diagram reproduced after Jancey (1974).

Anderberg (1973) attributes the attractiveness of clustering by single linkage to efficient computability. Another property which may explain a wide spread interest in the method is the relative ease with which its axiomatic characterization can be accomplished (Jardine & Sibson, 1968a,b, 1971; Cole & Wishart, 1970). No wonder then that the method has generated considerable interest, and also, that it gained implementations in the context of different formulations. One of these is *graph theory clustering.* A *graph* consists of points (nodes) and connecting lines (edges). A graph which shows only one path between any pair of nodes is said to be a spanning tree. When the sum length of the edges is minimal the graph is called a *minimum spanning tree*, designated variously as a 'shortest simply connected tree' (Kruskal, 1956) or 'Prim network' (Cavalli-Sforza & Edwards, 1967). Groups found by single linkage clustering have been shown to be the groups that dismember the minimum spanning tree by removing long edges (cf. Gower & Ross, 1969; Zahn, 1971; Rohlf, 1973). Whereas the ecological applications are not too wide spread (e.g. Thalen, 1971; van Groenewoud & Ihm, 1974; Pakarinen, 1976; Feoli, 1975), the applications in numerical taxonomy are numerous (e.g. Estabrook, 1966; Wirth, Estabrook & Rogers, 1966; Jardine & Sibson, 1971; Sneath & Sokal, 1973; Briane, Lazare, Roux & Sastre, 1974; McNeill, 1975; Ojaveer, Mullat & Võhandu, 1975).

Graphs are frequently used in vegetation analysis to show patterns of association between species. Graphs of this sort, known as *plexus diagrams*, are spanning trees in which only certain short edges are retained. By removing the long edges, the clustered appearance of points is enhanced. Examples of plexus diagrams are numerous (e.g. De Vries, Baretta & Hamming, 1954; McIntosh, 1962, 1973; Agnew, 1961; Juhász-Nagy, 1963; Looman, 1963; van Groenewoud, 1964; Beals & Cope, 1964), some reviewed by Greig-Smith (1964; also Whittaker, 1967; Shimwell, 1971; Kershaw, 1973; McIntosh, 1973). In some applications, matrix rearrangements are used rather than graphs to identify groups of similar individuals (e.g. Sokal & Sneath, 1963; Daniel & Holubičková, 1972). In any case, the groups recognized depend on the threshold similarity chosen. Other similar methods, utilizing correlations, have been suggested (e.g. Godron, Guillerm. Romane & Sabato-Pizzini, 1969).

d. Fixed group-number methods

These too, similar to the other clustering methods, attempt to subdivide a sample into a number of groups in such a way that the partitions optimize some group property. In one strategy, the problem amounts to sorting individuals into groups when the number of groups is specified a priori. Anderberg (1973) and Cormack (1971) review the broader literature and comment on different techniques. The method of Jancey (1966a) is an example. It works through successive iterative cycles, fitting centroids (equal in number to the number of groups desired) in such a manner that the between groups centroid distances are maximized in the final configuration. The algorithm can be summarized as follows;

1. Assume that k groups are to be formed.
2. Choose k pivotal points if not given a priori. The pivotal points may be centroids in k partitions of the data.
3. Assign each sample point to the cluster of the pivotal point which lies nearest to it.
4. Compute k new centroids and use these as new pivotal points.
5. Continue with step *3*, repeating the procedure until group contents no longer change with assignments.

Jancey's method incorporates a device to bypass local minima, and at the same time, to provide for fast convergence through permitting a new pivotal point to be located, some distance from a centroid, on the opposite side from the previous pivotal point. Golder & Yeomans (1973), in tackling a similar problem with a similar method, have addressed the question of cluster stability experimentally. They have reanalysed data with different values of k and looked for consistency in the clustering pattern. Ball & Hall (1965) described yet another iterative algorithm, with the same purpose, which has been applied to vegetation data by del Moral (1975).

Gower (1974) described another similar method under the title *maximal predictive classification.* This method uses dichotomous variables. Described in terms of our terminology, Gower's method seeks partitions of a sample of n quadrats into k groups which maximize the quantity $W_k - B_k$ at different values of k. If we designate by $\bar{X}_i$ the descriptor of group i, i.e. a p-valued vector of presence/absence scores for p species in the ith group, and by X_{ij}

the descriptor of quadrat j of group i, i.e. the p-valued vector of presence/absence scores in quadrat j of group i, then the following interpretations of W_k and B_k are appropriate:

W_k – the number of matches of the quadrat descriptors with their respective group descriptors,

B_k – the number of matches of the quadrat descriptors with descriptors of other classes.

The computational problem at even small values of k would be enormous, or simply impossible, if all different k partitions in a sample of n quadrats were examined. Gower (1974) chooses to adopt an iterative relocation algorithm (cf. Cormack, 1971) and to perform the computation at only selected values of k. The solution obtained is locally optimal in satisfying the objectives. The method may prove to be useful when the goal is to construct reference categories (types) for identification purposes. However, the algorithm is defined only for dichotomous (presence/absence) variables.

In a second strategy, the problem differs in as much as it allows the number of groups to change in the clustering process. Rubin's (1966, 1967) hill-climbing algorithm is an example. The clustering here begins with input of an initial class structure which may be derived through arrangement of individuals into a given number of classes randomly or non-randomly by some other method. A new partition is obtained by moving an individual out of its class to join another class, or to form its own class, to produce a new partition. The new partition is accepted if it is better than the previous partition, evaluated by use of a preference relation which need not be a numerical measure. The process is iterated until no move can improve on the partition. Provisions are made to move more than one individual at a time, to overcome local maxima, or to allow retrogression and restarting with a new initial arrangement. Whereas Rubin (1967) claims that the method works well in well-structured data, an optimal solution may not be reached.

e. A note on information clustering

The treatment of classification methods, in which the clustering parameter is a measure of information divergence, is not to be construed as an indication of sharp differences in clustering strategy. As a matter of fact, information divergences may be associated with any of the strategies so far discussed. They can serve as measures of group separation in average linkage clustering and in other centroid clustering methods. Information divergence can also serve as a basis for additive partitioning, in a similar manner as the sum of squares, or as the clustering parameter in the single linkage methods.

Ecological and related applications, methodology, advantages and shortcomings, programming aspects, etc., are treated in numerous papers (e.g. Macnaughton-Smith, 1965; Williams, Lambert & Lance, 1966; Lance & Williams, 1966b; Wallace & Boulton, 1968; Webb, Tracey, Williams & Lance, 1967a; Orlóci, 1968a, 1969a,b, 1970a,b, 1971a,b, 1972a,b, 1975b; Dale, 1971; Dale, Lance & Albrecht, 1971; Goldstein & Grigal, 1972; Dale & Anderson, 1972; Goodall, 1973a,b; Williams, 1973; Godron, 1975). The algorithms, whether divisive or agglomerative, always attempt to maximize the component of information between groups and to minimize the component within the group. In this, they rely on the information inequalities of Section 2.7. Since many of the information measures are directly identifiable as forms of the *minimum discrimination information statistics* (Kullback, 1959), or related to it, different authors were tempted to apply information measures probabilistically. Such attempts are made peculiar by the fact that clustering procedures maximize (or minimize) the information value. Because of this, it is doubtful that the sampling distribution of maxima (or minima) could still be approximated by the chi square distribution which provides the probabilities. Different information functions weight the data differently. This, in some cases, may be detrimental to the influence of rare species. The problem in this is not unique to information analysis and it does not need to be a disadvantage (Crawford & Wishart, 1967).

Of the many information theoretical formulations, it will be sufficient to consider the application of some which we gave in Chapter 2, while we follow Orlóci's (1970a,b) summary of the procedures:

1. $I(A;B)$ of formula (2.36). This divergence is computed from count or frequency data. It represents a simple sum of p individual bits of information, each contributed by a different species. Such a simple summation of terms may, of course, be objectionable if the species are not independent. We can build a clustering algorithm about $I(A;B)$, similarly as in sum of squares clustering, in which fusions are selected to minimize $I(A;B)$.
2. $I(A;B)$ of formula (2.37). This function is computed from counts or frequencies. No assumptions are required about species independence. The clustering algorithm is similar to sum of squares clustering.
3. $r(F_h;F_i)$ of formula (2.32). When the data represent quantities on some arbitrary scale, such as the data in Table 2-1, the coherence coefficient is a convenient measure of resemblance between the entities. The coherence coefficient can be analyzed in average linkage clustering or single linkage clustering. Although formula (2.32) defines the coherence coefficient for species, it can be defined also for pairs of quadrats.

Algorithms, not based on the minimum discrimination information measure, but on a *multiple of entropy of order one* (formula 2.24), which minimize the within group information, have been discussed by, e.g., Lambert & Williams (1966), Williams, Lambert & Lance (1966), Lance & Williams (1966b), Williams & Lance (1968) and Orlóci (1969a) in ecological classifications, and by Macnaughton-Smith (1965) elsewhere. Comments are offered by Field (1969) on the use of the multiple of entropy to analyse heterogeneous communities, by Bottomley (1971) regarding significance tests, and by Tracey (1968; Cormack, 1971) regarding distribution properties and stopping rules.

f. Probabilistic relocations

We have already considered Goodall's similarity index (S) in Chapter 2. We have seen that it measures the complement of the probability that any pair of quadrats drawn from a homogeneous population at random could be as similar as the pair whose similarity was actually measured. Goodall (1970b) has determined that S has a rectangular distribution when the sampling is random and the population is homogeneous. This fact allows testing the null hypothesis that N quadrats in a group are in fact random individuals from a common homogeneous population. The test may rely on the divergence

$$2I(\mathbf{F};\mathbf{F}^\circ) = 2\sum_j f_j \ \ln f_j/f^\circ, \ j = 1, ..., k$$

in which a k-valued observed frequency distribution $\mathbf{F}$, in which $N(N-1)/2$ similarity values are condensed, is compared with a k-valued, $N(N-1)/2$ totalled equidistribution $\mathbf{F}^\circ$. When the null hypothesis is true, $2I(\mathbf{F};\mathbf{F}^\circ)$ is approximately distributed as a chi square variate with $k-1$ degrees of freedom. A different test could be based on the maximum similarity (max$S$) in the group. For the sampling distribution of max$S$, Goodall (1970b) defined the α probability point S_α,

$$\ln S_\alpha = \frac{\ln(1-\alpha)}{N(N-1)/2}.$$

When max$S$ exceeds S_α it indicates that the two quadrats (j, k) which correspond to max$S$ can be regarded as falling into the same natural cluster. The methodological implication is clear: regard j, k as a nucleus for a homogeneous group.

Goodall's (1968) affinity index (A) gives the probability that a randomly chosen quadrat, whose parent population is homogeneous, will be at least as similar to the mean vector $\overline{\mathbf{X}}$ (norm) of its own population as an observed similarity $A(j, \overline{\mathbf{X}})$. A small index value means large similarity of j to $\overline{\mathbf{X}}$. Following Goodall (1970b), we can write for the α probability point A_α of the sampling distribution of the minimum affinity index,

$$\ln(1-A_\alpha) = \frac{\ln(1-\alpha)}{M},$$

in which M is the number of individuals (outsiders) in the sample not in a group. If quadrat j of the outsiders has affinity $A(j, \overline{\mathbf{X}}) = \min A < A_\alpha$ then j can be added to the group. This is of course equivalent to accepting the null hypothesis that the group and quadrat j have identical parent populations. To counterbalance the effect of M on the test, Goodall (1970b) suggests a corrected value for the 0.05 probability, which we give as $A^*_{0.05}$ in

$$\ln A^*_{0.05} = -3.00 - 0.444 \ln M \ .$$

Goodall's (1968) deviant index D is a measure which gives the probability that any randomly chosen quadrat can fall at least as far away from the mean vector $\overline{\mathbf{X}}$ of its own homogeneous population, as the observed value $D(j, \overline{\mathbf{X}})$. This is a direct expression of similarity. The α probability point in the distribution of min$D$ is given by Goodall (1970b) as D_α in

$$\ln(1-D_\alpha) = \frac{\ln(1-\alpha)}{N}.$$

If $D(j, \bar{\mathbf{X}}) = \min D < D_\alpha$ then the quadrat j showing the least similarity to $\bar{\mathbf{X}}$ is removed from the group. Such a decision would imply that we reject the null hypothesis, stating that j and $\bar{\mathbf{X}}$ come from different homogeneous populations. Goodall (1970b) has found that the actual probability level tends to fall below the nominal α value in the tests, and suggested the corrected $D^*_{0.05}$ probability point,

$$\ln D^*_{0.05} = -2.9957 - 1.0398 \ln N + 0.1059 (\ln N)^2 \quad .$$

How could we incorporate these indices in a clustering procedure? Goodall (1970b) describes three possibilities for Q-clustering:

1. *Aggregative procedure.* A group nucleus is formed of two quadrats (j, k) for which $S(j, k) > S_\alpha$. In this, S is Goodall's (1964, 1966a) probabilistic similarity index. The group is enlarged step-by-step to a size up to which the hypothesis of homogeneity is tenable in tests based on the affinity index.
2. *Rejective procedure.* The group is subjected to one of two tests. The first test employs the similarity index S to identify subsets of quadrats whose similarities with other quadrats in the group are sufficiently low to regard them as members of different populations. The second test uses the deviant index to locate quadrats responsible for heterogeneity in the group. In either case, once heterogeneity is demonstrated, the quadrats responsible are removed to *purify* the group.
3. *Combined agglomerative and divisive procedure.* The application of the deviant index and the similarity index alternate, to respectively eliminate outliers, or to find cluster nuclei which are built up by adding quadrats to the initial groups using the affinity index as the agglomeration parameter.

In either of these procedures, once a homogeneous group of maximum size is formed, its member quadrats are removed from the sample, and the quadrats still remaining are reanalysed similarly as in successive clustering.

These methods of cluster analysis require large samples for improved accuracy. They also incorporate a definition of homogeneity that implies multidimensional random variation in uncorrelated variables. Since such type of variation is tied up with conditions which are not frequent in vegetation surveys, applications may be restricted. Transformations are however available to pass from correlated species to uncorrelated variables. One possibility is to subject the data to component analysis and use the components to redescribe the quadrats. The suitability of Goodall's (1970b) methods to handle variables of nominal, ordinal, interval and ratio scales is an attractive feature. Another is a reasonably light computational load, not incommensurable with the computational difficulty in most of the clustering techniques. In spite of these advantages, the probabilistic methods did not attract much following. The reasons for this may lie in the weaknesses already mentioned, and in the mathematical difficulty in the original descriptions fragmented between numerous short articles.

Gibson (1959), Lazarsfeld (1950) and Lazarsfeld & Henry (1968) pursue similar objectives with their methods as Goodall (1970b) with his. They assume that homogeneous clusters exist in the data in which the (binary) variables are uncorrelated. A novel methodological aspect is the determination of parameters such as class number, class sizes, probabilities of variables in the clusters, as solutions in a system of equations. These methods, nevertheless, present formidable difficulties for computation, and they are unlikely to have success in handling more than a few groups in any one analysis. These leave the classificatory objectives better approached by other methods.

B. R-clustering methods

a. Successive clustering

This group includes techniques in which one cluster is completely formed before the formation of another cluster begins. Carlson (1970) and Cormack (1971) review the methods in general. We shall discuss applications in vegetation studies. The computations normally start with the entire sample and establish homogeneous groups, one at a time, by narrowing group content. A group is considered homogeneous when only a small proportion, say α, of the species correlations (association) remains significant in the group (e.g. Goodall, 1953; Hopkins, 1957; Yarranton, 1973), or when the ratio of average distances within the cluster and with outsiders is held below a certain critical value (e.g. Frey & Võhandu, 1967; Frey & Groenewoud, 1972). Goodall's (1953) method is summarized below:

1. Find the most frequent species showing significant association with other species. Measure association as a χ^2 of a 2 x 2 contingency table.
2. Group together all quadrats which contain the pivot species.
3. Analyse this group of quadrats to identify the next pivot species as in step 1 and continue at step 2. Stop when a group is formed in which the number of significant associations is reduced below an acceptable level.
4. Remove this group from the sample and reanalyse the remaining quadrats starting at step 1.
5. When the clustering is completed, recombine groups in pairs. Retain a combined group as final if no significant associations re-emerge. Otherwise, retain the uncombined groups.

Goodall's algorithm could be associated with definitions of homogeneity of different kinds. Yarranton's (1973) suggestion could be followed. He asserts, while relying on the results of Erdős & Rényi (1960, 1961), that in a homogeneous group (community, association, type, etc.) the expected number of species with 0, 1, 2, etc., significant associations follow a hypergeometric distribution. Yarranton (1973) uses the Poisson distribution to approximate this. He measures the divergence of the observed distribution of the number of significant associations from the Poisson, with the observed mean number as parameter, as a chi square quantity. This facilitates a statistical test on group homogeneity.

A quite unique feature of the methods is a reliance on standard statistical tests to establish homogeneous groups, and also, on large samples for accuracy. The practice of relying on standard statistical distributions, such as χ^2, may however be open to questions since the sample statistic may have a sampling distribution different from the assumed standard, partly because of the statistic used which may represent a minimum rather than an average, for which the standard distributions are formulated.

Goodall's (1953) method has become the object of several modifications (e.g. Williams & Lambert, 1959; Crawford & Wishart, 1967). We shall discuss these in the sequel. We already considered Goodall's (1966c, 1970a,b, 1973b) other probabilistic clustering methods in which tests of homogeneity are based on firmer grounds. In these, while fusions or subdivisions are conditional on statistical tests of homogeneity, the probabilities are defined on the basis of the sampling distribution of derived indices, such as the similarity index (Goodall, 1968) or the deviant index (Goodall, 1966e). Goodall (1970b) determined sampling distributions for these under the assumption that the sampling is random and the sampled population is homogeneous. What is meant by homogeneity? It means here, that species values (presence, number, yield, etc.) are randomly assorted between the population units (stands, quadrats, etc.). Clearly, such a definition must mean scale dependence, since whether a multispecies population will or will not appear homogeneous depends on the ground pattern of species and the size of the quadrats used in the sampling. Such definition of homogeneity must also mean lack of species correlations. Other definitions can however be conceived to allow species to be correlated in a homogeneous population, provided that the same pattern of species correlations characterize the entire population.

By this we mean that the same type of species correlations would arise irrespective of the stratal origin of the sampling units.

b. Simultaneous clustering

Another group of the R-clustering techniques is exemplified by the methods of *association analysis* (Williams & Lance, 1958; Williams & Lambert, 1959) and its variants (Crawford & Wishart, 1967; Pielou, 1969b; Hill, Bunce & Shaw, 1975). These can serve the objective of constructing dichotomous keys (cf. Osborne, 1963; Ivimey-Cook, 1968) for sorting quadrats between vegetation types. The dichotomies correspond to the possession or lack of certain strategically chosen species.

Association analysis is a monothetic, subdivisive, hierarchical method. The division criterion is basically a sum of squares, defined as the sum of p mean square contingency values of species. The clustering is such that each subdivision maximizes the between groups sum of squares on a single species. To illustrate association analysis we shall consider an example following Williams & Lambert (1959). They sampled a *Callunetum* based on a rectangular grid of 44 x 14 l-metre quadrats placed 5 m apart in both directions. The record for one quadrat was lost but all of the remaining 615 were used. The data include presence scores for five species: *A Calluna vulgaris,*

B Molinia caerulea, C Erica teralix, D Erica cinerea, E Pteridium aguilinum.

The objective in association analysis is to subdivide a sample of quadrats hierarchically into groups. Let N designate the number of quadrats and p the number of species in the sample. The steps in one version include:

1. Compute a $p \times p$ matrix **S** of mean square contingency values. An element of **S** is $S_{hi} = \chi^2_{hi}/N$, where χ^2_{hi} measures association between species h and i as a chi square of a 2 x 2 contingency table. Williams & Lambert use Yates' correction in the formula for χ^2_{hi}. It has been shown that the S_{hi} values represent squared correlation coefficients with a domain $0 \leqslant S_{hi} \leqslant 1$. The **S** matrix for the Williams & Lambert data is given by

$$S = \begin{bmatrix} - & 0.08343 & 0.07424 & - & - \\ 0.08343 & - & 0.1525 & 0.02052 & 0.1116 \\ 0.07424 & 0.1525 & - & 0.00787 & 0.02289 \\ - & 0.02052 & 0.00787 & - & 0.01125 \\ - & 0.1116 & 0.02289 & 0.01125 & - \end{bmatrix} .$$

A dash in the off-diagonal positions indicates non-significant χ^2; these are taken as zeros. Self comparisons are ignored (dash replacing values in the principal diagonal positions).

2. Compute the column (row) sums in **S**. For this we have,

$$\mathbf{Q} = (0.15767 \quad 0.36805 \quad 0.2575 \quad 0.03964 \quad 0.14574) .$$

These quantities are sums of squares that are loaded on the individual species vectors.

3. Choose the species with the highest sum of squares to divide the sample into two groups of quadrats. The species with maximum Q is *Molinia caerulea* (B). When Williams & Lambert (1959) inspected their data, they found that 474 quadrats contained *Molinia caerulea.* These quadrats were put in group B. The remaining 141 quadrats were placed in group b.

4. Recompute the χ^2 matrix within the two groups. For group B the mean square contingency values are given in

$$S_B = \begin{bmatrix} - & \mathrm{I} & 0.052321 & - & \mathrm{I} \\ \mathrm{I} & \mathrm{I} & \mathrm{I} & \mathrm{I} & \mathrm{I} \\ 0.052321 & \mathrm{I} & - & - & - \\ - & \mathrm{I} & - & - & \mathrm{I} \\ \mathrm{I} & \mathrm{I} & \mathrm{I} & \mathrm{I} & - \end{bmatrix}$$

and for group b in

$$S_b = \begin{bmatrix} - & \mathrm{I} & - & 0.03078 & - \\ \mathrm{I} & \mathrm{I} & \mathrm{I} & \mathrm{I} & \mathrm{I} \\ - & \mathrm{I} & - & - & \mathrm{I} \\ 0.03078 & \mathrm{I} & - & - & \mathrm{I} \\ - & \mathrm{I} & \mathrm{I} & \mathrm{I} & - \end{bmatrix} .$$

Symbol I in these indicates indeterminate values.
5. Choose species to subdivide the groups. In group B, the choice is either species A or C, since the corresponding maximum Q values are equal. In group b, the choice is either species A or D. Here too, the Q values are equal. There is clearly an ambiguity in both cases which we resolve by reference to the Q values in the previous S matrix. These indicate subdivision of group B based on *Erica tetralix*, and subdivision of group b based on *Calluna vulgaris.* When the quadrats were inspected, it was found that 250 of 474 in group B contained *Erica tetralix.* This group is labeled BC. The group of the remaining 244 quadrats of group B is labeled by Bc. Similarly, group b was subdivided into group bA of 105 quadrats that contained *Calluna vulgaris*, and group ba of 36 quadrats which did not contain that species. The results are summarized in Fig. 4-5.

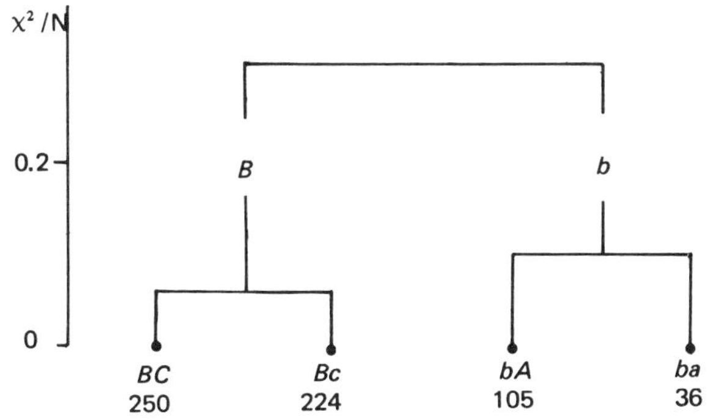

Fig. 4-5. Results from association analysis. See explanations in the main text.

In samples of large numbers of species, the subdivisions are not carried through all the species before the clustering is terminated. The stopping rule (Lance & Williams, 1965) may specify a given α probability point of χ^2, maximum χ^2, a χ^2 value proportionate to sample size, or group size (cf. Grunow, 1964) which, when obtained, will terminate group formation. Association analysis has gained broad acceptance for several reasons:
1. The data for input, consisting of presence scores of species in quadrats, are easily obtained.
2. The algorithm lends itself to efficient programming and it is easily adapted to large data sets.
3. Once the species to subdivide the sample are identified, assignment of new quadrats to groups is effortless.

Three main variants of the method are in circulation: one for classification of quadrats (*normal analysis*), another for the classification of species (*inverse analysis*) and a third for *nodal analysis* which concentrates entries in blocks in the data table, to serve as a basis for recognizing groups (noda) in the sample (Lambert & Williams, 1962).

Hill, Bunce & Shaw (1975) examined association analysis and found that it satisfies most of their criteria for a suitable classification:

1. It can produce classes in heterogeneous data which are ecologically meaningful.
2. The classification is open-ended and it is easy to assign new quadrats to established groups.
3. The method is computationally efficient.

But they disqualify association analysis on one criterion, namely, on freedom from committing serious misclassifications.

Gittins (1965) considers the case of collecting suitable data and the practical convenience of the classification produced a definite advantage in association analysis. Harberd (1960), Ivimey-Cook & Proctor (1966) and Ivimey-Cook (1972) regard susceptibility to misclassifications, due to chance absences of species from quadrats, a serious weakness. Problems with chaining are noted by Noy-Meir, Tadmor & Orshan (1970), and in general with the method, by Greig-Smith (1964), Lange, Stenhouse & Offler (1965), Cormack (1971), Shimwell (1971), Kershaw (1973), and Coetzee & Werger (1975). Lange (1966) explains the implications of the changing pattern of species association on sampling for association analysis, and Crawford & Wishart (1966) consider the methods performance in extremely unstable environments. Greig-Smith, Austin & Whitmore (1967; Ward, 1970) combine ordination with association analysis to correct for misclassifications.

The measurement of species association by χ^2 is not without problems (cf. Emden, 1972), and the interpretation of χ^2 is difficult in the context of association analysis. To overcome the problem of cumulative errors in summing χ^2 quantities, Pielou (1969b) suggests another division criterion. Given two subgroups A and a of quadrats, distinct in that A contains and a lacks species A, the divergence of the subgroups is measured by Pielou (1969b) as the quantity,

$$\chi^2_{Aa} = \sum_h \sum_i \frac{(f_{hi} - f^\circ_{hi})^2}{f^\circ_{hi}},$$

where the summation is over the two subgroups, $i = A,\ a$, and over all p species, $h = 1, \ldots, p$, excluding A. The value of f_{hi} specifies the number of occurrences of species h in subgroup i and $f^\circ_{hj} = f_{h.}f_{.i}/f_{..}$ in which $f_{h.}$ is the number of occurrences of species h in the union group of A and a, $f_{.i}$ the number of occurrences of all species, except A, in group i, and $f_{..} = f_{1.} + \ldots + f_{p.}$. A division of the sample into A and a is accepted if χ^2_{Aa} is at least as large as the chi square for division on any of the remaining species. Pielou (1969b) provides a numerical example to illustrate the method.

The *group analysis* of Crawford & Wishart (1967) is another method which relies on successive subdivisions, using species presence or absence, but the selection criterion for species to subdivide the sample is the quantity,

$$\mu_i'^2 = 4(D_i - P_i \sum_j S_j)^2, \quad j = 1, \ldots, N\ .$$

When S_j is defined as the number of species in quadrat j, expressed as a proportion of the total number of species in the sample, then D_i can be defined as the sum of S_j for those quadrats in which species i occurs. P_i is the proportion of quadrats in which species i occurs. The sample is subdivided on the species for which $\mu_i'^2$ is largest. Subsequent subdivisions of the groups follow a similar rule. The clustering stops when $\sum_j S_j/N$ exceeds an arbitrary threshold value. $\mu_i'^2$ is a frequency related quantity, and as defined, it gives extra weight to frequent species occurring in species rich quadrats. Whereas Crawford & Wishart (1967) consider this an advantage, others disagree (see Goodall, 1973a). A modification has been proposed by Crawford & Wishart (1968) and a similar method, but with single linkage clustering, by Carmichael, George & Julius (1968). The single linkage version was offered to recognize natural clusters, for which it was used by Stringer (1973) in a vegetation analysis.

Hill, Bunce & Shaw (1975) suggested *indicator species analysis* as a substitute for association analysis. Their method retains the obvious advantages of association analysis, but avoids the high misclassification rate. The steps in the analysis are as follows:

1. Order the n quadrats in the sample according to co-ordinates on the first (Y) ordination axis. Although the ordination method used is reciprocal ordering (see Section 3.6), indicator species analysis may be based on ordinations other than this method.
2. Split the set of n quadrats at the center of gravity (mean value) on Y. Label these groups as L and R indicating the left hand and right hand side of the dichotomy. The groups contain n_L and n_R quadrats, respectively.
3. Determine indicator values for species according to

$$I_h = \left| n_{Lh}/n_L - n_{Rh}/n_R \right| ,$$

where I_h is the value for species h, n_{Lh} the number of quadrats containing species h on the left hand side, and n_{Rh} the number of quadrats containing species h on the right hand side of the dichotomy. A value $I_h = 1$ indicates that species h is a perfect indicator; for this to happen, species h must be present in all quadrats on one side and absent from all quadrats on the opposite side of the dichotomy.
4. Identify the five indicator species with highest I_h. The choice of five and not one, two, etc., is entirely a matter of convenience. If only one species were used, the method would differ little from association analysis.
5. Assign a score of -1 to left side indicators and $+1$ to right side indicators.
6. Construct an indicator score for each quadrat by adding up the scores of indicator species which it contains. The quadrat scores may thus vary from -5 to 5 inclusive.
7. Select an indicator threshold. This is the maximum indicator score of a quadrat for inclusion in the minus (–) group. The threshold value should yield (–) and plus (+) groups which coincide as completely as possible with the left (L) and right (R) groups of the original division.
8. Assign quadrats with indicator scores not in excess of the indicator threshold to the (–) group and the others to the (+) group. Declare misclassifica-

tion if a quadrat's group affiliations based on the L/R dichotomy and on the $-/+$ dichotomy conflict. In other words, if a quadrat occurs on the left hand side of the ordination dichotomy and on the plus side of the indicator dichotomy, or in the reverse, it is misclassified.
9. To resolve minor misclassifications define a narrow zone of difference about the center of gravity on Y and assign the quadrats in it to group R if the indicator score is minus, irrespective of initial affiliations to groups L or R.
10. Repeat the stept to further subdivide the R and L groups. Continue until subdivisions are no longer considered useful, then stop.
This method produces a series of dichotomous divisions each characterized by a set of indicator species with +1 or −1 scores and an indicator threshold value. To key out the group affiliation of a new quadrat at any dichotomy, its indicator score is calculated and compared to the indicator threshold which then puts it in the left or right group.

Recognizing a possible underlying relationship between quadrats and species, different classification methods were suggested to extract from a sample clusters, called *noda*, characterized by unique combinations of species and quadrats. To Williams & Dale (1965), a nodum means 'a set of points and the set of axes in which the points constitute a galaxy'. The method proposed by Lambert & Williams (1962) represents an example. In this, quadrats are clustered based on species correlations (associations) and species on the basis of quadrat resemblances (associations). Either clustering process may modify the other. Another example is the method of Tharu & Williams (1966; Webb, Tracey, Williams & Lance, 1967a,b). This method can produce noda at any level in the analysis.

The objective in the Tharu & Williams (1966) method, described as *concentration analysis*, is to find a series of inosculate subdivisions in the data that maximally concentrate the presence scores in the cells of a 4-way partition. To be more explicit, let us state that the group, whose subdivision is contemplated in the analysis, contains p species and n quadrats. Let us assume that whereas a normal division (of association analysis) would yield one group of A quadrats and one group of α quadrats, inverse division would produce one group of B and another group of β species. If we perform both divisions, the normal and the inverse, they will dismember the $p \times n$ data matrix into four blocks with

AB	$B\alpha$
$A\beta$	$\alpha\beta$

number of cells in each block. Note that $AB + B\alpha + A\beta + \alpha\beta = pn$, the total number of cells in the table. Now, if occurrences (non-empty cells) were actually counted in the four blocks, we would obtain the values,

a	b
c	d .

In this, a represents the number of occupied cells in block *1*, b in block *2*, c in block *3* and d in block *4*. We shall refer to these as *observed concentrations.* Should species occurrence in the quadrats be entirely the consequence of chance, we would expect the population values for a, b, c, d to be in the proportions of ABk, $B\alpha k$, $A\beta k$ and $\alpha\beta k$ where $k = (a + b + c + d)/np$. We thus have two frequency distributions,

$$\mathbf{F}_1 = (a \;\; b \;\; c \;\; d)$$

$$\mathbf{F}_1^{\circ} = (ABk \;\; B\alpha k \;\; A\beta k \;\; \alpha\beta k) \quad .$$

The total divergence of the observed **F** from the expected **F**° (indicating minimal concentration) is measured by the chi square quantity,

$$\chi_1^2 = \frac{(a-ABk)^2}{ABk} + \frac{(b-B\alpha k)^2}{B\alpha k} + \frac{(c-A\beta k)^2}{A\beta k} + \frac{(d-\alpha\beta k)^2}{\alpha\beta k} \quad .$$

For the absences, we have

$$\mathbf{F}_{\mathrm{o}} = (AB-a \;\; B\alpha-b \;\; A\beta-c \;\; \alpha\beta-d)$$

$$\mathbf{F}_{\mathrm{o}}^{\circ} = [AB(1-k) \; B\alpha(1-k) \;\; A\beta(1-k) \;\; \alpha\beta(1-k)]$$

$$\chi_{\mathrm{o}}^2 = \frac{[AB-a - AB(1-k)]^2}{AB(1-k)} + \frac{[B\alpha-b - B\alpha(1-k)]^2}{B\alpha(1-k)}$$

$$+ \frac{[A\beta-c - A\beta(1-k)]^2}{A\beta(1-k)} + \frac{[\alpha\beta-d - \alpha\beta(1-k)]^2}{\alpha\beta(1-k)} \quad .$$

The total then is

$$\chi^2 = \chi_1^2 + \chi_{\mathrm{o}}^2 = \frac{(a-ABk)^2}{ABk(1-k)} + \frac{(b-B\alpha k)^2}{B\alpha k(1-k)}$$

$$+ \frac{(c-A\beta k)^2}{A\beta k(1-k)} + \frac{(d-\alpha\beta k)^2}{A\beta k(1-k)} \quad .$$

Tharu & Williams (1966) describe a partitioning of the total chi square into components of which

$$\chi_{A/\alpha}^2 = \frac{[(a+c)\alpha - (b+d)A]^2}{k(1-k)A\alpha pn}$$

gives a component in normal division, and

$$\chi^2_{B|\beta} = \frac{[(a+b)\beta - (c+d)B]^2}{k(1-k)B\beta pn}$$

gives a component in inverse division. These have special utility in that the largest of the two indicates the division that maximizes concentration. If $\chi^2_{A|\alpha} > \chi^2_{B|\beta}$ then a normal division into A and α is indicated and if $\chi^2_{A|\alpha} < \chi^2_{B|\beta}$ then an inverse division into B and β is indicated. Based on trials several inosculate divisions are tested and the best implemented. The entire analysis is then performed on the new groups recursively. A group is considered final, and not subdivided further, if one of two things occurred: percentage presence exceeds a specified limit in the group, or the group attains a critical minimum size.

Dale & Anderson (1973) applied concentration analysis to subartic vegetation, but they were disappointed with the results which lacked a simple group structure which characterized the results of other classificatory analyses. They turned to Macnaughton-Smith's (1965) *two-parameter model* which they incorporated in an inosculate, subdivisive procedure (cf. Dale & Webb, 1975). The rationale of the two-parameter model is briefly stated below with reference to presence data. Extensions to quantitative data are considered by Dale & Anderson (1973).

Within each group each quadrat j is associated with a measure of *stand productivity* (α_j) and each species h with a measure of *species performance* (β_h). If it is assumed that these properties interact in a multiplicative manner, it will make sense to express the odds of species h occurring in quadrat j by

$$\alpha_j\beta_h = p_{hj}/(1-p_{hj}) \ ,$$

and its probability by

$$p_{hj} = \alpha_j\beta_h/(1+\alpha_j\beta_h) \ . \qquad (4.7)$$

In the context of this model, occurrence in a quadrat is assumed to be entirely a consequence of h's performance and j's productivity. Dale & Anderson (1973) suggest that this model may be suitable to absorb a competition term when two species are compared,

$$p_{hj}/p_{ij} = (\beta_h + \alpha_j\beta_h\beta_i)/(\beta_i + \alpha_j\beta_h\beta_i) \ .$$

How to render these formulations operational? The α_i and β_h can be estimated from the data. For this, Macnaughton-Smith (1965) uses an iterative procedure with starting values defined according to

$$\alpha_j = \left[\frac{x_{.j}}{m-x_{.j}} \ \frac{mn-x_{..}}{x_{..}}\right]^{1/2}$$

$$\beta_h = \left[\frac{x_{h.}}{n-x_{h.}} \; \frac{mn-x_{..}}{x_{..}} \right]^{1/2},$$

where $x_{.j}$ is the total number of occurrences in quadrat j, $x_{h.}$ the number of quadrats in which species h occurs, $x_{..}$ the grand total of occurrences, m the number of species in the group and n the number of quadrats. Dale & Anderson (1973) use the initial values of α_j and β_h without iterations, anticipating that iterations to more accurate values would increase the computational load out of proportion with the accuracy gained. Macnaughton-Smith (1965) iterates until the αs and βs stabilize and then measures the fit of the observations to the model in group A based on

$$I_A = \sum_h x_{h.} \ln \beta_h + \sum_j x_{.j} \ln \alpha_j - \sum_h \sum_j \ln (1+\alpha_j\beta_h), \; h=1, ..., m;$$
$$j=1, ..., n \; .$$

Now, A is subdivided into two subgroups B and C with n_B and n_C quadrats. The I_B and I_C are computed. Macnaughton-Smith (1965) regards A final if the maximum value of $\Delta = I_B + I_C - I_A$ is such that $2\Delta < \chi^2_{0.05,\, m-3}$. This is the 0.05 probability point of the chi square distribution with $m-3$ degrees of freedom. Such a decision rule is of course arbitrary and its interpretation as a statistic in probabilistic terms is open to question.

Dale & Anderson (1973) have used the two-parameter model in this way: A normal analysis is performed to obtain the quadrat groups. The quadrat groups are ordinated by a suitable method based on an expression of divergence in terms of Δ. The groups are ordered on the first ordination axis. Species then are listed so that all species having β maximum within a given quadrat group fall adjacent, listed in order of descending β values. The result is a data table with both quadrats and species ordered. The appearance of the table is one of the traditional type in which the entries are concentrated in blocks. Dale & Anderson (1973) subjected the results to scrutiny for ecological informativeness and for resemblance of group contents. They concluded that the method compares favorably with other methods which they would have considered under the circumstances. They warn, however, of shortcomings in the method and problems with the interpretation.

C. Gradient and other ordination-based methods

Open model factor analysis (Cattell, 1965), component analysis (Chapter 3) and other methods of ordination are commonly used to serve classificatory objectives. In some applications, individuals are collected in a common group if their co-ordinates are similar on all ordination axes, or characters are assigned to the same group when they show high affinity to a given ordination axis (e.g. Dagnelie, 1960; Crovello, 1968c; Ornduff & Crovello, 1968; Noy-Meir, 1970; Fresco, 1971). Some applications rely on class recognition by inspection of scatter diagrams in one dimension (e.g. Brown & Curtis,

1955; Looman, 1963) or in two or more dimensions (e.g. Gittins, 1965; Fabbro, Feoli & Saulti, 1975). The success with such inspections can be greatly enhanced by viewing stereograms. The construction of stereograms is described in Section 3.1.

Noy-Meir (1971b, 1973b; also Noy-Meir & Whittaker, 1977) advocates classifications via an eigenvalue and vector analysis coupled with varimax rotation of axes (cf. Kaiser, 1958; Harman, 1967). A similar rotation has been tried earlier by Ivimey-Cook & Proctor (1967) who found that in the rotated positions, the axes were more easily interpreted as identifiers of vegetation types. The objective in Noy-Meir's (1971b) method is to extract ordination axes which project from the zero origin in the direction of high-density regions (clusters or noda) in sample space. Noy-Meir (1971b) described the method as *nodal ordination.* The axes are eigenvectors of a matrix of non-centered vector scalar products. Noy-Meir (1971b) and others (e.g. Ivimey-Cook & Proctor, 1967; Hinneri, 1972; Pignatti & Pignatti, 1975) find advantages in non-centering and also in axis rotation. After rotation, the axes tend to be better aligned each with a different group of quadrats or group of species which have high positive scores.

The pre-analysis treatment of the data in nodal analysis may or may not include normalizations or other adjustments on which the types of groups recognized would depend. The distinctness of groups is measured by the *coefficient of conjunction.* The number of axes estimated influence the number of groups recognized. As to actual numbers, decisions are arbitrary. Criteria include replicability of groups in different samples, or the amount of sum of squares accounted for in total by the eigenvalues.

It is appropriate to say that the success of nodal analysis by ordination will depend on the data structure. It can be postulated that axes will be identifiable as clusters without ambiguity, only if the clusters are orthogonal subspaces of the sample. Such clusters are shown in Fig. 4-6. The further the data structure is removed from this kind, which is likely to happen in actual samples, the less reliable the classification becomes.

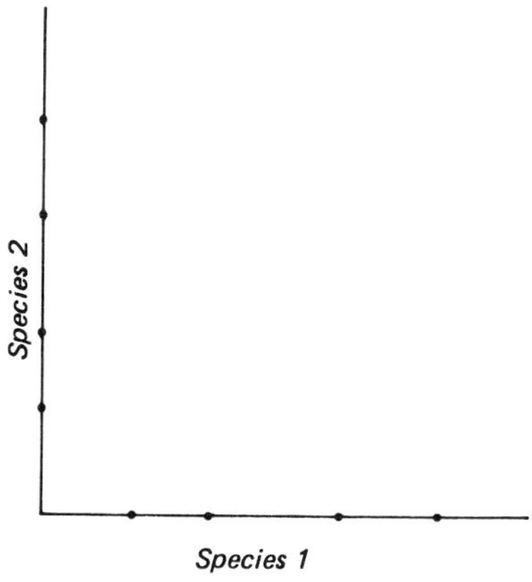

Fig. 4-6. Diagram indicating orthogonal clusters in a two-species sample. Points represent quadrats. Explanations are given in the main text.

Nodal ordination produces not only nodal groups, but also a non-nodal group of residuals. The members of this have to be disposed at the end of the clustering process. This may require further analyses. Nodal ordination can nevertheless be a useful method, as it has been shown by Noy-Meir (1971b, 1973b), in producing results consistent with common sense expectations from knowledge of the vegetation in the sampling site.

We note the conceptual similarity of choosing a position vector, which represents the centroid (or nearly the centroid) of a sheaf of position vectors, to serve as ordination axes, and of defining classification units as groups of position vectors that cluster most closely about the individual ordination axes. While the former is peculiar to position vectors ordination (see Section 3.4), the latter is implicit in the method of Yarranton, Beasleigh, Morrison & Shafi (1972). Their method will yield classification units with different degrees of overlap, depending on the threshold correlation at which clusters of position vectors are recognized.

Noy-Meir (1973b) and Lambert, Meacock, Barrs & Smartt (1973) studied the classificatory potential of ordinations along similar lines as suggested by Goodall (1964) and Dale (1975). The idea that ordinations are useful tools to give guidance in classificatory problems appears to be a sound one, even though it has been questioned. Carlson (1970) summarized the opposing arguments:

1. Ordinations produce axes and not clusters.
2. Axis rotation may dismember groups.
3. The number of groups found may be restricted by the number of axes extracted. (This applies mainly to those methods which define a one to one correspondence between ordination axes and groups.)

We may draw attention to one further point. Since scatter diagrams represent potentially distorted projections of point clusters, classes recognized from scatter diagrams may not be reliable. This should be understood in such a way that clusters of points that are spatially separated may not appear distinct in a scatter diagram.

Noda, as high density phases in sample space, may be detected by methods quite different from what we have considered so far. Imagine a curved surface (G) which shows the density changes of points in sample space. This surface would have several local maxima and minima corresponding to the high or low density phases. Now, if we think of maxima as centers of clusters, we may define a cluster as those points which are located on the slopes of G around a given maximum. Methods were suggested by Ihm (1965), Hill, Silvestri, Ihm, Farchi & Lanciani (1965) which can detect density maxima of this sort at different levels of resolution and identify the associated clusters of points. The algorithm is summarized here following Katz & Rohlf (1973):

1. Define a density function,

$$f(\mathbf{X}) = \sum_j e^{-d^2(\mathbf{X};\mathbf{X}_j)/w^2}, \quad j=1,\dots,n \tag{4.8}$$

Here $\mathbf{X}$ and $\mathbf{X}_j$ are p-valued vectors; they define respectively an arbitrary point and the point image of quadrat j on S. The value of $f(\mathbf{X})$ is higher when point $\mathbf{X}$ falls in a high density phase and lower when it lies in a low density phase. $d(\mathbf{X};\mathbf{X}_j)$ is the Euclidean distance of points $\mathbf{X}$ and $\mathbf{X}_j$ (formula (2.1)) and w is an arbitrary number.

2. Select a level of resolution for detecting maxima. This requires specification of a value for w. The effect of w on $f(\mathbf{X})$ is such that if it is sufficiently large only a single maximum, coinciding with the sample centroid, will be found and if it is sufficiently small as many maxima will be found as there are points in the sample. Clearly, by assigning different numerical values to w, few or many clusters can be detected.

3. Select a point $\mathbf{X}_j$ and move it up slope in the direction of the local maximum. Analytically, this is done by the function $\mathbf{X}: = \mathbf{X} + \Delta\mathbf{X}$ implemented recursively with the starting value of $\mathbf{X}$ defined as $\mathbf{X}_j$ and the final value by $\mathbf{X}_j^*$. In fact, $\mathbf{X}_j^*$ holds the co-ordinates of a local maximum. The value of $\Delta\mathbf{X}$ is defined by

$$\Delta\mathbf{X} = \frac{\partial f(\mathbf{X})}{\partial \mathbf{X}} \; \frac{S}{\| \partial f(\mathbf{X})/\partial \mathbf{X} \|} \quad ,$$

for the ith element of which we have

$$\frac{\partial f(X_i)}{\partial X_i} = \sum_{j=1}^{n} 2 f(\mathbf{X}) \, (X_{ij} - X_i)/w^2 \quad .$$

It is clear from these, that $f(\mathbf{X})$ need not be defined as in formula (4.8), but it may be one of many different functions. It has been suggested that S, the step size, be chosen to be equal to about $w/2$. When the maximum is passed (i.e. $\Delta\mathbf{X} < 0$) S is further reduced to allow reaching a point close to the maximum.

4. Repeat step 3, taking quadrats one at a time ($j = 1, ..., n$), and store the $\mathbf{X}_j^*$ vectors.

5. Quadrats for which the computations yield an identical $\mathbf{X}^*$ vector (within given tolerance limits) are represented by points in the neighbourhood of a local maximum, and thus they are considered to be members of the same cluster.

6. Repeat the analysis starting at step 3 for different values of w.

By increasing w, adjacent local maxima will fuse and be replaced by a new maximum near their centroid. In the process, a hierarchical classification is produced. This will not necessarily be nested since some quadrats that are members in a common cluster at one level may shift to different clusters at a higher or lower level. The method has been shown to be insensitive to outliers (cf. Katz & Rohlf, 1973).

Katz & Rohlf (1973) used the name *function-point cluster analysis* to designate gradient clustering. Related techniques were suggested by Buttler (1969; *gravitational clustering*) and Patrick & Fisher (1969; *cluster mapping*). Anderberg (1973) comments on some of the methodology. Another related technique is due to Rohlf (1970). In this, changing shape and size trends of clusters are taken into account in decisions concerning assignments. The procedure requires specification of the functional form of cluster shapes that are permissible. This technique, according to Rohlf (1970), will more likely be successful if the clusters follow a relatively simple and consistent shape trend.

D. Predictive clustering

While classifications are applied deterministically in most vegetation studies, probabilistic applications have been introduced (e.g. Goodall, 1970b). We discussed these in some detail. Since these rely on statistical tests for group homogeneity, they assume implicitly or explicitly that the groups recognized are samples from disjoint parent populations.

Other somewhat related applications involve estimation as the classification objective. When estimation is understood in the sense that from sample class values the unknown population class values are estimated, there may be serious difficulties with the estimation, because the classes are not constructed in a random sorting process, but rather, they combine individuals of maximum similarity. It follows then that the sample class values are likely to be biased. Macnaughton-Smith (1965) has considered this problem and suggested a solution:

1. Derive a classification for the first random half of the sample.
2. Impose the derived classification upon the second half.

Since the two halves have no individuals in common, membership in one or the other being the consequence of chance alone, the class values in the second half will hopefully represent unbiased estimates of the population class values. Imposition of an existing classification on the individuals of another sample may be accomplished on the basis of given diagnostic species, when the first random sample is classified in terms of some species-based monothetic classification (e.g. Williams & Lambert, 1959; Hill, Bunce & Shaw, 1975), or on the basis of some multivariate assignment function when the first classification is polythetic (see Chapter 5 for details).

Yet another predictive method has been described by Macnaughton-Smith (1963). We shall present a summary of this in the context of a vegetation study, seeking groups of quadrats within which species values are maximally predictive of certain conditions in their environment. Let us consider an example. We have completed a vegetation survey in which each of N quadrats, randomly chosen, were described by data indicating the presence or absence of p species and the possession or lack of visible signs of recent burning. Let the entries in the table

		Species	
		+	–
Burning	+	a_h	b_h
	–	c_h	d_h

specify the number of quadrats a_h that show signs of burning and possess species h, the number of quadrats b_h that show signs of burning but lack species h, the number of quadrats c_h that show no signs of burning and possess species h, and lastly, the number of quadrats d_h which neither show signs of burning nor possess species h. The association of burning and species can be measured by some suitable function, such as chi square,

$$\chi_h^2 = \frac{(|a_h d_h - b_h c_h| - N/2)^2}{(a_h+b_h)(c_h+d_h)(a_h+c_h)(b_h+d_h)} .$$

The first partition divides the N quadrats into two groups, one with and the other without species h for which the χ_h^2 value is largest, assuming that this value is also larger than the α probability point ($\chi^2_{\alpha;1}$) of the chi square distribution with 1 degree of freedom. The groups are reanalysed independently until group size drops below 8, or the maximum value of chi square drops to a value not larger than $\chi^2_{\alpha;1}$. At this stage, there are k groups with N_j quadrats and a species list S_j in the jth group. The species lists will overlap between the groups. If the proportion of quadrats that showed signs of burning in group j is p_j, then the joint predictive value of species in list S_j is taken to be proportional to p_j. Thus the $p_1, ..., p_k$ values, when ordered according to descending magnitude, rank the different lists (S_j, $j = 1, ..., k$) according to predictive value from high likelihood of burning to low or none.

E. Methods for large samples

The classification techniques which we described so far are, in the main, designed to provide optimal solution to specific classificatory problems without particular concern for computational efficiency in applications involving large samples. Even if we had access to a large and fast computer, the vastness of core space required may force us to look for procedures of high computational efficiency. Baum & Lefkovitch (1972) described one method which could be helpful. Others are offered by, e.g., Crawford & Wishart (1968), Frey (1969b) and Janssen (1975). Still other, potentially helpful methods include association analysis (Williams & Lambert, 1959), group analysis (Crawford & Wishart, 1967), predictive classification (Gower, 1974) and indicator species analysis (Hill, Bunce & Shaw, 1975).

The program package by Wishart (1969a) offers an unified algorithm for a diversity of clustering methods. One method in the package can perform hierarchical clustering on very large numbers of individuals. Another program package by Parks (1969, 1970) performs a weighted centroid cluster analysis on very large samples. Park's program also performs component analysis.

4.10 Evaluation of clustering results

A. Recognition of types

Classifications normally use devices which monitor efficiency at any stage during the clustering process. Such a device may be a relative measure of heterogeneity between the groups, often a ratio of sum of squares or information, or some other quantity such as the fusion similarity. Considering Fig. 4-2 as our example, classification efficiency at the 2-type level can be determined via the following analysis of sums of squares:

Source of sum of squares	*Sum of squares*	*% of total*
Between groups	1.81 – 0.06 – 0.14 = 1.61	89
Within groups	0.06 + 0.14 = 0.20	11
Total	1.81	100 .

There are few measures on which a similar additive analysis could be performed. Information is one of them, but average similarity is not. The additive property of the sum of squares and information will hold true regardless of the number of groups involved.

When a dendrogram is given, and the goal is to recognize vegetation types, the dendrogram may be intercepted at a level corresponding to a specified classification efficiency, and the intercepted stems may be used to delineate vegetation types. The criterion for choosing groups, however, need not specify classification efficiency, but simply, it may specify the number of vegetation types to be formed, in which case, the dendrogram is intercepted at a level incorporating the required number of stems.

Clustering methods often incorporate stopping rules that prevent the algorithm from producing new fusions or further subdivisions once an a priori established level of homogeneity (or heterogeneity) is reached. In some cases, a subdivision would not be implemented if the value of some measure such as χ^2 (e.g. Goodall, 1953; Williams & Lambert, 1959; Noy-Meir, Tadmor & Orshan, 1970), information I (e.g. Macnaughton-Smith, 1965) or the variance ratio F (e.g. Edwards & Cavalli-Sforza, 1965) dropped below or at least reached the vicinity of a specified threshold value depending on the clustering strategy. Similar stopping rules could be associated with agglomerative clustering procedures. Whether the clustering is subdivisive or agglomerative, the groups formed minimize heterogeneity within the groups, or maximize the discontinuity among the groups. Due to minimization (or maximization), stopping rules that refer χ^2, I or F to standard statistical distributions, may be ineffective since the sample value may have a sampling distribution different from what we would expect if the groups had a random basis for their existence. To this problem Goodall's (1970b, 1973a,b) procedures offer a plausible solution. His derivation of tests points up the value of a combined heuristic and axiomatic appraoch in the solution of distribution problems.

There are classifications where no hierarchy is constructed. Jancey's (1966a) method is a good example of this. In his method, the number of groups to be formed is specified, and the dissections will stop only after all individuals are allocated to the groups. In Jancey's (1966a) method, groups will be formed regardless of the existence or non-existence of natural groups in the sample. In Jancey's (1974) other method groups will be formed only if the data structure is discontinuous. In other words, the groups must actually exist as density phases (i.e. noda) in sample space before they are recognized as discrete entities.

B. Representation of types

After the types have been recognized, their vegetational and environmental characteristics have to be described. Where hierarchical relationships are to be demonstrated, dendrogram representations are appropriate. Dendrograms provide information about pairwise similarities of types and conglomerates of types. But in such a representation only a fraction of the information, residing in the matrix of resemblance values that relate the types, is retained by the fusion pattern traceable in a dendrogram. Dendrograms can nevertheless provide information about the degree of distinctness of the groups. If most fusions occur at a lower level (with low within-group heterogeneity), and large groups persist and fuse only at the higher levels, we conclude that the groups are sharply distinct.

Types are often displayed as distinct clusters of points or as ellipses (ellipsoids) in two (or three) dimensions (e.g. Gittins, 1965; Jancey, 1966b; Orlóci, 1967a, 1968c) in a reference system of ordination axes. Since projections into lower dimensional spaces are involved, the success of such representations will depend on the complexity of the actual group structure in the sample.

To display ecological information, types may be mapped in a reference system of environmental variables (e.g. Whittaker, 1956). When, however, interest centers on displaying information about the shape and orientation of point clusters, representing types, or where the clusters are known to be extremely variable in these properties, ordinations within types may be revealing (Greig-Smith, 1971a,b).

Users classifying the vegetation of an area often choose mapping of the ground pattern of types as a means for presentation of classification results. Bradfield & Orlóci (1975) describe an example. Mueller-Dombois & Ellenberg (1974) treat the idea of vegetation mapping, describe general methods and give examples.

Plexus diagrams (networks or spanning trees) with noda representing types is another obvious possibility (see McIntosh, 1973) to present the clustering results. Such diagrams, just as dendrograms, are two dimensional maps of a potential multidimensional configuration, and thus they do not convey the full information in the resemblance matrix.

C. Comparison of types

Comparisons via statistical tests of appropriately stated hypotheses can reveal useful information about the classification results. Circular arguments should however be avoided, and also, efforts should not be wasted in tediously showing significant differences where they are expected to occur with complete regularity springing directly from the maximizations or minimizations in the clustering algorithm.

To avoid circularity, group comparisons should be based on variables other than those which formed the groups. For example, when the groups are represented by vegetation types, delimited on the basis of species composition, a comparison may be based on environmental variables (e.g. Mukkattu, 1974). Materials for such a comparison are given in Table 4-1 and Fig. 4-7.

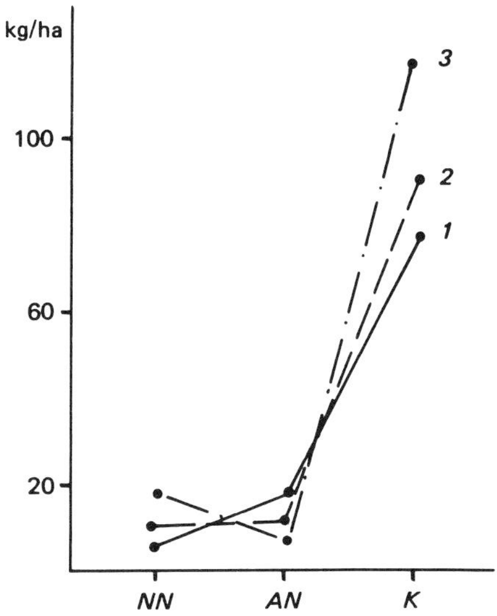

Fig. 4-7. Soil chemical profiles of vegetation types (*1, 2, 3*). Vertical scale indicates mean values within the types. *Legend to symbols: NN* – nitrate nitrogen; *AN* – ammoniacal nitrogen; *K* – potassium. All elemental in kg/ha units following Table *4-1*.

Let X_{jmh} define in symbolic terms a data element in Table 4-1, and let *j, m, h* identify respectively a quadrat in which the soil sample is taken, vegetation type in which the quadrat is taken, and variable. For example, X_{213} = 43.2 is the quantity of elementary potassium (variable *3*) in the second quadrat of type *1*. The values are summed within rows and within columns, and the totals are given in the marginal cells. The symbol $X_{jm.}$ represents the *j*th row total in type *m*, and symbol $X_{.mh}$ the *h*th column total in the same type. The total in type *m* is $X_{.m.}$ and the total within the *h*th column $X_{..h}$. The grand total for the table is designated by $X_{...}$, the number of replicates in type *m* by n_m, and in all types by $n_1 + ... + n_k = N$.

Table 4-1. Some soil chemical properties in three vegetation types. Replicates are soil samples from quadrats. Entries in table represent quantities in kg/ha units

Type	*Replicate*	*Elementary Nitrate N*	*Variable Elementary Ammonical N*	*Elementary K*	*Total*
1	*1*	7.5	20.1	42.0	69.6
	2	10.9	20.4	43.2	74.5
	3	10.8	17.9	40.4	69.1
	4	12.2	18.0	84.3	114.5
	5	5.1	15.4	69.5	90.0
	6	6.2	12.1	99.8	118.1
	7	3.0	21.0	110.4	134.4
	8	6.6	16.1	101.4	124.1
	9	12.9	13.8	87.7	114.4
	10	4.0	18.0	115.2	137.2
Total		79.2	172.8	793.9	1045.9
2	*1*	15.2	11.0	94.7	120.9
	2	11.7	13.5	70.8	96.0
	3	9.1	7.8	87.5	104.4
	4	7.0	10.9	125.6	143.5
	5	8.9	15.1	71.0	95.0
	6	11.7	15.3	133.5	160.5
	7	10.1	13.8	77.7	101.6
	8	6.2	16.0	70.3	92.5
Total		79.9	103.4	731.1	914.4
3	*1*	20.6	9.6	121.9	152.1
	2	20.8	4.1	101.0	125.9
	3	17.1	5.8	83.3	106.2
	4	15.1	12.1	121.4	148.6
	5	16.3	8.9	105.4	130.6
	6	14.9	11.9	128.8	155.6
	7	17.3	5.8	103.6	126.7
	8	13.7	8.7	163.3	185.7
	9	15.1	10.0	126.5	151.6
Total		150.9	76.9	1055.2	1283.0
Grand total		310.0	353.1	2580.2	3243.3

Symbol k designates the number of types and p the number of variables.

The analysis to be outlined requires the assumption that the populations from which the samples derive are multivariate normal. This means, in terms of our earlier spatial analogy, that the points are from linear clusters of uniform density. Under this assumption we can stipulate different hypotheses about the k types:

H_0 : The types are indistinguishable in terms of the p environmental variables.

H_{01} : The point clusters, which represent the types, are indistinguishable based on their size, shape, or orientation.

H_{02} : The type mean vectors estimate the same population mean vector. The general hypothesis of type indistinguishability H_0 has two components H_{01} and H_{02}. It is quite logical to begin with H_{01} because the test on H_{02} requires the assumption that H_{01} is true. The criterion that describes the shape, size and orientation of linear point clusters is the covariance matrix. Let $\mathbf{S}_m$ represent the mth sample covariance matrix of the variables. This matrix has $p \times p$ elements. A characteristic element of $\mathbf{S}_m$ is S_{him}, relating the ith and hth variables in type m. Because the covariance matrix completely describes a linear cluster, less its location, a test on H_{01}, against an alternative hypothesis which negates H_{01}, is equivalent to a test on the null hypothesis

$$H_{01} : E(\mathbf{S}_1) = \ldots = E(\mathbf{S}_k)$$

which stipulates that the k population covariance matrixes are equal, in contrast to the alternative that at least some of them are not. The letter E indicates expectation. The test criterion is MC^{-1} which we calculate according to

$$M = \sum_m (n_m - 1) \ln |\mathbf{S}| - \sum_m (n_m - 1) \ln |\mathbf{S}_m| , m = 1, \ldots, k$$

and

$$C^{-1} = 1 - \frac{2p^2 + 3p - 1}{6(p+1)(k-1)} \left[\sum_m \frac{1}{n_m - 1} - \frac{1}{\sum_m (n_m - 1)} \right],$$
$$m = 1, \ldots, k \, .$$

In these expressions $|\mathbf{S}|$ is the determinant of the pooled covariance matrix whose elements are defined by

$$S_{hi} = \frac{\sum_m (n_m - 1) S_{him}}{\sum_m (n_m - 1)} , \quad m = 1, \ldots, k$$

where S_{him} is the covariance of variable h and i in type m, and $|\mathbf{S}_m|$ is the determinant of the mth covariance matrix $\mathbf{S}_m$; n_m represents the number of samples taken from type m; k is the number of types; and p the number of variables. We accept H_{01} at α probability if the condition

$$MC^{-1} < \chi^2_{\alpha; (k-1) p (p+1)/2}$$

is satisfied. Otherwise we reject H_{01}.

Based on the data in Table 4-1 and the program TSTCOV in the Appendix, we have the following numerical results:

$$k = 3 \quad n_1 = 10 \quad n_2 = 8 \quad n_3 = 9 \quad p = 3$$

$$\mathbf{S}_1 = \begin{bmatrix} 12.5218 & -1.0784 & -52.3609 \\ -1.0784 & 8.5351 & -26.6524 \\ -52.3609 & -26.6521 & 840.434 \end{bmatrix}$$

$$|\mathbf{S}_1| = 53538.4$$

$$\mathbf{S}_2 = \begin{bmatrix} 8.3270 & -1.1854 & 7.9813 \\ -1.1854 & 7.8564 & -12.6739 \\ 7.9813 & -12.6739 & 634.510 \end{bmatrix}$$

$$|\mathbf{S}_2| = 39020.0$$

$$\mathbf{S}_3 = \begin{bmatrix} 6.2525 & -3.9396 & -30.7371 \\ -3.9396 & 7.7628 & 34.7115 \\ -30.7371 & 34.7115 & 515.228 \end{bmatrix}$$

$$|\mathbf{S}_3| = 10549.9$$

$$\mathbf{S} = \begin{bmatrix} 9.2085 & -2.0633 & -27.5532 \\ -2.0633 & 8.0797 & -2.1207 \\ -27.5532 & -2.1207 & 671.971 \end{bmatrix}$$

$$|\mathbf{S}| = 40718.9$$

$$M = 24 \ln 40718.9 - 9 \ln 53538.4 - 7 \ln 39020.0 - 8 \ln 10549.9$$
$$= 8.6396$$

$$C^{-1} = 1 - \frac{18 + 8}{(6)(4)(2)} \left[\frac{1}{9} + \frac{1}{7} + \frac{1}{8} - \frac{1}{24} \right] = 1 - 0.1827 = 0.8173 \ .$$

From these we have $MC^{-1} = 7.061$. Because $\chi_{0.05;\,12} = 21.03$, the condition

$$MC^{-1} < \chi^2_{0.05;12}$$

is true, and H_{01} is accepted. We thus declare indistinguishability on shape, size and orientation in the k clusters. The question remains: are the mean vectors distinguishable?

The test on the equality of covariance matrices clears the way for a test on the type mean vectors about which we hypothesized under H_{02}. This hypothesis can be restated as

$$H_{02} : \ E(\overline{\mathbf{X}}_1) = \ldots = E(\overline{\mathbf{X}}_k) \ ,$$

where $\overline{\mathbf{X}}_m = (\overline{X}_{m1} \ldots \overline{X}_{mp})$ is the mean vector of type m whose elements $\overline{X}_{m1}, \ldots, \overline{X}_{mp}$ represent the variate means. The letter E indicates expectation. H_{02} is tested against an alternative hypothesis which negates H_{02}. The procedure is based on the multivariate analysis of variance which partitions the sums of squares and cross products. We define the cross products of variable h and i in the total sample, irrespective of types, as

$$S_{hi} = \sum_m \sum_j X_{jmh} X_{jmi} - X_{..h} X_{..i}/N, \quad m = 1, \ldots, k; \; j = 1, \ldots, n_m \; .$$

Any such cross product can be partitioned into a between (H) and within (E) component:

$$H_{hi} = \sum_m X_{.mh} X_{.mi}/n_m - X_{..h} X_{..i}/N, \; m = 1, \ldots, k \qquad \text{and}$$

$$E_{hi} = \sum_m \left[\sum_j (X_{jmh} X_{jmi}) - X_{.mh} X_{.mi}/n_m \right], \; m = 1, \ldots, k; \; j = 1, \ldots, n_m \; .$$

To test the hypothesis of equal type mean vectors we extract the largest eigenvalue λ of $\mathbf{HE}^{-1}$ where $\mathbf{H}$ is a $p \times p$ matrix of elements H_{hi} and $\mathbf{E}$ is a $p \times p$ matrix of elements E_{hi}. The test criterion, $\theta = \lambda/(1 + \lambda)$, is referred to the appropriate Heck (1960) chart with α probability and parameters $s = \min(k-1, p)$, $m = (\,|\,k-p-1\,|\,-1)/2$, and $n = (N-k-p-1)/2$. H_{02} is accepted at α probability if the condition,

$$\theta < \theta_{\alpha;\, s,m,n} \, ,$$

is satisfied. Otherwise, H_{02} is rejected.

Based on the data in Table 4-1 the following results emerge by an analysis based on program EMV in the Appendix:

$$\overline{\mathbf{X}}_1 = \begin{bmatrix} 7.9 \\ 17.3 \\ 79.4 \end{bmatrix} \qquad \overline{\mathbf{X}}_2 = \begin{bmatrix} 10.0 \\ 12.9 \\ 91.4 \end{bmatrix} \qquad \overline{\mathbf{X}}_3 = \begin{bmatrix} 16.8 \\ 8.5 \\ 117.2 \end{bmatrix}$$

$$\mathbf{S} = \begin{bmatrix} 617.101 & -412.991 & 995.941 \\ -412.991 & 555.647 & -1610.02 \\ 995.941 & -1610.02 & 23113.2 \end{bmatrix}$$

$$\mathbf{H} = \begin{bmatrix} 396.096 & -363.471 & -1657.22 \\ -363.471 & 361.733 & -1559.13 \\ 1657.22 & -1559.13 & 6985.90 \end{bmatrix}$$

$$\mathbf{E} = \begin{bmatrix} 221.005 & -49.5202 & -661.277 \\ -40.5202 & 193.913 & -50.8975 \\ -661.277 & -50.8975 & 16127.300 \end{bmatrix}$$

$$\mathbf{E}^{-1} = \begin{bmatrix} 0.00555111 & 0.00147857 & 0.000232282 \\ 0.00147857 & 0.00555505 & 0.0000781584 \\ 0.000232282 & 0.0000781584 & 0.0000717777 \end{bmatrix}$$

$$\mathbf{HE}^{-1} = \begin{bmatrix} 2.04630 & -1.30392 & 0.182549 \\ -1.84497 & 1.3517 & -0.168066 \\ 8.51681 & -5.66470 & 0.764574 \end{bmatrix}$$

$$\begin{aligned} \lambda &= 12.1929 \\ \theta &= 12.1929/13.1929 = 0.924 \\ s &= \min(2,3) = 2 \\ m &= (|3 - 3 - 1| - 1)/2 = 0 \\ n &= (27 - 3 - 3 - 1)/2 = 10 \;. \end{aligned}$$

Since the chart value $\theta_{0.05;2,0,10} = 0.375$ is less than the calculated value θ, we reject H_{02} and declare that at least two of the type mean vectors are significantly different.

In the next step, after rejection of H_{02}, we could examine mean vectors in pairs to determine which are significantly different and which are not, and we could also consider the question of which variables are contributing most strongly to the differences. We shall limit the subsequent discussions to tests concerned with the hypotheses:

$$\begin{aligned} H_{021} &: E(\overline{\mathbf{X}}_1) = E(\overline{\mathbf{X}}_2) \\ H_{022} &: E(\overline{\mathbf{X}}_1) = E(\overline{\mathbf{X}}_3) \\ H_{023} &: E(\overline{\mathbf{X}}_2) = E(\overline{\mathbf{X}}_3) \;. \end{aligned}$$

The test criterion is

$$T^2_{mz} = \frac{n_m n_z}{n_m + n_z} (\overline{\mathbf{X}}_m - \overline{\mathbf{X}}_z)' \, \mathbf{S}^{-1} (\overline{\mathbf{X}}_m - \overline{\mathbf{X}}_z)$$

for any m or z where n_m and n_z indicate the number of samples from types m and z, $\overline{\mathbf{X}}_m$ and $\overline{\mathbf{X}}_z$ are type mean vectors, and $\mathbf{S}$ is the pooled covariance matrix for the two types. The quantity,

$$F = (n_m + n_z - p - 1) \; T^2 \,/((n_m + n_z - 2)p),$$

has the variance ratio distribution with degrees of freedom p and $n_m + n_z - p - 1$. For type *1* and *2* the criterion becomes:

$$T_{12}^2 = \frac{80}{18}(7.9 - 10.0 \quad 17.3 - 12.9 \quad 79.4 - 91.4)$$

$$\begin{bmatrix} 0.133227 & 0.0354857 & 0.00557475 \\ 0.0354857 & 0.133321 & 0.00187579 \\ 0.00557475 & 0.00187579 & 0.00172266 \end{bmatrix} \begin{bmatrix} 7.9 - 10.0 \\ 17.3 - 12.9 \\ 79.4 - 91.4 \end{bmatrix}$$

$$= 12.37 \text{ and } F_{12} = \frac{14}{48} \, 12.37 = 3.61 \quad .$$

To explain further these results, we note that in this particular example we use the inverse of the pooled covariance matrix **S** which we computed in the test for equality of covariance matrices. We do use the same inverse because it is convenient, and also, justified. Firstly, the inverse of **S** is already available since it had to be computed before the determinant of **S** could be found. Secondly, we have theoretical justification springing from the test of H_{01}. When we accepted H_{01} we implied that pooling $\mathbf{S}_1$, $\mathbf{S}_2$, $\mathbf{S}_3$ is justified because they are all estimates of a common population covariance matrix.

Similar computations yield the remaining values of T_{13}^2 and T_{23}^2. The complete set of results are summarized in the following table:

Types	*F*	*Degrees of freedom*	$F_{0.05;\, p, n_m + n_z - p - 1}$	*Hypothesis* H_{02} *rejected* (R)
1,2	3.61	3,14	3.34	R
1,3	28.0	3,15	3.29	R
2,3	11.3	3,13	3.41	R .

We conclude that all pairs of mean vectors differ significantly. The difference is smallest between types *1* and *2*, and greatest between *1* and *3*.

Some further comments should be made about these results. Longhand calculations, even in the present example where only three variables are involved and only three groups, are not really recommended. The computations should be performed by a computer. Regarding precision, it is logical that we retain not more than one decimal when we give the mean vectors. This is the precision in the measurements themselves. However, in the intermediate results, such as in the inverse of any matrix, we retain six significant digits. This is the computer's precision. In the test quantities such as F and χ^2 the number of decimals retained coincide with the numbers given in a relevant statistical table.

In the example above, the hypothesis of equal population covariance matrices is found to be tenable. From this we conclude that the point clusters which represent the different populations in sample space have identical shape, size and orientation. It is logical then to perform tests on the equality of mean vectors to reveal if there are in fact differences in locations. It is quite conceivable, however, that a test may lead to rejection of the null hypotheses of identical covariance matrices. If it did, we would have no justification to carry out tests on mean vectors in the way we have done.

To consider the implications, let it be assumed that the k population covariance matrices are in fact unequal. While a direct consequence of this is that we cannot proceed testing hypotheses about mean vectors as we have done, we can still extract information from the sample about the populations in a different approach of reasoning. Firstly, we note that differences between covariance matrices may take on a variety of forms. They may involve cluster size or orientation. Since these have different significance in interpretations of the materials studied (cf. Reyment, 1969), it may be profitable to identify more closely in what respect the covariance matrices are actually different. We could for instance assert that the difference is only in cluster size, but the orientation is the same, and then go about testing the assertion as a hypothesis. Equivalently, we may hypothesize that the ith eigenvector β_{im} of the population covariance matrix Σ_m, estimated by the eigenvector $\mathbf{b}_{im}$ of the sample covariance matrix $\mathbf{S}_m$, is in fact also an eigenvector of the population covariance matrix Σ_z estimated by $\mathbf{S}_z$. The test criterion is due to Anderson (1963; also Reyment, 1969):

$$\chi^2 = (n_z-1)\left[\lambda_{iz}\, \mathbf{b}'_{im}\, \mathbf{S}_z^{-1}\, \mathbf{b}_{im} + \lambda_{iz}^{-1}\, \mathbf{b}'_{im}\, \mathbf{S}_z \mathbf{b}_{im} - 2\right] .$$

The sampling distribution of χ^2 approximates that of a chi square variate with $p-1$ degrees of freedom, assuming that n_z is large and that the population distribution from which the sample is taken is multivariate normal. In the formula, n_z is the size of sample z, λ_{iz} the ith eigenvalue of $\mathbf{S}_z$ and $\mathbf{b}_{im}$ the ith eigenvector of $\mathbf{S}_m$. Symbol p indicates the number of variables. If we find that $\chi^2 < \chi^2_{\alpha;\, p\text{-}1}$, we accept the hypothesis and declare the ith principal axis of Σ_z parallel to the ith principal axis of Σ_m. The test is likely to be conservative. This means that we may reject some hypotheses based on the nominal value of χ^2 which we would accept should the actual value be available, since the nominal χ^2 tends to be larger than the actual value.

Similar tests may be performed on a number of $\mathbf{b}_{im}$, $i = 1...t$. If parallelism is shown, it is concluded that the difference is in size and not in orientation. This would indicate less severe conditions of covariance inequality than what would be the case if we concluded that the clusters are also oriented differently. A size difference of clusters is attributable to different degrees of variation in the population. However, when a difference in orientation is shown, it may not be unreasonable to suggest that the p variates in the different populations respond with different intensity to influences of specific underlying factors. We refer to Reyment (1969) for examples and to Anderson (1963) for exposition of the theory.

Apparently, the problem of testing two-sample hypotheses such as,

$$H_0 : E(\bar{\mathbf{X}}_1) = E(\bar{\mathbf{X}}_2); \quad H_1 : E(\bar{\mathbf{X}}_1) \neq E(\bar{\mathbf{X}}_2)$$

for which we have used T^2, has in fact been addressed by Fisher and Scheffé (cf. Anderson, 1958; also Reyment, 1962) under the assumption of unequal population covariance matrices. When $n_1 = n_2 = n$, the test statistic recommended is

$$F = \frac{(n-1)p}{n-p}\, T^2 \quad \text{with } p \text{ and } n-p \text{ degrees of freedom}$$

where

$$T^2 = n\, \bar{\mathbf{Y}}'\mathbf{S}^{-1}\, \bar{\mathbf{Y}}$$

and $\bar{\mathbf{Y}} = \bar{\mathbf{X}}_1 - \bar{\mathbf{X}}_2$. A characteristic element of $\mathbf{S}$ is

$$S_{hi} = \sum_j (Y_{hj} - \bar{Y}_h)(Y_{ij} - \bar{Y}_i)/(n-1), \quad j = 1, ..., n .$$

In this,

$$Y_{hj} = X_{hj1} - X_{hj2} \quad ,$$

i.e. the difference of the corresponding elements in the sample data matrices for groups. The columns in each are arbitrarily labelled ($j = 1, ..., n$). Generalizations to $n_1 \neq n_2$, and also to more than two groups, have been described (cf. Anderson, 1958).

The methods of comparison described for testing differences between types is computationally very complex. A simple method is available if the user is prepared to accept the assumption that the pairwise similarity of quadrats within the types is a random variable whose sampling distribution is normal. Under this particularly severe assumption, Mountford (1971) suggests to test the null hypothesis of no difference in average similarity between types using the statistic,

$$t = \left[\frac{(N-n-1)\,\Delta^2/V_\Delta^2}{S^2-\Delta^2/V_\Delta^2} \right]^{1/2},$$

if the quadrat data are different from the data based on which the types were defined. The data in Table 3-1 represent an example for such a situation. For the handling of the contingency when the test has to be based on similarity values which were maximized in the clustering process, Mountford (1971) suggests the criterion,

$$b = \frac{\Delta}{V_\Delta S}\quad .$$

If the null hypothesis is true, t is a t-variate with $N-n-1$ degrees of freedom. Some probability points in the sampling distribution of b are given by Mountford (1971). The symbols in these formulae accord with the following definitions:

$n \;=\; n_A + n_B$

n_A = number of quadrats in type A

n_B = number of quadrats in type B

$N \;=\; n(n-1)/2$

Δ = difference in average similarity = $\bar{S}_A + \bar{S}_B - 2\bar{S}_{AB}$

$\bar{S}_A$ = average similarity of quadrats in type A

$= \sum_j \sum_k S_{jkA}/[n_A(n_A-1)/2],\; j=1, ..., n_A - 1;\quad k = j+1, ..., n_A$

S_{jkA} = similarity of quadrats j and k in type A

$\bar{S}_B$ = average similarity of quadrats in type B

$= \sum_j \sum_k S_{jkB}/[n_B(n_B-1)/2],\; j=1, ..., n_B-1;\quad k=j+1, ..., n_B$

S_{jkB} = similarity of quadrats j and k in type B

$\bar{S}_{AB}$ = average similarity of quadrats in type A with quadrats in type B

$$= \sum_j \sum_k S_{jA,kB}/(n_A n_B), \; j=1,\dots,n_A; \; k=1,\dots,n_B$$

$S_{jA,kB}$ = similarity between quadrat j of type A and quadrat k of type B

V_Δ^2 = variance of the difference

$$= 2(n-1)(n-2)/[n_A(n_A-1)n_B(n_B-1)]$$

$V_\Delta = (V_\Delta^2)^{1/2}$

S^2 = sum of squares of similarities

$$= \sum_j \sum_k (S_{jk} - \bar{S} - \hat{g}_j - \hat{g}_k)^2, \; j=1,\dots,n-1; k=j+1,\dots,n$$

S_{jk} = similarity of quadrats j and k in the union group $(A+B)$

$\bar{S}$ = average similarity of quadrats in the union group $(A+B)$

$S = (S^2)^{1/2}$

$$\hat{g}_j = \left[\sum_k S_{jk} - (n-1)\bar{S}\right]/(n-2), \; k=1,\dots,n \text{ but } k \neq j; \; j=1,\dots,n \; .$$

In the context of the data in Table 3-1 we can state the null hypothesis as

$$H_o : E(\bar{S}_M + \bar{S}_G - 2\bar{S}_{MG}) = 0 \; .$$

This is equivalent to saying that there is no tendency for higher similarities inside types M and G than among them. H_0 is contrasted in the test with the alternative hypothesis,

$$H_1 : E(\bar{S}_M + \bar{S}_G - 2\bar{S}_{MG}) > 0 \; .$$

This is equivalent to stating that the high similarity values occur within the groups.

H_0 calls for a comparison of type M with type G. Let the **S** matrix be defined as a cross product matrix, given by

$$S_{(M+G)} = \left[\begin{array}{c|c} S_M & S_{MG} \\ \hline S_{GM} & S_G \end{array}\right]$$

$$= \left[\begin{array}{ccc|cc} - & 39.4341 & 47.0039 & 22.0879 & 19.3330 \\ 39.4341 & - & 38.8828 & 18.2515 & 15.9789 \\ 47.0039 & 38.8828 & - & 21.7693 & 19.0553 \\ \hline 22.0879 & 18.2515 & 21.7693 & - & 8.96148 \\ 19.3330 & 15.9789 & 19.0553 & 8.96148 & - \end{array}\right] .$$

The definition of **S** need not however be a cross product matrix. The subsequent numerical values are computed automatically in program MTFDT in which matrix $\mathbf{S}_{(M+G)}$ serves as input. After $n_M = 3$ and $n_G = 2$ are specified, the following output is produced:

$$n = 5, \; \bar{S}_M = 41.7736, \; \bar{S}_{MG} = 19.4127$$

$$N = 10, \; \bar{S}_G = 8.96148, \; \bar{S} = 25.0758$$

$$\hat{\mathbf{g}} = (\hat{g}_1 \; \hat{g}_2 \; \hat{g}_3 \; \hat{g}_4 \; \hat{g}_5)$$

$$= (9.18521 \; 4.08134 \; 8.80268 \; -9.74436 \; -12.3249)$$

$$S^2 = 79.6159, \; V_\Delta^2 = 2, \; t = 5.71 \quad .$$

With $N-n-1 = 4$ degrees of freedom, the probability of a more extreme t is less than 0.0024. It is concluded that the best course to be followed is to declare significance on the difference between types M and G based on their similarities of soil chemical properties. However, in no way can the user verify the validity of the assumption that the S_{jk} in fact have a normal sampling distribution. This may of course undermine confidence in the results.

Comparisons may be performed between types for many reasons. One common reason would call for combining types in order to reduce their number. The variables used in the comparison may be the same as those which formed the types or they may be different. In any case, if types are to be combined, the comparison should probably be treated as a deterministic one. The reason is that the user, by recognizing vegetation types as different units in the first place, implies that they represent different multispecies populations. Because of this, to combine them is truly an arbitrary step, to say the least, not compatible with statistical thinking. While we recognize this, we should also note that the user is at liberty of applying the various measures of group comparison in the present chapter or in Chapter 2 without probabilistic connotations.

D. Predictive use of classifications

A specific kind of test is needed when there are two classifications of the same set of quadrats and we intend to use one to predict the properties in the other. For example, we may recognize classes on the basis of species abundance, and we may wish to use these classes to predict given environmental properties. This we would do in such a way that we could make statements about the environmental conditions in a new quadrat once its affiliation with a vegetation class has been established. The question is how reliably can a given vegetational classification be used for such a prediction?

This question can, of course, be answered only with local relevance based on a survey of both the vegetation and environment. Let us assume that such a survey is completed and that subsequent analyses of the survey data yield two classifications of the same set of N quadrats. Let us further assume that the vegetational classification consists of k classes, and that there are t classes in the classification derived from environmental data. We wish to test the hypothesis that the k vegetation classes have no predictive value for the conditions in the t environmental classes.

Table 4-2. Comparison of classifications. Entries represent quadrat counts in classes of vegetation (A, B, C, D), in classes of soil texture (a, b, c, d), and in common between the two.

Soil textural class	*Vegetation class*				
	A	B	C	D	*Total*
a	4	1	19	8	32
b	5	27	3	4	39
c	8	16	6	20	50
d	12	2	1	1	16
Total	29	46	29	33	137

To illustrate the procedure, we shall consider the data in Table 4-2. In this table the same set of 137 quadrats are sorted according to two sets of classificatory criteria. The first set A, B, C, D signifies vegetation classes, and the second set a, b, c, d represents soil textural classes. The entries in the body of the table signify the number of quadrats which are common to the respective classes. In the marginal cells the values indicate the total number of quadrats in the individual classes.

The degree to which the soil textural classes are unpredictable from the vegetation classes is entirely dependent on the degree to which the two classifications are independent. The criterion, particularly suited for testing independence in such a contingency table, is the *mutual information*, given by

$$2I(\textit{vegetation; soil texture}) = 2 \sum_h \sum_i f_{hi} \ln \frac{f_{hi} f_{..}}{f_{h.} f_{.i}} \tag{4.9}$$

The summations are taken from $h = a$ to d and $i = A$ to D or $h = 1, \ldots, n$ and $i = 1, \ldots, c$ in the $n \times c$ contingency table. The remaining symbols are defined as follows: f_{hi} element in hi cell, $f_{h.}$ hth row total, $f_{.i}$ ith column total, $f_{..}$ grand total for the table. The working formula to be used in the computations accords with

$$2I(\textit{vegetation; soil texture}) = 2[\sum_h \sum_i f_{hi} \ln f_{hi} + f_{..} \ln f_{..}$$

$$- \sum_h f_{h.} \ln f_{h.} - \sum_i f_{.i} \ln f_{.i}]$$

$$= (2)((2)(4 \ln 4) + 19 \ln 19 + (2)(8 \ln 8) + 5 \ln 5 + 27 \ln 27 + 3 \ln 3$$

$$+ 16 \ln 16 + 6 \ln 6 + 20 \ln 20 + 12 \ln 12 + 2 \ln 2 + 137 \ln 137$$

$$- 32 \ln 32 - 39 \ln 39 - 50 \ln 50 - 16 \ln 16 - (2)(29 \ln 29)$$

$$- 46 \ln 46 - 33 \ln 33)$$

$$= (2)(346.868 + 674.037 - 493.745 - 486.805)$$

$$= 80.706$$

This quantity is referred to the χ^2 distribution with 9 degrees of freedom. On observing that $\chi^2_{0.05;\,9} = 16.919$ is less than $2I$, we reject the hypothesis of independence, and accept the alternative hypothesis that the two classifications are related. What we conclude from the test about the strength of the relationship between the two classifications is that the probability of obtaining an I-divergence of the magnitude as ours, when the population classifications indeed are independent, is less than 5%. This is a rather weak statement, and we prefer in addition something quantitative in terms of a relative scale. For this purpose, we may compute the coherence coefficient. In the example, Rajski's metric is defined by

$$d(vegetation; soil\ texture) = 1 - \frac{I(vegetation; soil\ texture)}{I(vegetation, soil\ texture)}$$

$$= 1 - \frac{\sum_h \sum_i f_{hi} \ln f_{hi} f_{..}/(f_{h.} f_{.i})}{- \sum_h \sum_i f_{hi} \ln f_{hi}/f_{..}},$$

with summations $h = a, \ldots, d$; $i = A, \ldots, D$, where $a, \ldots, d$ and $A, \ldots, D$ indicate respectively the rows and columns in the contingency table (Table 4-2). The corresponding numerical value is given by

$$d(vegetation; soil\ texture) = 1 - \frac{40.355}{327.169} = 0.877 \quad .$$

From this, we derive the coherence coefficient,

$$r(vegetation; soil\ texture) = (1 - d^2)^{1/2} = (1 - 0.877^2)^{1/2} = 0.481 \quad .$$

This indicates a reasonably high average affinity between the two classifications. We conclude, in general, that on the average the vegetation class affiliation of quadrats and their membership in a soil class are quite strongly related, and with regard to our original question, that the predictive value of the vegetational classification for soil texture is reasonably high.

One further question flows from what we have said so far: which vegetation class has a high predictive value for which soil textural class? To answer this, we partition $2I$ (*vegetation; soil texture*) into components specific to A, B, C, or D and look at them individually. The relevant results for A are the following:

$$2I(A; soil\ texture) = 2 \sum_h f_{hA} \ln \frac{f_{hA} f_{..}}{f_{h.} f_{.A}}$$

$$= 2 \sum f_{hA} \left[\ln \frac{f_{hA}}{f_{.A}/n} - \ln \frac{f_{h.}}{f_{..}/n} \right].$$

The working formula is given by

$$2I(A; soil\ texture) = 2[\sum_h f_{hA} \ln f_{hA} + f_{.A} \ln f_{..} - \sum f_{hA} \ln f_{h.} - f_{.A} \ln f_{.A}]$$

$$= (2)(4 \ln 4 + 5 \ln 5 + 8 \ln 8 + 12 \ln 12 + 29 \ln 137 - 4 \ln 32 - 5 \ln 39 - 8 \ln 50 - 12 \ln 16 - 29 \ln 29)$$

$$= (2)(60.047 + 142.679 - 96.748 - 97.652) = 16.652.$$

This is a χ^2 with 3 degrees of freedom. Since $\chi^2_{0.05;3} = 7.815$ is exceeded by the value $2I$, we declare significance on $2I$ conditional on vegetation class A. The test indicates that $2I$ (A; *soil texture*) significantly differs from zero. Although this may be an interesting fact, it does not tell us much about the extent to which A can predict soil texture relative to B, C, or D. We can, of course, compute A's specific share in Rajski's metric and the corresponding coherence value, and use these as a measure for A's weakness or strength of prediction. These are given by

$$d(A; soil\ texture) = 1 - \frac{I(A; soil\ texture)}{I(vegetation, soil\ texture)} = 1 - \frac{8.327}{327.169} = 0.975$$

$$r(A; soil\ texture) = (1 - 0.975^2)^{1/2} = 0.222\ .$$

Similar calculations yield components specific to B, C, D. The results are given below:

$$2I(B; soil\ texture) = (2)(27 \ln 27 + 16 \ln 16 + 2 \ln 2 + 46 \ln 137 - \ln 32 - 27 \ln 39 - 16 \ln 50 - 2 \ln 16 - 46 \ln 46)$$

$$= (2)(134.735 + 226.319 - 170.519 - 176.118) = 28.834$$

$$d(B; soil\ texture) = 1 - \frac{14.417}{327.169} = 0.956$$

$$r(B; soil\ texture) = (1 - 0.956^2)^{1/2} = 0.293$$

$$2I(C;\ soil\ texture) = (2)(19 \ln 19 + 3 \ln 3 + 6 \ln 6 + 29 \ln 137 + - 19 \ln 32 - 3 \ln 39 - 6 \ln 50 - \ln 16 - - 29 \ln 29)$$

$$= (2)(69.991 + 142.679 - 103.084 - 97.652)$$

$$= 23.868$$

$$d(C;\ soil\ texture) = 1 - \frac{11.934}{327.169} = 0.964$$

$$r(C;\ soil\ texture) = (1 - 0.964^2)^{1/2} = 0.266$$

$$2I(D;\ soil\ texture = (2)(8 \ln 8 + 4 \ln 4 + 20 \ln 20 + 33 \ln 137 - 8 \ln 32 - 4 \ln 39 - 20 \ln 50 - \ln 16 - 33 \ln 33)$$

$$= (2)(82.095 + 162.359 - 123.393 - 115.385)$$

$$= 11.352$$

$$d(D;\ soil\ texture) = 1 - \frac{5.676}{327.169} = 0.983$$

$$r(D;\ soil\ texture) = (1 - 0.983^2)^{1/2} = 0.184$$

The partial results are summarized below:

Source	*Component of information 2I*	*Degrees of freedom (DF)*	$\chi^2_{0\cdot05;DF}$	*Component of Rajski's metric*	*Coherence coefficient*
A	16.652	3	7.815	0.975	0.222
B	28.834	3		0.956	0.293
C	23.868	3		0.964	0.266
D	11.352	3		0.983	0.184
Total	80.706			0.877	0.481

Several comments are in order regarding these results. Firstly, we should point out that the sum of the individual components of $2I$ should be equal to $2I$(*vegetation; soil texture*). Secondly, the one complement of the sum of one complements of the components of Rasjki's metric is equal to d(*vegetation; soil texture*), i.e. $1 - (1 - 0.975) + (1 - 0.965) + (1 - 0.964) + (1 - 0.983)) = 0.877$. We further note that all $2I$ values represent significant divergences. On this basis we can give the following characterizations:

1. Vegetation class B is the best single predictor for soil texture. When a quadrat is identified as a member in vegetation class B, it most likely is also a

member of soil texture class *b*. This is a consequence of the frequencies in Table 4-2.

2. Vegetation class *C* is the second best predictor for soil texture. When a quadrat is identified as a member in *C*, it most likely is also a member in soil texture class *a*.

3. Vegetation class *A* is the third ranking predictor for soil texture. When a quadrat is identified as a member in *A*, it most probably is also a member in soil texture class *d*.

4. The least reliable predictor is vegetation class *D*. A quadrat identified as a member of *D* most probably is a member in soil texture class *c*.

If we use the coherence coefficient to measure how much reliance can be placed on a classification as a predictor of conditions indicated by another classification, we may also use Rajski's metric as our heuristic measure for uncertainty in the prediction. In these terms, when two classifications are identical, the prediction of one based on the other is without uncertainty, and thus their mutual information, i.e. $2I$ (*vegetation; soil texture*) in the example, is maximal, the value of Rajski's metric is zero, the value of the coherence coefficient is one. At the other extreme, when the two classifications are completely different, it is impossible to make predictions from one about the other, the mutual information is zero, Rajski's metric is one, and the coherence coefficient is zero.

In the foregoing example we assumed that the classification is completed and the remaining objective is to measure its predictive value for soil textural classes. We may of course inject a predictive function $f(Y \mid X)$, e.g. formula (4-9), into the clustering algorithm itself, and seek fusions (or subdivisions) which maximize $f(Y \mid X)$. The following may be considered in this connection.

1. Define three sets of variables; Z including the classificatory variables (e.g. species abundance); X, a predictor variable with k states (e.g. *A, B, C, D* vegetation types); and Y including the predicted variable with r states (e.g. *a, b, c, d* soil textural classes).

2. Implement trial fusions (or subdivisions) of the sample of n quadrats into k classes and accept the solution which maximizes $f(Y \mid X)$.

Workable algorithms for this have been described and examples given by, e.g., Macnaughton-Smith (1965), Williams, Haydock, Edye & Ritson (1971) and Beeston & Dale (1975).

E. *Classification as a means for scaling*

We shall consider a situation in which parts of an object are described separately, but the descriptions are recombined when the objects are analysed in a cluster analysis or ordination. The object could be a vegetation stand and its parts may be the different strata such as the tree layer, shrub layer, herb layer or the ground cover of bryophytes.

When treating relevés that record species in multistratal communities, we may (1) combine the records of the same species from different strata into a single record and subject the combined record to analysis. But by so doing, the analysis will lose information about strata. We may also (2) retain a separate record for the same species in each stratum as if it were for different species. But this would increase the number of species-level categories. As yet in another option (3), the strata may be treated as separate objects and their record analysed in a classification to derive new descriptors for the stands in the manner as suggested by Norris & Dale (1971) in their work on the classification of soil profiles. We present their method in the context of multidimensional scaling.

The objective is to replace the relevés of a multi-layered stand by new relevés of t-descriptors which, while not stratified, still retain information about stratal structure. The procedure is rather simple. We shall assume that we have records from all strata in each stand. The records for n stands and p species are fragmented into $t \times n$ new relevés of p species. An example is described below. The original records, representing cover/abundance estimates, are given in Table 4-3. We can break up this table into 15 partial records and give them in Table 4-4. The two-digit column numbers in the first row of Table 4-4 identify the stratum (first digit) and relevé (second digit) in the original record (Table 4-3). For example, *24* identifies the record in the second stratum, fourth relevé.

Table 4-3. Sample data. See explanations in the main text

Stratum	*Species*	*Relevé* 1	2	3	4	5
I.	*1. Alnus rubra*	1	0	6	0	1
	2. Populus trichocarpa	5	1	1	7	8
	3. Thuja plicata	0	5	0	0	0
II.	*4. Acer circinatum*	4	6	8	5	7
	3. Thuja plicata	4	0	0	0	3
	1. Alnus rubra	0	5	0	0	1
III.	*5. Disporum oreganum*	5	5	6	5	5
	6. Polystichum munitum	2	1	1	1	1
	3. Thuja plicata	1	0	2	1	1

Table 4-4. Partial relevé records extracted from Table 4-3

Species	*11**	*12*	*13*	*14*	*15*	*21*	*22*	*23*	*24*	*25*	*31*	*32*	*33*	*34*	*35*
	1**	2	3	4	5	6	7	8	9	10	11	12	13	14	15
1	1	0	6	0	1	0	5	0	0	1	0	0	0	0	0
2	5	1	1	7	8	4	0	0	0	3	0	0	0	0	0
3	0	5	0	0	0	0	0	0	0	0	1	0	2	1	1
4	0	0	0	0	0	4	6	8	5	7	0	0	0	0	0
5	0	0	0	0	0	0	0	0	0	0	5	5	6	5	5
6	0	0	0	0	0	0	0	0	0	0	2	1	1	1	1

*Subscript; **Record number

The next step is to subject the 15 partial records to a cluster analysis. We could use any technique, but we choose sum of squares clustering (Section 4-9). When this is performed it yields the fusions shown in Fig. 4-8. We

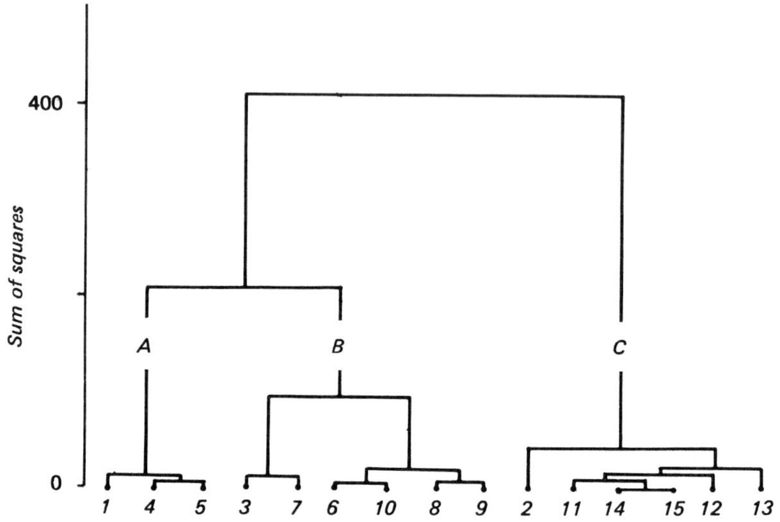

Fig. 4-8. Sum of squares clustering of 15 partial relevé records listed in Table *4-4.* See explanations in the main text.

are free to choose how many groups are to be recognized. But it should not be forgotten that the number of groups recognized will influence the final outcome in scaling. In any event the partial records should not be fragmented into too many groups nor into too few. In the example 3 or 4 groups appear reasonable. To have some idea about clustering efficiency, we perform an analysis of sum of squares for 3 and 4 groups:

Source	*Sum of squares*	*Efficiency %*
Between 3 groups	271.57	$\frac{271.57}{411.73} = 66$
Within 3 groups	140.16	
Total	411.73	

Source	*Sum of squares*	*Efficiency %*
Between 4 groups	322.40	$\frac{322.40}{411.73} = 78$
Within 4 groups	89.33	
Total	411.73	

Should we go for an even higher number of groups, efficiency would benefit, but computational economy would suffer. We choose three groups. These are as follows:

Group label	*Partial relevé records in group*					
A	*1*	*4*	*5*			
B	*3*	*7*	*6*	*10*	*8*	*9*
C	*2*	*11*	*14*	*15*	*12*	*13* .

Using the group labels of Fig. 4-8 as descriptors of partial relevé records, we redescribe the relevés of Table 4-3 in Table 4-5. The information in Table 4-5 summarizes the similarities of the strata within and between stands at the

Table 4-5. Redescription of relevés based on the group labels of Fig. 4-8 in Table 4-3

Stratum	*Relevé* *1*	*2*	*3*	*4*	*5*
I	*A*	*C*	*B*	*A*	*A*
II	*B*	*B*	*B*	*B*	*B*
III	*C*	*C*	*C*	*C*	*C*

chosen level of classification efficiency. This, with the three groups, is 66%. The new data in the form of group labels are readily available for analysis in an information theory method, and, if quantified, also in other methods of classification or ordination.

One possibility for metric scaling, i.e. quantification of group labels in Table 4-5, is in the form of transition matrices. Given three group symbols (A, B, C) we can construct a 3 x 3 matrix, for any relevé. For the first relevé, we have

$$\mathbf{F}_1 = \begin{bmatrix} 0 & 1 & 0 \\ 0 & 0 & 1 \\ 0 & 0 & 0 \end{bmatrix} .$$

The elements in $\mathbf{F}_1$ give the frequency with which a given symbol is followed by another symbol of a specified kind. For instance, the last element in the second row is the frequency by which symbol *B* is followed by symbol *C* in relevé *1* of Table 4-5. The first row (column) of $\mathbf{F}_1$ corresponds to symbol *A*, the second to symbol *B*, and the third to symbol *C*.

To completely describe *n* relevés, *n* transition matrices are required. Transition matrices are comparable by their elements between relevés as if they were scores on new variables. Examples for analysis of such matrices are described by Norris & Dale (1971) and Dale, Macnaughton-Smith, Williams & Lance (1970). The divergence of two relevés is measured through comparison of their transition matrices $\mathbf{F}_m$ and $\mathbf{F}_z$ in the form

$$\Delta I_{mz} = I_{mz} - I_m - I_z$$

where $I_m = -\sum_i \sum_j f_{ijm} \ln f_{ijm}/f_{..m}$; I_z is similarly defined and $I_{mz} = -\sum_i \sum_j (f_{ijm} + f_{ijz}) \ln (f_{ijm} + f_{ijz})/(f_{..m} + f_{..z})$. The summations are taken from i or $j = 1$ to r with r indicating the number of group symbols (3 in the example). Here f_{ijm} or f_{ijz} is the ijth element in $\mathbf{F}_m$ or $\mathbf{F}_z$ and $f_{..m}$ or $f_{..z}$ is the total of all elements in $\mathbf{F}_m$ or $\mathbf{F}_z$. Alternative manipulations with transition matrices to measure divergence are readily conceived based on formulations which we treated in Section 2-6.

F. Further on deterministic comparisons

Comparisons should be treated as statistical problems if differences in class contents between the classifications are in part attributable to sampling variation in the states of random variables. Some of the preceding examples were of this kind. The problem of comparison may however arise in yet another context. To illustrate this, we shall consider the case in which a given sample of entities, such as N quadrats, are subjected to cluster analyses by the use of different methods. Since the description of the quadrats remains unchanged in the different analyses, differences revealed in class contents are entirely the consequence of differences in clustering algorithms. The comparison of classifications has to be deterministic in this case, since the clustering algorithms presumably differ from one another by design and not by chance.

An example is outlined below. It involves three vegetation maps given as Figs. 4,5 in a paper by Ivimey-Cook, Proctor & Rowland (1975). The maps are images of three classifications of the same 340 contiguous quadrats based on three different methods: association analysis (Williams & Lambert, 1960, 1961), a method of information analysis (Williams, Lambert & Lance, 1966), and a method of group analysis (Crawford & Wishart, 1967). We observe that identical symbols within a given map identify quadrats of the same type. The consistency with which types in one map tend, on the average, to have quadrats in common with types in the others is a basis to measure their similarity. With classification consistency understood in this particular sense, we can quantify the similarities of the maps based on the information conveyed by the joint frequencies of the type symbols. The joint frequencies for the two maps (a, b) in Fig. 4 of the original paper are displayed in Table 4-6 in this book. Numbers in the table replace the group symbols in the original maps.

We use the coherence coefficient (formula (2.28)) to measure similarity. The individual terms of information accord with the formulations given in Section 2-6:

$$I(\mathbf{F}_a) = f_{a.} \ln f_{a.} - \sum_i f_{ai} \ln f_{ai} = 952.032$$

$$I(\mathbf{F}_b) = f_{b.} \ln f_{b.} - \sum_j f_{bj} \ln f_{bj} = 851.388 \ .$$

The summations are for all row (i) and column (j) totals in Table 4-6. $I(\mathbf{F}_a)$ and $I(\mathbf{F}_b)$ measure the total information in the marginal distributions $\mathbf{F}_a$, $\mathbf{F}_b$ whose elements are $(f_{a1} \; ... \; f_{a19})$ and $(f_{b1} \; ... \; f_{b13})$. The grand totals are given by $f_{a.} = f_{b.} = \sum_i f_{ai} = \sum_j f_{bj}$. The information in the joint distribution of the group symbols is

$$I(\mathbf{F}_a, \mathbf{F}_b) = f_{a.} \ln f_{a.} - \sum_i \sum_j f_{ai,bj} \ln f_{ai,bj} = 1283.81 \quad .$$

The summations are for all cells in Subtable 1 of Table 4-6. The symbol $f_{ai,bj}$ signifies the joint frequency of type symbol i in map a and type symbol j in map b. From these results we can derive the coherence coefficient for the maps:

$$r(\mathbf{F}_a; \mathbf{F}_b) = 0.804 \quad .$$

Table 4-6. Joint distribution of type symbols in maps referred to in the main text. Subtables 1, 2 and 3 compare respectively maps a/b, a/c and b/c. Numerals 1-19, 1-13 and 1-14 along margins replace the type symbols in maps. Entries in body of table are joint frequencies. Blanks indicate zeros.

Subtable 1

	Type symbols in map b													*Total*
Type symbols in map a	*1*	*2*	*3*	*4*	*5*	*6*	*7*	*8*	*9*	*10*	*11*	*12*	*13*	$\mathbf{F}_a$
1	1	5	7		1		1						1	16
2		7			1	1	1							10
3	2		4	9			5		5			1		26
4					1					1		5		7
5				1	4	1								6
6							6							6
7	8			1		2		2						13
8			1	2					11			8		22
9	1		10	12			1		3					27
10							2			11		7		20
11	1					16		13						30
12			8	2									15	25
13					5						9			14
14					2		11				8			21
15	10		1	10		9	1	1	3					35
16		1									7			8
17					1					2	3	6		12
18					1				6					7
19										33		2		35
Total $\mathbf{F}_b$	23	13	31	37	16	29	28	16	28	47	27	29	16	340

Subtable 2

	Type symbols in map c														Total
	1	*2*	*3*	*4*	*5*	*6*	*7*	*8*	*9*	*10*	*11*	*12*	*13*	*14*	F_a
Type symbols in map a *1*	5	1	8	1									1		16
2	6		1	2	1										10
3		7		5								9	1	4	26
4				1		3				2				1	7
5		1		1	2					1		1			6
6				5	1										6
7		3			2		8								13
8		5		2				5	1	1		2	2	4	22
9	1							16	6		4				27
10				3						14			2	1	20
11		6			21		3								30
12								11			12			2	25
13				8	4	2									14
14				16	4	1									21
15		13			12		8					2			35
16				2		6									8
17				3	1	1				7					12
18		4			3										7
19				1		1				33					35
Total F_c	12	40	9	50	51	14	19	21	18	58	16	14	6	12	340

Subtable 3

	Type symbols in map c														Total
	1	*2*	*3*	*4*	*5*	*6*	*7*	*8*	*9*	*10*	*11*	*12*	*13*	*14*	F_b
Type symbols in map b *1*	1	8			2		11	1							23
2	9		2	2											13
3		1	6	1			1	8	2		6	2		4	31
4		10		2	2		3	5	7		3	3		2	37
5				6	7					1		1		1	16
6		7		1	20							1			29
7	2			18	3					1		2	2		28
8		2			11		3								16
9		12		1	2		1	7				5			28
10				3						44					47
11				13	4	10									27
12				3		4			1	12			4	5	29
13			1						8		7				16
Total F_c	12	40	9	50	51	14	19	21	18	58	16	14	6	12	340

Similar calculations yield values for maps *a,c* and *b,c*:

$$r(\mathbf{F}_a;\mathbf{F}_c) = 0.780$$

$$r(\mathbf{F}_b;\mathbf{F}_c) = 0.746 \quad .$$

The r values indicate a high similarity of the maps. No probabilistic interpretation is appropriate, however, since the differences measured between maps reflect dissimilarities in clustering algorithm.

Several methods were described to measure similarity between classifications (see review by Rohlf, 1974). Of these, we shall briefly discuss the method proposed by Rand (1971). In Rand's method, pairs of individuals are counted. Since members of a pair are either in the same class or in different classes in the different classifications, the counts can be derived from the joint frequencies. Firstly,

$$A = \sum_i \sum_j f_{ai,bj}(f_{ai,bj} - 1)/2$$

happens to count the number of pairs X, Y such that X and Y are affiliates of the same class in classification $\mathbf{F}_a$ and also in classification $\mathbf{F}_b$. Summations are taken over all classes in $\mathbf{F}_a$ and $\mathbf{F}_b$. Symbol $f_{ai,\ bj}$ signifies the number of individuals in class i in F_a affiliated with class j in F_b. Secondly,

$$B = \sum_i \sum_j (f_{ai,bj} \sum_{k>i} \sum_{m>j} f_{ak,bm}) \quad .$$

gives the number of pairs whose members are in different classes in $\mathbf{F}_a$ and in $\mathbf{F}_b$. The summations are over all classes except those specified by $k \leqslant i$ and $m \leqslant j$. Rand's similarity index expresses the sum of the two counts as a ratio of all distinct pairs in a sample of N individuals,

$$R(\mathbf{F}_a;\mathbf{F}_b) = (A+B)/[N(N-1)/2] = 1-B/[N(N-1)/2] \ .$$

The numerical values for maps a, b, c in Figs. 4,5 of the original paper by Ivimey-Cook, Proctor & Rowland (1975) are given below:

$$R(\mathbf{F}_a;\mathbf{F}_b) = 0.908$$

$$R(\mathbf{F}_a;\mathbf{F}_c) = 0.891$$

$$R(\mathbf{F}_b:\mathbf{F}_c) = 0.884 \quad .$$

These indicate somewhat higher similarities than the coherence coefficient.

The limits for $R(\mathbf{F}_a;\ \mathbf{F}_b)$ are zero and unity. The limit values can be achieved under only rather unusual circumstances. For $R(\mathbf{F}_a;\ \mathbf{F}_b)$ to be zero, all the joint frequencies must be zero, i.e. $\mathbf{F}_a$ and $\mathbf{F}_b$ must not have any individuals in common. This can happen if $\mathbf{F}_a$ and $\mathbf{F}_b$ represent classifications of two disjoint sets of individuals. This is different from the coherence coefficient $r(\mathbf{F}_a;\ \mathbf{F}_b)$ of formula (2.28) which can have a zero value under much less restrictive conditions; namely, when the joint frequencies are proportional to the table marginal totals. Significantly, for the Rand index to have a unit value, all frequencies must be concentrated in a single cell in $\mathbf{F}_a$ and in $\mathbf{F}_b$. The coherence coefficient is different, for it to have a unit value the frequencies need not be so concentrated, but there must be for every cell in $\mathbf{F}_a$ a cell in $\mathbf{F}_b$ with identical contents.

G. *Comparison of dendrograms*

Dendrogram properties have been subject to intensive studies (e.g. Sneath & Sokal, 1973; Phipps, 1975, 1976a,b, and references therein) and it is quite obvious that the methods to describe them may differ considerably both in general approach and in choice of similarity index. We can recognize a major dichotomy in approach for comparing dendrograms. Whereas in one, the methods require description of dendrogram structure, i.e. the hierarchial pattern of fusions, on the basis of a dissimilarity matrix **T**, in the other, the methods handle matrices with elements measuring actual fusion similarities.

A description of the similarity levels at which fusions occur is essential to the method of Sokal & Rohlf (1962). An element t_{hi} of **T** in this method represents the dissimilarity value at which individuals h and i are fused in the dendrogram. However, such a description may prove to be too restrictive because it may give too much weight to quantitative differences among dendrograms which otherwise may be quite similar in terms of their fusion topology. There is much to be said for the Phipps (1971) method (cf. Farris, 1969a; Williams & Clifford, 1971; Rohlf, 1974) which describes dendrograms by their fusion topology. Such a description is based on counts, symbolically represented by t_{11}, t_{12}, ..., t_{hi}, ... in which the general element t_{hi} specifies the number of fusions that must occur before individuals h and i unite in a common group. Counts of this sort can provide unique descriptions for dendrograms and a suitable basis for comparisons.

Let the count t_{mhi} represent the h,i element in matrix $\mathbf{T}_m$ of group m. We shall refer to $\mathbf{T}_m$ as the mth dissimilarity matrix. Dendrograms whose dissimilarity matrices are equivalent are said to have identical structure. Considering dissimilarity matrices, they are characterized by certain common properties:

1. The elements are positive integer numbers, real numbers or zero, and symmetric. Because of the symmetry property, it is sufficient to give the counts in one half of **T** above or below the principal diagonal.
2. The elements are not independent.
3. The number of degrees of freedom in a dendrogram is less than the actual number of fusions.
4. With a decrease in the number of individuals, the number of possible dendrograms is reduced and the differences between the dendrograms become restricted to less and less variation.

The topological matrices $\mathbf{T}_m$ and $\mathbf{T}_o$ below describe respectively the dendrogram structures in Fig. 4-9:

$$\mathbf{T}_m = \begin{bmatrix} 0 & 1 & 4 & 4 & 3 \\ & 0 & 4 & 4 & 3 \\ & & 0 & 1 & 2 \\ & & & 0 & 2 \\ & & & & 0 \end{bmatrix} \quad \text{and } \mathbf{T}_o = \begin{bmatrix} 0 & 1 & 2 & 4 & 4 \\ & 0 & 2 & 4 & 4 \\ & & 0 & 3 & 3 \\ & & & 0 & 1 \\ & & & & 0 \end{bmatrix}.$$

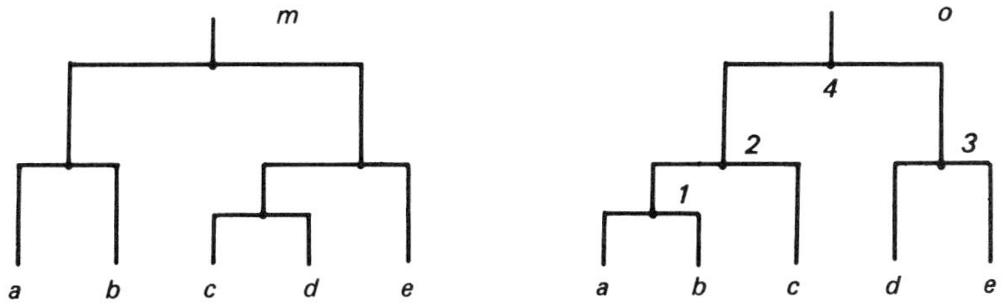

Fig. 4-9. Sample dendrograms described by topology matrices in the main text.

The problem we wish to address with these amounts to measuring the congruence of two dendrograms without isolating particular levels. If we isolated the levels, we would of course be facing once again the problem of comparing non-hierarchical classifications for which we could use similar methods as in the preceding sections.

Sokal & Rohlf (1962), who pioneered the subject, suggested the use of what they called *cophenetic correlation coefficient* to measure dendrogram similarity. Farris (1969a) studied the algebraic properties of this coefficient and its usefulness as an optimality criterion in different classificatory strategies. The cophenetic correlation coefficient relies on the cross product of standardized elements in the dissimilarity matrices,

$$r(\mathbf{T}_m;\mathbf{T}_o) = \sum_h \sum_{i>h} A_{mhi} A_{ohi} \,,$$

where the summation is taken from $h = 1$ to $N(N-1)/2-1$, and

$$A_{mhi} = (t_{mhi} - \bar{t}_m) \Big/ \left[\sum_h \sum_{i>h} (t_{mhi} - \bar{t}_m)^2 \right]^{1/2} ,$$

where t_{mhi} is an element of $\mathbf{T}_m$ and N is the number of objects classified. The symbol $\bar{t}_m$ is the mean of all elements in $\mathbf{T}_m$. A_{ohi} is similarly defined as A_{mhi}. The numerical values to yield $r(\mathbf{T}_m;\, \mathbf{T}_o)$ are determined in steps:

$$\begin{aligned} Q_{om} &= \sum_h \sum_{i>h} t_{mhi} t_{ohi} \\ &= (1)(1) + (4)(2) + \dots + (2)(1) \\ &= 84 \\ Q_m &= \sum_h \sum_{i>h} t_{mhi}^2 \\ &= 1^2 + 4^2 + \dots + 2^2 \\ &= 92 \end{aligned}$$

$$Q_o = \sum_h \sum_{i>h} t_{ohi}^2$$
$$= 1^2 + 2^2 + ... + 1^2$$
$$= 92$$

$$G_m = \sum_h \sum_{i>h} t_{ohi}$$
$$= 1 + 4 + ... + 2$$
$$= 28$$

$$G_o = \sum_h \sum_{i>h} t_{ohi}$$
$$= 1 + 2 + ... + 1$$
$$= 28$$

$$r(\mathbf{T}_m; \mathbf{T}_o) = \{Q_{mo} - G_m G_o/[N(N-1)/2]\}/\{[Q_m - G_m^2/(N(N-1)/2)][Q_o - G_o^2/(N(N-1)/2)]\}^{1/2}$$
$$= (84 - (28)(28)/10) / ((92 - 28^2/10)(92 - 28^2/10))^{1/2}$$
$$= 5.6/13.6$$
$$= 0.412 \quad .$$

The limits of r are +1 and −1 with 0 indicating no linear relationship between the elements of $\mathbf{T}_m$ and $\mathbf{T}_o$. Some may be tempted to interpret r not as a mathematical index, but rather as a statistic, using transformations to t and referring t to the standard sampling distribution of Student's t. This, however, would not work.

The advantages and disadvantages in applying the cophenetic correlation coefficient in dendrogram comparisons are discussed by, e.g., Sneath (1966a), Williams & Clifford (1971), Sneath & Sokal (1973) and Rohlf (1974, and references therein). Comparisons of dendrograms based on the measurement of similarity between the **T** matrices continue to be used. The measures, however, differ from the cophenetic correlation coefficient. Johnson (1967) prefers rank correlation measures and Jardine & Sibson (1968a, 1971) apply Euclidean and other metrics or metric-related formulations. Jackson (1969) uses a coefficient not unlike the rank correlation, and Farris (1973) relies on a technique which uses cluster descriptor variables. Williams & Clifford (1971) count nodes that separate individuals to derive their measure of dendrogram similarity, and Gower (1971b) relies on ordination scatter diagrams.

The choice of a similarity measure is of course dependent on previous decisions for what is to be done with the similarity values once calculated. Should the purpose be to develop ideas about the average resemblance of dendrograms, we could use almost any scale-free measure of similarity.

Should a measure serve as input in further analyses, such as in ordination or classification, our choice would be confined to similarity functions of specific kinds, consistent with the ordination or classification algorithm. One measure that is scale-free and easily converts into distance is Jardine & Sibson's $\alpha(\mathbf{T}_m; \mathbf{T}_o)$. This is defined by

$$\cos \alpha(\mathbf{T}_m; \mathbf{T}_o) = \frac{d^2(\mathbf{T}_m; \mathbf{O}) + d^2(\mathbf{T}_o; \mathbf{O}) - d^2(\mathbf{T}_m; \mathbf{T}_o)}{2d(\mathbf{T}_m; \mathbf{O})\, d(\mathbf{T}_o; \mathbf{O})} \quad .$$

In this,

$$d^2(\mathbf{T}_m; \mathbf{O}) = \sum_h \sum_{i>h} t^2_{mhi}$$

$$d^2(\mathbf{T}_o; \mathbf{O}) = \sum_h \sum_{i>h} t^2_{ohi}$$

$$d^2(\mathbf{T}_m; \mathbf{T}_o) = \sum_h \sum_{i>h} (t_{mhi} - t_{ohi})^2 \quad ,$$

with summations over all cells in the upper half of the **T** matrices. For the two dendrograms in Fig. 4-9, based on values in $\mathbf{T}_m$ and $\mathbf{T}_o$, we have

$$d^2(\mathbf{T}_m; \mathbf{O}) = Q_m = 92$$

$$d^2(\mathbf{T}_o; \mathbf{O}) = Q_o = 92$$

$$d^2(\mathbf{T}_m; \mathbf{T}_o) = (1-1)^2 + (4-2)^2 + \ldots + (2-1)^2$$

$$= 16$$

$$\cos \alpha(\mathbf{T}_m; \mathbf{T}_o) = \frac{92 + 92 - 16}{(2)((92)(92))^{1/2}}$$

$$= 168/184$$

$$= 0.913 \quad .$$

The two vectors $\overline{\mathbf{T}_m, \mathbf{O}}$ and $\overline{\mathbf{T}_o, \mathbf{O}}$ deviate by a mere 24°. This indicates a high similarity of $\mathbf{T}_m$ and $\mathbf{T}_o$.

Jackson (1969) devised a method which uses coefficients, not unlike the rank correlation, when measuring dendrogram similarity. After completing the computations which, as we shall see, are rather involved, we can make a statement how much a pair of hierarchical classifications (m, o) of the same elements, or more directly, the matrices $(\mathbf{T}_m, \mathbf{T}_o)$ which describe them, differ from one another. Jackson's method involves two coefficients,

$$g(-,s) = [R(-,s) + R(O,s)] \,/\, [R(-,s) + R(+,s) + R(O,s)]$$

$$g(+,s) = R(-,s)/[R(-,s) + R(+,s) + R(O,s)] \quad ,$$

which measure the discrepancy of $\mathbf{T}_m$ and $\mathbf{T}_o$. Although Jackson (1969) defines the elements of $\mathbf{T}_m$ and $\mathbf{T}_o$ as similarities, they could also be defined as dissimilarities. Jackson's method avoids direct comparison between the individual elements, t_{mhi} and t_{ohi}, by introducing the discriminant.

$$f[X] = f\,[(t_{mhi} - t_{mjk})/(t_{ohi} - t_{ojk})] \ .$$

The quantity $R(-, s)$ counts the comparisons for which $X > 0$, $R(+, s)$ counts the comparisons for which $X < 0$ and $R(0, s)$ the number of comparisons for which $X = 0$, undefined or infinity. No t_{mhi} less than s in $\mathbf{T}_m$, or t_{omi} less than some function of $s, f(s)$, in $\mathbf{T}_o$ are considered. Values for s, or the form of $f(s)$, are arbitrarily chosen.

Designating the joint frequency of a pair of corresponding dissimilarities in $\mathbf{T}_m$ and $\mathbf{T}_o$ by $f_{mj,ok}$, and the sum of the joint frequencies by $t = N(N-1)/2$, which is the number of elements in $\mathbf{T}_m$ or in $\mathbf{T}_o$, we can define the following quantities:

$$\begin{aligned}
g(-,s) &= [R(-,s) + R(0,s)]/[t(t-1)/2] = \\
&\quad [t(t-1)/2 - R(+,s)]/[t(t-1)/2] \\
g(+,s) &= R(-,s)/[t(t-1)/2] \\
R(+,s) &= \sum_h \sum_i f_{mh,oi} \left[\sum_{j>h} \sum_{k<i} f_{mj,ok} \right], \ h = 1, \ldots, n_m; \\
&\quad i = 1, \ldots, n_o \\
R(-,s) &= \sum_h \sum_i f_{mh,oi} \left[\sum_{j>h} \sum_{k<i} f_{mj,ok} \right], \ h = 1, \ldots, n_m; \\
&\quad i = 1, \ldots, n_o \\
R(0,s) &= t(t-1)/2 - R(-,s) - R(+,s) \ ,
\end{aligned}$$

where n_m and n_o are the numbers of distinct dissimilarity values in $\mathbf{T}_m$ and $\mathbf{T}_o$. Jackson (1969) refers to $R(+, s)$ and $R(-, s)$ respectively as the number of positive and the number of negative transitions, and to $R(0, s)$ as the number of ambiguous transitions. $g(-, s)$ is the ratio of the number of negative plus ambiguous transitions to the number of all distinct comparisons between the dissimilarity values in $\mathbf{T}_m$ or $\mathbf{T}_o$. It thus measures a discrepancy between $\mathbf{T}_m$ and $\mathbf{T}_o$. $g(+, s)$ is also a measure of discrepancy, but without considering ambiguous transitions. By varying s (or $f(s)$), it is possible to discover whether the small or large dissimilarity values contribute most to discrepancy.

We now outline a numerical example. The dendrograms to be compared are those given in Fig. 4-9. The number of units classified (N) is 5. The dissimilarity values were given as elements in $\mathbf{T}_m$ and $\mathbf{T}_o$. The joint frequencies are given in the table below:

Values in $\mathbf{T}_m$ \ Values in $\mathbf{T}_o$	*1*	*2*	*3*	*4*
1	1		1	
2	1		1	
3				2
4		2		2

We set s equal to zero and calculate the following quantities:

$$R(+,0) = (1)(7) + (1)(4) + (1)(6) + (1)(4) = 21$$

or

$$R(+,0) = (2)(4) + (2)(2) + (2)(4) + (1)(1) = 21$$

$$R(-,0) = (1)(3) + (1)(2) + (2)(2) = 9$$

or

$$R(-,0) = 1(1) + 2(4) = 9$$

$$R(0,0) = (10)(9)/2 - 9 - 21 = 15$$

$$g(-,0) = (9+15)/45 = (45-21)/45 = 0.533$$

$$g(+,0) = 9/45 = 0.200 \quad .$$

Since the number of ambiguous comparisons is large, $g(-, 0)$ indicates a much larger discrepancy than $g(+, 0)$. While the latter, as a scale-free measure of dissimilarity, lies rather close in size to $\cos \alpha(\mathbf{T}_m ; \mathbf{T}_o)$, $g(-, 0)$ is rather similar in size to the cross product $r(\mathbf{T}_m ; \mathbf{T}_o)$.

When the matrices $\mathbf{T}_m$ and $\mathbf{T}_o$ are compared on an element by element basis, we can deduce ideas from the comparison about the average similarity of the associated dendrograms. However, as Farris (1973) points out, characterization of the similarity (or dissimilarity) of two dendrograms by a single number – be it a correlation, distance, cosine or something else – does not provide a very reliable ground for interpretations, considering that similarities in the lower level clusters can swamp out marked dissimilarities at the higher levels. To remedy this problem, Farris (1973) suggests to examine the disagreement between dendrograms with respect to the extent to which the clusters of one are fragmented among the clusters of the other. The method involves counting steps, through which individuals of one classification can be regrouped in the classes of the other, similarly, as has been suggested by Webb, Tracey, Williams & Lance (1967a).

To count fragments, Farris (1973) devised cluster descriptor variables. These characterize membership in the clusters of a dendrogram. The number of fragments into which the clusters of dendrogram m disintegrate among the clusters of dendrogram o is measured by the number of extra steps taken in dendrogram o by the cluster descriptor variables of dendrogram m. The algorithm can be briefly stated as follows:

1. Describe cluster contents in one dendrogram, say m in Fig. 4-9, based on

a set of binary descriptor variables X_{1m}, X_{2m}, X_{3m}. (There are three clusters in dendrogram m.) The descriptor variables are specified by the column vectors in the table below:

Unit	*Cluster descriptors in dendrogram m* X_{1m}	X_{2m}	X_{3m}
a	1	0	0
b	1	0	0
c	0	1	1
d	0	1	1
e	0	0	1 .

A 1 indicates presence of a unit in a given cluster and 0 indicates absence. Cluster *1* consists of units *a* and *b*, cluster *2* of units *c* and *d* and cluster *3* of *c*, *d* and *e*. Trivial clusters such as the individual units by themselves or the entire sample of all units are not considered.

2. Determine the number of extra steps for each X_i of dendrogram m in dendrogram o. This accords with the formula,

$$E_i = \left[\sum_h X_{iL_k} - X_{iU_k} \right] - 1 \ .$$

X_{iL_k} denotes the state taken on by the lower node of the kth stem of dendrogram o in X_i; this state is set equal to the minimum value in X_i, corresponding to a unit in the cluster at the lower node of the kth stem of o. X_{iU_k} is similarly defined for the upper node of the kth stem. We note that some L_k will correspond to units a, b, c, d or e while two U_k will be the top node for the entire sample. In the example,

Stem k	*Nodes** L_k	U_k	*State of Lower/Upper node in* X_{1m}	X_{2m}	X_{3m}
	in dendrogram o				
1	*a*	*1*	1/1	–	–
2	*b*	*1*	1/1	–	–
3	*1*	*2*	1/0	–	–
4	*c*	*2*	–	1/0	1/0
5	*d*	*3*	–	1/0	1/1
6	*e*	*3*	–	–	1/1
7	*2*	*4*	–	–	–
8	*3*	*4*	–	–	1/0 .

* Labels identify clusters in Fig. 4-9.

The extra steps are the column totals according to formula E,

$$(E_1 \ E_2 \ E_3) = (0 \ 1 \ 1) \ .$$

The 0/0 scores are indicated by a dash in the table. The third score 1/0 under $\mathbf{X}_{1m}$ comes about this way: The cluster corresponding to $L_3 = 1$ in dendrogram o is (a, b). The lowest score for a, b in $\mathbf{X}_{1m}$ is equal to 1. The cluster corresponding to $U_3 = 2$ is (a, b, c) in dendrogram o. The lowest score for a, b, c in $\mathbf{X}_{1m}$ is 0. The third score under $\mathbf{X}_{1m}$ in the table thus is 1/0. The other scores are similarly determined. The sum of the differences of the elements in the L/U pairs minus 1 measures the extra steps. This sum indicates the actual magnitude of fragmentation of the clusters of dendrogram m among the clusters of dendrogram o.

3. Calculate the maximum number of extra steps which each cluster descriptor variable of m may take in dendrogram o. This number is $N_{mi}-1$, where N_{mi} is the number of units in cluster i of dendrogram m. We thus have,

	Cluster in dendrogram m		
	1	*2*	*3*
Maximum number of extra steps	1	1	2

4. Divide the extra steps by the corresponding maximum number of extra steps to obtain the individual values of the *cluster distortion coefficient:*

	Cluster in dendrogram m		
	1	*2*	*3*
Distortion coefficient	0	1	0.5

.

We note the following regarding the Farris (1973) method:

1. We compare dendrograms (m, o) without a need for topology matrices or other dissimilarity matrices to describe them.
2. A dissimilarity is measured in a way that tells us how much the clusters of dendrogram m are fragmented in the clusters of dendrogram o. We obtain for each cluster in dendrogram m a distortion coefficient with respect to dendrogram o.
3. The comparison is $m \leftarrow o$ which is not the same as $m \rightarrow o$. To complete the analysis, Rohlf (1974) suggests to repeat in the reverse $m \rightarrow o$.
4. An overall distortion coefficient is the average of the individual distortion coefficients. In our example, this has a value $1.5/3 = 0.5$. The distortion coefficient ranges form 0 to 1.

The average distortion measured by the Farris (1973) method falls far in value from $1 - \cos\alpha\,(\mathbf{T}_m ; \mathbf{T}_o)$, but close to $1 - r(\mathbf{T}_m ; \mathbf{T}_o)$ and also to $g\,(-,0)$ in the example. The measures which indicate moderate distortion do in fact describe reality : the two dendograms differ only by one unit, c, shifting position. From these we conclude that the main advantage of the method is that it not only detects distortion, but it also identifies the clusters subjected to distortion.

We shall briefly describe other methods for comparing dendrograms. One was suggested by Gower (1971b). This method uses ordination scatter diagrams. Principal axes analysis (Section 3.4) is performed on matrices $\mathbf{T}_m$ and $\mathbf{T}_o$. The classifications are then compared by superimposing the resulting scatter diagrams with centroids matched and diagrams rotated to a position of greatest coincidence of corresponding points. The method of Williams & Clifford (1971) is based on counting nodes that separate pairs of individuals. Let these be denoted as before by symbols t_{mjk} and t_{ojk} in dendrograms m and o. A measure of difference between m and o is given by

$$\delta_{mo} = 2 \sum_{j<k} \left| t_{mjk} - t_{ojk} \right| / [N(N-1)/2]$$

$$= 2D/C\,, \quad j = 1, \ldots, n-1$$

where j and k identify the entities classified. When the classification is truncated, so that the basic entities are groups rather than individuals, some or many of the t_{mjk} and t_{ojk} values will be zero. Williams & Clifford (1971) suggest counting as g_m the t_{mjk} that are zero in classification m but the associated t_{ojk} are not zero in classification o, and as g_o the cases which are zero in o but not in m. They suggest to use the total,

$$G = g_m + g_o\ ,$$

as a measure of the total number of transfers between groups.

It is quite obvious from our account that , while numerous and substantially different, the methods suggested are in the main deterministic. This is so by convenience, due to the fact that the elements in $\mathbf{T}_m$ or $\mathbf{T}_o$ are not completely independent. Let us define $\mathbf{T}_m$ and $\mathbf{T}_o$ as topology matrices (Phipps, 1975). The fact that fusion counts are not completely independent has a consequence of fundamental importance. Non-independence means that, for example, when three objects A, B, C give rise to fusions $A + B$ and $A + B + C$ with a topology matrix,

$$\mathbf{T} = \begin{array}{c|ccc} & A & B & C \\ \hline A & - & 1 & 2 \\ B & & - & 2 \\ C & & & - \end{array}\ ,$$

once t_{AB} is determined to be one, the value of t_{AC} or t_{BC} is also determined. Similar non-independence exists between certain elements in any larger topology matrix irrespective of its actual order. When the objective is to compare dendrogram topologies, the non-independence of fusion counts removes from consideration certain statistics, such as, for instance, χ^2, which assume that the elements are independent. The problem is that because the fusion counts are not independent the sampling distribution of a statistical index will probably not conform to the stipulated theoretical distribution, and the true distribution will be unknown under the circumstances. When the sampling distribution is unknown we can tell in no way whether an observed value of the statistic is small or large from the point of view of accepting or rejecting null hypotheses.

When the sampling distribution of a statistic cannot be derived from basic principles it still can be determined on empirical grounds (cf. Orlóci & Beshir, 1976). For example, we may generate a large number of dendrograms under the assumption that the null hypothesis of a single, homogeneous parent population for the samples is true, compute the statistic for each pair of dendrograms, and then, use the distribution of the computed values as the sampling distribution of the statistic. We derive, in the following example, such a distribution for

$$I(m;o) = \sum_h \sum_i \left[t_{mhi} \ln \frac{2t_{mhi}}{t_{mhi} + t_{moi}} + t_{ohi} \ln \frac{2t_{ohi}}{t_{mhi} + t_{ohi}} \right] \tag{4.10}$$

where m and o are two dendrograms with topology matrices $\mathbf{T}_m, \mathbf{T}_o$, and characteristic elements t_{mhi}, t_{ohi}. We proceed like this:

1. Sort individuals at random between quadrats based on the observed quantities of species.
2. Analyse the data into a dendrogram m.
3. Describe the dendrogram by a topology matrix $\mathbf{T}_m$.
4. Repeat steps 1 through 3 a large number of times, designated by E, but not in excess of the possible maximum number of distinct dendrograms that can be derived from the data under the chosen fusion strategy.
5. Compute $I(m; o)$ values for all distinct pairs of topology matrices. There will be $E(E-1)/2$ such values.
6. Order the $I(m; o)$ values in a sense of increasing magnitude and determine probability points by counting. An α probability point $I(m; o)_\alpha$ is such that the probability of a larger or equal $I(m; o)$ value is α.

Once probability points are determined for $I(m;o)$, we can use these points to test the null hypothesis that the two parent populations, corresponding to the sample topology matrices $\mathbf{T}_m$ and $\mathbf{T}_o$, are identical. Because the empirical distribution of $I(m; o)$ is generated under the assumption that the null hypothesis is true, we accept the null hypothesis with α probability if the condition

$$I(m; o) < I(m; o)_\alpha$$

is true, and reject it if

$$I(m;o) \geqslant I(m;o)_\alpha$$

is true.

Let us assume that the objects to be classified are species with total counts of individuals, given in Table 4-7, estimated to the nearest thousand

Table 4-7. Estimated numbers of individuals within survey site

Species	*Total*
1. *Poa compressa*	1994000
2. *Andropogon scoparius*	473000
3. *Artemisia campestris*	48000
4. *Melilotus alba*	37000
5. *Artemisia caudata*	21000
6. *Panicum virgatum*	12000
7. *Equisetum hyemale*	8000

in the survey site. We can use programs SSSIM2 and SSSIM3 to compute an empirical sampling distribution for $I(m;o)$ based on these species quantities. We note that there were 25 quadrats in the sample from which the data derived. After the computation of 30 dendrograms for the seven species and 435 values for $I(m;o)$, in accord with steps 1 to 6, and after counting the ordered values, we find the following probability points:

α	.975	.95	.90	.10	.05	.025
$I(m;o)_\alpha$	.638	.750	1.007	8.065	8.627	9.136

An entry $I(m;o)_\alpha$ in this table corresponds to the limit such that the probability of a larger or equal value in the empirical distribution of $I(m;o)$ is α. Let us assume that an analysis of samples from two soil types in the survey site for Table 4-7 yields dendrograms for the seven species with topology matrices,

$$\mathbf{T}_1 = \begin{bmatrix} 0 & 1 & 2 & 3 & 2 & 2 & 3 \\ & 0 & 2 & 3 & 2 & 2 & 3 \\ & & 0 & 2 & 1 & 1 & 2 \\ & & & 0 & 1 & 1 & 1 \\ & & & & 0 & 1 & 2 \\ & & & & & 0 & 2 \\ & & & & & & 0 \end{bmatrix}$$

$$\mathbf{T}_2 = \begin{bmatrix} 0 & 1 & 2 & 1 & 1 & 4 & 5 \\ & 0 & 2 & 1 & 1 & 3 & 4 \\ & & 0 & 1 & 1 & 3 & 4 \\ & & & 0 & 1 & 3 & 4 \\ & & & & 0 & 3 & 4 \\ & & & & & 0 & 2 \\ & & & & & & 0 \end{bmatrix}.$$

The information divergence (formula (4.10)) of $\mathbf{T}_1$ and $\mathbf{T}_2$ is $I(1;2) = 5.954$. We now wish to test the null hypothesis of a common parent population for the topology matrices. In other words, we want to decide whether 5.954 is a large enough number to declare significant the difference of $\mathbf{T}_1$ and $\mathbf{T}_2$. We choose $\alpha = 0.05$ as the rejection probability. Because the value $I(1;\ 2) = 5.954$ is less than $I(m;\ o)_\alpha = 8.627$ we accept the null hypothesis and declare $\mathbf{T}_1$ and $\mathbf{T}_2$ statistically indistinguishable. But when we consider the topology matrices,

$$\mathbf{T}_3 = \begin{bmatrix} 0 & 1 & 2 & 3 & 1 & 2 & 1 \\ & 0 & 2 & 3 & 1 & 2 & 1 \\ & & 0 & 2 & 1 & 1 & 1 \\ & & & 0 & 2 & 1 & 2 \\ & & & & 0 & 2 & 1 \\ & & & & & 0 & 2 \\ & & & & & & 0 \end{bmatrix}$$

and

$$\mathbf{T}_4 = \begin{bmatrix} 0 & 1 & 2 & 3 & 4 & 5 & 6 \\ & 0 & 2 & 3 & 4 & 5 & 6 \\ & & 0 & 2 & 3 & 4 & 5 \\ & & & 0 & 2 & 3 & 4 \\ & & & & 0 & 2 & 3 \\ & & & & & 0 & 2 \\ & & & & & & 0 \end{bmatrix},$$

we have to reject the null hypothesis. In this case $I(3,\ 4) = 11.547$, which far exceeds the α rejection limit.

H. Comparison of dendrogram structure and input data

Cluster analysis produces dendrograms from matrices of resemblance (**D**) through transformations. We have seen that we can use different methods to describe a dendrogram. The matrix $\mathbf{T}_m$ served for this purpose. Its element t_{mhi} may define the number of nodes counted between two entities (the units classified) in dendrogram m. The elements of $\mathbf{T}_m$ may also define the threshold similarity or dissimilarity at which pairs of entities fuse or separate. In any case, we may think of t_{mhi} as the distance of entities h and i in dendrogram m. Jardine & Sibson (1968a) describe such a distance as an ultrametric (see Section 2.4). Using $d(h,\ i)$, the distance of entity h from i in sample space, and t_{mhi}, their distance in the dendrogram, the function (cf. Jardine & Sibson, 1968a; Ward, 1963),

$$\Delta_k\,(\mathbf{D};\mathbf{T}) = \left[\sum_h \sum_i \left| d(h,i) - t_{mhi} \right|^{1/k}\right]^k \Big/ \left[\sum_h \sum_i d^{1/k}(h,i)\right]^k \quad (4.11)$$

with summation over all distinct pairs ($h < i$), gives a possible family of measures which we may use to determine how well the dendrogram fits the data. When $k = 1$, $\Delta_l(\mathbf{D};\ \mathbf{T})$ is the normalized mean distance. When $k = 1/2$, $\Delta_{1/2}(\mathbf{D};\ \mathbf{T})$ is the normalized root mean square. We give an example. We have the original distances which gave rise to the dendrogram in Fig. 4-2,

$$\mathbf{D} = \begin{bmatrix} 0.00 & 0.35 & 1.08 & 1.28 & 1.41 \\ & 0.00 & 0.88 & 1.10 & 1.30 \\ & & 0.00 & 0.24 & 0.51 \\ & & & 0.00 & 0.32 \\ & & & & 0.00 \end{bmatrix}.$$

The dendrogram distances (fusion levels) accord with

$$t_{muz} = \left[\frac{N_A + N_B}{N_A\ N_B}\ Q_{AB} \right]^{1/2}$$

which is the centroid distance of group A which includes quadrat u, and group B which includes quadrat z. The symbols are identical to those used in Section 4.9. For quadrats $u = 2$ and $z = 5$,

$$t_{m25} = \left[\frac{2+3}{(2)(3)}\ 1.61 \right]^{1/2}$$

$$= 1.16\ .$$

Other values can be similarly derived:

$$\mathbf{T}_m = \begin{bmatrix} 0.00 & 0.35 & 1.16 & 1.16 & 1.16 \\ & 0.00 & 1.16 & 1.16 & 1.16 \\ & & 0.00 & 0.24 & 0.41 \\ & & & 0.00 & 0.41 \\ & & & & 0.00 \end{bmatrix}.$$

These and formula (4.11) with k defined as 1/2 give us

$$\Delta_{1/2}(\mathbf{D};\mathbf{T}) = ((0.35 - 0.35)^2 + (1.08 - 1.16)^2 + \dots + (0.32 - 0.41)^2)^{1/2}/(0.35^2 + 1.08^2 + \dots + 0.32^2)^{1/2}$$

$$= 0.15$$

which indicates a relatively small amount (15%) of distortion.

Formula (4.11) assumes that d and t are in the same units. This is not always so. To handle different units, we may define a new quantity $\rho^2\ (\mathbf{D};\ \mathbf{T})$, as we have done in Section 1.7, and stress, as $\sigma(\mathbf{D};\ \mathbf{T}) = 1 - \rho^2(\mathbf{D};\ \mathbf{T})$. In the present example, $\rho^2(\mathbf{D};\ \mathbf{T}) = 0.890$ and $\sigma(\mathbf{D};\ \mathbf{T}) = 0.11$ or 11%. Other measures can be readily conceived, and in fact, several were suggested (e.g. Jardine & Sibson, 1968a, 1971; Farris, 1969a; Jackson, 1969; Rohlf, 1970; Cunningham & Ogilvie, 1972).

I. Reallocation

The purpose in reallocation is to rectify misclassifications which occur in the course of cluster analysis. To detect misclassified individuals, the classes are re-examined for their membership. There are many methods available to accomplish this task, among them, e.g., Goodall (1970b), Bradfield & Orlóci (1975) and del Moral (1975). We discuss several in Section 4.9 and in the next chapter.

4.11 Factors influencing choice

We can agree with Sokal (1966) and Rohlf (1970) that, in need of sufficient information about data structures, method selection lacks solid foundations. It is however well known that in the presence of strongly discontinuous data different clustering methods tend to produce similar results (cf. Sneath, 1966a; Muir, Hardie, Inkson & Anderson, 1970). But even then, inconsistent results can occur because of differences in resemblance functions (cf. Sammon, 1969; Wishart, 1971), differences in the threshold fusion similarity used, or differences in the clustering strategy which happened to be elected (cf. Austin, 1970; Cormack, 1971; Clifford & Williams, 1973).

The pivotal point in overcoming many of the problems with method selection is the clarity with which clustering optimality can be defined for when a precise definition is found an optimal classification can be formulated axiomatically or on the basis of trial and error (cf. Rubin, 1966; Friedman & Rubin, 1967; Hartigan, 1967; Rohlf, 1970). Any definition of optimality will no doubt reflect the user's philosophy on clustering and his or her ideas regarding approach, data structure, and objectives. Not surprisingly then, in applications different properties are stressed by different users which a clustering method should possess. To some of the users axiomatic properties are important (e.g. Jardine & Sibson, 1968a,b, 1971; Sibson, 1970), while to others the empirical properties, revealed through experiments, have a decisive role to play (e.g. Moss, 1968; Crovello, 1968a,b; Rand, 1971; Cunningham & Ogilvie, 1972). Still others stress utility (e.g. Williams, 1971; Sneath & Sokal, 1973).

Jardine & Sibson (1968a,b, 1971) view a classification as a transformation and describe the resulting hierarchical dendrogram as a pair $S = (p, d^*)$, where P is a set (of objects to be classified) and d^* is an ultrametric (see Section 2.4). The latter measures the level at which two objects are fused in the dendrogram. They assert that the transformation from resemblance values (d) to dendrogram (ultrametric d^*) must be such that certain fundamental conditions are not violated:

1. Unique results should be obtained from given data, i.e. the transformations should be well-defined.
2. Small changes in the data should produce small changes in the resulting dendrogram, i.e. the transformations should be continuous.
3. If the resemblance measure d is already an ultrametric it should not be changed by the transformation.
4. The transformation should be independent of the scale used by the resemblance function.

5. The transformation should be independent of the labelling of the objects.
6. The dendrograms produced from a P of extended or restricted contents should be consistent.
7. The transformation should impose minimum distortion on the resemblance values.

We have already considered the last point, and we have seen that the distortion can be measured. Whether a method satisfies the remaining conditions can be determined from careful consideration of its properties, if it is represented by a well-defined algorithm, or empirically through experiments (e.g. Kuiper & Fisher, 1975).

It is quite obvious that the Jardine & Sibson (1968a,b) conditions are very restrictive, and they can hardly be fully satisfied by any classificatory algorithm not from the family of single link (nearest neighbour) clustering. Yet, the single link methods are not those that are always the most useful in vegetation analysis. Common sense thus dictates not to overemphasize the axiomatic properties, but rather to adapt a set of criteria that allow for reasonable flexibility and diversity of choice. Hill, Bunce & Shaw (1975) suggest plausible criteria:

1. The clustering method should be successful in classifying heterogeneous vegetation data.
2. The resulting classes should be interpretable in ecological terms.
3. The classification should be open-ended to allow assignment of new samples of vegetation to classes without the need for further heavy computations.
4. The method should not be prone to serious misclassifications.

These authors believe that while the subdivisive methods tend to fail on criterion 4, the agglomerative methods tend to be weak on satisfying criteria 2 and 3.

One should really look for criteria that allow a diversity of choice, and especially, allow the choice to be dominated by the objectives to be achieved. We thus stress local objectives – whether prediction, problem recognition, data reduction or something else – under which a method is selected to perform clustering on a given set of data. We could also consider the availability of suitable computer programs and avoid straitjacketing the analysis by programs just because they were available. Consideration of computional efficiency may of course lead to preference for specialized methods which we considered in Section 4.9.

The clustering strategy must be well-defined with regard to the objectives. If the user is not sufficiently familiar with the data to formulate a clustering strategy, exploratory analyses may be performed based on the combined use of several methods. The first method could be, for instance, average linkage clustering and the second, clustering by neighbourhoods to put the results of average linkage clustering to test. Williams (1971) suggests initial assessment of group structure based on the use of, what he terms, weakly clustering methods (Jancey's, 1966a method is an example). He also suggests that the more intensely clustering methods (e.g. sum of squares clustering) may be used where cluster boundaries are to be sharpened.

Preference, at one time, gravitated toward the monothetic classifications in vegetation surveys. The disadvantages soon manifested themselves in sub-optimal divisions. The tendency in preference is now more toward the polythetic methods, including both probabilistic and deterministic procedures. Goodall's (1970a,b) probabilistic methods show considerable promise in that they rely on tests of homogeneity as a prelude to subdivisions or fusions.

Among the non-hierarchical techniques Jancey's (1966a) method and some of the nearest neighbour techniques, such as the method of clustering by neighbourhoods, may prove satisfactory. Whereas Jancey's method builds spherical clusters, the nearest neighbour techniques are suitable to trace out disjoint clusters of different shape or size. They are, however, sensitive to aberrant individuals and it may be well advised to identify and eliminate them before the clustering begins. We have considered methods that can help in this respect. If hierarchical clustering is required, and if cluster shape is to be made spherical, sum of squares clustering is likely to be successful. But this method restricts the definition of resemblance to a Euclidean distance. The agglomerative method of average linkage clustering gives greater flexibility in the measure of resemblance used, but it may also lead to spherical clusters.

A suitable clustering method must produce output not influenced by the order in which the objects are presented for analysis. In this respect, an inspection of the device in the algorithm which resolves ambiguities is important. Inspection of the clustering strategy, whether fusions or fissions are allowed in twos, threes, etc., may also be revealing. Scale independence is an important requirement; the clustering results must not be influenced by a scale factor in the data or resemblance function. Some schemes of evaluation, such as the one suggested by Jardine & Sibson (1968a,b, 1971) and Sibson (1970), stipulate that existing clusters must be preserved. This requirement, if satisfied, would exclude the subdivisive methods from consideration, knowing that these methods can irreparably dismember groups before detecting their existence. Optimality is often mentioned as a requirement (e.g. Jardine & Sibson, 1971; Sneath & Sokal, 1973). An optimal method will form clusters of objects only if justified by the data structure.

The methods to be preferred are those whose output gives the best possible fit to the data. Methods to measure this kind of fit were already discussed. Related to the idea of fit is the idea of clustering stability. This means that the effect of small changes in the data – such as changes in the number of variables (cf. Crovello, 1968a,b, 1969), in the sampling method, or in accumulating computer rounding errors – must yield small changes in the clustering results. The clustering is required sometimes to be monotone (with no reversals) in fusion levels (cf. Jardine & Sibson, 1971; Rohlf, 1970).

The properties required from a clustering method may not always be complementary. As a matter of fact, it is not too difficult to imagine a situation where they may be outright conflicting. In one such case, the ideal objectives may not square with optimal axiomatic properties. For example, to create a dichotomous key to facilitate the identification of vegetation groups based on presence or absence of single species, subdivisive methods will be used in spite of a distinct possibility that the cluster preservation and stability properties will be both violated. To more clearly see these aspects, experiments with different techniques are advised.

When we contemplate clustering by a given method, we must consider the implications regarding the data structure analysed, the resemblance measure applied, the clustering (fusion or fission) strategy followed, and the objectives achieved. It is a known fact that data are rarely completely continuous nor distinctly discontinuous. Outliers tend to promote the appearance of a discontinuous structure, while intermediate points, scattered between high density phases, tend to have the opposite effect. The analysis cannot be fully successful in putting into focus certain aspects of data structure when such data points are present, and yet, it may be highly successful when they are removed. In attempting to overcome the difficulties with outliers, or with intermediates, an iterative preliminary analysis may be performed on the entire sample to identify extraneous sample points. After removal of these, the data may be reanalysed. We have already considered methods for detection of extreme data points in Chapter 1.

The choice of a resemblance function will depend on the data, and also, on the clustering strategy. Certain types of data will dictate the use of certain types of resemblance functions. When the data consists of cover/abundance estimates in terms of symbols such as *r, +, 1,* etc., the choice may fall on information theory functions which, as we have seen in Chapter 2, are well-suited to handle mixed data types. Williams (1971) suggests the use of Euclidean distance (without adjustments) for continuous data with no extreme outliers, information functions for highly skewed binary data, and the so-called Canberra metric when the data contains few zeros (with no zero matches) and occasional extreme outliers that should be prevented from dominating the analysis. When probabilistic interpretations are required, Goodall's (1964, 1968, 1970b) indices may be applied in practically any type of data. The clustering strategy itself may, however, dictate the resemblance function to be used, such as in sum of squares clustering where the distance must be Euclidean.

Not uncommonly, data sets are very large, far in excess to what could be analysed based on a specific clustering method which would otherwise be desired. In cases of this sort, methods may be changed at different stages in the analysis. First level groups may be established by use of a 'quick and dirty' method. These groups could then be reanalysed based on optimal methods. The difficulty that has to be adequately resolved is that results from different methods are not always comparable. In spite of the difficulties, a mixed method strategy may still be preferred in the face of the alternative of not performing any cluster analysis on the data.

The dominance of results by one or a few variables, because of scale problems in measurements or sheer quantity, is often a source for difficulties, particularly in those methods that rely on monothetic subdivisions. It is known that inclusion or exclusion of such variables can completely alter the course of the analysis. The solutions suggested rely on preliminary analyses (see Chapter 1), or on clustering strategies flexible enough to permit recourse to a previous point in the procedure, to initiate a new analysis, to reveal and then to remove the problematic variables.

Cluster size may also be a source for problems in some strategies. Once a cluster has reached a certain critical size it may tend to draw individuals to itself. The opposite can also be true. We have commented on this already in Section 4.9. The recognition of this fact may lead to pursuance of strategies in which a cluster may be removed from the analysis as soon as its distinctness is established. The analysis may then be continued on the residual sample.

In one extreme, the objectives may dictate that the sample be dissected no matter what the properties of the data structure. However, a classification need not be rigidly prescribed, but it may be pursued only if groups actually exist as distinct clusters of points. In such a case, preliminary analyses of the data can reveal whether or not a more thorough search for clusters would be profitable (cf. Hill, 1969; Gnanadesikan & Wilk, 1969).

While the available methods are numerous, and the choice between them difficult, the task of classifying vegetation is somewhat simplified by not having to choose variables arbitrarily. In vegetation studies, the variables are normally well-defined and they are also applicable uniformly to all species in all quadrats of the sample. This cannot be said of variables and specimens in taxonomy where the variables are usually ill-defined, frequently specific to particular organs or individuals, and often logically or otherwise dependent (cf. Dale, 1968; Goodall, 1970a; Jardine & Sibson, 1971).

4.12 Classification: success or failure?

If no specific objective is stipulated, other than to achieve a dissection, then the question of success or failure need not be asked, for in such a case success would simply mean that the algorithm did the job: it subdivided the sample into groups. But whether or not the algorithm is appropriately used is another question. Appropriateness has to do with both axiomatic properties and suitability to achieve the stated objectives.

Intuitively, the approach in which all possible divisions of a sample of N individuals into k groups are inspected is appealing. The number of all possible divisions cannot however be simply enumerated, considering that this number accords with the combinatorial

$$G = \frac{1}{k!} \sum_{m=0}^{k} (-1)^{k-m} \frac{k!}{m!\,(k-m)!} m^N .$$

Clustering by the Edwards & Cavalli-Sforza (1965) algorithm is a case in point. While this algorithm is suitable to find an optimal solution that will maximize the sum of squares between groups through enumerating all possible subdivisions of the sample, or subsets thereof, into two, it can succeed with its objective only in samples of relatively small size. When, for instance, there is a sample of 25 individuals, to find the first optimal split into two would require examination of $2^{24} - 1 = 16{,}777{,}215$ alternative divisions. When $N = 40$, there would be more than 32,000 times as many divisions to inspect. The magnitude of the computation can be overwhelming, particularly when complex calculations have to be performed at each division and when the operations have to be performed over and over again at different levels.

When the question of success or failure comes up for evaluation, the answer cannot be divorced from the objective. We consider these:

1. *Prediction.* It is not too difficult to accept the idea that the predictive function is important. Clustering results, however, always have some predictive value for certain properties of the population from which the sample is taken. But the different classifications may be differently efficient in predicting population properties, or they may utilize different aspects of the information in the sample as a basis for prediction. Goodall (1966a,b,c,d,e, 1973a) sees the evaluation of success in this case as a problem in measuring the departure of the predicated values from their expectation. Examples for this are given by Orlóci (1972b) and also in this chapter. A prediction, to be reliable, must be unbiased. An unbiased prediction requires that suitable data be analysed by the use of suitable methods which can produce predicted values whose expectation is in fact the target population value. Random sampling is a precondition for unbiased prediction. The further the sampling scheme departs from a random design, the more difficult to know which population's values are being predicted.

Reliance on new data helps avoid circular arguments and enhance reliability. This is exactly what Macnaughton-Smith (1965) tried to do when he superimposed a classification on a new sample before using class values as predictors of certain population values. Regarding this point, an illustration should follow. When we decide to rely on vegetation types as predictors of soil conditions, we render the success of the prediction conditional on how well quadrat membership in the vegetation types can indicate quadrat membership in soil types. In measuring this, we relied on a contingency table (Table 4.2) and a function ($2I$) of association. Should the sampling be non-random, or the distribution of $2I$ depart from the assumed kind (χ^2), our conclusions about the value of vegetation types as predictors of soil conditions would have been biased.

2. *Exploratory analysis leading to problem recognition and clarification.* A frequently met objective of classifications is to suggest relationships and principles in the problem recognition or hypothesis formulation phase of an ecological study. Problem recognition is an indirect outcome of classification, and as such, it is not an objective based on which success could be measured. It is important, though, to choose a classification algorithm which does not unduly restrict the type of problems that can be recognized, does not lead to recognition of false problems, and yet, which reveals information from the sample regarding those aspects of the population with respect to which clear statements of problem are sought.

3. *Data reduction.* Since this involves passing from data to a classification, information losses along the way are inevitable. We examined this question in connection with ordinations in Section 3.8. Explicit measurements could be made in the following way:

a. We may measure stress in the fusion topology of the dendrogram, or among the classes if the clustering is non-hierarchical, with regard to the similarity (or distance) matrix on which the classification is based. This is an approach which we used in a preceding section on dendrograms.

b. We may rely on a set of external variables, such as soil characteristics when the classification is based on vegetation, to determine how much of the predictive value of the original species data is retained in the fusion topology of the classification (cf. Hall, 1970; Williams, Lambert & Lance, 1966; Goodall, 1973a). The degree of success could then be measured as, for instance, a canonical correlation of species and environmental variables, or as a canonical correlation of the group means of species and environmental variables.

Reliance on any single property in a complex system, such as in a classification, to evaluate success or failure may not be a wise undertaking. Evaluations should rather be based on several criteria, encompassing axiomatic properties as well as applicational aspects, such as the suitability of the method to achieve the objectives. It is true that any attempt to measure success based on how well an objective is achieved, or how much ecological information is supplied, would be weakened if the method of classification were chosen without an understanding of its fundamental properties. It is also true, however, that even if the fundamental properties are understood, it is impossible to tell which method to choose without the understanding of the objectives to be achieved.

What has been said in the last two sections should be of help to bring to attention some aspects of clustering which may help with method selection. The suggestions made are not, however, claimed to contain all the relevant points. The user should experiment with the different methods to learn about their potential, and then decide which are best under the circumstances. It is standard practice to select a method and subject the data to cluster analysis. Even if the choice is fortunate, the information provided by the results may be too one-sided. Here, just as in the case of ordinations, reliance on results from different methods is likely to be beneficial (cf. Johnson & Holm, 1968; Crawford, Wishart & Campbell, 1970; Phipps, 1970; Grigal & Goldstein, 1971). In this way, not only new information may come to light about the data structure, but also, experience may be gained about methods to benefit future users.

Chapter 5

IDENTIFICATION

In Chapter 4 we have considered methods whose objective is to partition a sample of objects into groups. In the present chapter we proceed on the assumption that the groups are formed and described, a measure of affinity given and decision rules specified. The remaining objective amounts to finding the group that is most likely to represent a parent population for a given individual. The problem thus is one in identification. We shall consider several methods.

5.1 Generalized distance

In statistical applications of the generalized distance it is assumed that the population distribution is multivariate normal. This implies, as we have seen already, that the groups can be represented by linear clusters of points in sample space. Such a cluster is ellipsoidal in shape, has uniform density, but possibly with an ascending density gradient toward the centroid. A cluster of this sort can be completely described by a mean vector and covariance matrix.

Let us assume that the points represent quadrats embedded within a reference system of species. With n_m quadrats and p species in group m, and with sample mean vectors,

$$\bar{\mathbf{X}}_m = \begin{bmatrix} \bar{X}_{1m} \\ \cdot \\ \bar{X}_{pm} \end{bmatrix},$$

and covariance matrix,

$$S_m = \begin{bmatrix} S_{11m} & \cdots & S_{1pm} \\ \cdot & \cdots & \cdot \\ S_{p1m} & \cdots & S_{ppm} \end{bmatrix},$$

the generalized distance of a new quadrat,

$$\mathbf{X} = \begin{bmatrix} X_1 \\ \cdot \\ X_p \end{bmatrix},$$

from group m can be calculated according to

$$d(\mathbf{X}, \overline{\mathbf{X}}_m) = [(\mathbf{X} - \overline{\mathbf{X}}_m)' \mathbf{S}_m^{-1} (\mathbf{X} - \overline{\mathbf{X}}_m)]^{\frac{1}{2}} \tag{5.1}$$

where $\mathbf{S}_m^{-1}$ is the inverse of $\mathbf{S}_m$. Based on formula (5.1), we can test two sets of hypotheses:

1. $H_{01}: E(\overline{\mathbf{X}}_m) = \mathbf{X}$

   $H_{11}: E(\overline{\mathbf{X}}_m) \neq \mathbf{X}$

2. $H_{02}: E(\overline{\mathbf{X}}_m) = E(\mathbf{X})$

   $H_{12}: E(\overline{\mathbf{X}}_m) \neq E(\mathbf{X})$.

The null hypothesis H_{01} puts the expectation of the sample mean vector $\overline{\mathbf{X}}_m$ at point $\mathbf{X}$. The question then to be answered amounts to determining the probability that a sample mean vector in group m may fall as far away from a standard, represented by $\mathbf{X}$, as it actually does. Under the null hypothesis,

$$F_m = \frac{(n_m - p)\ n_m}{p(n_m - 1)}\ d^2(\mathbf{X}, \overline{\mathbf{X}}_m) \tag{5.2a}$$

may be treated as a variance ratio and referred to the variance ratio distribution with p and $n_m - p$ degrees of freedom, assuming that the parent population of $\overline{\mathbf{X}}_m$ is multivariate normal. The null hypothesis (H_{02}) stipulates that $\overline{\mathbf{X}}_m$ and $\mathbf{X}$ are both samples from a multivariate normal population. Under the null hypothesis,

$$F_m = \frac{n_m(n_m - p)}{(n_m + 1)p(n_m - 1)}\ d^2(\mathbf{X}, \overline{\mathbf{X}}_m) \tag{5.2b}$$

has the F distribution with p and $n_m - p$ degrees of freedom. If the computed value F_m satisfies the condition

$$F_m < F_{\alpha; p, n_m - p} \tag{5.3}$$

at given probability α, we declare group m a likely parent population for $\mathbf{X}$. If more than one group satisfies this condition, $\mathbf{X}$ may be assigned to either group or to the group from which its distance is smallest.

Let us consider an example. We want to determine if either one of three *forest compartments* could represent a parent population for a given vegetation quadrat. The compartments are described on the basis of two species in random samples of 24, 31 and 20 quadrats respectively. The compartment mean vectors are given by

$$\overline{\mathbf{X}}_1 = \begin{bmatrix} \overline{X}_{11} \\ \overline{X}_{21} \end{bmatrix} = \begin{bmatrix} 2.1 \\ 4.8 \end{bmatrix} \qquad \overline{\mathbf{X}}_2 = \begin{bmatrix} \overline{X}_{12} \\ \overline{X}_{22} \end{bmatrix} = \begin{bmatrix} 3.9 \\ 4.6 \end{bmatrix}$$

$$\bar{\mathbf{X}}_3 = \begin{bmatrix} \bar{X}_{13} \\ \bar{X}_{23} \end{bmatrix} = \begin{bmatrix} 2.2 \\ 1.1 \end{bmatrix}$$

and the covariance matrices by

$$\mathbf{S}_1 = \begin{bmatrix} 1.043 & 0.042 \\ 0.042 & 0.931 \end{bmatrix} \qquad \mathbf{S}_2 = \begin{bmatrix} 1.503 & 0.165 \\ 0.165 & 0.265 \end{bmatrix}$$

$$\mathbf{S}_3 = \begin{bmatrix} 0.974 & 0.707 \\ 0.707 & 1.050 \end{bmatrix}.$$

A newly acquired vegetation quadrat is specified by its data vector,

$$\mathbf{X} = \begin{bmatrix} X_1 \\ X_2 \end{bmatrix} = \begin{bmatrix} 2 \\ 1 \end{bmatrix} .$$

We compute the affinity of $\mathbf{X}$ and the compartments in terms of formula (5.1). In the first step we determine the inverses of $\mathbf{S}_1$, $\mathbf{S}_2$, and $\mathbf{S}_3$ which are found to be,

$$\mathbf{S}_1^{-1} = \begin{bmatrix} 0.961 & -0.043 \\ -0.043 & 1.076 \end{bmatrix} \qquad \mathbf{S}_2^{-1} = \begin{bmatrix} 0.714 & -0.445 \\ -0.445 & 4.051 \end{bmatrix}$$

$$\mathbf{S}_3^{-1} = \begin{bmatrix} 2.008 & -1.352 \\ -1.352 & 1.862 \end{bmatrix} .$$

Based on these results we proceed to the calculation of the $d^2(\mathbf{X}, \bar{\mathbf{X}}_m)$ values for the different compartments.

Compartment 1:

$$(\mathbf{X} - \bar{\mathbf{X}}_1)' = (2 - 2.1 \quad 1 - 4.8) = (-0.1 \quad -3.8)$$

$$d^2(\mathbf{X},\bar{\mathbf{X}}_1) = (-0.1 \quad -3.8) \begin{bmatrix} 0.961 & -0.043 \\ -0.043 & 1.076 \end{bmatrix} \begin{bmatrix} -0.1 \\ -3.8 \end{bmatrix} = 15.514 .$$

The value of the test statistic for H_{01} (formula (5.2a)) is given by

$$F = \frac{(24-2)\,(24)\,(15.514)}{(2)\,(24-1)} = 178.07 .$$

Since F far exceeds the $\alpha = 0.05$ probability point, $F_{0.05;2,22} = 3.44$, we conclude (formula (5.3)) that the chances of $\bar{\mathbf{X}}_1$ being a sample from a population whose mean vector is the standard $\mathbf{X}$ are too small to allow assignment of $\mathbf{X}$ to $\bar{\mathbf{X}}_1$. The test on H_{02} leads to a similar conclusion. The value of the test statistic in this case (formula (5.2b)) is given by

$$F = \frac{(24)(24-2)(15{\cdot}514)}{(24+1)(2)(24-1)} = \frac{8191.392}{1150} = 7.12 \ .$$

F exceeds the $\alpha = 0.05$ probability points, $F_{0.05;2,22} = 3.44$, and the condition in formula (5.3) remains unsatisfied.

Compartment 2:

$$(\mathbf{X} - \bar{\mathbf{X}}_2)' = (2 - 3.9 \quad 1 - 4.6) = (-1.9 \ -3.6)$$

$$d^2(\mathbf{X},\bar{\mathbf{X}}_2) = (-1.9 \ -3.6) \begin{bmatrix} 0.714 & -0.445 \\ -0.445 & 4.051 \end{bmatrix} \begin{bmatrix} -1.9 \\ -3.6 \end{bmatrix} = 48.991 \ .$$

The value of the test statistic for H_{01} is given by

$$F = \frac{(31-2)(31)(48.991)}{(2)(31-1)} = 734.05,$$

and for H_{02} by

$$F = \frac{(31)(29)(48.991)}{(32)(2)(30)} = 22.94 \ .$$

At $\alpha = 0.05$ probability, $F_{0.05;2,29} = 3.33$ and the condition in formula (5.3) remains unsatisfied for both.

Compartment 3:

$$(\mathbf{X} - \bar{\mathbf{X}}_3)' = (2 - 2.2 \quad 1 - 1.1) = (-0.2 \ -0.1)$$

$$d^2(\mathbf{X}, \bar{\mathbf{X}}_3) = (-0.2 \ -0.1) \begin{bmatrix} 2.008 & -1.352 \\ -1.352 & 1.863 \end{bmatrix} \begin{bmatrix} -0.2 \\ -0.1 \end{bmatrix} = 0.0449 \ .$$

The value of the test statistic for H_{01},

$$F = \frac{(20-2)(20)(0.0449)}{2(20-1)} = 0.47,$$

is less than $F_{0.05;2,18} = 3.55$ which leaves the condition in formula (5.3) satisfied. We conclude that assignment of $\mathbf{X}$ to the third compartment is justified on a probabilistic basis. For this particular assignment the average misclassification probability is ((24)P(F>178.07) + (31)P(F>734.05))/55 $\simeq 0$. For the test on H_{02}, we have

$$F = \frac{(20)(0.0449)(18)}{(21)(2)(19)} = 0.02 \ .$$

At $\alpha = 0.05$ probability, $F_{0.05;2,18} = 3.55$, and the condition in formula (5.3) is satisfied. This also indicates that compartment 3 is a likely parent population for quadrat **X**. The average misclassification probability is $((24)P(F > 7.12) + (31)P(F > 22.94))/ 55 < 0.002$. We note the following:

1. Irrespective of the number of reference groups, we deal with the groups individually. We compute the distance $d(\mathbf{X}, \bar{\mathbf{X}}_m)$ on the basis of the mth covariance matrix $\mathbf{S}_m$; in this respect we differ from other users who recommend pooling the group covariance matrices and then computing the generalized distance based on the pooled covariance matrix. While pooling has practical advantages in simplifying the computations, it also puts additional constraints on the method, manifested in the assumption that the population group covariance matrices are equal.
2. The distance $d(\mathbf{X},\bar{\mathbf{X}}_m)$ can be associated with a standard statistical distribution when the samples are drawn at random from multivariate normal populations. Otherwise this distance measure will serve only as a deterministic expression of absolute separation.
3. Because an inverse of the covariance matrix is incorporated in the formula for distance $d(\mathbf{X}, \bar{\mathbf{X}}_m)$, a standardization occurs which is not applied directly to the species, but rather indirectly to their linear compound. This is obvious from the equivalent expression of formula (5.1),

$$d(\mathbf{X}, \bar{\mathbf{X}}_m) = \left[\sum_i Y_i^2/\lambda_i \right]^{1/2}, \ i = 1, \dots, t \qquad (5.4)$$

where t indicates the number of non-zero eigenvalues of the covariance matrix, λ_i the ith eigenvalue, and Y_i the ith component score for quadrat **X**. This form has been used by Bradfield & Orlóci (1975) in classification and mapping.

Let us consider the last point on the basis of a numerical example. The data are given in Table 5-1. It can be seen from Fig. 5-1 that the data define a linear cluster of points of reasonably even density. Such a cluster can be described reasonably well by a mean vector,

$$\bar{\mathbf{X}}_m = \begin{bmatrix} 5 \\ 5 \end{bmatrix},$$

Table 5-1. Co-ordinates for the points in Fig. 5-1

Species	*Quadrat*														
i	*1*	*2*	*3*	*4*	*5*	*6*	*7*	*8*	*9*	*10*	*11*	*12*	*13*	*14*	*15*
X_{1m}	2	3	2	3	4	3	3	4	5	6	5	5	4	4	5
X_{2m}	2	2	3	3	3	4	5	4	3	4	4	5	5	6	6

	16	*17*	*18*	*19*	*20*	*21*	*22*	*23*	*24*	*25*	$\bar{X}_{im}$
X_{1m}	6	7	6	7	8	7	6	5	7	8	5
X_{2m}	5	5	6	6	7	7	7	7	8	8	5

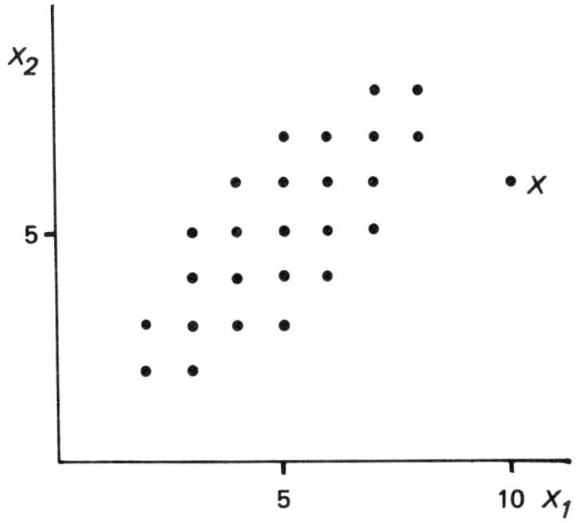

Fig. 5-1. Scatter diagram for quadrats of a hypothetical sample. X_1 and X_2 represent species. Point X signifies an external quadrat.

and covariance matrix,

$$\mathbf{S}_m = \begin{bmatrix} 3.167 & 2.417 \\ 2.417 & 3.167 \end{bmatrix} .$$

The inverse of $\mathbf{S}_m$ is

$$\mathbf{S}_m^{-1} = \begin{bmatrix} 0.756 & -0.577 \\ -0.577 & 0.756 \end{bmatrix} .$$

The separation of individual

$$\mathbf{X} = \begin{bmatrix} 10 \\ 6 \end{bmatrix}$$

from group m is

$$d^2(\mathbf{X}, \bar{\mathbf{X}}_m) = (5 \quad 1) \begin{bmatrix} 0.756 & -0.577 \\ -0.577 & 0.756 \end{bmatrix} \begin{bmatrix} 5 \\ 1 \end{bmatrix} = 13.886$$

or

$$d(\mathbf{X}, \bar{\mathbf{X}}_m) = 3.726 \ .$$

The same generalized distance may be derived from the component scores which we compute for **X** based on the formulae,

$$Y_{1X} = (\mathbf{X} - \bar{\mathbf{X}}_m)'\mathbf{b}_1 = (5 \quad 1) \begin{bmatrix} 0.707 \\ 0.707 \end{bmatrix} = 4.242$$

$$Y_{2X} = (\mathbf{X} - \bar{\mathbf{X}}_m)'\mathbf{b}_2 = (5 \quad 1) \begin{bmatrix} -0.707 \\ 0.707 \end{bmatrix} = -2.828 \ ,$$

where $\mathbf{b}_1$, $\mathbf{b}_2$ hold the component coefficients of $\mathbf{S}_m$. From the component scores and eigenvalues $\lambda_1 = 5.583$, $\lambda_2 = 0.750$, we obtain, according to formula (5.4),

$$d(\mathbf{X}, \bar{\mathbf{X}}_m) = \left[\frac{Y_{1X}^2}{\lambda_1} + \frac{Y_{2X}^2}{\lambda_2}\right]^{1/2} = \left[\frac{4.242^2}{5.583} + \frac{(-2.828)^2}{0.750}\right]^{1/2} = 3.726 \ .$$

This is the same quantity as the one derived by formula (5.1). According to formula (5.2a), we have

$$F_m = \frac{(25 - 2)(25)(3{\cdot}726)^2}{(2)(25 - 1)} = 166.31,$$

and according to formula (5.2b),

$$F_m = \frac{(25)(23)(3.726)^2}{(25 + 1)(2)(25 - 1)} = 6.40 \ .$$

In either case, the hypothesis that quadrat **X** is from a population represented by group m must be rejected.

It is revealing to note that when we plot points, based on the component scores in Table 5-2, we obtain a point cluster exactly like the one in Fig. 5-1. When, however, the component scores are standardized the point cluster becomes circular in shape (Fig. 5-2).

Table 5-2. Component scores of quadrats based on data in Table 5-1

Component	*Quadrat* 1	2	3	4	5	6	7
Y_{1m}	−0.866	−0.722	−0.722	−0.577	−0.433	−0.433	−0.289
Y_{2m}	0.000	−0.144	0.144	0.000	−0.144	0.144	0.289
	8	*9*	*10*	*11*	*12*	*13*	*14*
Y_{1m}	−0.289	−0.289	0.000	−0.144	0.000	−0.144	0.000
Y_{2m}	0.000	−0.289	−0.289	−0.144	0.000	0.144	0.289
	15	*16*	*17*	*18*	*19*	*20*	*21*
Y_{1m}	0.144	0.144	0.289	0.289	0.433	0.722	0.577
Y_{2m}	0.144	−0.144	−0.289	0.000	−0.144	−0.144	0.000
	22	*23*	*24*	*25*	λ_i		
Y_{1m}	0.433	0.289	0.722	0.866	5.583		
Y_{2m}	0.144	0.289	0.144	0.000	0.750		

5.2 Discriminant function

The method just described involved computation of the distance of a point **X** from the centroids of each reference group. Each group is regarded as a different case and its own covariance matrix is used in the definition of distance. The procedure is very tedious, though, because each group's covariance matrix has to be inverted, and when a new **X** is given, the complex operations (formula (5.1)) have to be repeated all over again. The computations can however be somewhat simplified by sacrificing simplicity in underlying assumptions.

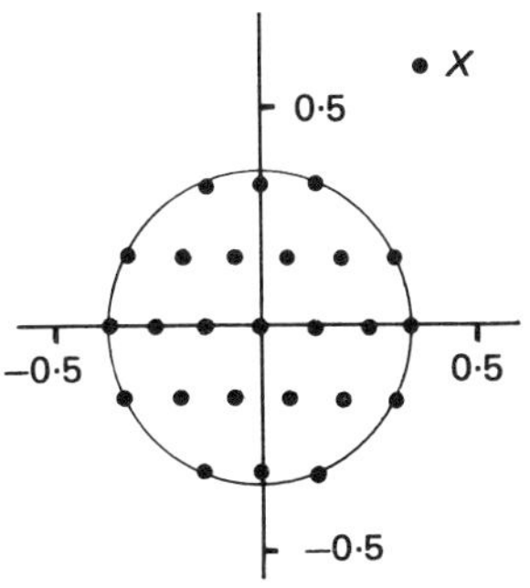

Fig. 5-2. Scatter diagram from standardized component scores. Explanations are given in the main text.

If the reference groups represent random samples from normal populations, whose covariance matrices are equal, identification may proceed according to Fisher's (1936; Rao, 1952) discriminant function or some of its variants (cf. Anderson, 1958; Reyment, 1972; Ottestad, 1975b; Fischer, Horánszky, Kiss & Szőcs, 1974). Discriminant functions project points onto lines connecting the groups in the direction of maximum difference. In this way, discriminant functions allow examination of th mutual nearness of groups in a comparable manner. Each discriminant function is specific to a pair of groups in most cases, but to several groups in at least one variant (Ottestad, 1975b).

We shall give a brief account of discriminant analysis using formulations that utilize the idea of distance. Let $\bar{\mathbf{X}}_1, ..., \bar{\mathbf{X}}_k$ and $\mathbf{S}_1, ..., \mathbf{S}_k$ represent k group mean vectors and associated covariance matrices, similarly as in the preceding example. Let the k population covariance matrices be equal (to justify pooling the sample group covariances matrices). Let a pooled covariance matrix be given by the weighted average,

$$\mathbf{S}_{mz} = [(n_m - 1)\mathbf{S}_m + (n_z - 1)\mathbf{S}_z] / (n_m + n_z - z) ,$$

and $n_1, ..., n_k$ represent sample sizes. Considering two groups, m and z, their squared generalized distance is

$$\begin{aligned} d^2(\bar{\mathbf{X}}_m;\bar{\mathbf{X}}_z) &= (\bar{\mathbf{X}}_m - \bar{\mathbf{X}}_z)'\mathbf{S}_{mz}^{-1}(\bar{\mathbf{X}}_m - \bar{\mathbf{X}}_z) \\ &= (\bar{\mathbf{X}}_m - \bar{\mathbf{X}}_z)'\mathbf{a}^{(mz)} \\ &= \sum_i a_i^{(mz)}(\bar{X}_{mi} - \bar{X}_{zi}), \qquad i = 1, ..., p \end{aligned}$$

Symbol $\mathbf{a}^{(mz)}$ stands for a vector of p discriminant coefficients $a_1^{(mz)}, ..., a_p^{(mz)}$. Here p is the number of species. The p coefficients are specific to the pair (mz) that are compared. The important point to be made is that, based on the use of the discriminant coefficients in $a_i^{(mz)}$ and the co-ordinates $(\mathbf{X} - \bar{\mathbf{X}}_m) = [(X_1 - \bar{X}_{m1}) ... (X_p - \bar{X}_{mp})]$, which describe the species in a quadrat $(\mathbf{X})$ whose compartmental affiliation is in question, the discriminant function, when given in the form of

$$t^2(\mathbf{X};\bar{\mathbf{X}}_m)^{(mz)} = \sum_i a_i^{(mz)}(X_i - \bar{X}_{mi}), \qquad i = 1, ..., p \qquad (5.5)$$

is still measuring distance. But, unlike formula (5.1), in this a distance is measured from the projection of the tip of the mth mean vector to the projection of the tip of $\mathbf{X}$ on the line in the direction of which the overlap between the density contours of the groups is minimized. The geometric picture is explained elsewhere (e.g. Cooley & Lohnes, 1971; Sneath & Sokal, 1973). Note that different distances are measured by formulae (5.1) and (5.5) as shown in Fig. 5-3. It is quite obvious that formula (5.5) takes on a new numerical form for each new $\mathbf{X}$ and m or z.

How to know to which group **X** should be assigned? As one possibility, we may decide to make an assignment only if there is a reasonably high probability that a relevé (as a point) could lie from the centroid of its population as far as the measured distance of **X** from that group's centroid from which its distance is smallest. If this group is m, to find this probability, we would have to define an F quantity associated with $t^2(\mathbf{X};\bar{\mathbf{X}}_m)^{(mz)}$ which could then be referred to the appropriate distribution with ν_1 and ν_2 degrees of freedom. An assignment would not be performed at α probability of a Type I error if $P(F > F_{\alpha;\nu_1,\nu_2}) < \alpha$. Alternatively, we could rely on the Bayes classification rule according to which X is assigned to group m if the relation,

$$
\begin{aligned}
&-1/2[(\mathbf{X}-\bar{\mathbf{X}}_m)'\,\underset{mz}{\mathbf{S}^{-1}}(\mathbf{X}-\bar{\mathbf{X}}_m)-(\mathbf{X}-\bar{\mathbf{X}}_z)'\,\underset{mz}{\mathbf{S}^{-1}}(\mathbf{X}-\bar{\mathbf{X}}_z)]\\
&= \quad [\mathbf{X}-1/2(\bar{\mathbf{X}}_m+\bar{\mathbf{X}}_z)]'\,\underset{mz}{\mathbf{S}^{-1}}(\bar{\mathbf{X}}_m-\bar{\mathbf{X}}_z)\\
&> \ln\frac{P_z\ \mathrm{L}(m|z)}{P_m\ \mathrm{L}(z|m)}
\end{aligned}
\tag{5.6}
$$

holds true for all groups $z = 1, \ldots, k$ (excluding m). Anderson (1958) describes the derivations and also gives references. In formula (5.6), $P_m = n_m/n$ and $P_z = n_z/n$ are estimates of the relative sizes of the populations from which the contents of groups m and z represent random samples; $n = n_m + n_z$. $\mathrm{L}(m \mid z)$ and $\mathrm{L}(z \mid m)$ measure the cost of misassignments (cf. Anderson, 1958). If we assume that population sizes are equal, and the cost of a misassignment is the same for all groups, **X** is assigned to group m if the value on the left-hand side of formula (5.6) is consistently larger than zero. This is the same as if we would make the assignment of **X** to group m conditional on $t(\mathbf{X};\, \mathbf{X}_m)^{(mz)}$ of formula (5.5) being the smallest distance among those actually measured. Such an assignment is of course entirely deterministic.

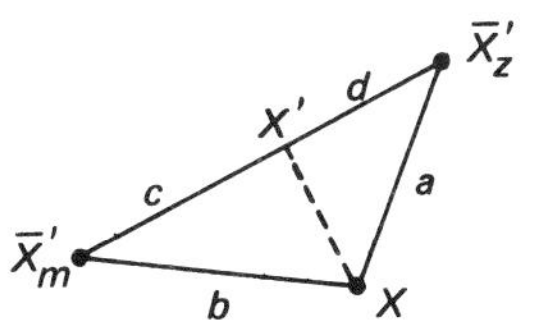

Fig. 5-3. Distances measured by formula (5.1) between two mean vectors ($\bar{\mathrm{X}}_m$, $\bar{\mathrm{X}}_z$) and an outsider (X) are indicated by letters a and b. Letters c and d indicate distances measured by formula (5.5). $\bar{X}'_m$ and $\bar{X}'_z$ indicate points marking the projections of the tips of the mth and zth mean vector on the best discriminating line between groups m and z. Symbol X' indicates the projection of X on the same line. See description of method in the main text.

When the number of groups is k, the preceding method requires $k(k-1)/2$ discriminant functions to complete an assignment. Ottestad (1975b) suggests modifications that allow the use of a single discriminant function even if k is larger than 2. While Ottestad's modifications simplify the computation, they do this with potential losses of precision on two different counts. Firstly, a single linear function is likely to be a less efficient discriminator among $k > 2$ groups than among $k = 2$ groups. Naturally, exceptions can be conceived. Secondly, Ottestad's method involves inspection of group means of the variates to determine which of the variates order the groups in the same sense or in the opposite sense. Since it may not be possible to assign all variables to only two subsets, according to the ordering which their means impose on the groups, some variables cannot be used in the analysis.

In the Ottestad method, just as in Fisher's, the computational problem amounts to finding discriminant coefficients on the basis of which a discriminant function can be constructed. Let the data vector $\mathbf{X}' = (X_1 \ldots X_p)$ describe a quadrat whose group affiliation is to be established. Let some variable on the same set of p species describe each of the k reference groups based on k mean vectors,

$$\bar{\mathbf{X}} = \begin{bmatrix} \bar{X}_{11} & \ldots & \bar{X}_{1k} \\ \cdot & \ldots & \cdot \\ \bar{X}_{p1} & \ldots & \bar{X}_{pk} \end{bmatrix},$$

and pooled covariance matrix $\mathbf{S}$. A general element $\bar{X}_{im}$ in $\bar{\mathbf{X}}$ is the mean value of species i in group m, and S_{hi} in $\mathbf{S}$ the pooled covariance of species h and i. When $h = i$, S_{hi} is a pooled variance. To determine numerical values for the discriminant coefficients we solve for $a_1, \ldots, a_p$ in p simultaneous equations,

$$\begin{bmatrix} S_{11} \ldots S_{1p} \\ \cdot \;\; \ldots \;\; \cdot \\ S_{p1} \ldots S_{pp} \end{bmatrix} \begin{bmatrix} a_1 \\ \cdot \\ a_p \end{bmatrix} = \begin{bmatrix} d_1 \\ \cdot \\ d_p \end{bmatrix}.$$

The peculiarity of Ottestad's method is the definition of the $d_1, \ldots, d_p$ values. While in Fisher's method d_i is the difference of two co-ordinates $\bar{X}_{im} - \bar{X}_{iz}$, in Ottestad's modification it is defined by

$$d_i = (V_i)^{1/2} \tag{5.7}$$

where

$$V_i = \frac{1}{k-1} \sum_m n_m (\bar{X}_{mi} - \bar{X}_i)^2 \;\; , \; m = 1, \ldots, k$$

This is the between groups sum of squares. k is the number of groups, n_m the size of group m; $n = n_1 + \ldots + n_k$; $\bar{X}_{mi}$ the mean and X_{mi} the total of the ith species in group m. $\bar{X}_i$ is the grand mean of species i in the sample.

The discriminant coefficients are defined by the relation,

$$\mathbf{a} = \mathbf{S}^{-1}\mathbf{d}\,. \tag{5.8}$$

We adopt a decision rule as before: assign **X** to group m if

$$t^2(\mathbf{X};\overline{\mathbf{X}}_m) = \sum_i a_i(X_i - \overline{X}_{im}), \quad i = 1, \ldots, p \tag{5.9a}$$

is the smallest value among all k computed values of $t^2(\mathbf{X};\overline{\mathbf{X}}_z)$.

We shall use the data in Table 4-1 to illustrate identification by Ottestad's generalized discriminant function. Let the soil chemical data in,

Quadrat	*Nitrate* *N*	*Ammoniacal* *N*	*Elementary* *k*
$\mathbf{X}'$	20	10	125 ,

describe a quadrat **X** whose group identity is sought. We wish to determine which of the three types, described in Table 4-1, represents the most likely parent population for **X**. The mean vectors of the types are given in the table,

Type	*Nitrate* *N*	*Ammoniacal* *N*	*Elementary* *k*
$\overline{\mathbf{X}}'_1$	7.92	17.28	79.39
$\overline{\mathbf{X}}'_2$	9.93	12.93	91.39
$\overline{\mathbf{X}}'_3$	16.77	8.54	117.24 .

We proceed through the following steps:

1. Derive the pooled covariance matrix and its inverse,

$$\mathbf{S} = \begin{bmatrix} -\ 9.20853 & -2.06334 & -27.5532 \\ -\ 2.06334 & 8.07972 & -\ 2.12072 \\ -27.5532 & -2.12072 & 679.971 \end{bmatrix}$$

$$\mathbf{S}^{-1} = \begin{bmatrix} 0.132981 & 0.0354032 & 0.00549897 \\ 0.0354032 & 0.133293 & 0.00185030 \\ 0.00549897 & 0.00185030 & 0.00169925 \end{bmatrix}.$$

2. Calculate the between group root mean squares of the variables (formula (5.7)),

$$\mathbf{d} = \begin{bmatrix} 14.0730 \\ 13.4487 \\ 59.1012 \end{bmatrix}.$$

For example, using an expanded form for V_i, we have

$$d_1 = (1/2(79.2^2/10 + 79.9^2/8 + 150.9^2/9 - 310.0^2/27))^{1/2}$$
$$= 14.0730 \; .$$

The other d_i are similarly calculated.
3. Decide on the sign of d_1, d_2, d_3. This is a peculiar step in Ottestad's algorithm. It is suggested that the user inspect the table for type means of species to get an idea how the means order the types. In our table *nitrate N* and *elementary k* order the types in an identical sense, but *ammoniacal N* orders them in an opposite sense. Accordingly, positive signs are assigned to d_1 and d_3 and a negative sign to d_2. We thus have,

$$\mathbf{d} = \begin{bmatrix} 14.0730 \\ -13.4487 \\ 59.1012 \end{bmatrix}.$$

When the differences in ordering are not as clear cut as in our case, arbitrary steps may be taken as prescribed by Ottestad (1975b) to resolve the problem.
5. Compute coefficients according to formula (5.8) to obtain,

$$\mathbf{a} = \mathbf{S}^{-1}\mathbf{d} = \begin{bmatrix} a_1 \\ a_2 \\ a_3 \end{bmatrix} = \begin{bmatrix} 1.72032 \\ -1.18504 \\ 0.15293 \end{bmatrix} .$$

These give us the discriminant functions with the appropriate type means substituted for $\bar{\mathbf{X}}_m$ in formula (5.9a):

$$t^2(\mathbf{X}; \bar{\mathbf{X}}_1) = 1.72032\,(X_1 - 7.92) - 1.18504\,(X_2 - 17.28) + 0.15293\,(X_3 - 79.39)$$

$$t^2(\mathbf{X}; \bar{\mathbf{X}}_2) = 1.72032\,(X_1 - 9.93) - 1.18504\,(X_2 - 12.93) + 0.15293\,(X_3 - 91.39)$$

$$t^2(\mathbf{X}; \bar{\mathbf{X}}_3) = 1.72032\,(X_1 - 16.77) - 1.18504\,(X_2 - 8.54) + 0.15293\,(X_3 - 117.24) \qquad (5.9b)$$

After substituting $X_1 = 20, X_2 = 10$ and $X_3 = 125$ in formula (5.9b), we get the distance scores,

Quadrat	*Distance from type* 1	2	3
X'	36.3837	25.9358	5.01321 .

Assignment of **X** to *Type 3* is indicated.

Having decided on the assignment, the question of how reliable is the decision arises. We can answer this by evaluating the goodness of the discriminant function on the basis of a method suggested by Ottestad (1975b). We have to measure the overlap between the clusters of points representing the types. For this, we define

$$y_{mj} = a_1 X_{1jm} + \ldots + a_p X_{pjm} \tag{5.10}$$

as the discriminant score of quadrat j of the sample in group m, and

$$\begin{aligned} OL_{mz} &= [\min(y_{mj}; j = 1, ..., n_m) - \max(y_{zj}; j = 1, ..., n_z)]/ \\ &\quad [\max(y_{mj}; j = 1, ..., n_m) - \min(y_{zj}; j = 1, ..., n_z)] \\ &= (a - b)/(c - d) \end{aligned} \tag{5.11}$$

as a measure of overlap. We assume that $(y_{\bar{X}_m} = a_1 \bar{X}_{1m} + ... + a_p \bar{X}_{pm}) \geqslant (y_{\bar{X}_z} = a_1 \bar{X}_{1z} + ... + a_p \bar{X}_{pz})$ in which $y_{\bar{X}_m}$ is the discriminant score of the centroid of group m and $y_{\bar{X}_z}$ is the same for group z. The graphical illustration is given in Fig. 5-4. The limits of OL_{mz} are -1 and $+1$. The lower limit

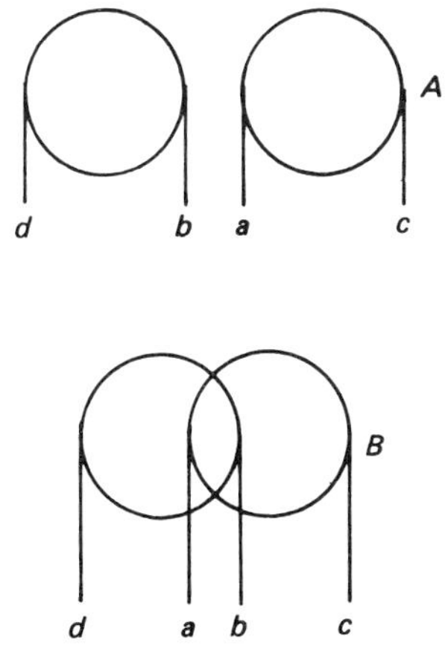

Fig. 5-4. Diagramatic representation of clusters between which the discrimination is sharp (*A*) or less sharp (*B*). Letters *a, b, c, d* indicate minima and maxima for the clusters on the best discriminating line. The method is described in the main text.

will be realized when there is zero discrimination, i.e. the two groups coincide ($a = d$ and $b = c$). The upper limit is realized when the discrimination is perfect, i.e. $(a = c) > (b = d)$. But this requires all replicates within a group to be identical and between groups different. Consequently, a negative value of OL_{mz} indicates overlap while a positive value indicates a gap between *Types m* and *z*. Using the same example as before, we have for the three types of Table 4.1:

Types	*Symbols in formula* (5.11)				OL_{mz}
m, z	*a*	*b*	*c*	*d*	$(a-b)/(c-d)$
2,1	2.45632	19.2505	27.5959	−4.49384	−0.52
3,1	30.2035	19.2505	46.3699	−4.49384	0.22
3,2	30.2035	27.5959	46.3699	2.45632	0.06

The discrimination power of the derived function increases with increasing values of OL_{mz}. It declines with decreasing values. We conclude that the discriminant function (formula (5.9b) is quite powerful between *Types 1* and *3*, somewhat less between *Types 2* and *3*, and not at all reliable between *Types 1* and *2*. We interpret the latter in the sense that if we were to choose between *Type 1* or *Type 2* as likely parent populations for **X**, we may not rely with confidence on the discriminant function as a basis for making the decision.

5.3 Rank order

Identification based on the generalized distance is burdened by the restrictive assumption of multivariate normality and a rather involved computational procedure. If nothing else, these aspects could explain its infrequent use in vegetation ecology. Rather than rely on methods in which identifications are based on the actual magnitude of a distance, the ecologist may consider using other methods which rely on rank order in a manner as suggested by Goodall (1968, 1970b). We have described several methods which use the order-related functions of Goodall (1970b) in Section 4.9. A variant is outlined below in connection with a numerical example:

1. Assuming that a sample of quadrats has been drawn from a vegetation type m whose sample mean vector is $\bar{\mathbf{X}}_m$, and that the elements of $\bar{\mathbf{X}}_m$ represent sample means $\bar{X}_{1m}, ..., \bar{X}_{tm}$ for species, compute the dissimilarity between a quadrat j in type m and the type mean vector $\bar{\mathbf{X}}_m$ with respect to species h as the quantity,

$$\delta_{hjm} = |X_{hjm} - \bar{X}_{hm}| \ .$$

Similarly, for the external quadrat **X** whose type affiliation is sought, compute

$$\delta_{hXm} = |X_{hX} - \bar{X}_{hm}| \ .$$

2. Compute the affinity of **X** and vegetation type m with respect to species h as the quantity,

$$p_{hXm} = P(\delta \geqslant \delta_{hXm}) \ .$$

This is the proportion of quadrats in type m which are at least as far away from the centroid of type m as the external quadrat **X** with respect to species h.

3. Determine the total affinity A_{Xm} of **X** by combining the proportions p_{hXm} for the t species as a simple product to obtain,

$$\chi^2{}_{Xm} = -2 \ln (p_{1Xm} \cdots p_{tXm})$$

and refer χ^2_{Xm} to the chi square distribution to obtain

$$A_{Xm} = P(\chi^2 \geqslant \chi^2{}_{Xm})$$

with $2t$ degrees of freedom. The affinity A_{Xm} in this case measures a probability that any member of type m is at least as dissimilar to the common type mean vector $\bar{\mathbf{X}}_m$ as **X** is to $\bar{\mathbf{X}}_m$. Since we form a simple product of probabilities, we must assume that the individual probabilities are additive, or equivalently, that the species are uncorrelated. Zero terms in the product will cause indeterminacy. To avoid this, the zeros should be omitted. When all terms are zero, A_{Xm} is set to 0.

4. Determine through similar manipulations the affinity of **X** to each vegetation type which may represent a parent population. Following this, identify m as the parent population for **X** if A_{Xm} is the highest affinity among all the affinity values. Alternatively, make the identification conditional on the actual size of A_{Xm} being in excess of a given threshold value α.

We shall base the illustrations on data given below:

Type A

Species	*Quadrat* 1	2	3	4	5	$\bar{X}_{hA}$
X_1	5	4	5	3	3	4
X_2	1	2	0	1	1	1
X_3	2	3	2	2	1	2

$$\mathbf{X} = \begin{bmatrix} 2 \\ 2 \\ 3 \end{bmatrix}.$$

At this point, we could replace the species by sets of component scores to remove the effect of species correlations from the measurement of affinities. However, for the purpose of illustration it will be simpler if we proceed with the data as presented. We begin with computing the δ_{hjA} values,

$$\begin{bmatrix} \delta_{11A} & \delta_{12A} & \delta_{13A} & \delta_{14A} & \delta_{15A} \\ \delta_{21A} & \delta_{22A} & \delta_{23A} & \delta_{24A} & \delta_{25A} \\ \delta_{31A} & \delta_{32A} & \delta_{33A} & \delta_{34A} & \delta_{35A} \end{bmatrix} = \begin{bmatrix} 1 & 0 & 1 & 1 & 1 \\ 0 & 1 & 1 & 0 & 0 \\ 0 & 1 & 0 & 0 & 1 \end{bmatrix}.$$

We calculate the δ_{hXA} values next,

$$\begin{bmatrix} \delta_{1XA} \\ \delta_{2XA} \\ \delta_{3XA} \end{bmatrix} = \begin{bmatrix} 2 \\ 1 \\ 1 \end{bmatrix}.$$

From these we obtain the p_{hXA} values,

$$\begin{bmatrix} p_{1XA} \\ p_{2XA} \\ p_{3XA} \end{bmatrix} = \begin{bmatrix} 0 \\ 2/5 \\ 2/5 \end{bmatrix}.$$

We find the affinity index, after omitting the single zero term and reducing the degrees of freedom from $2t$ to $2(t - 1)$ so that $\chi^2_{XA} = -2 \ln 4/25 = 3.665$ with 4 degrees of freedom, to be $A_{XA} = P(\chi^2 \geqslant 3.665) \approx 0.55$. When we set the critical α to 0.05 the condition $A_{XA} > \alpha$ is satisfied and *Type A* is declared a likely parent population for **X**. When several reference types are given, the assignment of **X** to one of them is based on the criteria which were discussed under item 4 in the preceding enumeration.

5.4 Information

Identifications may be based on the I-divergence information of formula (2.22) when the data are categorical. When a reference class is described as a p-valued probability distribution,

$$\mathbf{P}_j = (P_{j1} \ldots P_{jp}),$$

we can use the quantity,

$$2I(\mathbf{F}; \mathbf{P}_j) = 2 \sum_i f_i \ \ln f_i/(f_. P_{ji}), i = 1, \ldots, p \ ,$$

to evaluate the null hypothesis that the

$$\mathbf{F} = (f_i \ldots f_p)$$

represents a random sample from a population completely described by $\mathbf{P}_j$, against the alternative hypothesis that it does not. Under the null hypothesis

$2I(\mathbf{F}; \mathbf{P}_j)$ may be referred to the chi square distribution with $p - 1$ degrees of freedom. We note that the totals for distribution **F** and $\mathbf{P}_j$ are respectively $f.$ and 1.

Let the frequency distribution **F** = (58 62 70) describe a new quadrat whose class affiliation is in question. In **F**, the elements define counts of species occurrences in 80 sampling units. Let the distributions $\mathbf{P}_1$ = (0.260 0.300 0.440) and $\mathbf{P}_2$ = (0.318 0.290 0.392) specify the relative proportions of the same three species in two vegetation types which can be considered as possible parent populations for **F**. Our decision regarding the class affiliation of **F** may be formulated this way:

We assign **F** to Type 1 with α probability if the condition

$$2I(\mathbf{F};\mathbf{P}_1) < 2I(\mathbf{F};\mathbf{P}_2)$$

is true, provided that the conditions

$$2I(\mathbf{F};\mathbf{P}_1) < \chi^2_{\alpha,p-1} \text{ and}$$
$$2I(\mathbf{F};\mathbf{P}_2) \geqslant \chi^2_{\alpha,p-1}$$

are true also. In these expressions, $\chi^2_{\alpha,p-1}$ is the α probability point of the chi square distribution with $p - 1$ degrees of freedom. If these conditions are reversed, an assigment of **F** to type 2 is indicated. Furthermore, when $2I$ exceeds the corresponding chi square probability point in both types, no assignment may be made. Conversely, when they both are smaller than the stipulated probability point, an assignment to either type may be justified. In the example, we have the numerical values: $2I(\mathbf{F}; \mathbf{P}_1) = 4.187$ and $2I(\mathbf{F}; \mathbf{P}_2) = 1.203$. Since both of these quantities are smaller than the specified probability point $\chi^2_{0.05,2} = 5.991$, we conclude that an assignment to either type can be justified.

5.5 Bayesian analysis

In the preceding method, we measured the divergence of **F** from each reference type (or population in general) separately. If we followed suggestions by Machol & Singer (1971), or Baum & Lefkovitch (1972), we could have performed what they described as a simultaneous Bayesian analysis on **F**, $\mathbf{P}_1, \mathbf{P}_2$. The simultaneous comparison is based on the likelihood function,

$$L(\mathbf{F}; \mathbf{P}_1, \mathbf{P}_2) = \prod_i \left[\frac{P_{i1}}{P_{i2}}\right]^{f_i} \left[\frac{1-P_{i1}}{1-P_{i2}}\right]^{n-f_i}, \quad i = 1, ..., p \qquad (5.12)$$

where the product is for all p species in the sample. Symbol n signifies the number of sampling units in which occurrences f_i of species are counted in each quadrat. We also note that although the likelihood ratio, as formulated, assumes an underlying binomial distribution, extensions are possible to multinomial distributions in general (cf. Machol & Singer, 1971).

In the context of the example outlined in the previous section, and with n defined as 80, we have

$$\begin{aligned} L(\mathbf{F};\mathbf{P}_1,\mathbf{P}_2) &= ((0.260/0.318)^{58}((1-0.260)/(1-0.318))^{80-58}) \\ &\quad ((0.300/0.290)^{62}((1-0.300)/(1-0.290))^{80-62}) \\ &\quad ((0.440/0.392)^{70}((1-0.440)/(1-0.392))^{80-70}) \\ &= 0.461303 \quad . \end{aligned}$$

Machol & Singer (1971) observe that for function L to be less than unity, the frequency f_i in $\mathbf{F}$ should be high when the relative frequency P_{i2} is high and low when P_{i2} is low, provided that P_{i1} is low when P_{i2} is high and P_{i2} is low when P_{i1} is high. This is equivalent, of course, to saying that when L is less than unity, $\mathbf{F}$ is more similar to $\mathbf{P}_2$ than to $\mathbf{P}_1$, or when L is larger than unity, $\mathbf{F}$ is more similar to $\mathbf{P}_1$ than to $\mathbf{P}_2$. Based on these observations, Machol & Singer (1971) derive a simple decision rule which they feel formalizes what the taxonomists do intuitively when they contemplate assignments: assign $\mathbf{F}$ to the first population ($\mathbf{P}_1$) if L is 'considerably' greater than unity, or assign it to the second population ($\mathbf{P}_2$) if L is 'considerably smaller than unity. The meaning of 'considerably greater' or 'considerably smaller' is however not explicitly defined, but L would be regarded to indicate assignment to $\mathbf{P}_2$ if it were less than 0.01 and to $\mathbf{P}_1$ if it were greater than 1.00. In these terms, the value of 0.461303 is not sufficiently indicative to allow a clear cut decision in assigning $\mathbf{F}$ to one or the other of the reference groups.

While this procedure requires the knowledge of a priori probabilities of species presence in the reference types, the method used by Baum & Lefkovitch (1972) assumes that not only the a priori probabilities of species are known but also probabilities of the vegetation types in which the species occur. Giving probability $P(m)$ to *Type m* in our case means that we have estimated the proportionate representation of *Type m* (in quadrat-sized grid units) in the survey site from where quadrat $\mathbf{F}$ originates. In the example, we have two types ($m = 1, 2$). We shall take these to have probabilities $P(1) = 0.3$ and $P(2) = 0.7$ respectively. It is important to note that should we have $k > 2$ quadrats the $P(1), ..., P(k)$ values should still add up to unity. We shall proceed on the assumption that there are in fact k reference types, the mth of which is described by p 2-valued distributions,

$$\mathbf{P}(m) = \begin{bmatrix} \mathbf{P}(1,m) \\ \cdot \\ \mathbf{P}(p,m) \end{bmatrix} = \begin{bmatrix} P_{11m} & P_{12m} \\ \cdot & \cdot \\ P_{p1m} & P_{p2m} \end{bmatrix}.$$

A characteristic element P_{i1m} in the ith distribution ($\mathbf{P}_{im}$) gives the a priori probability of species i in a randomly chosen quadrat of *Type m*. In the example, $P_{311} = 0.440$. Since we are dealing with presence/absence, $P_{i2m} = 1 - P_{i1m}$. If we designate by f_{i1} the number of occurrences of species i in the n sampling units of quadrat **F** and define f_{i2} as $n - f_{i1}$, the p-valued vector

$$\mathbf{F} = \begin{bmatrix} \mathbf{F}(1) \\ \cdot \\ \mathbf{F}(P) \end{bmatrix} = \begin{bmatrix} f_{11} & f_{12} \\ \cdot & \cdot \\ f_{p1} & f_{p2} \end{bmatrix}$$

gives a complete description of quadrat **F**. We can give the quantity,

$$P(\mathbf{F}(i) \mid m) = P_{i1m}^{\;f_{i1}} \; P_{i2m}^{\;n-f_{i1}} \frac{n!}{f_{i1}\,!\;(n-f_{i1})\,!},$$

as the a priori probability of finding f_i units occupied by species i of n observed sampling units in a random quadrat of *Type m*. In other words, $P(\mathbf{F}(i) \mid m)$ is the a priori probability of a distribution $(f_{i1}\;\; n - f_{i1})$ in *Type m*. The joint a priori probability of p such distributions, i.e. the likelihood of a quadrat such as **F** in *Type m*, is

$$P(\mathbf{F}|m) = \prod_i P(\mathbf{F}(i)|m), \qquad i = 1, ..., p \; .$$

Bayes theorem (cf. Rényi, 1970) tells us that the a posteriori probability that we have in fact located a stand of *Type m*, when we select quadrat **F**, is

$$P(m\,|\mathbf{F}) = P(m)\; P(\mathbf{F}\,|m) / \sum_j P(j)\; P(\mathbf{F}\,|j)\,, \qquad j = 1, ..., k$$

in which m identifies the mth and j identifies the jth of k types. We shall show how to calculate this quantity. As the first step, present the sample estimates for species probabilities in the types:

$$\mathbf{P}(1) = \begin{bmatrix} 0.260 & 0.740 \\ 0.300 & 0.700 \\ 0.440 & 0.560 \end{bmatrix}$$

$$\mathbf{P}(2) = \begin{bmatrix} 0.318 & 0.682 \\ 0.290 & 0.710 \\ 0.392 & 0.608 \end{bmatrix} .$$

Then present the observed frequencies in **F**,

$$\mathbf{F} = \begin{bmatrix} \mathbf{F}(1) \\ \mathbf{F}(2) \\ \mathbf{F}(3) \end{bmatrix} = \begin{bmatrix} 58 & 22 \\ 62 & 18 \\ 70 & 10 \end{bmatrix}$$

The a priori probabilities of **F**(1), **F**(2), **F**(3), conditional on *Type 1*, are given by

$$P(\mathbf{F}(1) \mid 1) = 0.260^{58}\ 0.740^{22}\ \frac{80!}{58!\ 22!}$$

$$= 0.260^{58}\ 0.740^{22}\ \frac{69!\ (4.182353399)\ (10^{20})}{58!\ 22!}$$

$$= (4.210576471)\ (10^{-18})$$

$$P(\mathbf{F}(2) \mid 1) = 0.300^{62}\ 0.700^{18}\ \frac{80!}{62!\ 18!}$$

$$= 0.300^{62}\ 0.700^{18}\ \frac{69!\ (4.182353399)\ (10^{20})}{62!\ 18!}$$

$$= (2.206850132)\ (10^{-18})$$

$$P(\mathbf{F}(3) \mid 1) = 0.440^{70}\ 0.560^{10}\ \frac{80!}{70!\ 10!}$$

$$= 0.440^{70}\ 0.560^{10}\ \frac{69!\ (4.182353399)(10^{20})}{69!\ 10!\ (70)}$$

$$= (5.497019503)\ (10^{-16}) .$$

The product of the $P(\mathbf{F}(i) \mid 1)$ values gives the a priori probability for **F** in *Type 1*,

$$P(\mathbf{F} \mid 1) = (5.107891676)(10^{-51}) .$$

Proceeding to *Type 2*, we have the a priori probabilities,

$$P(\mathbf{F}(1) \mid 2) = 0.318^{58}\ 0.682^{22}\ \frac{80!}{58!\ 22!}$$

$$= 0.318^{58}\ 0.682^{22}\ \frac{69!}{58!\ 22!}\ (4.182353399)(10^{20})$$

$$= (8.256932689)\ (10^{-14})$$

$$P(\mathbf{F}(2) \mid 2) = 0.290^{62}\ 0.710^{18}\ \frac{80!}{62!\ 18!}$$

$$= 0.290^{62}\ 0.710^{18}\ \frac{69!}{62!\ 18!}\ (4.182353399)\ (10^{20})$$

$$= (3.48190145)\,(10^{-19})$$

$$P(\mathbf{F}(3) \mid 2) = 0.392^{70}\;0.608^{10}\;\frac{80!}{70!\;\;10!}$$

$$= 0.392^{70}\;0.608^{10}\;\frac{69!\;\;(4.182353399)\;(10^{20})}{69!\;10!\;\;(70)}$$

$$= (3.851413410)\,(10^{-19}).$$

The product of the $P(\mathbf{F}(i) \mid 2)$ values gives the a priori probability of **F** in *Type 2*,

$$P(\mathbf{F} \mid 1) = (1.107274650)(10^{-50})\,.$$

From the partial results we can compute the a posteriori probabilities, first for *Type 1* given quadrat **F**,

$$P(1 \mid \mathbf{F}) = \frac{(0.3)\,(5.107891676)\,(10^{-51})}{(0.3)\,(5.107891676)\,(10^{-51}) + (0.7)\,(1.107274650)\,(10^{-50})}$$

$$= \frac{(1.532367503)\,(10^{-51})}{(9.283290053)\,(10^{-51})}$$

$$= 0.165067287\,,$$

and then, for *Type 2* given quadrat **F**,

$$P(2 \mid \mathbf{F}) = \frac{(0.7)\,(1.107274650)\,(10^{-50})}{(0.3)\,(5.107891676)\,(10^{-51}) + (0.7)\,(1.107274650)\,(10^{-50})}$$

$$= \frac{(7.750922550)\,(10^{-51})}{(9.283290053)\,(10^{-51})}$$

$$= 0.834932713\;.$$

Since $k = 2$, $P(2 \mid \mathbf{F}) = 1 - P(1 \mid F)$ is a check on the calculations. Once the a posteriori probabilities are determined the simplest decision rule would stipulate assignment of **F** to the type for which $P(m \mid \mathbf{F})$ is largest. This provides a maximum likelihood assignment, the kind often used in biology (e.g. Jardine, 1971). Accordingly, we identify *Type 2* in the example as the most likely parent type for **F**. After the assignment is implemented we would expect the probability of misclassification to be $P(1 \mid \mathbf{F})$, or if there were k types, $\sum_j P(j \mid \mathbf{F})$ over all types $j \neq m$ where m is the type to which the assignment is made. When we compare this assignment with those in the preceding section, we can see that these sharply identify *Type 2* as a likely parent population for **F**. Baum & Lefkovitch (1972) relied on another kind of Bayesian (minimax) decision rule with unequal costs of misclassification.

A major limitation of the method is obvious. We need to know the type and/or the species probabilities or we must have estimates for them. Other problems with the methods in this and the previous section are considered below:

1. Strictly speaking $2I(\mathbf{F};\, \mathbf{P}_j)$, $L(\mathbf{F};\, \mathbf{P}_1, \mathbf{P}_2)$ and $P(m \mid \mathbf{F})$ should be used as statistics only when there is good reason to assume that the species in **F** are non-interacting in the sampled populations. This may rarely be the case, and therefore, we should normally anticipate the actual but unknown value of $2I(\mathbf{F};\mathbf{P}_j), L(\mathbf{F};\mathbf{P}_1, \mathbf{P}_2)$ and $P(m \mid \mathbf{F})$ to be less than their observed value.
2. It directly follows from the presence of species interactions that the assignments based on $2I(\mathbf{F};\, \mathbf{P}_j)$, $L(\mathbf{F};\, \mathbf{P}_1, \mathbf{P}_2)$ or $P(m \mid \mathbf{F})$ are likely to be conservative; we may decide to reject the null hypothesis on the basis of an observed value when in fact the actual values of $2I(\mathbf{F};\mathbf{P}_j), L(\mathbf{F};\mathbf{P}_1,\ \mathbf{P}_2)$ or $P(m \mid \mathbf{F})$ may indicate that it should be accepted.
3. We cannot tell whether **F**, as a random sample, happened to incorporate all the species which actually occur in the sampling site. Furthermore, **F** may display a species list which is richer or poorer than the species list in the reference types. These raise problems on different counts. Firstly, zeros are introduced into **F** and $\mathbf{P}_j$. However, the zeros cannot be uniquely interpreted; some will indicate that certain species, present in the sampling site, were missed by sampling, while others signify actual absences in the survey site. Secondly, when the zeros indicate that species were actually missed, the nominal value of $2I(\mathbf{F};\mathbf{P}_j)$, $L(\mathbf{F};\, \mathbf{P}_1, \mathbf{P}_2)$ or $P(m \mid \mathbf{F})$ can be either smaller or larger than the actual value. Since we do not know in which way the bias occurs, we cannot tell its influence in the decisions to accept or reject the null hypothesis.
4. The problems which arise because of indeterminacy in the presence of zeros dictate that identifications be based on sets of shared species. But then it should be clearly understood that the analysis will have relevance with respect only to these species.

The problems raised in the foregoing comments are in no way unique to identifications which use information or probability. The fact of the matter is that zeros in the data can lead to similar problems in most methods of data analysis.

5.6 Matching properties

Gower's (1974) predictive classification has been described in Section 4.9. When such a procedure produced the reference classes, a simple method of identification is available. The species list of the quadrat to be identified is compared with the species list of each reference class. The quadrat is assigned to the reference class giving the highest number of matches.

5.7 Dichotomous keys

In the unique cases of association analysis (Williams & Lambert, 1959) and in indicator species analysis (Hill, Bunce & Shaw, 1975), a dichotomously branching key is produced directly in the clustering phase of the analysis. This key relies either on the presence and absence of given species at a given level of the hierarchy, or on the indicator score of a quadrat and the indicator threshold. The methods are described in Section 4.9.

5.8 Further remarks on identification

In this chapter, we have assumed that a reference classification exists whose classes, or more precisely, the descriptors of the classes are inviolably final. Accordingly, through the process of identification, we could increase class membership but could not alter the class descriptors. We may of course use identification methods to reallocate individuals among classes and follow each reallocation with recomputation of the class descriptors as Rubin (1967) suggested.

When we wish to identify an object, we have to choose an affinity measure before we can implement an analysis to accomplish the objective of identification. Our choice may depend on the method of cluster analysis which produced the reference groups. If, for instance, the groups were recognized based on average linkage, it may be quite logical under specific circumstances to use average linkage as a criterion for measuring affinity. Similarly, if the groups were recognized on the basis of sum of squares, the same criterion may serve as a measure for affinity in identifications.

It is quite possible, however, that we recognize reference groups using one set of criteria, and then redescribe the groups in terms of some new criteria which then provide a basis for identification. When such is the case, the criteria which produce the reference groups have no relevance in the identifications, and a measure of affinity can be more freely chosen. The following examples will help illuminate these points:

Case 1. N quadrats are selected at random in an area of vegetation. Each quadrat is described in terms of species composition. The sample is subdivided into k groups by the method of sum of squares clustering. If $\mathbf{X}$ represents a new quadrat from the same area, described in terms of the same set of species, the minimum value of the function

$$\frac{n_m}{n_m + 1} \sum_h (X_h - \bar{X}_{hm})^2, \ m = 1, ..., k$$

is a logical criterion for identifying the best fitting group for **X**.

Case 2. Types are recognized based on the method of sum of squares clustering. Each type is then mapped and resampled in the field for the yield of given tree species within given sized quadrats. An additional quadrat is chosen whose type affiliation is sought in terms of the yield data and some affinity measure which, in this case, need not be the sum of squares. The reason for this is that the sum of squares no longer is a relevant criterion since the types are redescribed by new characters different from those which formed the type limits. Here the user has more freedom than in *Case 1* to choose an affinity measure. When there is good reason to commit the analysis to a covariance structure, the generalized distance of formula (5.1) or (5.5) is a logical choice. When, however, there is no compelling reason to prefer one affinity measure over another, identifications based on the order-related resemblance measures of Goodall (1970b) or the order-free measures of information may have special appeal.

Our approach to identification, whether probabilistic or deterministic, will depend on the decision whether or not the descriptors of the reference types are population parameters, or simply just estimates of the population parameters. In vegetation surveys, the identification problem is likely to come up in a probabilistic sense. It is however conceivable that identification will have to be deterministic in mapping or other exercises where assignments are made irrespective of the nominal value of an individual's affinity to the reference class to which its affinity is the greatest.

Chapter 6

MULTIVARIATE ANALYSIS – A DISCUSSION

In this chapter we give an overview of the materials which we presented in the preceding chapters. While we focus on such points which concern the choice of method, we also offer comments on the present state of the Art and provide bibliographic information.

6.1 Choice of method

We have described methods which have at least one property in common: they are suitable to analyse multidimensional data and thus to reveal information about correlations. In the measurement of correlations we relied on different objective functions that are based on, or can be derived from, the covariance, information or probability. We characterized the methods as 'multivariate' to signify that they deal with several correlated variables simultaneously.

Multivariate methods are characterized by certain revealing, common properties. We follow Szőcs (1973) in summarizing the major features in an attempt to place the methods in a broader context:

1. Multivariate methods use *multidimensional data* (Section 1.5) in which individuals, such as quadrats of vegetation, are described by p-valued vectors in which the elements are measurements on p variables, such as p distinguishable species. It is assumed that the individuals in the population are discrete units that can be recognized without ambiguity. It is also assumed, while allowing correlations between them, that the variables are distinct, measurable properties of objects (Section 1.6).
2. Multivariate methods do not, as a rule, associate different variables with different a priori values of importance. As a matter of fact, variables are normally treated alike in the same analysis. If different importance is to be attached to different variables, the data are manipulated to introduce differential weights prior to the analysis (Section 1.6).
3. Multivariate methods are concerned with correlated variables in the broad sense (Chapter 2). Should variables in a sample be independent, no real advantage could be gained by applying multivariate analysis.
4. Multivariate methods stipulate or assume the functional form taken by variate correlations (Section 3.2). This form is very often linear (Section 3.4), although as we have seen, there is an expanding armoury of new techniques that can handle non-linearly correlated variables (Sections 3.5, 3.6). Among these, the potentially most applicable are those that do not restrict non-linearity to a particular form.
5. Multivariate methods may or may not require assumptions about the underlying data structure (Section 3.2). If a specific type of data structure is assumed, it is most often one which is generated when the underlying dis-

tribution of the variables is multivariate normal.

6. The multivariate methods which we described in this book are *parametric* in the broad sense that they utilize actual measurements. Sometimes, they are also parametric in the statistical sense that they assume a certain type of probability distribution in the data. The distribution most often assumed is the p-dimensional normal distribution (Section 1.6).

7. Multivariate methods normally assume a data structure that does not incorporate the time dimension. This is definitely so with the methods which we described.

8. Multivariate methods are extremely time consuming to compute. They should not normally be contemplated without access to modern computational equipment.

To appreciate what methods of this sort can do for the ecologist, it is sufficient to consider what data analysis would be like if applications had to rely entirely on the univariate techniques. Firstly, variables would be analysed individually. Secondly, variable correlations would be ignored. And thirdly, all conclusions would be limited to a single variable at a time.

When we analyse variables individually, efficiency suffers. Consider, for example, a comparison between several diagnostic categories such as the different vegetation types in an area. We may wish to compare the vegetation types under the null hypothesis that they are indistinguishable on the basis of p environmental variables. Should we decide to follow a univariate approach in the testing, we would proceed as follows:

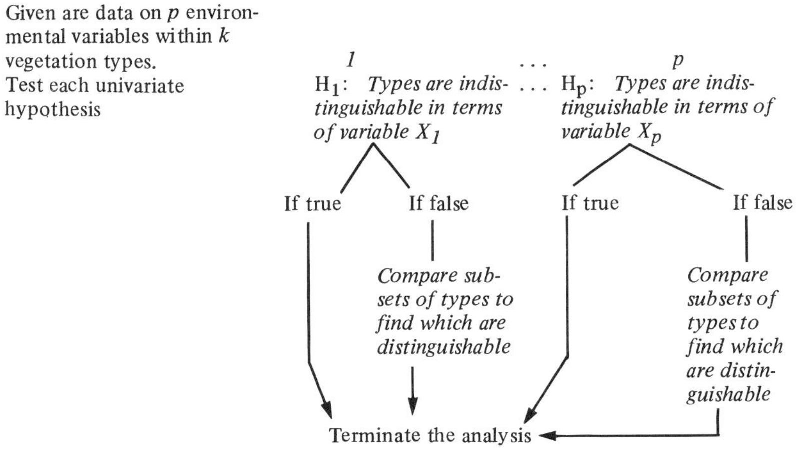

Given are data on p environmental variables within k vegetation types.
Test each univariate hypothesis
1
...
p
H1: Types are indistinguishable in terms of variable X1
...
Hp: Types are indistinguishable in terms of variable Xp
If true
If false
If true
If false
Compare subsets of types to find which are distinguishable
Compare subsets of types to find which are distinguishable
Terminate the analysis

It is quite obvious that the univariate test will have to be repeated for each variable separately. When the overall test on a single variable in the k types leads to rejection of the null hypothesis, the analysis is continued with comparison of the types in subsets to discover which of them differ significantly and which do not. Considering k types, the last step may at worst require $k(k-1)/2$ paired comparisons on each variable. An essential feature of these tests, and a reason for their potential inefficiency, is their reliance on a single variable at a time, even though the problem itself is typically multivariate. A more parsimonious analysis may, in the first place, subject all k types to a single test on a *multivariate null hypothesis* based on the use of all p variables simultaneously, in the same manner as illustrated in Section 4.9. Only if the null hypothesis is rejected should testing continue with more specific hypotheses – some of which may be univariate.

The *logistic advantage* of a multivariate approach is quite obvious. But besides the increased logistic efficiency, there are also other advantages to be gained as a direct consequence of analysing variables as correlated entities in a group. The *power* of the analysis increases, and in turn, the conclusions of the analysis become broad in relevance. Clearly then, what makes the multivariate methods so well suited for use in vegetation studies is the important fact that they are efficient, can convey information about correlations, and their results have wide relevance.

The multivariate techniques, however, just as any other method of data analysis, cannot be utilized to their fullest advantage, unless they are embedded in a general procedure characterized by strict internal consistency. The statements –

1. Definition of the problem in workable terms.
2. Data collection based on a statistical sampling design.
3. Data analysis based on a multivariate method.
4. Statistical inference.

– exemplify a procedure for the *embedding* of a multivariate method. We realize, of course, that the parts of this procedure are interrelated. The choice of a method of analysis must follow constraints imposed on it by decisions regarding the problem, sampling design, and method of inference.

The *mode of implementation* of a multivariate method depends on local circumstances. We may reason along two basically different lines in this connection. If all population units can be located and measured, or the sample has not been derived by random sampling, the approach necessarily is deterministic, and the multivariate methods must be applied as mathematical techniques with the sole objective of deriving precise descriptions of the materials at hand. The question of statistical estimation and inference does not arise. If, however, the sample has been randomly chosen, the approach will have to be statistical involving an estimation of unknown population parameters and tests of null hypotheses about them. However, it should be clear that in either case the underlying mathematical technique may be precisely the same, only the manner of implementation different.

It must be understood that when we speak of consistent procedures, we do not necessarily think of the formal procedures of mathematical statistics which preclude investigation of certain problems because their circumstances do not conform with conditions stipulated by theory. But we have in mind procedures of considerably more flexibility which are capable of being adopted to problems of any sort, broad or narrow, general or specific. *Our emphasis is on consistency in the sense of an orderly conduct of research programs from sampling to data analysis and inference.*

We must choose methods with a clear understanding of the objectives to be achieved. Poor judgement may result in failure to contribute significantly to the solution of the problem (Jeffers, 1967). We may of course recognize as objectives such broad aspects as summarization, multidimensional scaling, trend seeking, cluster recognition, identification, etc., or such specific aspects as may be involved when attempting to solve very specific problems. When summarization is accomplished on axes – and also in the case of multidimensional scaling, reciprocal ordering, or trend seeking – the assumption of a continuous data structure is implicit. Axes are sought, representing straight lines or curves, that can serve as a basis for simple, efficient representations. The methods that have primary relevance in this connection are known under the broad, collective term 'ordinations'. Class recognition is a matter for classification techniques which may dissect a perfectly continuous sample arbitrarily into groups, or delimit only groups separated by clear discontinuities. In certain cases, identifications may be sought. These entail the finding of classes which represent the most probable parent populations for unassigned individuals.

Once the relevant class of method is identified, a further choice within the class may depend on the data structure in the sample. Four broad categories may be recognized:

1. In the first category the data structures are linear and continuous. These mean that the points form an ellipsoidal cluster in sample space. Although such data structures are believed to be rare in vegetation samples, it is conceivable that a sample from a relatively narrow segment of the dominant environmental gradient may exhibit a nearly linear, continuous data structure.
2. The data structures in the second category are curved and continuous, with points arranged evenly within a single curved solid or on the surface of some manifold. It may be suggested that this type of structure occurs in samples taken from broad segments of the dominant environmental gradient.
3. The third category includes data structures which contain several linear clusters separated by discontinuities. This may be the case when a sample encompasses several short segments of several major gradients.
4. The fourth category contains data structures with points arranged in several curved clusters. Such curved and continuous data structures are probably the commonest in samples originating in vegetation surveys encompassing broadly different environmental conditions.

We understood, of course, that the foregoing categorization of data structure, while relevant in the ordination methods which we described, may have little relevance in other analyses. The reason being that in certain methods the concept of linearity, or as a matter of fact nonlinearity, may have no significance. Some classificatory techniques are examples in this connection in which only the continuity or discontinuity property is relevant.

When ordination is intended, several specific objectives may be served on which the choice of a particular method will have to depend. Some contingencies are given below:

Objective to be achieved	Data structure in sample	Method indicated
Data reduction through summarization of variation on axes	Linear, continuous	Component analysis
	Curved, continuous	Shepard & Carroll (1966), Noy-Meir (1974b), Sammon (1969), McDonald (1962, 1967), Phillips (1978)
	Linear or curved, disjoint	Sneath (1966b)*
Multidimensional scaling	As above, except that when linearity is suspected principal axes analysis is indicated. Other linear scaling techniques may be useful under specific circumstances. For instance, when it is essential to preselect the ordination poles the modified Bray & Curtis method is a logical choice. When there is concern about the computational difficulties with principal axes analysis, the use of other simple methods (e.g. position vectors ordination) may be appropriate. It is conceivable, that under certain circumstances, we may wish to do an ordination in which the ordination configuration is a given non-linear function of the actual data structure in the sample. When such is the case, we may consider applying some variants of the Kruskal method (Kruskal & Carmone, 1971).	
Trend seeking		
– a single species, several environmental variables	Linear or curved, continuous	Multiple regression analysis
– several species	Linear, continuous	Component analysis, principal axes analysis (Gower, 1966), orthogonal functions (Ottestad, 1975a).
	Curved, continuous	Shepard & Carroll (1966). Noy-Meir (1974b), Johnson (1973), Gauch, Chase & Whittaker (1974), Phillips (1978)
	Linear or curved, disjoint	Sneath (1966b)*
Reciprocal ordering		
– species scores given	Linear, continuous	Whittaker (1948), Williams (1952)
– species scores to be estimated from species abundance data	Linear, continuous	RQ-technique (Williams, 1952, Hatheway, 1971, Hill, 1973b)

* There is no sufficient evidence to reliably appraise this method in vegetation studies.

It should be clear that ordinations, in general, are suitable for handling continuous data structures, and conversely, a classificatory analysis is not a strictly logical objective in an ordination. We have quoted reasons for this in the preceding chapters. It will be sufficient here to stress one further point. If we have ordination objectives, and a discontinuous data structure in the sample, it may be wise to classify the sample first and then ordinate within the classes. Since we have to make decisions about the data structure, whose exact shape may not be known, it may happen that the method of ordination or classification which is chosen first is not the best method available. We may have to make several trials before a satisfactory method is found. 'Satisfactory' means that the method, besides satisfying certain norms of methodological acceptability, is also valid statistically and acceptable ecologically. The idea of subjecting the same data set to repeated analyses by different methods is generally advocated (e.g. Tukey & Wilk, 1966). The reason given is that in this way, different aspects of the data structure can be revealed, thereby supplying the user with possibly new, non-redundant information (cf. Crovello, 1970) about the materials under investigation.

In the group of classificatory analyses the methods (Chapter 4) can accomplish their task in different ways. The hierarchical techniques will normally produce classes irrespective of the existence or non-existence of discrete groups in the sample. To this extent, the hierarchical techniques are prone to produce dissections. However, we have seen already that while certain techniques will produce dissections, others can in fact find groups if groups actually exist or will signal the user if there are no groups that would satisfy given criteria. We cannot tell which technique is potentially most useful without first considering the local conditions of application, but we can suggest broad possibilities:

Method	*Characterization*
Q – clustering	Classification or dissection of a sample produces hierarchical or fixed-number group arrangements based on direct measurement of resemblance between objects.
– average linkage and other centroid techniques	Fusions depend on the average or centroid distance of groups unweighted by group size (Sokal & Michener, 1958; Gower, 1967b).
– minimum variance methods	Fusions depend on the squared centroid distance of groups weighted by group size which produce spherical, well-balanced groups (Ward, 1963; Edwards & Cavalli-Sforza, 1965; Orlóci, 1967a).
– single linkage methods	Fusions produce clusters of varying shape, separated by gaps (discontinuities) equal to a current state of the neighbourhood radius (Sneath, 1957b; Jardine & Sibson, 1971; Orlóci, 1976b).
– fixed group-number clustering	Subdivisions dissect the sample into a specified number of groups whose centroid distances (weighted or unweighted) are maximized (Jancey, 1966a); the positive matches between the descriptors of individuals and the descriptors of their respective clusters are maximized (Gower, 1974); or some other preference relation is maximized (Rubin, 1966).
– relocative clustering	Groups are produced aggregatively from objects for which the statistical hypothesis of a common parent population is tenable (Goodall, 1970b); groups are produced rejectively, for each of which the statistical hypothesis of a common parent population for the member objects is tenable (Goodall, 1970b); or groups are produced in a combined agglomerative and divisive procedure within which the statistical hypothesis of homogeneity holds true.
R – clustering	Classification or dissection of a sample produces hierarchical

Method	Characterization
	or non-hierarchical arrangements based on presence/absence, or indicator value of certain attributes.
– successive clustering	Subdivisions produce groups of objects, one at a time, in which variable correlations are minimized (Goodall, 1953).
– simultaneous clustering	The sample is subdivided into groups of objects based on the possession or lack of strategically selected attributes which maximize a function of association (Williams & Lambert, 1959), a function of indicator value (Hill, Bunce & Shaw, 1975) or other functions (e.g. Crawford & Wishart, 1967; Pielou, 1969b); some may maximize concentration (Lambert & Williams, 1962; Tharu & Williams, 1966). The clustering may also be inosculate, combining subdivisions based on presence and absence of attributes in objects to form groups, and ordinations to order the objects within the groups (Dale & Anderson, 1973).
Gradient and other ordination-based clustering	Groups are recognized based on affinities of variates or objects with ordination axes (e.g. Noy-Meir, 1971b, 1973b); based on locating sheaves of position vectors (Yarranton, Beasleigh, Morrison & Shafi, 1972); or through locating points in the neighbourhoods of local maxima on a density surface (Ihm, 1965; Katz & Rohlf, 1973).
Predictive clustering	Groups of objects are produced in which the states of certain variables are maximally predictive of the states of the same or other variables in the type in which the objects originate (Macnaughton-Smith, 1963, 1965).

The methods mentioned above are rich in technique which, under appropriate circumstances, can yield optimal solutions. There are nevertheless clustering methods whose aim is more fundamental in that they will sacrifice optimality to render the analysis of excessively large samples feasible in the face of the alternative of not performing any analysis on the samples.

Identification (Chapter 5) involves the measurement of affinity between a reference class and an external individual, and then, assignment of the individual to the class with which its affinity is greatest. The choice of an affinity measure is crucial. Two distinct cases must be recognized. In the first case, identifications are based on the variables and measures which formed the reference classes. For instance, when the reference classes are the product of average linkage clustering of species data, the identifications may rely on average linkage, as a measure of affinity, and the same kind of species data. In the second case, identifications are sought in terms of variables and measures which have not been used in the delimitation of the reference classes. An example in this connection includes several environmental classes to which new sampling units (quadrats) are assigned on the basis of vegetation data. While in the first case the affinity measure is predetermined, in the second it can be more freely chosen. We can make some points on technique using selected cases:

Methods	*Remarks*
Generalized distance	Underlying multivariate normal populations are assumed. This assumption may prove too restrictive.
Discriminant function	Same assumption applies as above plus the assumption of equal covariance matrices. The last is rarely warranted. The methods may handle groups in pairs (Fisher, 1936; Rao, 1952) or more than two groups at a time (Ottestad, 1975b).
Rank order	Identification is based on the probability of an object belonging to a reference group (e.g. Goodall, 1968, 1970b). Reference classes are assumed large and their exact contents known.
I-divergence information	Categorical variables are assumed. Identification is based on evaluation of the statistical hypothesis that the object in question comes from a population completely described by the frequency distribution of the reference class (e.g. Orlóci, 1972b).
Bayesian analysis	The likelihood of an object possessing a set of attribute states with given probabilities is expressed, and the likelihood of the type conditional on the occurrence of a given object is determined (Machol & Singer, 1971; Baum & Lefkovitch, 1972). Independent probabilities of species are assumed and the relative proportions of reference types (reference classes) may have to be known.
Matching properties	Attribute lists are compared. An object is identified as a member of the class with which it has matching properties in excess of a specified number (Gower, 1974).
Dichotomous keys	Presence or absence of attributes (Williams & Lambert, 1959), or possession or lack of indicator attributes (Hill, Bunce & Shaw, 1975) identifies, dichotomies in a hierarchical arrangement.

We have used in the text different formal procedures for testing hypotheses which rely on statistical indices of known sampling distributions including F, χ^2, etc. While such indices can provide elegant tests, their scope of application is rather limited for reasons of implicit assumptions about distributions. A method of testing statistical significance, however, need not be based on a standard statistical index. It is quite possible to proceed heuristically involving random simulations in which a sampling distribution is determined empirically. We have seen an example of this when we compared dendrograms in section 4.10.

6.2 Observations regarding data and method

When the ecologist decides to apply some multivariate method, the decision calls into question the fundamental aspects of understanding not only the immediate objectives, but also the broader circumstances of the entire application. Quite obviously, in this connection, Anderberg's (1973) comments have general relevance. Along similar lines we offer the following thoughts:

1. No method should be claimed to represent a general preference strategy in the analysis of vegetational data. If the analysis of any data set is limited to one method, some possibly important information will go undetected. Conversely, if different methods are used, they may reveal information about different aspects of the data structure (cf. Walker, 1974). Because of this, it may be best to go along with the suggestions of Tukey & Wilk (1966) who regard data analysis as an open-ended, interactive, iterative procedure in which the user exploits different avenues of approach in the hope of gaining greater insight into the data.

2. If the application of different methods leads to similar conclusions about a certain property of the data structure, the probability that the property exists, not as an artifact of the analysis, but in reality, is largely increased. Classifications represent a case in point. If conceptually different clustering algorithms tend to produce the same divisions in the data, the idea that the divisions are real is readily accepted. The converse is of course untrue, for not unexpectedly differences in subdivisions by different methods may mean that (i) the data structure is continuous, (ii) some of the methods perform dissections, or (iii) different methods reveal different aspects of the group structure.

3. The results of multivariate analysis are not necessarily validated by the fact that the method and equipment used are sophisticated in design or application. The validity will depend on the broader circumstances included in the sampling method, selection of variables, measuring instrument used, transformations, etc. What follows directly from these is that the ecologist should not feel a personal attachment to the method of analysis or to the results it produces, either because of respect for the method's originator or its successes in previous applications, to the extent that it would prevent critical examinations.

4. Axiomatic properties are important and any method of data analysis should be consistent in this respect. It is not uncommon, however, to find methods in use for which such properties are unknown. This is a disadvantage which renders a method's evaluation entirely dependent on the indirect means of comparing its results to known facts, or to some conceptual construct with which the results are expected to be concordant. In contrast to this situation, the methods that are well-known are capable of evaluation directly on axiomatic grounds, and also indirectly based on the examination of the results.

5. When the ecologist evaluates the results of an analysis, the main interest shifts to consistency with known facts and to correlation with known trends and classes in the variation of the environment or vegetation, in the context of new data. By no means should the validation of results depend on the data from which they are generated, or on the method of which they are an outcome. The latter point may of course be further qualified by saying that the method is unimportant as long as it is free of inconsistencies in the logic and free of mistakes in the data and arithmetic. Weaknesses in these will render the results useless, unworthy for further evaluation.

6. Results emitting from multivariate analyses are of two kinds. One kind includes the numerical values that are facts, and if not erroneous, undisput-

able. The sample mean, variance, etc. are examples. The other kind is what the user derives from the results through the mental process that puts numerical values into context and perspective. The latter are the basis of interpretations and decisions if further action is required.

7. Ordinations or classifications need not mean solutions to a specific problem. Often, they are means only of displaying the information in the data. The advantage is still obvious, for when information is displayed the need for revisions in the method may be indicated, or problems may be recognized for further investigation. When display alone is the objective, it is best to begin the analysis with methods that are simple and inexpensive to use.

8. It should not be forgotten that multivariate methods not only reveal information about the data structure, but often, they change the data structure in the course of the analyses and provide information about the new structure. Changes come about through standardizations or other adjustments of the data (e.g. Orlóci, 1967a) which can produce a new structure completely unlike the original. If such is the case, the interpretation of results is directly relevant to the new data structure, and only indirectly to the original via the functions which were operative in bringing about structural changes. Methods which use the correlation coefficient, single or double standardizations, condensations into frequency distributions, etc. are typical examples. Furthermore, the methods may be predisposed through their algorithm to respond to classes or trends of variation of a certain kind. The axes in some ordinations, for instance, respond only to linear trends, and certain clustering methods can produce only spherical clusters. These may or may not be regarded as advantages depending on the data and objectives.

9. It is a fact that multivariate analysis is often performed on samples from broad populations with sampling intensity being very low. This then means that certain trends or classes of variation which the ecologist may believe to exist in the population, may not receive sufficient representation in the sample. The implication is obvious: the results of the analysis will have limited relevance. Ecologists proposed many methods of overcoming the sampling problem through preferential selection of unity (cf. Westhoff & van der Maarel, 1973) or through the use of variants of restricted and stratified random sampling (e.g. Greig-Smith, 1964). However, no universally satisfactory solution has yet been found. If we choose preferential sampling, we may obtain a sample that will represent an imaginary population which we conceived and not the population in which we sample. If, on the other hand, we decide to incorporate a random element of choice, the sampling may become far too tedious or outright impractical.

With these thoughts in mind let us move on to the last section of the main text, the Prognosis.

6.3 Prognosis

The last few decades have seen the rise of formal statistical theories, some of which readily translated into useful methodologies in data analysis. Concurrently with these, new methods of data analysis evolved, in vegetation science and elsewhere, without much statistics, but with the objective, as Tukey & Wilk (1966) notes, to lay open the data to display the 'unexpected'.

With the new theories and methodologies, a novel and essentially objective outlook is emerging in vegetation studies to serve new purposes. The trend, in fact, seems to point to wide-spread future uses of methods founded upon mathematics and probability theory, in association with elaborate computer-based systems of sampling, data analysis and inference. The principal value of such systems, as noted by Shubik (1954) in connection with political behaviour, is that it should render the complex problems manageable, while also providing facilities for exact repetition, and also, precision and speed in the computations.

The availability of electronic data processing has lead and continues to lead to new directions in research. On the one hand, many of the conventional procedures of vegetation analysis are automated; in this way their efficiency is increased. On the other hand, new questions are asked about trends and classes of variation which would have been extremely impractical in the past. These have sometimes produced zealous, over-enthusiastic applications of methods, often under inappropriate circumstances (cf. Greig-Smith, 1964).

It appears that the strictly statistical, probabilistic aspects of vegetation analysis may have to be brought into closer focus in the future than they were in the past. Goodall (1970b) sees advantages to be gained through statements of clear-cut hypotheses and methods capable of performing tests on them. While we can whole-heartedly agree with this view, we may also add that with the hypothesis testing aspect vegetation science will experience a shift of its main thrust from the descriptive uses of multivariate techniques to statistical problem solving. Whereas hypothesis development, in Goodall's (1970b) words, '... can take place under more congenial conditions where requirements for full statistical rigour do not apply and intuition can be given free reign', hypothesis testing will have '... no place for intuition and demands that the special requirements for the testing method, usually including random sampling from a definite population, be met'. Goodall (1970b) foresees fruitful models and hypotheses in statistical ecology '... to be based on biological rather than purely mathematical premises'. He also believes that in 'statistical ecology in the past, mathematical and statistical considerations have too often played the role of master rather than servant' and that 'the subject is likely to mark time until its biological aspects resume their rightful place'.

We share the sentiments expressed by Pielou (1971) in her informal comments which we shall paraphrase to contrast 'soft science' and 'hard science':

An example of hard science is celestial mechanics. It provides astronomers with the means to make predictions about comets, like Halley's. The astronomer's predictions are precise. The comet proves it when it turns up at the right place on time. This is what we expect from a hard science. Its predictions are exact and confident. Ecology is a soft science. What this really means is that in ecology we cannot normally predict things with such high degrees of accuracy. The difficulties are rooted in the fact that good ecological data are difficult to obtain, that ecological data carry excessive random variation which can obscure the things about which we wish to make predictions, and that the ecological objects do not have the good behaviour, like comets. Furthermore, ecological problems are usually complex. They cannot be reasoned out as neatly as the orbital path of comets or the trajectory of a space module.

These thoughts should warn against overestimating the limits to which statistical and other mathematical methods can be put to useful service in plant ecology or in other domains of *soft science.* The crucial criterion is, of course, usefulness which can be tested only on an empirical basis. To this effect, the last word is that of the practising ecologists who are qualified only to answer the questions posed by Greig-Smith (1971a): Does a given multivariate method prove practical in use? Can the results be interpreted more readily, more exactly and with greater certainty than those obtained by other means?

6.4 Bibliographic notes

An exhaustive list of publications using mathematical methods in vegetation ecology and in related fields, up to 1962, has been compiled by Goodall (1962). Recent surveys appear in various textbooks, monographs and articles, including Sokal & Sneath (1963), Sneath & Sokal (1973), Greig-Smith (1964), Lambert & Dale (1964), Kershaw (1964), Williams & Dale (1965), McIntosh (1967b, 1973), Goodall (1970a), Crovello (1970), Crovello & Moss (1971), Shimwell (1971), Williams (1971), Blackith & Reyment (1971) and Szőcs (1972), among others. A series of essays by Juhász-Nagy (1966a,b, 1968), treating the conceptual framework, provides references on that topic. McIntosh (1970) establishes connections to and provides references on aspects of experimental plant biology on a conceptual basis, and Tukey (1962) interprets the general statistical (mathematical) framework in terms of an outlook less alien to data analysts in ecology than most of the conventional interpretations. Regarding general procedure, readers will find the commentaries of Sokal (1965), Lambert & Dale (1964), Williams & Dale (1965) and Greig-Smith (1971a) most instructive.

The underlying mathematical and/or statistical theory is explained by: Searle (1966) and Rektorys (1969) for mathematics; Sampford (1962) for sampling theory; Ostle (1963), Sokal & Rohlf (1969) and Dagnelie (1969) for univariate statistical methods; Rényi (1970) and Feller (1968, 1971) for probability theory; Kullback (1959) and Kullback, Kupperman & Ku (1962) for information theory; Seal (1964), Morrison (1967), Anderson (1958), Bartlett (1965), Dempster (1969) and Rao (1972) for multivariate statistical methods; and Bartlett (1960) and Pielou (1969a, 1974) for stochastic population processes and pattern. Computer programs and schemes for organizing data analysis are described by Cooley & Lohnes (1971), Grench & Thacher (1965), Williams & Lance (1965), Lance & Williams (1965, 1966a, 1967a, 1968), Wishart (1969a), Cole (1969), Rao (1971), Goldstein & Grigal (1972), Gauch (1971, 1977), Moore (1972), Hall (1965), Romane, Bacou, David, Godron & Lepart (1974), among others.

REFERENCES

(Numbers between square brackets indicate chapters of occurrence.)

Abramowitz, M. & I.A. Stegun. 1970. Handbook of Mathematical Functions. (9th printing). U.S. Department of Commerce, National Bureau of Standards, Applied Math. Series 55, Superintendent of Documents, U.S.A. Government Printing Office, Wash. D.C. 20402. [1]

Adam, P., H.J.B. Birks, B. Huntley & I.C. Prentice. 1975. Phytosociological studies at Malham Tarn Moss and Fen, Yorkshire, England. Vegetatio 30:117-132. [3, 4]

Agnew, A.D.Q. 1961. The ecology of Juncus effusus L. in North Wales. J. Ecol. 49:83-102. [3, 4]

Allen, T.F.H. 1971. Multivariate approaches to the ecology of algae on terrestrial rock surfaces in North Wales. J. Ecol. 59:803-826. [3]

Allen, T.F.H. & J.F. Koonce. 1973. Multivariate approaches to algal stratagems and tactics in systems analysis of phytoplankton. Ecology 54:1234-1246. [3, 4]

Allen, T.F.H. & S. Skagen. 1973. Multivariate geometry as an approach to algal community analysis. Br. phycol. J. 8:267-287. [3]

Anderberg, M.R. 1973. Cluster Analysis for Applications. New York, Academic Press. [1, 2, 3, 4, 6]

Anderson, A.J.B. 1971a. Similarity measure for mixed attribute types. Nature 232: 416-417. [2, 3]

Anderson, A.J.B. 1971b. Ordination methods in ecology. J. Ecol. 59:713-726. [3]

Anderson, D.J. 1965. Classification and ordination in vegetation science: controversy over a non-existent problem? J. Ecol., 53:521-526. [4]

Anderson, T.W. 1958. An Introduction to Multivariate Statistical Analysis. New York, Wiley. [4, 5, 6]

Anderson, T.W. 1963. Asymptotic theory for principal component analysis. Ann. Math. Statist. 34:122-148. [4]

Armstrong, J.S. 1967. Derivation of theory by means of factor analysis or Tom Swift and his electric factor analysis machine. Am. Stat. 21:17-21. [3]

Arno, S.F. & J.R. Habeck. 1972. Ecology of alpine larch (Larix lyallii Parl.) in the Pacific Northwest. Ecol. Monogr. 42:417-450. [3]

Auclair, A.N. & G. Cottam. 1973. Multivariate analysis of radial growth of black cherry (Prunus serotina Erhr.) in southern Wisconsin oak forests. Amer. Midl. Natur. 89:408-425. [3]

Austin, M.P. 1968. An ordination study of a chalk-grassland community. J. Ecol. 56:739-757. [3]

Austin, M.P. 1970. An applied ecological example of mixed-data classification. In: R.S. Anderssen & M.R. Osborne (eds.), Data Representation, pp. 113-117. Melbourne, Queensland Univ. Press. [3]

Austin, M.P. 1971. Role of regression analysis in plant ecology. Proc. Ecol. Soc. Aust. 6:63-75. [3]

Austin, M.P. 1972. Models and analysis of descriptive vegetation data. In: J.N.R. Jeffers (ed.), Mathematical Models in Ecology, pp. 61-86. London. Blackwell. [3]

Austin, M.P. 1976a. On non-linear species response models in ordination. Vegetatio 33:33-41. [3]

Austin, M.P. 1976b. Performance of four ordination techniques assuming different non-linear species response models. Vegetatio 33:43-49. [3]

Austin, M.P. 1977. Problems of non-linearity in ordination. Vegetatio 35:000-000. [3]

Austin, M.P., P.S. Ashton & P. Greig-Smith. 1972. The application of quantitative methods to vegetation survey. III. Re-examination of rain forest data from Brunei. J. Ecol. 60:305-324. [3, 4]

Austin, M.P. & P. Greig-Smith. 1968. The application of quantitative methods to vegetation survey. J. Ecol. 56:827-844. [1]
Austin, M.P. & I. Noy-Meir. 1971. The problem of non-linearity in ordination: Experiments with two-gradient models. J. Ecol. 59:763-773. [3]
Austin, M.P. & L. Orlóci. 1966. Geometric models in ecology. II. An evaluation of some ordination techniques. J. Ecol. 54:217-227. [2, 3]
Ayyad, M.A. & A.A. El-Ghonemy. 1976. Phytosociological and environmental gradients in a sector of the western desert of Egypt. Vegetatio 31:93-102. [4]
Bachacou, J. 1973. L'effet Guttman dans l'analyse de données phytosociologiques. Dept. de Biometrie et de Calcul Automatique, Station de Biometrie du C.N.R.F. (Mimeographed.) [3]
Bachacou, J. 1974. Analyse de données non lineaires a 2 gradients. Dept. de Biometrie et de Calcul Automatique, Statio de Biometrie du C.N.R.F. (Mimeographed.) [3]
Ball, G.H. & D.J. Hall. 1965. Isodata, a novel method of data analysis and pattern classification. SRI Stanford Research Institute, Menlo Park, Calif. 94025. (Mimeographed.) [4]
Bannister, P. 1966. The use of subjective estimates of cover-abundance as the basis for ordination. J. Ecol. 54:665-674. [1]
Bannister, P. 1968. An evaluation of some procedures used in simple ordination. J. Ecol. 56:27-34. [3]
Barkman, J.J. 1958. Phytosociology and Ecology of Cryptogamic Epiphytes Including a Taxonomic Survey and Description of Their Vegetation Units in Europe. Assen, Van Gorcum. [2]
Bartels, P.H., G.F. Bahr, D.W. Calhoun & G.L. Wied. 1970. Cell recognition by neighborhood grouping techniques in Ticas. Acta Cytol. 14:313-324. [2]
Bartlett, M.S. 1936. The square root transformation in analysis of variance. Suppl. J. R. Statist. Soc. 3:68-78. [1]
Bartlett, M.S. 1950. Test of significance in factor analysis. Brit. J. Math. Stat. Psych. 3:77-85. [3]
Bartlett, M.S. 1960. Stochastic Population Models in Ecology and Epidemiology. London, Methuen. [1, 6]
Bartlett, M.S. 1965. Multivariate Statistics. In: T.H. Waterman & H.J. Morowitz (eds.), Theoretical and Mathematical Biology, pp. 201-224. New York, Blaisdell. [6]
Basharin, G.P. 1959. On a statistical estimate for the entropy of a sequence of independent random variables. Theory Prob. Applic. 4:333-336. [2]
Batchelor, B.G. 1971. Improved distance measure for pattern recognition. Electronics 7:521-524. [2]
Baum, B.R. & L.P. Lefkovitch. 1972. A model for cultivar classification and identification with reference to oats (Avena). I. Establishment of the groupings by taximetric methods. Can. J. Bot. 50:121-130. [4, 5, 6]
Beals, E. 1960. Forest bird communities in the Apostle Islands of Wisconsin. Wilson Bull. 72:156-181. [2]
Beals, E.W. 1965a. Ordination of some corticolous cryptogamic communities in south-central Wisconsin. Oikos 16:1-8. [3]
Beals, E.W. 1965b. Species patterns in a Lebanese Poterietum. Vegetatio 13:69-87. [2, 3]
Beals, E.W. 1973. Ordination: Mathematical elegance and ecological naïveté. J. Ecol. 61:23-35. [3]
Beals, E.W. & J.B. Cope. 1964. Vegetation and soils in Eastern Indiana woods. Ecology 45:777-792. [4]
Becking, R.W. 1957. The Zürich-Montpellier School of Phytosociology. Bot. Rev. 23:411-488. [1, 2]
Beeston, G.R. & M.B. Dale. 1975. Multiple predictive analysis: a management tool. Proc. Ecol. Australia 9:172-181. [4]
Benninghoff, W.S. & W.C. Southworth. 1964. Ordering of tabular arrays of phytosociological data by digital computer. Tenth Int. Bot. Congress Abst., pp. 331-332. Edinburgh. [4]
Benzécri, J.P. 1969. Statistical analysis as a tool to make patterns emerge from data.

In: S. Watanabe (ed.), Methodologies of Pattern Recognition, pp. 35-74. New York, Academic Press. [3]
Blackith, R.E. & R.A. Reyment. 1971. Multivariate Morphometrics. New York, Academic Press. [4, 6]
Bonner, R.E. 1964. On some clustering techniques. Bull. I. S. I. 43:411-425. [2]
Borko, H., D.A. Blankenship & R.C. Burket. 1968. On-line information retrieval using associative indexing. RADC-TR-68-100. Griffiss Air Force Base, New York. (Mimeographed.) [4]
Bormann, R.H. 1953. The statistical efficiency of sample plot size and shape in forest ecology. Ecology 34:474-487. [1]
Bottliková, A., P. Daget, J. Drdoš, J.L. Guillerm, F. Romane & H. Ružičková. 1976. Quelques résultats obtenus par l'analyse factorielle et les profils écologiques sur des observations phyto-écologiques recueillies dans la vallée de Liptov (Tchécoslovaquie). Vegetatio 31:79-91. [3]
Bottomley, J. 1971. Some statistical problems arising from the use of the information statistic in numerical classification. J. Ecol. 59:339-342. [4]
Bourdeau, P.F. 1953. A test of random versus systematic ecological sampling. Ecology 34:499-512. [1]
Bouxin, G. 1975a. Ordination and classification in the savanna vegetation of the Akagera Park (Rwanda, Central Africa). Vegetatio 29:155-167. [3, 4]
Bouxin, G. 1975b. Ordination of quantitative and qualitative data in a savanna vegetation (Rwanda, Central Africa). Vegetatio 30:197-200. [3]
Bouxin, G. 1976. Ordination and classification in the upland Rugege Forest (Rwanda, Central Africa). Vegetatio 32:97-115. [3]
Bradfield, G.E. & L. Orlóci. 1975. Classification of vegetation data from on open beach environment in southwestern Ontario: cluster analysis followed by generalized distance assignments. Can. J. Bot. 53:495-502. [4, 5]
Braun-Blanquet, J. 1928. Pflanzensoziologie. Grundzuge der Vegetationskunde. 1. Auflage. Berlin, Springer-Verlag. [4]
Braun-Blanquet, J. 1951. Pflanzensoziologie. Grundzuge der Vegetationskunde. 2. Auflage. Wien, Springer-Verlag. [4]
Bray, J.R. & J.T. Curtis. 1957. An ordination of the upland forest communities of southern Wisconsin. Ecol. Monogr. 27:325-349. [1, 2, 3]
Briane, J.P., J.-J. Lazare, G. Roux & C. Sastre.1974. L'analyse factorielle des correspondances et l'arbre de longueur minimum; exemples d'application. Adansonia 14:111-137. [2, 3, 4]
Brisse, H. & G. Grandjouan. 1971. Adaptation d'une méthode de classification multivariable par similitudes à l'écologie végétale en milieu naturel I. Exposé de la méthode. Oecol. Plant. 6:163-187. [1, 2, 3]
Brown, R.T. & J.T. Curtis. 1952. The upland conifer-hardwood forests of northern Wisconsin. Ecol. Monogr. 22:217-234. [4]
Bryant, E.H., B. Crandall-Stotler & R.E. Stotler. 1972. A factor analysis of the distribution of some Puerto Rican liverworts. Can. J. Bot. 51:1545-1554. [3]
Buell, M.F., A.N. Langford, D.W. Davidson & L.F. Ohmann. 1966. The upland forest continuum in northern New Jersey. Ecology 47:416-432. [3]
Buttler, G.A. 1969. A vector field approach to cluster analysis. Pattern Recognition 1:291-299. [4]
Cain, S.A. 1938. The species-area curve. Amer. Midl. Nat. 19:573-581. [1]
Cajander, A.K. 1909. Über Waldtypen. Acta Forestalia Fennica. 1:1-175. [4]
Carlson, K.A. 1970. A multivariate classification of reformatory inmates. Ph.D. thesis. Univ. of Western Ontario, London, Canada. [4]
Carmichael, J.W., J.A. George & R.S. Julius. 1968. Finding natural clusters. Syst. Zool. 17:144-150. [4]
Carmichael, J.W. & P.H.A. Sneath. 1969. Taximetric maps. Syst. Zool. 18: 402-415. [4]
Cattell, R.B. 1965. Factor analysis: an introduction to essentials. Biometrics, 21:190-215, 405-435. [4]
Cattel, R.B. & M.A. Coulter. 1966. Principles of behavioural taxonomy and the mathematical basis of the taxonome computer program. Brit. J. Math. Statist. Psych. 19:237-269. [4]

Cavalli-Sforza, L.L. & A.W.F. Edwards. 1967. Phylogenetic analysis: models and estimation procedures. Evolution 21:550-570. [4]
Češka, A. 1966. Estimation of the mean floristic similarity between and within sets of vegetational relevés. Folia Geobot. Phytotax., Praha 1:93-100. [2]
Češka, A. 1968. Application of association coefficients for estimating the mean similarity between sets of vegetation relevés. Folia Geobot. Phytotax, Praha 3:57-64. [4]
Češka, A. & H. Roemer. 1971. A computer program for identifying species-relevé groups in vegetation studies. Vegetatio 23:255-276. [4]
Chandapillai, M.M. 1970. Variation in fixed dune vegetation at Newborough Warren, Anglesey. J. Ecol. 58:193-201. [3]
Chardy, P., M. Glemarec & A. Laurec. 1976. Application of inertia methods to benthic marine ecology: practical implications of the basic options. Estuar. Coast. Mar. Sci. 4:179-205. [1, 2, 3]
Clements, F.E. 1916. Plant succession an analysis of the development of vegetation. Carnegie Inst. Wash. Publ. 242:1-512. [4]
Clifford, H.T. & D.W. Goodall. 1967. A numerical contribution to the classification of the Poaceae. Aust. J. Bot. 15:499-519. [2]
Clifford, H.T. & W. Stephenson. 1975. An Introduction to Numerical Classification. New York, Academic Press. [4]
Clifford, H.T. & W.T. Williams. 1973. Classificatory dendrograms and their interpretation. Aust. J. Bot. 21:151-162. [4]
Clifford, H.T., W.T. Williams & G.N. Lance. 1969. A further numerical contribution to the classification of the Poaceae. Aust. J. Bot. 17:119-131. [2]
Coetzee, B.J. & M.J.A. Werger. 1975. On association-analysis and the classification of plant communities. Vegetatio 30:201-206. [4]
Cole, A.J. (ed.) 1969. Numerical Taxonomy. London, Academic Press. [6]
Cole, A.J. & D. Wishart. 1970. An improved algorithm for the Jardine-Sibson method of generating overlapping clusters. Comp. J. 13:156-163. [4]
Cole, L.C. 1949. The measurement of interspecific association. Ecology 30:411-424. [2]
Cole, L.C. 1957. The measurement of partial interspecific association. Ecology 38:226-233. [2]
Cooley, W.W. & P.R. Lohnes. 1971. Multivariate Data Analysis. New York, Wiley. [5, 6]
Cormack, R.M. 1971. A review of classification. J. Roy. Statis. Soc. Series A. 134:321-353. [1, 2, 4]
Cottam, G. 1949. The phytosociology of an oak woods in southwestern Wisconsin. Ecology 30:271-287. [1, 3]
Cottam, G. & J.T. Curtis. 1949. A method for making rapid surveys of woodlands by means of pairs of randomly selected trees. Ecology 30:101-104. [1]
Cottam, G., F.G. Goff & R.H. Whittaker. 1973. Wisconsin comparative ordination. In: R.H. Whittaker (ed.), Handbook of Vegetation Science, Vol. 5, pp. 193-221. The Hague, Junk. [3]
Cottam, G. & R.P. McIntosh. 1966. Vegetational continuum. Science 152:546-547. [4]
Cramér, H. 1946. Mathematical Methods of Statistics. Princeton, Princeton Univ. Press. [2, 4]
Crawford, R.M.M. & D. Wishart. 1966. A multivariate analysis of the development of dune slack vegetation in relation to coastal accretion at Tentsmuir, Fife. J. Ecol. 54:729-743. [2, 4]
Crawford, R.M.M. & D. Wishart. 1967. A rapid multivariate method for the detection and classification of groups of ecologically related species. J. Ecol. 55:505-524. [4, 6]
Crawford, R.M.M. & D. Wishart. 1968. A rapid classification and ordination method and its application to vegetation mapping. J. Ecol. 56:385-404. [4]
Crawford, R.M.M., D. Wishart & R.M. Campbell. 1970. A numerical analysis of high altitude scrub vegetation in relation to soil erosion in the eastern Cordillera of Peru. J. Ecol. 58:173-191. [2, 4]
Crovello, T.J. 1968a. The effect of change of number of OTU's in a numerical taxo-

nomic study. Bittonia 20:346-367. [4]
Crovello, T.J. 1968b. The effect of alteration of technique at two stages in a numerical taxonomic study. Univ. Kansas Sci. Bull. 47:761-786. [4]
Crovello, T.J. 1968c. Key communality cluster analysis as a taxonomic tool. Taxon 17:241-258. [4]
Crovello, T.J. 1969. Effects of change of characters and of number of characters in numerical taxonomy. Am. Midl. Nat. 81:68-86. [4]
Crovello, T.J. 1970. Analysis of character variation in ecology and systematics. Ann. Rev. Ecol. Syst. 1:55-98. [1, 3, 4, 6]
Crovello, T.J. & W.W. Moss. 1971. A bibliography on classification in diverse disciplines. Class. Soc. Bull. 2:29-45. [6]
Cuanalo de la C., H.E. & R. Webster. 1970. A comparative study of numerical classification and ordination of soil profiles in a locality near Oxford. J. Soil. Sci. 21:340-352. [3, 4]
Cunningham, K.M. & J.C. Ogilvie. 1972. Evaluation of hierarchical grouping techniques: a preliminary study. Comput. J. 15:209-213. [4]
Curtis, J.T. 1959. The Vegetation of Wisconsin: An Ordination of Plant Communities. Univ. Wisconsin Press, Madison. [3]
Curtis, J.T. & R.P. McIntosh. 1950. The interrelations of certain analytic and synthetic phytosociological characters. Ecology 31:434-455. [1, 3]
Curtis, J.T. & R.P. McIntosh. 1951. An upland forest continuum in the prarie-forest border region of Wisconsin. Ecology 32:476-496. [1, 3, 4]
Czekanowski, J. 1909. Zur differential Diagnose der Neandertalgruppe. Korrespbl. dt. Ges. Anthrop. 40:44-47. [2]
Dabinett, P.E. & A.M. Wellman. 1973. Numerical taxonomy of the genus Rhizopus. Can. J. Bot. 51:2053-2064. [4]
Daget, P., M. Godron & J.L. Guillerm. 1972. Profils écologiques et information mutuelle entre espèces et facteurs écologiques. In: E. van der Maarel & R. Tüxen (eds.), Grundfragen und Methoden in der Pflanzensoziologie. The Hague, Junk. [1, 2]
Dagnelie, P. 1960. Contribution à l'étude des communautés végétales par l'analyse factorielle. Bull. Serv. Carte phytogéogr. Sér. B. 5:7-71, 93-195. [2, 3, 4]
Dagnelie, P. 1965a. L'étude des communautés végétales par l'analyse statistique des liasons entre les espèces et les variables écologiques: principes fondamentaux. Biometrics 21:345-361. [2]
Dagnelie, P. 1965b. The study of plant communities by the statistical analysis of correlations between species and ecological variables: an example. Biometrics 21:890-907. [3]
Dagnelie, P. 1969. Théorie et Méthodes Statistiques. Vols. 1, 2. Gembloux, Duculot. [6]
Dagnelie, P. 1971. Some ideas on the use of multivariate statistical methods in ecology. In: G.P. Patil, E.C. Pielou & W.E. Walters (eds.), Statistical Ecology, Vol. 3, pp. 167-174. London, Pennsylvania State Univ. Press. [3]
Dagnelie, P. 1973. L'analyse factorielle. In: R.H. Whittaker (ed.), Handbook of Vegetation Science, Vol. 5, pp. 223-248. The Hague, Junk. [3]
Dale, M.B. 1968. On property structure, numerical taxonomy and data handling. In: V.H. Heywood (ed.), Modern Methods in Plant Taxonomy, pp. 185-197. London, Academic Press. [1]
Dale, M.B. 1971. Information analysis of quantitative data. In: G.P. Patil, E.C. Pielou & W.E. Waters (eds.), Statistical Ecology, Vol. 3, pp. 133-148. London, Penn. State Univ. Press. [4]
Dale, M.B. 1975. On objectives of methods of ordination. Vegetatio 30:15-32. [3, 4]
Dale, M.B. & D.J. Anderson. 1972. Qualitative and quantitative information analysis. J. Ecol. 60:639-653. [4]
Dale, M.B. & D.J. Anderson. 1973. Inosculate analysis of vegetation data. Aust. J. Bot. 21:253-276. [3, 4, 6]
Dale, M.B., G.N. Lance & L. Albrecht. 1971. Extensions of information analysis. Austral. Comput. J. 3:29-34. [4]
Dale, M.B., P. Macnaughton-Smith, W.T. Williams & G.N. Lance. 1970. Numerical clas-

sification of sequences. Austr. Comp. J. 2:9-13. [4]
Dale, M.B. & L. Quadraccia. 1973. Computer assisted tabular sorting of phytosociological data. Vegetatio 28:57-73. [4]
Dale, M.B. & L.J. Webb. 1975. Numerical methods for the establishment of association. Vegetatio 30:77-81. [4]
Daniel, M. & B. Holubičková. 1972. Interspecific relationships of gamasoid mites in the nests of Clethrionomys glareolus. Folia Parasitologica, Praha 19:67-86. [4]
Daubenmire, R. 1966. Vegetation: Identification of typal communities. Science 151: 291-298. [4]
Davidson, R.A. 1967. A cybernetic approach to classification preliminaries. Taxon 16:3-7. [4]
Dempster, A.P. 1969. Elements of Continuous Multivariate Analysis. Reading, Mass., Addison-Wesley. [6]
De Vries, D.M., J.P. Baretta & G. Hamming. 1954. Constellation of frequent herbage plants, based on their correlation in occurrence. Vegetatio 5-6:105-111. [3, 4]
Dokuchaev, V.V. 1899. On the Theory of Natural Zones. St. Petersburg. [4]
Ducker, S.C., W.T. Williams & G.N. Lance. 1965. Numerical classification of the Pacific forms of Chlorodesmis (Chlorophyta). Aust. J. Bot. 13:489-499. [4]
Edwards, A.W.F. & L.L. Cavalli-Sforza. 1964. Reconstruction of evolutionary trees. In: V.H. Heywood & J. McNeill (eds.), Phenetic and Phylogenetic Classification, pp. 67-76. Syst. Ass. Pub. 6. [2]
Edwards, A.W.F. & L.L. Cavalli-Sforza. 1965. A method for cluster analysis. Biometrics 21:362-375. [4, 6]
Ehrendorfer, F. 1954. Gedankungen zur Frage der Struktur und Anordnung der Lebensgemeinschaften. Angew. Pfl. Soziol., Wien, Festschr. Aichinger 1:151-157. [3]
Ellenberg, H. & D. Mueller-Dombois. 1966. Tentative physiognomic-ecological classifition of plant formations of the earth. Ber. geobot. Inst. Rübel 37:21-55. [1]
Emden, M.H. van. 1972. Interaction analysis; an application of information theory in phytosociology. In: E. van der Maarel & R. Tüxen (eds.), Grundfragen und Methoden in der Pflanzensoziologie, pp. 113-120. The Hague, Junk. [4]
Emlen, J.M. 1973. Ecology: An Evolutionary Approach. London, Addison-Wesley. [1]
Erdős, P. & A. Rényi. 1960. On the evolution of random graphs. Publ. Math. Inst., Hungarian Acad. Sci. 5:17-61. [4]
Erdős, P. & A. Rényi. 1961. On a classical problem of probability theory. Publ. Math. Inst., Hung. Acad. Sci. 6:215-220. [4]
Escofier-Cordier, B. 1969. L'analyse factorielle des correspondances. Série Rech. 13. Cah. Bur. univ. Rech. opér. Univ. Paris 13:25-59. [2, 3]
Estabrook, G.F. 1966. A mathematical model in graph theory for biological classification. J. Theoret. Biol. 12:297-310. [2, 3, 4]
Estabrook, G.F. 1967. An information theory model for character analysis. Taxon 16:86-97. [2]
Fabbro, A. del, E. Feoli & G. Sauli. 1975. An indirect gradient analysis of the bryophyte vegetation of the 'magredi' in Friuli. Giorn. Bot. Ital. 109:361-374. [3, 4]
Farris, J.S. 1969a. On the cophenetic correlation coefficient. Syst. Zool. 18:279-285. [4]
Farris, J.S. 1969b. A successive approximation approach to character weighting. Syst. Zool. 18:374-385. [1]
Farris, J.S. 1973. On comparing the shapes of taxonomic trees. Syst. Zool. 22:50-54. [4]
Farris, J.S., A.G. Kluge & M.J. Eckardt. 1970. A numerical approach to phylogenetic systematics. Syst. Zool. 19:172-189. [3]
Fekete, G. & Z. Szőcs. 1974. Studies on interspecific association processes in space. Acta Bot. Acad. Sci. Hung. 20:227-241. [1, 4]
Feller, W. 1968. An Introduction to Probability Theory and Its Applications. Vol. I. (3rd ed.), New York, Wiley. [6]
Feller, W. 1971. An Introduction to Probability Theory and Its Applications. Vol. II. (2nd ed.), New York, Wiley. [6]
Feoli, E. 1973a. An index for weighing characters in monothetic classifications. (Italian with English summary.) Giorn. Bot. Ital. 107:263-268. [1]

Feoli, E. 1973b. Un esempio di ordinamento di tipi fitosociologici mediante l'analisi delle componenti principali. Not. Fitosoc. 7:21-22,. [4]
Feoli, E. 1975. On the use of characteristic species combinations to compare vegetation types. (Italian with English summary.) Giorn. Bot. Ital. 109:87-96. [3, 4]
Feoli, E. 1977. On the resolving power of principal component analysis in plant community ordinations. Vegetatio 33:119-125. [3]
Feoli, E. & G. Bressan. 1972. Floristic affinity of bentonic vegetational types of Cala di Mitigliano (Massa Lubrense, Napoli). Giorn. Bot. Ital. 106:245-256. [4]
Feoli, E. & L. Feoli Chiapella. 1974. Analisi multivariata di rilievi fitosociologici delle faggete della Majella. Not. Fitosoc. 9:37-53. [3, 4]
Feoli Chiapella, L. & E. Feoli. 1977. A numerical phytosociological study of the summits of the Majella massive (Italy). Vegetatio 34:21-39. [3, 4]
Ferrari, T.J. & J. Mol. 1967. Factor analysis of causal models. Neth. J. Agric. Sci. 15:38-49. [3]
Field, J.G. 1969. The use of the information statistic in the numerical classification of heterogeneous systems. J. Ecol. 57:565-569. [4]
Field, J.G. & F.T. Robb. 1970. Numerical methods in marine ecology. 2. Gradient analysis of rocky shore samples from False Bay. Zool. Afric. 5:191-210. [3]
Finney, D.J. 1963. Probit Analysis. Rev. ed. London, Cambridge. Univ. Press. [1]
Fischer, J., A. Horánszky, N. Kiss & Z. Szőcs. 1974. A new variant of discriminant analysis and its application to distinguishing Festuca populations. Annal. Univ. Scient. Budapest 16:63-86. [5]
Fisher, R.A. 1936. The use of multiple measurements in taxonomic problems. Ann. Eugenics 7:179-188. [5, 6]
Fisher, R.A. 1940. The precision of discriminant function. Ann. Eugenics 10:422-429. [3]
Fisher, R.A., A.S. Corbet & C.B. Williams. 1943. The relation between the number of species and the number of individuals in a random sample of an animal population. J. Animal Ecol. 12:42-58. [2]
Fleiss, J.L. & J. Zubin. 1969. On the methods and theory of clustering. Multivariate Behaviour. Res. 4:235-250. [4]
Forgy, E.W. 1963. Detecting 'natural' clusters of individuals. Western Psychological Association, Santa Monica, Calif. (Mimeographed.) [4]
Forgy, E.W. 1965. Cluster analysis of multivariate data: efficiency versus interpretability of classifications. Biometrics 21:768-769. [4]
Forsythe, W.L. & O.L. Loucks. 1972. A transformation for species response to habitat factors. Ecology 53:1112-1119. [3]
Fosberg, F.R. 1967. A classification of vegetation for general purposes. In: G.F. Peterkin (ed.), A guide to the Check Sheet for I.B.P. Areas, No. 4, pp. 73-120. Oxford, Blackwell. [1]
Fraser, A.R. & M. Kováts. 1966. Stereoscopic models of multivariate statistical data. Biometrics 22:358-367. [3, 4]
Fraser, D.A.S. 1965. On information in statistics. Ann. Math. Stat. 36:890-896. [2]
Fresco, L.F.M. 1969. Factor analysis as a method in synecological research. Acta Bot. Neerl. 18:477-482. [3]
Fresco, L.F.M. 1971. Compound analysis, a preliminary report on a new numerical approach in phytosociology. Acta. Bot. Neerl. 20:589-599. [4]
Fresco, L.F.M. 1972. Eine direkte quantitative analyse von vegetationsgrenzen und gradienten. In: E. van der Maarel & R. Tüxen (eds.), Grundfragen und Methoden in der Pflanzensoziologie, pp. 99-112. The Hague, Junk. [3]
Frey, T. 1966. On the significance of Czekanowski's index of similarity. Applicationes Mathematicae 9:1-7. [2]
Frey, T. 1969a. Review of Estonian quantitative plant ecology. In: Plant Taxonomy, Geography and Ecology in the Estonian S.S.R., pp. 60-70. Tallin, Valgus. [2]
Frey, T. 1969b. A new cluster analysis program for large matrices. In: Quantitative Methods of Vegetation Study, pp. 43-46. (Russian with English summary.) Inst. of Zool. and Bot., Academy of Sciences, Tartu State Univ., Estonian S.S.R. [4]
Frey, T.E.A. 1973. The Finnish School and forest site-types. In: R.H. Whittaker (ed.),

Handbook of Vegetation Science, pp. 403-433. The Hague, Junk. [4]
Frey, T. & H. van Groenewoud. 1972. A cluster analysis of the D^2 matrix of white spruce stands in Saskatchewan based on the maximum-minimum principle. J. Ecol. 60:813-886. [4]
Frey, T. & L. Võhandu. 1967. A new method for the establishing of classification units. Can. Dep. For. Transl. 155. From: Eesti NSV Tead. Toim. Biol. 15:565-576. [4]
Friedman, H.P. & J. Rubin. 1967. On some invariant criteria for grouping data. J. Amer. Statist. Assoc. 62:1159-1178. [4]
Furnival, G.M. 1971. All possible regression with less computation. Technometrics 13:403-408. [1]
Gabriel, K.R. & R.R. Sokal. 1969. A new statistical approach to geographic variation analysis. Syst. Zool. 18:259-278. [3]
Gauch, H.G. 1971. Bray-Curtis Ordination Cornell Ecology Program 4. (Mimeographed.) Ecology and Systematics, Langmuir Laboratory, Cornell University, Ithaca, New York. [6]
Gauch, H.G. 1973a. The relationship between sample similarity and ecological distance. Ecology 54:618-622. [3]
Gauch, H.G. 1973b. A quantitative evaluation of the Bray and Curtis ordination. Ecology 54:829-836. [3]
Gauch, H.G. 1977. Ordiflex: A Flexible Computer Program for Four Ordination Techniques. Ecology and Systematics, Cornell University, Ithaca, New York. (Mimeographed.) [6]
Gauch, H.G. & G.B. Chase. 1974. Fitting the Gaussian curve to ecological data. Ecology 55:1377-1381. [3, 6]
Gauch, H.G., G.B. Chase & R.H. Whittaker. 1974. Ordination of vegetation samples by Gaussian species distributions. Ecology 55:1382-1390. [3]
Gauch, H.G. & R.H. Whittaker. 1972. Comparison of ordination techniques. Ecology 53:868-875. [3]
Gengerelli, J.A. 1963. A method for detecting subgroups in a population and specifying their membership. J. Psychology 55:457-468. [2]
Gibson, W.A. 1959. Three multivariate models: factor analysis, latent structure analysis, and latent profile analysis. Psychometrika 24:229-252. [4]
Gimingham, C.H., N.M. Pritchard & R.M. Cormack. 1966. Interpretation of a vegetational mosaic on limestone in the island Gotland. J. Ecol. 54:481-502. [3]
Gittins, R. 1965. Multivariate approaches to a limestone grassland community. III. A comparative study of ordination and association analysis. J. Ecol. 53:411-425. [2, 3, 4]
Gittins, R. 1968. Trend-surface analysis of ecological data. J. Ecol. 56:845-869. [3]
Gittins, R. 1969. The application of ordination techniques. In: I.H. Rorison (ed.), Ecological Aspects of the Mineral Nutrition of Plants, pp. 37-66. Oxford, Blackwell. [3]
Gnanadesikan, R. & M.B. Wilk. 1969. Data analytic methods in multivariate statistical analysis. In: P.R. Krishnaiah (ed.), Multivariate Analysis II., pp. 593-638. New York, Academic Press. [4]
Godron, M. 1968. Quelques applications de la notion de fréquence en écologie végétale. Oecol. Plant. 185-212. [2]
Godron, M. 1975. Préservation, classification et évolution des phytocénoses et des milieux. Biol. Contemp. Roma 2:6-14. [4]
Godron, M., J.L. Guillerm, F. Romane & L. Sabato-Pizzini. 1969. Sur L'interpretation des matrices de coefficients de corrélation en phytosociologie. Oecol. Plant. 4:15-26. [4]
Goff, F.G. & G. Cottam. 1967. Gradient analysis: the use of species and synthetic indices. Ecol. 48:793-806. [3]
Goff, F.G. & P.H. Zedler. 1972. Derivation of species succession vectors. Amer. Midl. Natur. 87:397-412. [3]
Golder, P.A. & K.A. Yeomans. 1973. The use of cluster analysis for stratification. Appl. Stat. 22:213-219. [4]

Goldstein, R.A. & D.F. Grigal. 1972. Computer programs for the ordination and classification of ecosystems. Ecol. Sci. Div. Publication No. 417. Oak Ridge Nat. Lab., Oak Ridge, Tennessee. [4, 6]
Good, I.J. 1965. Categorization of classification. In: Mathematics and Computer Science in Biology and Medicine, pp. 115-125. London, H.M.S.O. [4]
Goodall, D.W. 1952a. Quantitative aspects of plant distribution. Biol. Rev. 27:194-245. [1, 2]
Goodall, D.W. 1952b. Some considerations in the use of point quadrats for the analysis of vegetation. Aust. J. Sci. Res. Ser. B. 5:1-41. [1]
Goodall, D.W. 1953. Objective methods for the classification of vegetation I. The use of positive interspecific correlation. Aust. J. Bot. 1:39-63. [1, 4, 6]
Goodall, D.W. 1954a. Objective methods for the classification of vegetation. III. An essay in the use of factor analysis. Aust. J. Bot. 2:304-324. [1, 3]
Goodall, D.W. 1954b. Vegetational classification and vegetational continua. Angew. Pfl. Soziol., Wien. [3]
Goodall, D.W. 1961. Objective methods for the classification of vegetation. IV. Pattern and minimal area. Aust. J. Bot. 9:162-196. [1]
Goodall, D.W. 1962. Bibliography of statistical plant sociology. Excerpta Botanica Sect. B 4:16-322. [6]
Goodall, D.W. 1963. The continuum and the individualistic association. Vegetatio 11:297-316. [4]
Goodall, D.W. 1964. A probabilistic similarity index. Nature 203:1098. [2, 4]
Goodall, D.W. 1966a. A new similarity index based on probability. Biometrics 22:883-907. [2, 4]
Goodall, D.W. 1966b. Numerical taxonomy of bacteria – some published data reexamined. J. gen. Microbiol. 42:25-37. [2, 4]
Goodall, D.W. 1966c. Classification, probability and utility. Nature 211:53-54. [4]
Goodall, D.W. 1966d. Hypothesis-testing in classification. Nature 211:329-330. [4]
Goodall, D.W. 1966e. Deviant index: a new tool for numerical taxonomy. Nature 210:216. [4]
Goodall, D.W. 1967. The distribution of the matching coefficient. Biometrics 23:647-656. [2]
Goodall, D.W. 1968. Affinity between an individual and a cluster in numerical taxonomy. Biom. Prax. 9:52-55. [2, 4, 5, 6]
Goodall, D.W. 1969a. A procedure for recognition of uncommon species combination in sets of vegetation samples. Vegetatio 18:19-35. [1, 2, 4]
Goodall, D.W. 1969b. Simulating the grazing situation. In: F. Heinmets (ed.), Biomathematics, Vol. 1, pp. 211-236. New York, Marcel Dekker. [3]
Goodall, D.W. 1970a. Statistical plant ecology. Ann. Rev. Ecol. Syst. 1:99-124. [1, 4, 6]
Goodall, D.W. 1970b. Cluster analysis using similarity and dissimilarity. Biom. Prax. 11:34-41. [4, 6]
Goodall, D.W. 1973a. Numerical methods of classification. In: R.H. Whittaker (ed.), Handbook of Vegetation Science, Vol. 5, pp. 575-615. The Hague, Junk. [1, 2, 3, 4]
Goodall, D.W. 1973b. Sample similarity and species correlation. In: R.H. Whittaker (ed.), Handbook of Vegetation Science, Vol. 5, pp. 105-156. The Hague, Junk. [2,3, 4]
Goodall, D.W. 1975. Setting objectives for ecological research. Bull. Ecol. Soc. Aust. 5:3-8. [1]
Goodman, L.A. & W.H. Kruskal. 1954. Measures of association for cross classifications. J. Am. Stat. Ass. 49:732-764. [2]
Goodman, L.A. & W.H. Kruskal. 1959. Measures of association for cross classifications. II. Further discussion and references. J. Am. Stat. Ass. 54:123-163. [2]
Goodman, M.M. 1972. Distance analysis in biology. Syst. Zool. 21:174-186. [2]
Gounot, M. 1969. Méthodes d'Étude Quantitative de la Végétation. Paris, Masson. [4]
Gower, J.C. 1966. Some distance properties of latent root and vector methods used in multivariate analysis. Biometrika 53:325-338. [3]

Gower, J.C. 1967a. Multivariate analysis and multidimensional geometry. The Statist. 17:13-28. [3]
Gower, J.C. 1967b. A comparison of some methods of cluster analysis. Biometrics 23:623-637. [4, 6]
Gower, J.C. 1971a. A general coefficient of similarity and some of its properties. Biometrics 27:857-871. [1, 2, 3]
Gower, J.C. 1971b. Statistical methods of comparing different multivariate analyses of the same data. In: Hodson, F.R., D.G. Kendall & P. Tautu (eds.), Mathematics in the Archaeological and Historical Sciences, pp. 138-149. Edinburgh, Univ. Press. [4]
Gower, J.C. 1974. Maximal predictive classification. Biometrics 30:643-654. [4, 5, 6]
Gower, J.C. & G.J.S. Ross. 1969. Minimum spanning trees and single linkage cluster analysis. Appl. Stat. 18:54-64. [4]
Green, P.E. & V.R. Rao. 1969. A note on proximity measures and cluster analysis. J. Market. Res. 6:359-364. [2]
Greenstadt, J. 1960. The determination of the characteristic roots of a matrix by the Jacobi method. In: A. Ralston & H.S. Wilf (eds.), Mathematical Methods for Digital Computers, pp. 84-91. New York, Wiley. [3]
Greig-Smith, P. 1952a. The use of random and contiguous guadrats in the study of the structure of plant communities. Ann. Bot., Lond., 16:293-316. [1]
Greig-Smith, P. 1952b. Ecological observations on degraded and secondary forest in Trinidad, British West Indies. J. Ecology 40:283-315. [1]
Greig-Smith, P. 1957. Quantitative Plant Ecology. London, Butterworths. [1, 4]
Greig-Smith, P. 1964. Quantitative Plant Ecology. (2nd ed.) London, Butterworths. [1, 2, 3, 4, 6]
Greig-Smith, P. 1971a. Analysis of vegetation data: the user viewpoint. In: G.P. Patil, E.C. Pielou & W.E. Waters (eds.), Statistical Ecology, Vol. 3, pp. 149-166. London, Penn. State Univ. Press. [1, 3, 4, 6]
Greig-Smith, P. 1971b. Application of numerical methods to tropical forests. In: G.P. Patil, E.C. Pielou & W.E. Waters (eds.), Statistical Ecology, Vol. 3, pp. 195-204. London, Penn. State Univ. Press. [4]
Greig-Smith, P., M.P. Austin & T.C. Whitmore. 1967. The application of quantitative methods to vegetation survey. I. Association analysis and principal component ordination of rain forest. J. Ecol. 55:483-503. [3, 4]
Grench, R.E. & H.C. Thacher. 1965. Collected algorithms 1960-1963 from the communications of the Association for Computing Machinery. U.S. Dept. of Commerce, Springfield, Virginia. [3, 6]
Grigal, D.F. & R.A. Goldstein. 1971. An integrated ordination classification analysis of an intensively sampled oak-hickory forest. J. Ecol. 59:481-492. [1, 4]
Grigal, D.F. & L. F. Ohmann. 1975. Classification, description and dynamics of upland plant communities within a Minnesota wilderness area. Ecol. Monogr. 45:389-407. [1, 3]
Groenewoud, H. van. 1964. An analysis and classification of white spruce communities in relation to certain habitat features. Can. J. Bot. 43:1025-1036. [4]
Groenewoud, H. van. 1965. Ordination and classification of Swiss and Canadian coniferous forests by various biometric and other methods. Ber. geobot. Inst. ETH Stiftg. Rübel, Zürich 36:28-102. [2, 3]
Groenewoud, H. van & P. Ihm. 1974. A cluster analysis based on graph theory. Vegetatio 29:115-120. [4]
Grunow, J.O. 1964. Objective classification of plant communities. S. Afr. J. Agric. Sci. 7:171-172. [4]
Guinochet, M. 1973. Phytosociologie. Masson, Paris. [3]
Guttman, L. 1959. Metricizing rank-ordered or unordered data for a linear factor analysis. Sankhya 21:257-268. [3]
Hall, A.V. 1965. Studies of the South African species of Eulophia. J.S. Afr. Bot. Suppl. Vol. 5. [6]
Hall, A.V. 1967a. Methods for demonstrating resemblance in taxonomy and ecology. Nature 214:830-831. [4]
Hall, A.V. 1967b. Studies in recently developed group-forming procedures in taxo-

nomy and ecology. J. S. Afr. Bot. 33:185-196. [4]
Hall, A.V. 1970. A computer-based method for showing continua and communities in ecology. J. Ecol. 58:591-602. [4]
Hamdan, M.A. & C.P. Tsokos. 1971. An information measure of association in contingency tables. Information and Control 19:174-179. [2]
Hansell, R.I.C. 1973. The detection and estimation of character weighting in classifications. J. theor. Biol. 39:297-314. [1]
Hansell, R.I.C. & D.A. Chant. 1973. A method for estimating relative weights applied to characters by classical taxonomist. Syst. Zoology, 22:46-49. [1]
Hansell, R.I.C. & B. Ewing. 1973. The detection and estimation of character weighting in classifications. J. theoret. Biol. 39:297-314. [1]
Harberd, D.J. 1960. Association-analysis in plant communities. Nature 185:53-54. [4]
Harberd, D.J. 1962. Application of a multivariate technique to ecological survey. J. Ecol. 50:1-17. [2, 3]
Harman, H.H. 1967. Modern Factor Analysis (2nd ed.), Chicago, Univ. Chicago Press. [4]
Hartigan, J.A. 1967. Representation of similarity matrices by trees. J. Amer. Stat. Assn. 62:1140-1158. [4]
Hatheway, W.H. 1971. Contingency-table analysis of rain forest vegetation. In: G.P. Patil, E.C. Pielou & W.E. Waters (eds.), Statistical Ecology, Vol. 3, pp. 271-313. London, Penn. State Univ. Press. [3, 6]
Hawksworth, F.G., G.F. Estabrook & D.J. Rogers. 1968. Application of an information theory model for character analysis in the genus Arceuhobium (Viscaceae). Taxon 17:605-619. [2]
Heck, D.L. 1960. Charts of some upper percentage points of the distribution of the largest characteristic root. Ann. Math. Stat. 31:625-642. [4]
Hill, L.R., L.G. Silvestri, P. Ihm, G. Farchi & P. Lanciani. 1965. Automatic classification of staphylococci by principal component analysis and a gradient method. J. Bacteriol. 1393-1401. [3, 4]
Hill, M.O. 1969. On looking at large correlation matrices. Biometrika 56:249-254. [4]
Hill, M.O. 1973a. Diversity and evenness: a unifying notation and its consequences. Ecology 54:427-432. [2]
Hill, M.O. 1973b. Reciprocal averaging: an eigenvector method of ordination. J. Ecol. 61:237-249. [3, 6]
Hill, M.O. 1974. Correspondence analysis: A neglected multivariate method. Appl. Statist. 23:340-354. [3]
Hill, M.O., R.G.H. Bunce & M.W. Shaw. 1975. Indicator species analysis, a divisive polythetic method of classification, and its application to a survey of native pinewoods in Scotland. J. Ecol. 63:597-613. [4, 5, 6]
Hinneri, S. 1972. An ecological monograph on eutrophic deciduous woods in the SW Archipelago of Finland. Annal. Univ. Turku. Ser. A. 50:1-131. [3, 4]
Hodson, F.R., P.H.A. Sneath & J.E. Doran. 1966. Some experiments in the numerical analysis of archeological data. Biometrika 53:311-324. [4]
Hoerner, S. von. 1967. Least-squares fit of a Gaussian to radio sources. Astrophysical J. 147:467-470. [4]
Hole, F.D. & M. Hironaka. 1960. An experiment in ordination of some soil profiles. Proc. Soil. Sci. Soc. Am. 24:309-312. [3]
Holgate, P. 1971. Notes on the Marczewski-Steinhaus coefficient of similarity. In: G.P. Patil, E.C. Pielou & W.E. Waters (eds.), Statistical Plant Ecology, Vol. 3, pp. 181-193. London, Penn. State Univ. Press. [2]
Holzner, W. & F. Stockinger. 1973. Der Einsatz von Elektronenrechnern bei der pflanzensoziologischen Tabellenarbeit. Österr. Bot. Z. 121:303-309. [4]
Hopkins, B. 1957. Pattern in the plant community. J. Ecol. 45:451-463. [3, 4]
Hotelling, H. 1933. Analysis of a complex of statistical variables into principal components. J. Ed. Psych. 24:417-441, 498-520. [3]
Hughes, R.E. & D.V. Lindley. 1955. Application of biometric methods to problems of classification in ecology. Nature. 175:806-807. [2]
Hulett, G.K., R.T. Coupland & R.L. Dix. 1966. The vegetation of sand dune areas

within the grassland region of Saskatchewan. Can. J. Bot. 44:1307-1331. [3]
Hurlbert, S.H. 1969. A coefficient of interspecific association. Ecology 50:1-9. [2]
Ihm, P. 1965. Automatic classification in anthropology. In: D. Hymes (ed.), The Use of Computers in Anthropology, pp. 357-376. London, Mouton. [4, 6]
Ihm, P. & H. van Groenewoud. 1975. A multivariate ordering of vegetation data based on Gaussian type gradient response curves. J. Ecol. 63:767-777. [3]
Ivimey-Cook, R.B. 1968. Investigations into the phenetic relationships between species of Onosis L. Watsonia 7:1-23. [4]
Ivimey-Cook, R.B. 1972. Association analysis – some comments on its use. In: E. van der Maarel & R. Tüxen (eds.), Grundfragen und Methoden in der Pflanzensoziologie, pp. 89-97. The Hague, Junk. [4]
Ivimey-Cook, R.B. & M.C.F. Proctor. 1966. The application of association-analysis to phytosociology. J. Ecol. 54:179-192. [4]
Ivimey-Cook, R.B. & M.C.F. Proctor. 1967. Factor analysis of data from an East Devon heath: a comparison of principal component and rotated solutions. J. Ecol. 55:405-413. [3, 4]
Ivimey-Cook, R.B., M.C.F. Proctor & D.M. Rowland. 1975. Analysis of the plant communities of a heathland site: Aylesbeare Common, Devon, England. Vegetatio 31:33-45. [2, 4]
Jaccard, P. 1901. Distribution de la flore alpine dans le Bassin des Dranses et dans quelques régions voisines. Bull. Soc. vaud. Sci. nat. 37:241-272. [2]
Jackson, D.M. 1969. Comparison of classifications. In: A.J. Cole (ed.), Numerical Taxonomy, pp. 91-111. London, Academic Press. [4]
James, F.C. 1970. Geographic size variation in birds and its relationship to climates. Ecology 51:365-390. [3]
James, F.C. 1971. Ordinations of habitat relationships among breeding birds. The Willson Bull. 83:215-236. [3]
James, F.C. & H.H. Shugart.1974. The phenology of the nesting season of the American robin (Turdus migratorius) in the United States. The Condor 76:159-168. [3]
Jancey, R.C. 1966a. Multidimensional group analysis. Aust. J. Bot. 14:127-130. [4, 6]
Jancey, R.C. 1966b. The application of numerical methods of data analysis to the genus Phyllota Benth. in New South Wales. Aust. J. Bot. 14:131-149. [4]
Jancey, R.C. 1974. Algorithm for detection of discontinuities in data sets. Vegetatio 29:131-133. [4]
Janssen, J.G.M. 1972. Detection of some micropatterns of winter annuals in pioneer communities of dry sandy soils. Acta. Bot. Neerl. 21:609-616. [4]
Janssen, J.G.M. 1975. A simple clustering procedure for preliminary classification of very large sets of phytosociological relevés. Vegetatio 30:67-71. [4]
Jardine, N. 1970. Algorithms, methods and models in the simplification of complex data. Comput. J. 13:116. [4]
Jardine, N. 1971. Patterns of differentiation between human local populations. Phil. Trans. Roy. Soc. Lond. B. 263:1-33. [2, 5]
Jardine, N. & R. Sibson. 1968a. The construction of hierarchic and non-hierarchic classifications. Comput. J. 11:177-184. [2, 4]
Jardine, N. & R. Sibson. 1968b. A model for taxonomy. Math. Biosci. 2:465-482. [4]
Jardine, N. & R. Sibson. 1971. Mathematical Taxonomy. New York, Wiley. [1, 2, 3, 4, 6]
Jeffers, J.N.R. 1967. Two case studies in the application of principal component analysis. Statistician 17:29-43. [6]
Jeglum, J.K., C.F. Wehrhahn & M.A. Swan. 1971. Comparisons of environmental ordinations with principal component vegetational ordinations for sets of data having different degrees of complexity. Can. J. For. Res. 1:99-112. [3]
Jesberger, J.A. & J.W. Sheard. 1973. A quantitative study and multivariate analysis of corticolous lichen communities in the southern boreal forest of Saskatchewan. Can. J. Bot. 51:185-201. [3, 4]
Johnson, M.P. & R.W. Holm. 1968. Numerical taxonomic studies in the genum Sarcostemma R.Br. (Asclepiadaceae). In: V.H. Heywood (ed.), Modern Methods in Plant Taxonomy, pp. 199-217. London, Academic Press. [4]

Johnson, R. 1973. A study of some multivariate methods for the analysis of botanical data. Ph. D. thesis, Utah State Univ. Logan, Utah. [3, 6]
Johnson, S.C. 1967. Hierarchical clustering schemes. Psychometrika 32:241-254. [4]
Jolliffe, I.T. 1972. Discarding variables in a principal component analysis. I: Artificial data. Appl. Statist. 21:160-173. [1]
Jolliffe, I.T. 1973. Discarding variables in a principal component analysis. II. Real data. Appl. Statist., 22:21-31. [1]
Jones, K.S. 1970. Some thoughts on classification for retrieval. J. Document. 26:89-101. [4]
Juhász-Nagy, P. 1963. Investigations on the Bulgarian vegetation. Some hygrophilous plant communities (I-III). Acta Biologica Debrecina 2:47-70. [4]
Juhász-Nagy, P. 1964. Some theoretical models of cenological fidelity. I. Acta Biol. Debrecina 3:33-43. [2]
Juhász-Nagy, P. 1966a. Some theoretical problems of synbotany Part 1. Primary considerations on a conceptual network. Acta Biol. Debrecina 4:59-66. [6]
Juhász-Nagy, P. 1966b. Some theoretical problems of synbotany. Part 2. Preliminaries on an axiomatic model-building. Acta Biol. Debrecina 4:67-81. [6]
Juhász-Nagy, P. 1967a. On association among plant populations. I. Multiple and partial association: a new approach. Acta Biol. Debrecina 5:43-56. [2]
Juhász-Nagy, P. 1967b. On some 'characteristic areas' of plant community stands. In: Proceedings of the Colloquium on Information Theory, Bolyai Mathematical Society, Debrecen, Hungary. [3]
Juhász-Nagy, P. 1968. Some theoretical problems of synbotany. Part 3. The importance of methodology. Acta Biol. Debrecina 6:65-77. [6]
Kaesler, R.L. & J. Cairns. 1972. Cluster analysis of data from limnological surveys of the Upper Potomac River. Amer. Midl. Natur. 88:56-67. [4]
Kaiser, H.F. 1958. The varimax criteria for analytic rotation in factor analysis. Psychometrika 23:187-200. [4]
Katz, J.O. & F.J. Rohlf. 1973. Function-point cluster analysis. Syst. Zool. 22:295-301. [4, 6]
Kendall, M.G. 1966. Discrimination and classification. In: P.R. Krishnaiah (ed.), Multivariate Analysis, pp. 165-185. New York, Academic Press. [4]
Kendall, M.G. & A. Stuart. 1969. The Advanced Theory of Statistics. Vol. 1 (3rd ed.) London, Griffin. [1]
Kendall, M.G. & A. Stuart. 1973. The Advanced Theory of Statistics. Vol. 2. (3rd ed.) New York, Hafner. [1, 2, 3]
Kendall, M.G. & A. Stuart. 1976. The Advanced Theory of Statistics. Vol. 3. (3rd ed.) New York, Hafner. [1]
Kendrick, W.B. 1965. Complexity and dependence in computer taxonomy. Taxon 14:141-154. [1]
Kendrick, W.B. & J.R. Proctor. 1964. Computer taxonomy in Fungi Imperfecti. Can. J. Bot., 42:65-88. [1]
Kershaw, K.A. 1964. Quantitative and Dynamic Ecology. London, Arnold. [4, 6]
Kershaw, K.A. 1968. Classification and ordination of Nigerian savanna vegetation. J. Ecol. 56:467-482. [3]
Kershaw, K.A. 1973. Quantitative and Dynamic Plant Ecology. (2nd ed.). London, Arnold. [1, 3, 4]
Kershaw, K.A. & W.R. Rouse. 1973. Studies on lichen-dominated systems. V. A primary survey of a raised-beach system in northwestern Ontario. Can. J. Bot. 51:1285-1307. [3]
Kessell, S.R. & R.H. Whittaker. 1976. Comparisons of three ordination techniques. Vegetatio 32:21-29. [3]
Khinchin, A.I. 1957. Mathematical Foundations of Information Theory. New York, Dover. [2]
Knight, D.H. 1965. A gradient analysis of Wisconsin prairie vegetation on the basis of plant structure and function. Ecology 46:744-747. [1]
Knight, D.H. & O.L. Loucks. 1969. A quantitative analysis of Wisconsin forest vegetation on the basis of plant function and gross morphology. Ecology 50: 219-234. [1]

Kortekaas, W.M. & E. van der Maarel. 1973. A numerical classification of spartinetum vegetation. II. Comparison of the computer based numerical system with the system published in 'Prodrome des Groupements Végétaux d'Europ'. (Mimeographed.) [4]

Kortekaas, W.M., E. van der Maarel & W.G. Beeftink. 1976. A numerical classification of European Spartina communities. Vegetatio 33:51-60. [4]

Koterba, W.D. & J.R. Habeck. 1971. Grassland of the North Fork Valley, Glacier Park, Montana. Can. J. Bot. 49:1627-1636. [3]

Krajina, V.J. 1933. Die Pflanzengesellschaften des Mlynica-Tales in den Vysoke Tatry (Hohe Tatra). Mit besonderer Berucksichtigung der okologischen Verhaltnisse. Bot. Centralbl., Beih., Abt. 2, 50:744-957; 51:1-224. [4]

Krajina, V.J. 1960. Ecosystem classification of forests. Silva Fennica 105:107-110, 123-138. [4]

Krajina, V.J. 1961. Ecosystem classification of forests: summary. In: Recent Advances in Botany, pp. 1599-1603.Toronto, Univ. Toronto Press. [4]

Kruskal, J.B. 1956. On the shortest spanning subtree of a graph and the traveling salesman problem. Proc. Amer. Math. Soc. 7:48-50. [4]

Kruskal, J.B. 1964a. Multidimensional scaling by optimizing goodness of fit to a nonmetric hypothesis. Psychometrika 29:1-27. [3]

Kruskal, J.B. 1964b. Nonmetric multidimensional scaling: a numerical method. Psychometrika 29:115-129. [3]

Kruskal, J.B. & F. Carmone. 1971. How to use the M-D-SCAL (Version 5M) and other useful information. Bell Telephone Laboratories Murray Hill, New Jersey, U.S.A., and University of Waterloo, Waterloo, Ontario, Canada. [3, 6]

Kruskal, J.B. & J.D. Carroll. 1966. Geometric models and badness of fit functions. In: P.B. Krishnaiah, Multivariate Analysis, pp. 639-671. London, Academic Press. [3]

Krzanowski, W.J. 1971. A comparison of some distance measures applicable to multinomial data, using a rotational fit technique. Biometrix 27:1062-1068. [2]

Kubíková, J. & M. Rejmánek. 1973. Notes on some quantitative methods in the study of plant community structure. Preslia (Praha) 45:154-164. [1]

Kuiper, F.K. & L. Fisher. 1975. A Monte Carlo comparison of six clustering procedures. Biometrics 31:777-783. [4]

Kullback, S. 1959. Information Theory and Statistics. New York, Wiley. [2, 4, 6]

Kullback, S., M. Kupperman & H.H. Ku. 1962. Tests for contingency tables and Markov chains. Technometrics 4:573-608. [2, 6]

Lacoste, A. 1976. Relations floristiques entre les groupements prairiaux du Triseto-Polygonion et les Megaphorbiales (Adenostylion) dans les Alpes Occidentales. (English summaré) Vegetatio 31:161-176. [3]

Lacoste, A. & M. Roux. 1971. L'analyse multidimensionelle en phytosociologie et écologie. Application à des données de l'étage subalpin des Alpes maritimes. I. L'analyse des données floristiques. Oecol. Plant. 6:353-369. [3]

La France, C.R. 1972. Sampling and ordination characteristics of computer-simulated individualistic communities. Ecology 53:387-397. [3]

Lambert, J.M. & M.B. Dale. 1964. The use of statistics in phytosociology. Adv. Ecol. Res. 2:59-99. [1, 3, 4, 6]

Lambert, J.M., S.E. Meacock, J. Barrs & P.F.M. Smartt. 1973. AXOR and MONIT: two new polythetic-divisive strategies for hierarchical classification. Taxon 22:173-176. [4]

Lambert, J.M. & W.T. Williams. 1962. Multivariate methods in plant ecology. IV. Nodal analysis. J. Ecol. 50:775-802. [4, 6]

Lambert, J.M. & W.T. Williams. 1966. Multivariate methods in plant ecology. VI. Comparison of information-analysis and association-analysis. J. Ecol. 54:635-664. [4]

Lance, G.N. & W.T. Williams. 1965. Computer analysis for monothetic classification (Association analysis). Comput. J. 8:246-249. [2, 4, 6]

Lance, G.N. & W.T. Williams. 1966a. A generalized sorting strategy for computer classifications. Nature 212:218. [4, 6]

Lance, G.N. & W.T. Williams. 1966b. Computer programs for hierarchical polythetic classification ('Similarity analysis'). Comput. J. 9:60-64. [2, 3, 4]

Lance, G.N. & W.T. Williams. 1967a. Mixed-data classificatory programs. Agglomerative systems. Aust. Comput. J. 1:15-25. [2, 6]
Lance, G.N. & W.T. Williams. 1967b. A general theory of classificatory sorting strategies. I. Hierarchical systems. Comput. J. 9:373-380. [4]
Lance, G.N. & W.T. Williams. 1968. Mixed-data classificatory programs. II. Divisive systems. Aust. Comput. J. 1:82-85. [6]
Lange, R.T. 1966. Sampling for association analysis. Aust. J. Bot. 14:373-378. [4, 6]
Lange, R.T., N.S. Stenhouse & C.E. Offler. 1965. Experimental appraisal of certain procedures for the classification of data. Aust. J. Bio. Sci. 18:1189-1205. [4, 6]
Lausi, D. 1972. Die Logik der Pflanzensoziologischen vegetationsanalyse – Ein Deutungsversuch. In: E. van der Maarel & R. Tüxen (eds.), Grundfragen und Methoden in der Pflanzensoziologie, pp. 17-28. The Hague, Junk. [2]
Lawley, D.N. & A.E. Maxwell. 1963. Factor Analysis as a Statistical Method. London, Butterworths. [3]
Lazarsfeld, P.F. 1950. The logical and mathematical foundations of latent structure analysis. In: S.A. Stouffer, L. Guttman, E.A. Suchman, P.F. Lazarsfeld, S.A. Star & J.A. Clausen (eds.) Measurement and Prediction, pp. 316-413. Princeton, Princeton Univ. Press. [4]
Lazarsfeld, P.F.. & N.W. Henry. 1968. Latent Structure Analysis. Boston, Houghton Mifflin. [4]
Lee, A. 1968. Numerical taxonomy and influenza B virus. Nature 217:620-622. [4]
Lerman, I.C. 1969. On two criteria of classification. In: A.J. Cole (ed.), Numerical Taxonomy, pp. 114-128. London, Academic Press. [2]
Lerman, I.C. 1970. Les Bases de la Classification Automatique. Paris, Gauthier-Villars. [2]
Levandowsky, M. 1972. An ordination of phytoplankton populations in ponds of varying salanity and temperature. 53:398-407. [2]
Levandowsky, M. & D. Winter. 1971. Distance between sets. Nature 234:34-35. [2]
Lieth, H. & G.W. Moore. 1971. Computerized clustering of species in phytosociological tables and its utilization for field work. In: G.P. Patil, E.C. Pielou & W.E. Waters (eds.), Statistical Ecology, Vol. 1, pp. 403-422. London, Penn. State Univ. Press. [4]
Linfoot, E.H. 1957. An informational measure of correlation. Information and Control 1:85-89. [2]
Looman, J. 1963. Preliminary classification of grasslands in Saskatchewan. Ecol. 44:15-29. [3, 4]
Loucks, O.L. 1962. Ordinating forest communities by means of environmental scalars and phytosociological indices. Ecol. Monogr. 32:137-166. [3]
Lubke, R.A. & J.B. Phipps. 1973. Taximetrics of Loudetia (Gramineae) based on leaf anatomy. Can. J. Bot. 51:2127-2146. [4]
Maarel, E. van der. 1966. Over vegetatiestructuren, -relaties en systemen in het bijzonder in de duingraslanden van Voorne. Thesis, Univ. Utrecht. [2]
Maarel, E. van der. 1967. Variation in vegetation and species diversity along a local environmental gradient. Acta Bot. Neerl. 16:211-221. [3]
Maarel, E. van der. 1969. On the use of ordination models in phytosociology. Vegetatio 19:21-46. [2, 3, 4]
Maarel, E. van der. 1970. Vegetationsstruktur und Minimum-areal in Einem Dünen-Trockenrasen. In: R. Tüxen (ed.), Gesellschaftsmorphologie, pp. 218-219. The Hague, Junk. [1, 4]
Maarel, E. van der. 1971. Basic problems and methods in phytosociology. Vegetatio 22:275-283. [4]
Maarel, E. van der. 1972. Ordination of plant communities on the basis of their plant genus, family and order relationships. In: E. van der Maarel & R. Tüxen (eds.), Grundfragen und Methoden in der Pflanzensoziologie, pp. 183-206. The Hague, Junk. [3]
Maarel, E. van der. 1975. The Braun-Blanquet approach in perspective. Vegetatio 30: 213-219. [4]

Machol, R.E. & R. Singer. 1971. Bayesian analysis of generic relations in Agaricales. Nova Hedwigia 21:753-787. [5, 6]
Macnaughton-Smith, P. 1963. The classification of individuals by the possession of attributes associated with a criterion. Biometrics 19:364-366. [4, 6]
Macnaughton-Smith, P. 1965. Some statistical and other numerical techniques for classifying individuals. London, H.M.S.O. [4, 6]
Macnaughton-Smith, P., W.T. Williams, M.B. Dale & L.G. Mockett. 1964. Dissimilarity analysis: a new technique of hierarchical sub-division. Nature 202:1034-1035. [1, 4]
Mahalanobis, P.C. 1936. On the generalized distance in statistics. Nat. Inst. Sci. India Proc. 2:49-55. [2]
Major, J. 1961. On two trends in phytocoenology (Annotated translation of V.M. Ponyatovskaya 1959.) Vegetatio 10:373-385. [4]
Maka, J.E. 1973. A mathematical approach to defining spatially recurring species groups in a montane rain forest on Mauna Loa, Hawaii. Technical Report No. 31, U.S.I.B.P., University of Hawaii, Honolulu. (Mimeographed). [4]
Marczewski, E. & H. Steinhaus. 1958. On a certain distance of sets and the corresponding distance of functions. Coll. Math. 6:319-327. [2]
Margalef, D.R. 1958. Information theory in ecology. Yearbook of the Society for General Systems Research 3:36-71. [2]
Martin, J. 1969. Ordination methods in lichenocoenology. In: Quantitative Methods of Vegetation Study, Collected Conference Reports, pp. 83-86. (Russian with English summary.) Institute of Zoology and Botany, Academy of Sciences, Tartu State Univ., Estonian S.S.R. [3]
Maycock, P.F. 1963. The phytosociology of the deciduous forests of extreme southern Ontario. Can. J. Bot. 41:379-438. [3]
Mayr, E. 1963. Animal Species and Evolution. Cambridge, Mass., Harvard Univ. Press. [1]
McCabe, G.P. Jr. 1975. Computations for variable selection in discriminant analysis. Technometrics 17:103-109. [1]
McDonald, R.P. 1962. A general approach to nonlinear factor analysis. Psychometrika 27:397-415. [3]
McDonald, R.P. 1966. Application of non-linear factor analysis in ecology. Proc. Symp. Ecol. Soc. Aust., Armidale. [3]
McDonald, R.P. 1967. Numerical methods for polynomial models in nonlinear factor analysis. Psychometrika 32:77-112. [3]
McIntosh, R.P. 1957. The York woods, a case history of forest succession in southern Wisconsin. Ecol. 38:29-37. [3, 4]
McIntosh, R.P. 1958. Plant communities. Science 128:115-120. [3, 4]
McIntosh, R.P. 1962. Pattern in a forest community. Ecol. 43:25-33. [4]
McIntosh, R.P. 1967a. An index of diversity and the relation of certain concepts to diversity. Ecology 48:392-403. [2]
McIntosh, R.P. 1967b. The continuum concept of vegetation. Bot. Rev. 33:130-187. [3, 4, 6]
McIntosh, R.P. 1970. Community, competition and adaptation. Quarter. Rev. Biol. 45:259-280. [6]
McIntosh, R.P. 1972. Forests of Catskill Mountains, New York. Ecol. Monogr. 43:143-161. [3]
McIntosh, R.P. 1973. Matrix and plexus techniques. In: R.H. Whittaker (ed.), Handbook of Vegetation Science, Vol. 5, pp. 157-191. The Hague, Junk. [3, 4, 6]
McIntosh, R.P. & R.T. Hurley. 1964. The spruce-fir forests of the Catskill Mountains. Ecol. 45:314-326. [3]
McKay, R.J. 1976. Simultaneous procedures in discriminant analysis involving two groups. Technometrics 18:47-53. [1]
McNeill, J. 1972. The hierarchical ordering of characters as a solution to the dependent character problem in numerical taxonomy. Taxon 21:71-82. [1]
McNeill, J. 1974. The handling of character variation in numerical taxonomy. Taxon 23:699-705. [1]

McNeill, J. 1975. A generic revision of Portulacaceae, tribe Montieae, using techniques of numerical taxonomy. Can. J. Bot. 53; 789-809. [4]
Miller, G.A. & W.G. Madow. 1954. On the maximum likelihood estimate of the Shannon-Wiener measure of information. AFCRC-TR-54-75, Air Research and Development Command, Bolling Air Force Base, Washington, D.C. [2]
Moore, J.J. 1972. An outline of computer-based methods for the analysis of phytosociological data. In: E. van der Maarel & R. Tüxen (eds.), Grundfragen und Methoden in der Pflanzensoziologie, pp. 29-38. The Hague, Junk. [4]
Moore, J.J., P. Fitzsimons, E. Lambe & J. White. 1970. A comparison and evaluation of some phytosociological techniques. Vegetatio 20:1-20. [4]
Moore, J.J. & A. O'Sullivan. 1970. A comparison between the result of the Braun-Blanquet method and those of 'cluster analysis'. The use of a digital computer for re-arranging a phytosociological array according to the principals of Braun-Branquet. In: R. Tüxen (ed.), Gesellschaftsmorphologie, pp. 26-30. The Hague, Junk. [4]
Moore, J.J. & A.M. O'Sullivan. 1973. A phytosociological survey of the Irish Molinio-Arrhenatheretea using computer techniques. Dept. of Botany, University College, Dublin. [6]
Moral, R. del. 1975. Vegetation clustering by means of ISODATA: revision by multiple discriminant analysis. Vegetatio 29:179-190. [4]
Moravec, J. 1973. The determination of minimal area of phytocenoses. Folia Geobot. Phytotax. Praha, 8:23-47. [1]
Moravec, J. 1975. Die Anwendung von Stetigkeitsartengruppen zur numerischen Ordnung von pflanzensoziologischen Tabellen. Vegetatio 30:41-47. [1]
Morisita, M. 1959. Measuring of interspecific association and similarity between communities. Mem. Fac. Sci. Kyushu Univ. Ser. B. 3:65-80. [2]
Morrison, D.F. 1967. Multivariate Statistical Methods. London. McGraw-Hill. [3, 6]
Moss, W.W. 1966. The biological and systematic relationships of the martin mite, Dermanyssus prognephilus Ewing (Acari: Mesostigmata: Dermanyssidae). Ph. D. Thesis, Univ. Kansas. [3]
Moss, W.W. 1967. Some new analytic and graphic approaches to numerical taxonomy with an example from the Dermanyssidae (Acari). Syst. Zool. 16:177-207. [3]
Moss, W.W. 1968. Experiments with various techniques of numerical taxonomy. Syst. Zool. 17:31-47. [4]
Motyka, J. 1947. O zadaniach i metodach badan geobotanicznych (French summ.: Sur les buts et les méthodes des recherches géobotaniques). Annls. Univ. Mariae Curie-Sklodowska, Lubin, Sect. C. Suppl. 1:1-168. [4]
Mountford, M.D. 1962. An index of similarity and its application to classificatory problems. In: P.W. Murphy (ed.), Progress in Soil Zoology, pp. 43-50. London, Butterworths. [2]
Mountford, M.D. 1971. A test of the difference between clusters. In: G.P. Patil, E.C. Pielou & W.E. Waters (eds.), Statistical Ecology, Vol. 3, pp. 237-257. London, Penn. State Univ. Press. [4]
Mueller-Dombois, D. & H. Ellenberg. 1974. Aims and Methods of Vegetation Ecology. New York, Wiley. [1, 2, 4]
Muir, J.W., H.G.M. Hardie, R.H.E. Inkson & H.J.B. Anderson. 1970. The classification of soil profiles by traditional and numerical methods. Geoderma 4:81-90. [4]
Mukkattu, M.M. 1974. Classification of natural communities based on covariance and related functions. Can. J. Bot. 52:2341-2349. [4]
Neal, M.W. & K.A. Kershaw. 1973a. Studies on lichen-dominated systems. III. Phytosociology of a raised-beach system near Cape Henrietta Maria, northern Ontario. Can. J. Bot. 51: 1115-1125. [3]
Neal, M.W. & K.A. Kershaw. 1973b. Studies on lichen-dominated systems. IV. The objective analysis of Cape Henrietta Maria raised-beach systems. Can. J. Bot. 51:1177-1190. [3]
Needham, R.M. & K.S. Jones. 1964. Keywords and clumps. J. Document, 20:5-16. [4]

Norris, J.M. & J.P. Barkham. 1970. A comparison of some Cotswold beechwoods using multiple discriminant analysis. J. Ecol. 58:603-619. [1]
Norris, J.M. & M.B. Dale. 1971. Transition matrix approach to numerical classification of soil profiles. Soil Sci. Soc. Amer. Proc. 35:487-491. [4]
Noy-Meir, I. 1970. Component analysis of semi-arid vegetation in southeastern Australia. Ph. D. Thesis, Aust. Nat. Univ., Canberra. [4]
Noy-Meir, I. 1971a. Multivariate analysis of desert vegetation. II. Qualitative/quantitative partition of heterogeneity. Israel J. of Bot. 20:203-213. [1]
Noy-Meir, I. 1971b. Multivariate analysis of the semi-arid vegetation in southeastern Australia: nodal ordination by component analysis. In: N.A. Nix (ed.), Quantifying Ecology, Proc. Ecol. Soc. Aust. 6:159-193. [3, 4, 6]
Noy-Meir, I. 1973a. Data transformations in ecological ordination. I. Some advantages of non-centering. J. Ecol. 61:329-341. [2, 3, 4]
Noy-Meir, I. 1973b. Divisive polythetic classification of vegetation data by optimized division on ordination components. J. Ecol. 61:753-760. [3, 4, 6]
Noy-Meir, I. 1974a. Multivariate analysis of the semiarid vegetation in southeastern Australia. II. Vegetation catenae and environmental gradients. Aust. J. Bot. 22:115-140. [3]
Noy-Meir, I. 1974b. Catenation: quantitative methods for the definition of coenoclines. Vegetatio 29:89-99. [3, 6]
Noy-Meir, I. 1977. Some problems and recent developments in ordination. Vegetatio [3]
Noy-Meir, I. & M.P. Austin. 1970. Principal-component ordination and simulated vegetational data. Ecology 51:551-552. [3]
Noy-Meir, I., N.H. Tadmor & G. Orshan. 1970. Multivariate analysis of desert vegetation. I. Association analysis at various quadrat sizes. Israel J. Bot. 19:561-591. [1, 4]
Noy-Meir, I., D. Walker & W.T. Williams. 1975. Data transformations in ecological ordination. II. On the meaning of data standardization. J. Ecol. 63:779-800. [1, 2, 3]
Noy-Meir, I. & R.H. Whittaker. 1977. Recent developments in continuous multivariate techniques. In: R.H. Whittaker (ed.), Handbook of Vegetation Science, Vol. 5 (2nd ed.). (In press.) The Hague, Junk. [3, 4]
Numata, M. 1966. Some remarks on the method of measuring vegetation. Bull. Marine Lab., Chiba Univ. 8:71-77. [1]
Ochiai, A. 1957. Zoogeographic studies on the solenoid fishes found in Japan and its neighbouring regions. Bull. Jap. Soc. Sci. Fish. 22:526-530. [2]
Odum, E.P. 1950. Bird populations of the Highlands (North Carolina) Plateau in relation to plant succession and avian invasion. Ecology 31:587-605. [2]
Ojaveer, E., J. Mullat & L. Võhandu. 1975. A study of infraspecific groups of the Baltic east Coast autumn herring by two new methods based on cluster analysis. In: T. Frey, M. Kangur & K. Elberg (eds.), Productivity of Estonian Water Bodies, pp. 28-50. IBP Est. NC. Tiigi 61, 202400 Tartu, Estonian SSR. [4]
Onyekwelu, S.S.C. 1972. The vegetation of dune slacks at Newborough Warren. I. Ordination of the vegetation. J. Ecol. 60:887-896. [3]
Orlóci, L. 1966. Geometric models in ecology. I. The theory and application of some ordination methods. J. Ecol. 54:193-215. [1, 3, 4]
Orlóci, L. 1967a. An agglomerative method for classification of plant communities. J. Ecol. 55:193-206. [2, 4, 6]
Orlóci, L. 1967b. Data centering: a review and evaluation with reference to component analysis. Syst. Zool. 16:208-212. [3]
Orlóci, L. 1968a. Information Analysis in phytosociology: partition, classification and prediction. J. theor. Biol. 20:271-284. [4]
Orlóci, L. 1968b. Definitions of structure in multivariate phytosociological samples. Vegetatio 15:281-291. [2, 4]
Orlóci, L. 1968c. A model for the analysis of structure in taxonomic collections. Can. J. Bot. 46:1093-1097. [4]
Orlóci, L. 1969a. Information theory models for hierarchic and non-hierarchic classifi-

cations. In: A.J. Cole (ed.), Numerical Taxonomy, pp. 148-164. London, Academic Press. [2]
Orlóci, L. 1969b. Information analysis of structure in biological collections. Nature 223:483-484. [4]
Orlóci, L. 1970a. Analysis of vegetation samples based on the use of information. J. theor. Biol. 29:173-189. [4]
Orlóci, L. 1970b. Automatic classification of plants based on information content. Can. J. Bot. 48:793-802. [4]
Orlóci, L. 1971a. An information theory model for pattern analysis. J. Ecol. 59:343-349. [2, 4]
Orlóci, L. 1971b. Information theory techniques for classifying plant communities. In: G.P. Patil, E.C. Pielou & W.E. Waters (eds.), Statistical Ecology, Vol. 3, pp. 259-270. London, Penn. State Univ. Press. [4]
Orlóci, L. 1972a. On objective functions of phytosociological resemblance. Am. Midl. Nat. 88:28-55. [2, 4]
Orlóci, L. 1972b. On information analysis in phytosociology. In: E. van der Maarel & R. Tüxen (eds.), Grundfragen und Methoden in der Pflanzensoziologie, pp. 75-88. The Hague, Junk. [2, 4, 6]
Orlóci, L. 1973a. Ordination by resemblance matrices. In: R.H. Whittaker (ed.), Handbook of Vegetation Science, Vol. 5, pp. 249-286. The Hague, Junk. [2, 3, 4]
Orlóci, L. 1973b. Ranking characters by a dispersion criterion. Nature 244:371-373. [3]
Orlóci, L. 1973c. An algorithm for cluster seeking in ecological collections. Vegetatio 27:339-345. [4]
Orlóci, L. 1974a. Revisions for the Bray & Curtis ordination. Can. J. Bot. 52:1773-1776. [3]
Orlóci, L. 1974b. On information flow in ordination. Vegetatio 29:11-16. [3]
Orlóci, L. 1975a. Measurement of redundancy in species collection. Vegetatio 31:65-67. [1, 3]
Orlóci, L. 1975b. Partition of information: some formulae revisited. Aust. J. Bot. 23:977-979. [4]
Orlóci, L. 1976a. Ranking species by an information criterion. J. Ecol. 64:417-419. [1, 2, 3]
Orlóci, L. 1976b. TRGRPS – An interactive algorithm for group recognition with an example from Spartinetea. Vegetatio 32:117-120. [3, 4, 6]
Orlóci, L. & E. Beshir. 1976. A heuristic test for homogeneity in species populations. Vegetatio 31:141-145. [2, 4]
Orlóci, L. & M.M. Mukkattu. 1973. The effect of species number and type of data on the resemblance structure of a phytosociological collection. J. Ecol. 61:37-46. [1, 3]
Ornduff, R. & T.J. Crovello. 1968. Numerical taxonomy of Limnanthacea. Amer. J. Bot. 55:173-182. [4]
Osborne, D.V. 1963. Some aspects of the theory of dichotomous keys. New Phytol 62:144-160. [4]
Ostle, B. 1963. Statistics in Research. Ames, Iowa State Univ. Press. [6]
Ottestad, P. 1975a. Component analysis: an alternative system. Int. Stat. Rev. 43:83-108. [3]
Ottestad, P. 1975b. Discrimination analysis. Int. Stat. Rev. 43:301-315. [5, 6]
Pakarinen, P. 1976. Agglomerative clustering and factor analysis of south Finnish mire types. Ann. Bot. Fennici 13:35-41. [3, 4]
Parker-Rhodes, A.F. & D.M. Jackson. 1969. Automatic classification in the ecology of the higher fungi. In: A.J. Cole (ed.), Numerical Taxonomy, pp. 181-215. New York, Academic Press. [2]
Parks, J.M. 1969. Classification of mixed mode data by R-mode factor analysis and Q-mode cluster analysis on distance function. In: A.J. Cole (ed.), Numerical Taxonomy, pp. 216-219. London, Academic Press. [4]
Parks, J.M. 1970. Fortran IV Program for Q-mode Cluster Analysis on Distance Func-

tion with Printed Dendrogram. Comp. Contrib. 46. State Geol. Survey, Univ. of Kansas, Lawrence. [4]
Patrick, E.A. & F.P. Fischer. 1969. Cluster mapping with experimental computer graphics. IEEE. Transactions on Computers. C-18 11:987-991. [4]
Patten, B.C. (ed.) 1971, 1972, 1975. Systems Analysis and Simulation Ecology. Vols. 1-3. New York, Academic Press. [1]
Pemadasa, M.A., P. Greig-Smith & P.H. Lovell. 1974. A quantitative description of the distribution of annuals in the dune system at Aberffraw, Anglesey. J. Ecol. 62:379-402. [3]
Phillips, D.L. 1978. Non-linear ordination: Field and computer simulation testing of a new method. Vegetatio (in press). [3, 6]
Phipps, J.B. 1970. Studies in the Arundinelleae (Gramineae). X. Preliminary taximetrics. Can. J. Bot. 48:2333-2356. [4]
Phipps, J.B. 1971. Dendrogram topology. Syst. Zool. 20:306-308. [4]
Phipps, J.B. 1972. Studies in the Arundinelleae (Gramineae). XIII. Taximetrics of the loudetioid, tristachyoid and danthoniopsoid groups. Can. J. Bot. 50:935-948. [4]
Phipps, J.B. 1975. Dendrogram topology: nomenclature. Can. J. Bot. 53:2047-2049. [4]
Phipps, J.B. 1976a. The number of classifications. Can. J. Bot. 54:686-688. [4]
Phipps, J.B. 1976b. Dendrogram topology: capacity and retrieval. Can. J. Bot. 54:679-685 [4]
Pielou, E.C. 1966a. Shannon's formula as a measure of species diversity: its use and missuse. Amer. Natur. 100:463-465. [2]
Pielou, E.C. 1966b. Species-diversity and pattern-diversity in the study of ecological succession. J. Theor. Biol. 10:370-383. [2]
Pielou, E.C. 1966c. The measurement of diversity in different types of biological collections. J. Theor. Biol. 13:131-144. [2]
Pielou, E.C. 1967. A test for random mingling of the phases of a mosaic. Biometrics 23:657-670. [2]
Pielou, E.C. 1969a. An Introduction to Mathematical Ecology. New York, Wiley-Interscience. [1, 2, 3, 4]
Pielou, E.C. 1969b. Association tests versus homogeneity tests: their use in subdividing quadrats into groups. Vegetatio 18:4-18. [4, 6]
Pielou, E.C. 1971. Measurement of structure in animal communities. In: J.A. Wiens (ed.), Ecosystem Structure and Function, pp. 113-135. Corvallis, Oregon State University Press. [6]
Pielou, E.C. 1974. Population and Community Ecology: Principles and Methods. New York, Gordon and Breach. [1, 2]
Pietsch, W. & W.R. Muller-Stoll. 1968. Die Zwergbinsen-Gesellschaft der nackten Teichboden im ostlichen Mitteleuropa, Eleocharito-Caricetum bohemicae. Mitt. Flor.-soz. Arbeitsgem N. F. 13:14-47. [2]
Pignatti, E. & S. Pignatti. 1975. Syntaxonomy of the Sesleria varia-grasslands of the calcareous Alps. Vegetatio 30:5-14. [3]
Poissonet, P.S., J.A. Poissonet, M.P. Godron & G.A. Long. 1973. A comparison of sampling methods in dense herbaceous pasture. J. Range Management 26:65-67. [1]
Poldini, L. & E. Feoli. 1976. Phytogeography and syntaxomy of the Caricetum firmae S.L. in the Carnic Alps. Vegetatio 32:1-10. [4]
Précsényi, I. & Z. Szőcs. 1969. Modification of a method of continuum studies. (Magyar with English summary.) Bot. Közlem. 56:189-196. [3]
Prichard, N.M. & A.J.B. Anderson, 1971. Observations on the use of cluster analysis in botany with an ecological example. J. Ecol. 59:727-747. [4]
Prim, R.C. 1957. Shortest connection networks and some generalizations. Bell Syst. Tech. J. 36:1389-1401. [3]
Quadling, C. 1967. Evaluation of tests and grouping of cultures by a two-stage principal component method. Can. J. Microbiol. 13:1379-1400. [4]
Quastler, H. 1956. The status of information theory in biology. In: H.P. Yockey (ed.), Symposium on Information Theory in Biology, pp. 399-402. New York, Pergamon. [2]

Rahman, N.A. 1962. On the sampling distribution of the studentized Penrose measure of distance. Ann. Hum. Genet. 26:97-106. [2]
Rajski, C. 1961. Entropy and metric spaces. In: C. Cherry (ed.), Information Theory, pp. 41-45. London, Butterworths. [2]
Ramenski, L.G. 1930. Zur Methodik der vergleichenden Bearbeitung und Ordnung von Pflanzenlisten und anderen Objekten, die durch mehrere, verschiedenartig wirkende Faktoren bestimmt werden. Beitr. Biol. Pfl. 18:269-304. [3]
Ramenski, L.G. 1938. Introduction to the complex soil-geobotanical investigation of lands. (In Russian). Moscow, Selkhozgiz. [3]
Rand, W.M. 1971. Objective criteria for the evaluation of clustering methods. J. Amer. Stat. Assoc. 66:846-850. [4]
Rao, C.R. 1952. Advanced Statistical Methods in Biometric Research. New York, Wiley. [3, 5]
Rao, C.R. 1962. A note on a generalized inverse of a matrix with application to problems in mathematical statistics. J.R. Statist. Soc. B. 24:152-158. [2]
Rao, C.R. 1972. Recent trends of research work in multivariate analysis. Biometrics 28:3-22. [6]
Rao, M.R. 1971. Cluster analysis and mathematical programming. J. Amer. Statist. Assoc. 66:622-626. [6]
Rektorys, K. 1969. A Survey of Applicable Mathematics. Cambridge, Mass. M.I.T. Press. [6]
Rényi, A. 1961. On measures of entropy and information. In: J. Neyman (ed.), Proceedings of the 4th Berkeley Symposium on Mathematical Statistics and Probability, pp. 547-561. Berkeley, Univ. Calif. Press. [2]
Rényi, A. 1970. Probability Theory. New York, Elsevier. [5, 6]
Rescigno, A. & G.A. Maccacaro. 1961. The information content of biological classifications. In: C. Cherry (ed.), Information theory, pp. 437-445. London, Butterworths. [1]
Reyment, R.A. 1962. Observations on homogeneity of covariance matrices in paleontologic biometry. Biometrics 18:1-11. [4]
Reyment, R.A. 1969. A multivariate paleontological growth problem. Biometrics 25:1-8. [4]
Reyment, R.A. 1971. Multivariate normality in morphometric analysis. Math. Geol. 4:357-368. [1]
Reyment, R.A. 1972. The discriminant function in systematic biology. NATO Advanced Study Institute, Athens; Publication No. 137, pp. 311-335. Palaeontological Institution University of Uppsala. [5]
Rhodes, A.M., S.E. Malo, C.W. Campbell & S.G. Carmer. 1971. A numerical taxonomic study of the Avocado (Persea americana Mill.). Amer. Soc. Hort. Sci. 96:391-395. [4]
Rice, E.L. 1967. A statistical method for determining quadrat size and adequacy of sampling. Ecology 48:1047-1049. [1]
Rochow, J.J. 1972. A vegetational description of a Mid-Missouri forest using gradient analysis techniques. Amer. Midl. Natur. 87:377-396. [3]
Rogers, D.J. 1963. Taximetrics – new name, old concept. Brittonia 15:285-290. [1]
Rohlf, F.J. 1965. Character correlation in numerical taxonomy. Proc. XII. Int. Congr. Ent., London. [3]
Rohlf, F.J. 1968. Stereograms in numerical taxonomy. Syst. Zool. 17:246-255. [3, 4]
Rohlf, F.J. 1970. Adaptive hierarchical clustering schemes. Syst. Zool. 18:58-82. [3, 4]
Rohlf, F.J. 1972. An empirical comparison of three ordination techniques in numerical taxonomy. Syst. Zool. 21:271-280. [3]
Rohlf, F.J. 1973. Hierarchical clustering using the minimum spanning tree. Comput. J. 16:93-95. [4]
Rohlf, F.J. 1974. Methods of comparing classifications. Ann. Rev. Ecol. Syst. 5:101-113. [4]
Rohlf, F.J. 1975. Generalization of the gap test for the detection of multivariate outliers. Biometrics 31:93-101. [1, 2]
Rohlf, F.J. 1977. A note on the measurement of redundancy. Vegetatio 34:63-64. [1]

Rohlf, F.J. & R.R. Sokal. 1962. The description of taxonomic relationships by factor analysis. Syst. Zool. 11:1-16. [3]
Romane, F. 1972. Un example d'utilisation de l'analyse factorielle des correspondances en écologie végétale. In: E. van der Maarel & R. Tüxen (eds.), Grundfragen und Methoden in der Pflanzensoziologie, pp. 151-182. The Hague, Junk. [3]
Romane, F., A.M. Bacou, P. David, M. Godron & J. Lepart. 1974. An example of the organization of plant ecology data processing. C.N.R.S., L. Emberger, B.P. 5051, 34033 Montpellier Cedex, France. [6]
Rubin, J. 1966. An approach to organizing data into homogeneous groups. Syst. Zool. 15:169-182. [4, 6]
Rubin, J. 1967. Optimal classification into groups: an approach for solving the taxonomy problem. J. Theoret. Biol. 15:103-144. [4, 5]
Russel, P.F. & T.R. Rao. 1940. On habitat and association of species of anopheline larvae in south-eastern Madras. J. Mal. Inst. Ind. 3:153-178. [2]
Sammon, J.W. 1969. A nonlinear mapping for data structure analysis. IEEE Trans. Computers C – 18:401-409. [3, 4]
Sampford, M.R. 1962. An Introduction to Sampling Theory. Edinburgh, Oliver and Boyd. [1, 6]
Scott, J.T. 1974. Correlation of vegetation with environment: a test of the continuum and community-type hypotheses. In: B.R. Strain & W.D. Billings (eds.), Handbook of Vegetation Science, Vol. 6, pp. 87-109. The Hague, Junk. [4]
Seal, H.L. 1964. Multivariate Statistical Analysis for Biologists. London, Methuen. [2, 3, 6]
Searle, S.R. 1966. Matrix Algebra for the Biological Sciences. New York, Wiley. [6]
Shafi, M.I. & G.A. Yarranton. 1973. Vegetation heterogeneity during a secondary (postfire) succession. Can. J. Bot. 51:73-90. [4]
Shannon, C.E. 1948. A mathematical theory of communication. Bell System Tech. J. 27:379-423. [2]
Shepard, R.N. 1962. The analysis of proximities: Multidimensional scaling with an unknown distance function. I. Biometrika 27:125-140. [3]
Shepard, R.N. & J.D. Carroll. 1966. Parametric representation of nonlinear data structures. In: P.R. Krishnaiah (ed.), Multivariate Analysis, pp. 561-592. London, Academic Press. [3, 6]
Shimwell, D.W. 1971. The Description and Classification of Vegetation. Seattle, Univ. Wash. Press. [1, 4, 6]
Shubik, M. 1954. Readings in Game Theory and Political Behavior. Garden City, Doubleday. [6]
Sibson, R. 1969. Information radius. Z. Wahr. 14:149-160. [2]
Sibson, R. 1970. A model for taxonomy. II. Math. Biosci. 6:405-430. [4]
Silvestri, L.G. & L.R. Hill. 1964. Some problems of the taxometric approach. In: V.H. Heywood & J. McNeill (eds.), Phenetic and Phylogenetic Classification, pp. 87-104. Systematics Association, London. [4]
Sinha, R.N. & R.J. Lee. 1970. Maximum likelihood factor analysis of natural arthropod infestations in stored grain bulks. Res. Popul. Ecol. 12:51-60. [3]
Sinha, R.N., G. Yaciuk & W.E. Muir. 1973. Climate in relation to deterioration of stored grain. Oecologia 12:69-88. [3]
Smartt, P.F.M. & J.E.A. Grainger. 1974. Sampling for vegetation survey: some aspects of the behavior of unrestricted, restricted, and statified techniques. J. Biogeogr. 1:193-206. [1]
Smartt, P.F.M., S.E. Meacock & J.M. Lambert. 1974. Investigations into the properties of quantitative vegetation data. J. Ecol. 62:735-759. [1]
Sneath, P.H.A. 1957a. Some thoughts on bacterial classification. J. gen. Microbiol. 17:184-200 [1]
Sneath, P.H.A. 1957b. The application of computers to taxonomy. J. gen Microbiol. 17:201-226. [4, 6]
Sneath, P.H.A. 1961. Recent developments in theoretical and quantitative taxonomy. Syst. Zool. 10:118-139. [4]
Sneath, P.H.A. 1962. The construction of taxonomic groups. In: G.S. Ainsworth &

P.H.A. Sneath (eds.), Microbial Classification, pp. 289-332. Cambridge, Cambridge Univ. Press. [4]
Sneath, P.H.A. 1965. A method for curve seeking from scattered points. Comput. J. 8:383-391. [3]
Sneath, P.H.A. 1966a. A comparison of different clustering methods as applied to randomly spaced points. Class. Soc. Bull. 1:2-18. [4]
Sneath, P.H.A. 1966b. A method for curve seeking from scattered points. Comput. J. 8:383-391. [6]
Sneath, P.H.A. 1969. Recent trends in numerical taxonomy. Taxon 18:14-21. [1]
Sneath, P.H.A. & R.R. Sokal. 1973. Numerical Taxonomy. San Francisco, Freeman. [1, 2, 4, 5, 6]
Sobolev, L.N. & V.D. Utekhin. 1973. Russian (Ramensky) approaches to community systematization. In: R.H. Whittaker (ed.), Handbook of Vegetation Science, Vol. 5, pp. 75-103. The Hague, Junk. [3]
Sokal, R.R. 1958. Thurstone's analysical method for simple structure and a mass modification thereof. Biometrika 23:237-257. [3]
Sokal, R.R. 1961. Distance as a measure of taxonomic similarity. Syst. Zool. 10:70-79. [2]
Sokal, R.R. 1965. Statistical methods in systematics. Biol. Rev. 40:337-391. [6]
Sokal, R.R. 1966. Numerical taxonomy. Sci. Amer. 215:106-116. [4]
Sokal, R.R. 1974. Classification: purposes, principles, progress, prospects. Science 185:1115-1123. [4]
Sokal, R.R. & T.J. Crovello. 1970. The biological species concept: a critical evaluation. Amer. Nat. 104:127-153. [1]
Sokal, R.R. & H.V. Daly. 1961. An application of factor analysis to insect behavior. Univ. Kansas Sci. Bull. 42:1067-1097. [3]
Sokal, R.R., H.V. Daly & F.J. Rohlf. 1961. Factor analytical procedures in a biological model. Univ. Kansas Sci. Bull. 42:1099-1121. [3]
Sokal, R.R. & C.D. Michener. 1958. A statistical method for evaluating systematic relationships. Univ. Kansas Sci. Bull. 38:1409-1438. [2, 4, 6]
Sokal, R.R. & F.J. Rohlf. 1962. The comparison of dendrograms by objective methods. Taxon. 11:33-40. [4]
Sokal, R.R. & F.J. Rohlf. 1969. Biometry. San Francisco. Freeman. [1, 2, 6]
Sokal, R.R. & F.J. Rohlf. 1970. The intelligent ignoramus, an experiment in numerical taxonomy. Taxon 19:305-319. [3]
Sokal, R.R. & P.H.A. Sneath. 1963. Principles of Numerical Taxonomy. San Francisco, Freeman. [1, 2, 4, 6]
Sørensen, T. 1948. A method of establishing groups of equal amplitude in plant sociology based on similarity of species content. Biol. Skr., K. danske Videmsk. Selsk. 5:1-34. [1, 2, 4]
Spatz, G. 1972. Eine möglichkeit zum Einsatz der elektronischen Datenverarbeitung bei der pflanzensoziologischen Tabellenarbeit. In: E. van der Maarel & R. Tüxen (eds.), Grundfragen und Methoden in der Pflanzensoziologie, pp. 251-258. The Hague, Junk. [4]
Spatz, G. & J. Siegmund. 1973. Eine Methode zur tabellarischen Ordination, Klassifikation und Ökologischen Auswertung von pflanzensoziologischen Bestandsaufnahmen durch den Computer. Vegetatio 28:1-17. [4]
Stanek, W. & L. Orlóci. 1973. A comparison of Braun-Blanquet's method with sum-of-squares agglomeration for vegetation classification. Vegetatio 27:323-345. [2, 4]
Staniforth, R.J. & P.B. Cavers. 1976. An experimental study of water dispersal in Polygonum spp. Can. J. Bot. 54:2587-2596. [2]
Stephenson, W. & W.T. Williams. 1971. A study of the benthos of soft bottoms, Sek Harbour, New Guinea, using numerical methods. Aust. J. mar. Freshwat. Res. 22:11-34. [4]
Stockinger, J.J. & W.F. Holzner. 1972. Rationelle Methode zur Auswertung pflanzensoziologischer Aufnahmen mittels Elektronenrechner. In: E. van der Maarel & R. Tüxen (eds.), Grundfragen und Methoden in der Pflanzensoziologie, pp. 239-248. The Hague, Junk. [4]

Stout, B.B., J.M. Deschenes & L.F. Ohmann. 1975. Multi-species model of a deciduous forest. Ecology 56:226-231. [3]
Strahler, A.H. 1977. Response of woody species to site factors in Maryland: evaluation of sampling plans and binary measurement techniques. Vegetatio 35:1-19. [1]
Stringer, P.W. 1973. An ecological study of grasslands in Banff, Jasper and Waterton Lake National Parks. Can. J. Bot. 51:383-411. [4]
Swan, J.M.A. 1970. An examination of some ordination problems by use of simulated vegetational data. Ecology 51:89-102. [3]
Swan, J.M.A. & R.L. Dix. 1966. The phytosociological structure of upland forest at Candla Lake, Saskatchewan. J. Ecol. 54:12-40. [3]
Swan, J.M.A., R.L. Dix & C.F. Wehrhahn. 1969. An ordination technique based on the best possible stand-defined axes and its application to vegetation analysis. Ecology 50:206-212. [2, 3]
Switzer, P. 1971. Notes on the 'Notes on the Maczewski-Steinhous coefficient of similarity'. In: G.P. Patil, E.C. Pielou & W.E. Waters (eds.), Statistical Ecology, 3:190-191. London, Penn. State Univ. Press. [2]
Szőcs, Z. 1971. The beechwoods of the Vértes Mountains. II. An investigation of the interspecific correlations. (Magyar with English summary). Bot. Közlem 58:47-52. [3]
Szőcs, Z. 1972. The beechwoods of the Vértes Mountains. III. An investigation of similarity of stands. Ann. Univ. Scient. Budapest 14:179-184. [3, 6]
Szőcs, Z. 1973. On the botanical application of some multivariate analyses. II. General characterization. (Magyar with English summary) Bot. Közl. 60:29-34. [6]
Tansley, A.G. 1939. The British Islands and their Vegetation. Cambridge, Cambridge Univ. Press. [4]
Thalen, D.C.P. 1971. Variation in some saltmarsh and dune vegetation in the Netherlands with special reference to gradient situation. Acta Bot. Neerl. 20:327-342. [4]
Tharu, J. & W.T. Williams. 1966. Concentration of entries in binary arrays. Nature 211:549. [4, 6]
Torgerson, W.S. 1952. Multidimensional scaling: I. Theory and method. Psychometrika 17:401-419. [3]
Torgerson, W.S. 1958. Theory and Methods of Scaling. New York. Wiley. [1]
Tracey, J.G. 1968. Investigation of changes in pasture composition by some classificatory methods. J. Appl. Ecol. 5:639-648. [4]
Trass, H. & N. Malmer. 1973. North European approaches to classification. In: R.H. Whittaker (ed.), Handbook of Vegetation Science, Vol. 5, pp. 529-574. The Hague, Junk. [1]
Tukey, J.W. 1962. The future of data analysis. Ann. Math. Statist. 33:1-67. [6]
Tukey, J.W. & M.B. Wilk. 1966. Data analysis and statistics: an expository overview. AFIPS Conf. Proc. Fall Joint Comput. Conf. 29:695-709. [1, 6]
Underwood, R. 1969. The classification of constrained data. Syst. Zool. 18:312-317. [1]
Vasilevich, V.I. 1969a. Some problems of discrimination of natural groups. In: Quantitative Methods of Vegetation Study, Collected Conference Reports, pp. 31-34. (Russian with English Summary.) Institute of Zoology and Botany, Academy of Sciences, Tartu State Univ., Estonian S.S.R. [4]
Vasilevich, V.I. 1969b. Statistical Methods in Geobotany. (In Russian). Leningrad, Nauka. [4]
Vasilevich, V.I. 1973. A coenoquant as a minimal spatial entity in vegetation cover. (Russian with English summary). Bot. J. Leningrad 58:1241-1252. [1]
Vogle, R.J. 1966. Vegetational continuum. Science 152:546. [4]
Wali, M.K. & V.J. Krajina. 1973. Vegetation environment relationships of some subboreal spruce zone ecosystems in British Columbia. Vegetatio 26:237-381. [3]
Walker, B.H. 1974. Some problems arising from the preliminary manipulation of plant ecological data for subsequent numerical analysis. J.S. Afr. Bot. 40:1-13. [6]
Walker, B.H. 1975. Vegetation-site relationships in the Harvard Forest. Vegetatio 29:169-178. [3]
Walker, B.H. & C.F. Wehrhahn. 1971. Relationships between derived vegetation gra-

dients and measured environmental variables in Saskatchewan wetlands. Ecology 52:86-95. [3]
Walker, J., I. Noy-Meir, D.J. Anderson & R.M. Moore. 1972. Multiple pattern analysis of a woodland in South Central Queensland. Aust. J. Bot. 20:105-118. [3]
Wallace, C.S. & D.M. Boulton. 1968. An information measure for classification. Comput. J. 11:185-194. [4]
Ward, J.H. Jr. 1963. Hierarchical grouping to optimize an objective function. J. Amer. Static. Ass. 58:236-244. [4, 6]
Ward, J.H. Jr. & M.E. Hook. 1963. Application of an hierarchical grouping procedure to a problem of grouping profiles. Ed. Psych. Measure. 23:69-81. [4]
Ward, S.D. 1970. The phytosociology of Calluna-Arctostaphylos heath in Scotland and Scandinavia. I. Dinnet Moor, Aberdeenshire. J. Ecol. 58:847-863. [4]
Watt, K.E. 1968. Ecology and Resource Management. New York, McGraw-Hill. [1]
Webb, L.J., J.G. Tracey, W.T. Williams & G.N. Lance. 1967a. Studies in the numerical analysis of complex rain-forest communities. I. A comparison of methods applicable to site/species data. J. Ecol. 55:171-191. [4]
Webb, L.J., J.G. Tracey, W.T. Williams & G.N. Lance. 1967b. Studies in the numerical analysis of complex rain-forest communities. II. The problem of species-sampling. J. Ecol. 55:525-538. [4]
Webb, L.J., J.G. Tracey, W.T. Williams & G.N. Lance. 1970. Studies in the numerical analysis of complex rain-forest communities. V. A comparison of the properties of floristic and physiognomic-structural data. J. Ecol. 58:203-232. [1]
Werger, M.J.A. 1973. On the use of association analysis and principal component analysis in interpreting a Braun-Blanquet phytosociological table of a Dutch grassland. Vegetatio 28:129-144. [4]
West, N.E. 1966. Matrix cluster analysis of montane forest vegetation of the Oregon Cascades. Ecol. 47:975-980. [4]
Westhoff, V. & E. van der Maarel. 1973. The Braun-Blanquet approach. In: R.H. Whittaker (ed.), Handbook of Vegetation Science, pp. 617-726. The Hague, Junk. [1, 2, 4, 6]
Westman, W.E. 1975. Edaphic climax pattern of the Pygmy Forest Region of California. Ecol. Monogr. 45:109-135. [3]
Whittaker, R.H. 1948. A vegetation analysis of the Great Smoky Mountains. Ph. D. Thesis, University of Illinois, Urbana. [3, 6]
Whittaker, R.H. 1952. A study of summer foliage insect communities in the Great Smoky Mountains. Ecol. Monogr. 22:1-44. [2, 3]
Whittaker, R.H. 1956. Vegetation of the Great Smoky Mountains. Ecol. Monogr. 26:1-80. [3, 4]
Whittaker, R.H. 1962. Classification of natural communities. Bot. Rev. 28:1-239. [4]
Whittaker, R.H. 1967. Gradient analysis of vegetation. Biol. Rev. 42:207-264. [3, 4]
Whittaker, R.H. 1972. Convergences of ordination and classification. In: E. van der Maarel & R. Tüxen (eds.), Grundfragen und Methoden in der Pflanzensoziologie, pp. 39-54. The Hague, Junk. [4]
Whittaker, R.H. 1973a. Direct gradient analysis: techniques. In: R.H. Whittaker (ed.), Handbook of Vegetation Science, Vol. 5, pp. 7-31. The Hague, Junk. [3]
Whittaker, R.H. 1973b. Approaches to classifying vegetation. In: R.H. Whittaker (ed.), Handbook of Vegetation Science, Vol. 5, pp. 323-354. The Hague, Junk. [4]
Whittaker, R.H. & H.G. Gauch. 1973. Evaluation of ordination techniques. In: R.H. Whittaker (ed.), Handbook of Vegetation Science, Vol. 5, pp. 287-321. The Hague, Junk. [2, 3]
Whittaker, R.H. & H.G. Gauch. 1977. Evaluation of ordination techniques. In: R.H. Whittaker (ed.), Handbook of Vegetation Science, Vol. 5 (2nd ed.). (In press.) The Hague, W. Junk. [3]
Wilkins, D.A. & M.C. Lewis. 1969. An application of ordination to genecology. New Phytol. 68:861-871. [3]
Wilks, S.S. 1963. Multivariate statistical outliers. Sankhyā Ser. A 25:407-26. [1]
Williams, E.J. 1952. Use of scores for the analysis of association in contingency tables. Biometrika 39:274-289. [2, 3, 6]

Williams, W.T. 1967. Numbers, taxonomy and judgement. Bot. Rev. 33:379-386. [4]
Williams, W.T. 1969. The problem of attribute-weighting in numerical classification. Taxon 18:369-374. [1]
Williams, W.T. 1971. Principles of clustering. Ann. Rev. Ecol. Syst. 2:303-326. [4, 6]
Williams, W.T. 1973. Partition of information. Aust. J. Bot. 21:277-281. [4]
Williams, W.T. & H.T. Clifford. 1971. On the comparison of two classifications of the same set of elements. Taxon 20:519-522. [4]
Williams, W.T., H.T. Clifford & G.N. Lance. 1971. Group-size dependence: a rational for choice between numerical classifications. Comput. 14:157-162. [4]
Williams, W.T. & M.B. Dale. 1962. Partition correlation matrices for heterogeneous quantitative data. Nature 196:602. [1]
Williams, W.T. & M.B. Dale. 1965. Fundamental problems in numerical taxonomy. Adv. Bot. Res. 2:35-68. [2, 4, 6]
Williams, W.T., M.B. Dale & P. Macnaughton-Smith. 1964. An objective method of weighting in similarity analysis. Nature 201:426. [1]
Williams, W.T., K.P. Haydock, L.A. Edye & J.B. Ritson. 1971. Analysis of a fertility trial with Droughtmaster cows. Aust. J. Agric. Res. 22:979-991. [4]
Williams, W.T. & J.M. Lambert. 1959. Multivariate methods in plant ecology. I. Association-analysis in plant communities. J. Ecol. 47:83-101. [1, 2, 4, 5, 6]
Williams, W.T. & J.M. Lambert. 1960. Multivariate methods in plant ecology. II. The use of an electronic digital computer for association-analysis. J. Ecol. 48:689-710. [2, 4]
Williams, W.T. & J.M. Lambert. 1961. Multivariate methods in plant ecology. III. Inverse association-analysis. J. Ecol. 49:717-729. [2, 4]
Williams, W.T., J.M. Lambert & G.N. Lance. 1966. Multivariate methods in plant ecology. V. Similarity analyses and information-analysis. J. Ecol. 54:427-445. [4]
Williams, W.T. & G.N. Lance. 1958. Automatic subdivision of associated populations. Nature 182:1755. [4]
Williams, W.T. & G.N. Lance. 1965. Logic of computer-based intrinsic classifications. Nature 207:159-161. [4, 6]
Williams, W.T. & G.N. Lance. 1968. The choice of strategy in the analysis of complex data. Statistician 18:31-43. [4]
Williams, W.T., G.N. Lance, M.B. Dale & H.T. Clifford. 1971. Controversy concerning the criteria for taxonometric strategies. Comput. J. 14:162-165. [4]
Williams, W.T., G.N. Lance, L.J. Webb, J.G. Tracey & M.B. Dale. 1969. Studies in the numerical analysis of complex rain-forest communities. III. The analysis of successional data. J. Ecol. 57:515-535. [2]
Wirth, M., G.F. Estabrook & D.J. Rogers. 1966. A graph theory model for systematic biology with an example for the Oncidiinae (Orchidaceae)[1] Syst. Zool. 15:59-69. [4]
Wishart, D. 1969a. An algorithm for hierarchical classifications. Biometrics 22:165-170. [4, 6]
Wishart, D. 1969b. Numerical classification method for deriving natural classes. Nature 22:97-98. [4]
Wishart, D. 1971. A generalized approach to cluster analysis. Part of Ph. D. Thesis, University of St. Andrews, Scotland. [4]
Yarranton, G.A. 1966. A plotless method of sampling vegetation. J. Ecol. 54:229-237. [1]
Yarranton, G.A. 1967a. Principal components analysis of data from saxicolous bryophyte vegetation at Steps Bridge, Devon. I. A quantitative assessment of variation in the vegetation. Can. J. Bot. 45:93-115. [3]
Yarranton, G.A. 1967b. Principal component analysis of data from saxicolous bryophyte vegetation at Steps Bridge, Devon. II. An experiment with heterogeneity. Can. J. Bot. 45:229-247. [3]
Yarranton, G.A. 1967c. Organismal and individualistic concepts and the choice of methods of vegetation analysis. Vegetatio 15:113-116. [3]
Yarranton, G.A. 1969. Plant ecology: a unifying model. J. Ecol. 57:245-250. [3]
Yarranton, G.A. 1970. Towards a mathematical model of limestone pavement vegeta-

tion. III. Estimation of the determinants of species frequency. Can. J. Bot. 48:1387-1404. [3]

Yarranton, G.A. 1973. A graph theoretical test of phytosociological homogeneity. Vegetatio 28:283-298. [4]

Yarranton, G.A., W.J. Beasleigh, R.G. Morrison & M.I. Shafi. 1972. On the classification of phytosociological data into nonexclusive groups with a conjecture about determining the optimum number of groups in a classification. Vegetatio 24:1-12. [4, 6]

Zahn, C.T. 1971. Graph-theoretical methods for detecting and describing Gestalt clusters. IEEE Transp. Comput. C-20, 1:68-86. [4]

Zar, J.H. 1974. Biostatistical Analysis. Englewood Cliffs, Prentice-Hall. [1]

APPENDIX

1. Introduction

The computer programs in this Appendix supplement the examples in the main text. The programs use the BASIC computer language. This is a conversational language designed to solve mathematical problems from a teletype console. The fundamentals of BASIC are described by J. G. Kemeny and T. E. Kurtz in their manual "Basic" published at Dartmouth College, Computing Center (1968). An adaptation to the PDP-10 Time-Sharing System, for which the present programs were written, is found in the manual "Advanced Basic for the PDP-10" by Digital Equipment Corporation, Maynard, Massachusetts, U.S.A.(1974).

A typical sequence of commands to run a BASIC program from a teletype console on the PDP-10 includes:

```
.R BASIC /
(READY)
OLD RANK /
(READY)
RUN /
```

The words between parentheses indicate computer response. The symbol / signifies the return key on the teletype which echoes as a carriage return and line feed. Upon receiving the RUN command the computer begins processing the program. It is assumed that the program, RANK in this example, exists as a disk file.

To open a disk file for a program the following commands give a typical example:

```
.R BASIC /
(READY)
NEW RANK /
(READY)
SAVE /
(READY)
10 PRINT "PROGRAM NAME --- RANK" /
.
. -Type all statements in program-
.
1170 END /
REPLACE /
(READY)
```

The SAVE command opens file RANK on disk. The REPLACE command causes the computer to transfer program RANK from user core to disk.

2. Brief Description of Programs

RANK - Species are ranked based on a sum of squares criterion. The method is described in Chapter 1.

SPVAR - Species are ranked according to specific variance. The method is described in Chapter 1.

STRESS - When run after program RANK this program computes stress analysis in a manner described in Chapter 1.

STRESC - Stress analysis is computed for count data following steps as outlined in Chapter 1.

EUCD - Euclidean distances are computed from raw observations or from normalized data according to formula (2.1) or (2.2) as described in Chapter 2.

OBLIK - This program should be run after CORRS has already been run with options z=1 and v=1. Euclidean distances are computed based on oblique reference axes. The method is explained in Chapter 2.

GEND - A generalized distance is computed for two quadrat vectors. The description of method is given in Chapter 2.

CORRS - Vector scalar products are computed in accordance with the descriptions given in Chapter 2.

PINDEX - A version of Goodall's similarity index (Chapter 2) is computed.

HYPD - The Calhoun distance is computed for species. The method and an example are described in Chapter 2.

INFC - Information quantities are computed as chosen depending on the option selected by the user. The definitions are given in Chapter 2 and the options specified in the listing of the program.

RANKIN - Information is partitioned into components based on which species are ranked. A description and an example are given in Chapter 2.

PCAR - The R-algorithm of component analysis is computed as outlined in Chapter 3.

PCAD - The D or Q-algorithm of component analysis is computed as described in Chapter 3. The input data include Euclidean distances generated in program EUCD, or vector scalar products computed in program CORRS.

OFORD - Scores are determined for quadrats to serve as ordination co-ordinates on orthogonal functions. The method is described in Chapter 3 including an example (under ordination by orthogonal functions).

PVO - Position vectors ordination (described in Chapter 3) is computed in this program. Vector scalar products (from program CORRS) represent the input data.

BCOAX - The Bray and Curtis ordination is computed in this program with an option to perpendicularize the co-ordinate axes as described in Chapter 3. File COOR of program CORRS serves as input.

RQT - The RQ-algorithm is computed for reciprocal ordering. Counts of individuals represent input data. The technique is described in Chapter 3.

STEREO - Left and right sets of co-ordinates are produced for stereograms. The method and an example are described in Chapter 3.

ALC - Average linkage clustering is performed on the data. A similarity matrix serves as input. The procedure is described in Chapter 4.

SSA - Sum of squares clustering (Chapter 4) is computed. Euclidean distances are used for input data.

TRGRPS - This program performs a deterministic test on the hypothesis that a sample divides along natural discontinuities into a specified number of groups. Descriptions of the method and an example are given in Chapter 4.

TSTCOV - This program computes a test on the equality of covariance matrices. The method is described in Chapter 4.

EMV - A test on the equality of mean vectors is computed (see descriptions in Chapter 4).

MTFDT - The hypothesis that there is no difference in average similarity between two groups is tested based on Mountford's method. An example and the method are described in Chapter 4.

SSSIM2 - This is a simulation program. It generates dendrograms and determines their topology matrices for input in SSSIM3. The procedure is described in Chapter 4.

SSSIM3 - An empirical distribution is determined for I(m;o) from input originating in program SSSIM2. The method is described in Chapter 4.

3. Program Directory

4. A Note Regarding Programs

The programs were written with a dual purpose in mind. Firstly, they should help those who would like to apply the methods, but are unable to write their own programs. Secondly, they should also help to give a clear and explicit statement of the algorithms. To avoid obscuring the computational steps, certain parts of the programs do not provide for the simplest and most efficient computations. However, the inefficiencies are rather inconsequential in computations with small data sets. Those who wish to use the programs with excessively large volumes of data can have the programs revised to suit better their own specific needs.

5. Programs From Chapter 1

```
RANK            22:31           20-DEC-76

00010 PRINT "PROGRAM NAME --- RANK"
00020 REM---    THIS PROGRAM RANKS SPECIES ACCORDING TO A SUM OF
00030 REM SQUARES CRITERION. THE DATA FOR INPUT ARE STORED IN DISK
00040 REM FILE RAWD AS P SETS OF N NUMBERS.  P INDICATES
00050 REM THE NUMBER OF SPECIES AND N THE NUMBER OF QUADRATS.
00060 REM SPECIES MUST NOT HAVE ZERO VARIANCE.
00070 PRINT "=============================================="
00080 FILES RAWD,STAD,ORD
00090 REM---FILES STAD AND ORD ARE WRITTEN IN THE PROGRAM. STAD HOLDS
00100 REM   THE STANDARDIZED DATA AND ORD THE SPECIES RANKS FOR
00110 REM   INPUT IN PROGRAM STRESS.
00120 SCRATCH #2,3
00130 DIM R(5,5),X(5,10),D(10,5),Y(5),K(5)
00150 REM---EXPLANATION OF ARRAY SYMBOLS:
00170 REM          R - A P*P ARRAY
00180 REM          X - A P*N ARRAY
00190 REM          D - THE TRANSPOSE OF ARRAY X
00200 REM          K, Y - P-VALUED VECTORS
00220 MAT K=ZER
00230 REM---READ DATA
00240 PRINT "NUMBER OF SPECIES P";
00250 INPUT P
00260 PRINT "NUMBER OF QUADRATS N";
00270 INPUT N
00280 FOR I=1 TO P
00290 FOR K=1 TO N
00300 READ #1,X(I,K)
00310 NEXT K,I
00330 REM---CENTER, STANDARDIZE DATA WITHIN ROWS (SPECIES)
00340 PRINT "TYPE 1 IF STANDARDIZATION REQUIRED ELSE TYPE 0";
00350 INPUT Q
00360 FOR I=1 TO P
00370 LET M=S=0
00390 FOR K=1 TO N
00400 LET M=M+X(I,K)
00410 LET S=S+X(I,K)**2
00420 NEXT K
00430 LET S=(S-M**2/N)
00440 FOR K=1 TO N
00450 LET X(I,K)=X(I,K)-M/N
00460 IF Q=0 THEN 480
00470 LET X(I,K)=X(I,K)/SQR(S)
00480 WRITE #2,X(I,K)
00490 NEXT K,I
00510 REM---GENERATE R MATRIX
00520 PRINT "TYPE 1 IF PRINTING OF R MATRIX REQUIRED"
00530 PRINT "ELSE TYPE 0";
00540 INPUT Q
00550 PRINT "TYPE 1 IF PRINTING OF SUM OF SQUARED VECTOR"
```

```
00560 PRINT "PROJECTIONS REQUIRED ELSE TYPE 0";
00570 INPUT Z
00580 MAT D=TRN(X)
00590 MAT R=X*D
00600 FOR I=1 TO P
00610 LET A9=A9+R(I,I)
00620 NEXT I
00630 LET A=0
00640 LET M=0
00650 IF Q=0 THEN 720
00660 PRINT
00670 PRINT
00680 PRINT "R MATRIX OR RESIDUAL"
00690 MAT PRINT R;
00700 REM---COMPUTE SUM OF SQUARED PROJECTIONS OF VECTORS
00710 REM   ON A GIVEN SPECIES VECTOR
00720 FOR I=1 TO P
00730 LET S=0
00740 IF R(I,I)=0 THEN 840
00750 FOR K=1 TO P
00760 LET S=S+(R(I,K)**2)/R(I,I)
00770 NEXT K
00780 IF Z=0 THEN 810
00790 PRINT "THE SUM OF SQUARED PROJECTIONS OF SPECIES VECTORS"
00800 PRINT "ON SPECIES"I"IS"S
00810 IF M>=S THEN 840
00820 LET M=S
00830 LET L=I
00840 NEXT I
00850 IF L=0 THEN 1050
00860 LET A=A+1
00870 PRINT "SPECIES"L"RANK"A"SPECIFIC SUM OF SQUARES"M;
00880 PRINT "PER CENT"100*M/A9
00890 LET K(L)=A
00900 REM---COMPUTE RESIDUAL OF R
00910 FOR I=1 TO P
00920 LET Y(I)=R(I,L)/SQR(R(L,L))
00930 NEXT I
00940 FOR I=1 TO P
00950 FOR K=I TO P
00960 LET R(I,K)=R(I,K)-Y(I)*Y(K)
00970 LET R(K,I)=R(I,K)
00980 NEXT K
00990 LET R(L,I)=0
01000 LET R(I,L)=0
01010 NEXT I
01020 LET L=0
01030 IF A=P THEN 1100
01040 GO TO 640
01050 FOR I=1 TO P
01060 IF K(I)>0 THEN 1090
01070 LET A=A+1
01080 LET K(I)=A
01090 NEXT I
01100 FOR L=1 TO P
```

```
01110 WRITE #3,K(L)
01120 PRINT "SPECIES"L"RANK"K(L)
01130 NEXT L
01140 PRINT
01150 PRINT "CONTINUE WITH PROGRAM STRESS IF STRESS ANALYSIS";
01160 PRINT " IS REQUIRED"
01170 END

RAWD            22:39          20-DEC-76

01000 45,2,9,26,3,5,2,90,31,16,18,92,32,48,73,80,95,13,92,78
01010 3,40,5,83,68,27,2,17,1,23,10,61,11,3,32,2,39,2,8,6
01020 9,53,99,21,49,81,72,6,90,62

.R BASIC

READY
OLD RANK
READY
RUN

RANK            22:41          20-DEC-76

PROGRAM NAME --- RANK
===============================================
NUMBER OF SPECIES P ?5
NUMBER OF QUADRATS N ?10
TYPE 1 IF STANDARDIZATION REQUIRED ELSE TYPE 0 ?0
TYPE 1 IF PRINTING OF R MATRIX REQUIRED
ELSE TYPE 0 ?0
TYPE 1 IF PRINTING OF SUM OF SQUARED VECTOR
PROJECTIONS REQUIRED ELSE TYPE 0 ?0

SPECIES 2 RANK 1 SPECIFIC SUM OF SQUARES 17027.2 PER CENT 46.2184
SPECIES 3 RANK 2 SPECIFIC SUM OF SQUARES 9061.53 PER CENT 24.5965
SPECIES 5 RANK 3 SPECIFIC SUM OF SQUARES 6281.31 PER CENT 17.0499
SPECIES 4 RANK 4 SPECIFIC SUM OF SQUARES 2956.14 PER CENT 8.0241
SPECIES 1 RANK 5 SPECIFIC SUM OF SQUARES 1514.55 PER CENT 4.11108

SPECIES 1 RANK 5
SPECIES 2 RANK 1
SPECIES 3 RANK 2
SPECIES 4 RANK 4
SPECIES 5 RANK 3

CONTINUE WITH PROGRAM STRESS IF STRESS ANALYSIS IS REQUIRED

TIME:  1.72 SECS.

READY
```

```
SPVAR          12:32          20-DEC-76

00010 PRINT "PROGRAM NAME -- SPVAR"
00020 REM--- THIS PROGRAM DETERMINES SPECIFIC VARIANCES FOR P SPECIES.
00030 REM THE SPECIES ARE RANKED BY MULTIPLE CORRELATION. DATA ARE
00040 REM READ FROM DISK FILE COR9 WHICH CONTAINS THE UPPER HALF OF A
00050 REM CROSS PRODUCTS MATRIX INCLUDING ALSO THE SUMS OF SQUARES IN
00060 REM THE PRINCIPAL DIAGONAL POSITIONS. THE CROSS PRODUCTS AND
00070 REM SUM OF SQUARES CAN BE GENERATED IN PROGRAM CORRS.
00100 PRINT "================================"
00110 FILES COR9
00120 DIM S(5,5),R(5,5),Y(5),V(5)
00130 DIM Q(5),Z(5)
00140 REM---EXPLANATION OF ARRAY SYMBOLS:
00150 REM           S, R -  P*P ARRAYS
00160 REM           Y, V, Q, Z, -  P-VALUED VECTORS
00170 PRINT "NUMBER OF SPECIES P";
00180 INPUT P
00190 PRINT "TYPE 1 IF PRINTING OF INTERMEDIATE RESULTS "
00200 PRINT "IS REQUIRED ELSE TYPE 0";
00220 INPUT G
00230 LET M2=P+1
00240 MAT Z=CON
00250 FOR H=1 TO P
00260 FOR I=H TO P
00270 READ #1,R(H,I)
00280 LET R(I,H)=R(H,I)
00290 NEXT I
00310 NEXT H
00320 IF G=0 THEN 360
00330 PRINT "CROSS PRODUCTS MATRIX"
00340 MAT PRINT R;
00360 PRINT
00370 PRINT "SUMS OF SQUARES (SS), SPECIFIC SS (SPSS),"
00380 PRINT "COMMON SS (CSS) AND SQUARED MULTIPLE"
00390 PRINT "CORRELATION (R^2) IN P-SPECIES SAMPLE"
00400 FOR C=1 TO P
00410 IF Z(C)=0 THEN 720
00420 MAT S=R
00430 LET O=0
00440 FOR M=1 TO P
00450 IF Z(M)=0 THEN 680
00460 IF M=C THEN 680
00470 LET O=O+1
00480 FOR H=1 TO P
00490 IF Z(H)=0 THEN 510
00500 LET Y(H)=S(H,M)/SQRT(S(M,M))
00510 NEXT H
00520 IF G=0 THEN 560
00540 PRINT "Y CO-ORDINATES"
00550 MAT PRINT Y;
00560 FOR H=1 TO P
00570 FOR I=1 TO P
00580 LET S(H,I)=S(H,I)-Y(H)*Y(I)
00590 NEXT I,H
```

```
00600 FOR H=1 TO P
00610 LET S(M,H)=0
00620 LET S(H,M)=0
00630 NEXT H
00640 IF G=0 THEN 680
00660 PRINT "RESIDUAL "O
00670 MAT PRINT S;
00680 NEXT M
00690 LET V(C)=1-S(C,C)/R(C,C)
00700 PRINT "SPECIES"C"SS"R(C,C)"SPSS"S(C,C)"CSS"R(C,C)-S(C,C);
00710 PRINT "R^2"1-S(C,C)/R(C,C)
00720 NEXT C
00730 LET M=0
00740 FOR I=1 TO P
00750 IF Z(I)=0 THEN 790
00760 IF M>V(I) THEN 790
00770 LET M=V(I)
00780 LET C=I
00790 NEXT I
00800 PRINT "MAX R^2"V(C)"SPEC"C
00810 IF M2=2 THEN 840
00820 PRINT
00830 PRINT "SPECIES"C"REMOVED"
00840 LET M2=M2-1
00850 LET Q(M2)=C
00860 LET Z(C)=0
00880 FOR I=1 TO P
00890 LET Y(I)=R(C,I)/SQR(R(C,C))
00900 NEXT I
00910 FOR I=1 TO P
00920 FOR J=I TO P
00930 LET R(I,J)=R(I,J)-Y(I)*Y(J)
00940 LET R(J,I)=R(I,J)
00950 NEXT J,I
00960 FOR I=1 TO P
00970 LET R(I,C)=0
00980 LET R(C,I)=0
00990 NEXT I
01000 IF M2=1 THEN 1055
01010 IF G=0 THEN 1050
01030 PRINT "RESIDUAL CROSS PRODUCTS"
01040 MAT PRINT R;
01050 GO TO 400
01055 PRINT
01060 PRINT "SUMMARY OF R^2 VALUES"
01080 FOR I=1 TO P
01090 PRINT "SPECIES "I;V(I)
01100 NEXT I
01110 END

COR9            12:38            20-DEC-76

01000 6896.9,-5611.9,-1470.1,-2527.6,-5445.8,9022.9,594.1
01010 2913.6,5646.8,7462.9,581.4,-2913.8,3576.4,686.2,9881.6
```

```
.R BASIC

READY
OLD SPVAR
READY
RUN

SPVAR          12:41          20-DEC-76

PROGRAM NAME -- SPVAR
==============================
NUMBER OF SPECIES P ?5
TYPE 1 IF PRINTING OF INTERMEDIATE RESULTS
IS REQUIRED ELSE TYPE 0 ?0

SUMS OF SQUARES (SS), SPECIFIC SS (SPSS),
COMMON SS (CSS) AND SQUARED MULTIPLE
CORRELATION (R^2) IN P-SPECIES SAMPLE
SPECIES 1 SS 6896.9 SPSS 1514.55 CSS 5382.35 R^2 0.780401
SPECIES 2 SS 9022.9 SPSS 3495.24 CSS 5527.66 R^2 0.612626
SPECIES 3 SS 7462.9 SPSS 3672.12 CSS 3790.78 R^2 0.507949
SPECIES 4 SS 3576.4 SPSS 1795.22 CSS 1781.18 R^2 0.498037
SPECIES 5 SS 9881.6 SPSS 2232.79 CSS 7648.81 R^2 0.774045
MAX R^2 0.780401 SPEC 1

SPECIES 1 REMOVED
SPECIES 2 SS 4456.58 SPSS 3495.24 CSS 961.349 R^2 0.215714
SPECIES 3 SS 7149.54 SPSS 3672.12 CSS 3477.42 R^2 0.486383
SPECIES 4 SS 2650.08 SPSS 1795.22 CSS 854.855 R^2 0.322577
SPECIES 5 SS 5581.59 SPSS 2232.79 CSS 3348.8 R^2 0.599972
MAX R^2 0.599972 SPEC 5

SPECIES 5 REMOVED
SPECIES 2 SS 4191.83 SPSS 3495.24 CSS 696.59 R^2 0.166178
SPECIES 3 SS 4175.07 SPSS 3672.12 CSS 502.943 R^2 0.120463
SPECIES 4 SS 2342.81 SPSS 1795.22 CSS 547.587 R^2 0.233731
MAX R^2 0.233731 SPEC 4

SPECIES 4 REMOVED
SPECIES 2 SS 3635.01 SPSS 3495.24 CSS 139.774 R^2 3.84521E-2
SPECIES 3 SS 3818.97 SPSS 3672.12 CSS 146.848 R^2 3.84521E-2
MAX R^2 3.84521E-2 SPEC 3

SPECIES 3 REMOVED
SPECIES 2 SS 3495.24 SPSS 3495.24 CSS 0 R 2 0
MAX R^2 0 SPEC 2

SUMMARY OF R^2 VALUES
SPECIES  1  0.780401
SPECIES  2  0
SPECIES  3  3.84521E-2
SPECIES  4  0.233731
SPECIES  5  0.599972

TIME:  2.31 SECS.
```

```
READY


STRESS          20:53          22-DEC-76

00010 PRINT "PROGRAM NAME --- STRESS"
00020 REM---  RUN THIS PROGRAM AFTER PROGRAM RANK TO COMPUTE
00030 REM STRESS ANALYSIS.
00040 PRINT "=============================================="
00050 FILES STAD, ORD, DIST
00060 SCRATCH #3
00070 DIM X(5,10), R(5)
00100 REM---EXPLANATION OF ARRAY SYMBOLS:
00110 REM           X - A P*N ARRAY WHERE P REPRESENTS THE NUMBER
00120 REM               OF SPECIES AND N THE NUMBER OF QUADRATS
00130 REM           R - A P-VALUED VECTOR
00160 REM---READ DATA
00170 PRINT "NUMBER OF SPECIES P";
00180 INPUT P
00190 PRINT "NUMBER OF QUADRATS N";
00200 INPUT N
00210 LET A=N
00220 LET N=P
00230 LET P=A
00240 FOR K=1 TO N
00250 READ #2,R(K)
00260 NEXT K
00270 FOR K=1 TO N
00280 LET R1=R(K)
00290 FOR J=1 TO P
00300 READ #1,A
00310 LET X(R1,J)=A
00320 NEXT J,K
00330 REM---COMPUTE DISTANCES BASED ON P SPECIES
00340 LET N1=0
00350 FOR K=1 TO P-1
00360 FOR J=K+1 TO P
00370 LET A=0
00380 FOR H=1 TO N
00390 LET A=A+(X(H,K)-X(H,J))^2
00400 NEXT H
00410 WRITE #3,A
00420 NEXT J,K
00430 REM---COMPUTE DISTANCES USING REDUCED NUMBER OF SPECIES
00440 LET N1=N1+1
00450 LET S1=S2=S3=S4=S5=0
00460 RESTORE #3
00470 FOR K=1 TO P-1
00480 FOR J=K+1 TO P
00490 READ #3,A
00500 LET B=0
00510 FOR H=1 TO N1
00520 LET B=B+(X(H,K)-X(H,J))^2
00530 NEXT H
```

```
00540 LET S1=S1+A
00550 LET S2=S2+B
00560 LET S3=S3+A^2
00570 LET S4=S4+B^2
00580 LET S5=S5+A*B
00590 NEXT J,K
00600 REM---COMPUTE CORRELATION AND STRESS
00610 LET A=(P*(P-1))/2
00620 LET S5=S5-S1*S2/A
00630 LET S3=S3-S1^2/A
00640 LET S4=S4-S2^2/A
00650 PRINT "NUMBER OF SPECIES IN REDUCED SET "N1
00660 LET B=S5/SQR(S3*S4)
00670 PRINT "CORRELATION"B,"STRESS  "100*(1-B*B)
00680 PRINT
00690 IF N1=N THEN 710
00700 GO TO 440
00710 END

STAD          20:54          22-DEC-76

01000 22.1,-20.9,-13.9,3.1,-19.9,-17.9,-20.9,67.1,8.1,-6.9
01010 -44.1,29.9,-30.1,-14.1,10.9,17.9,32.9,-49.1,29.9,15.9
01020 -23.9,13.1,-21.9,56.1,41.1,9.99999E-2,-24.9,-9.9,-25.9
01030 -3.9,-7.4,43.6,-6.4,-14.4,14.6,-15.4,21.6,-15.4,-9.4
01040 -11.4,-45.2,-1.2,44.8,-33.2,-5.2,26.8,17.8,-48.2,35.8
01050 7.8

ORD           20:54          22-DEC-76

01000 5,1,2,4,3

.R BASIC

READY
OLD STRESS
READY
RUN

STRESS        20:55          22-DEC-76

PROGRAM NAME --- STRESS
=========================================================
NUMBER OF SPECIES P ?5
NUMBER OF QUADRATS N ?10

NO OF SPECIES IN REDUCED SET  1
CORRELATION 0.728787     STRESS  46.8869

NO OF SPECIES IN REDUCED SET  2
CORRELATION 0.728298     STRESS  46.9582

NO OF SPECIES IN REDUCED SET  3
CORRELATION 0.886349     STRESS  21.4386
```

```
NO OF SPECIES IN REDUCED SET  4
CORRELATION 0.920366      STRESS  15.2926

NO OF SPECIES IN REDUCED SET  5
CORRELATION 1.            STRESS -5.96046E-6

TIME:  2.22 SECS.

READY


STRESC          14:04          01-MAR-77

00010 PRINT "PROGRAM NAME --- STRESC"
00020 REM--- STRESS IS COMPUTED BASED ON COUNT DATA USING K,
00030 REM THE MAXIMUM LIMIT FOR COUNTS, AS A VARIABLE.  THE
00040 REM DATA ARE STORED IN DISK FILE RAWD AS P SETS OF N
00050 REM NUMBERS. P SPECIFIES THE NUMBER OF SPECIES AND N
00060 REM THE NUMBER OF QUADRATS.
00070 PRINT "=================================================="
00080 FILES RAWD, DIST
00090 SCRATCH #2
00100 DIM X(5,10)
00110 REM---EXPLANATION OF ARRAY SYMBOL:
00120 REM            X - A P*N ARRAY
00130 REM---READ DATA
00140 PRINT "NUMBER OF SPECIES P";
00150 INPUT P
00160 PRINT "NUMBER OF QUADRATS N";
00170 INPUT N
00180 PRINT
00190 PRINT " K","CORRELATION"," STRESS"
00200 FOR I=1 TO P
00210 FOR J=1 TO N
00220 READ #1,X(I,J)
00230 NEXT J,I
00240 REM---COMPUTE DISTANCES FROM ORIGINAL COUNTS
00250 LET N1=N*(N-1)/2
00260 LET S1=S2=0
00270 FOR J=1 TO N-1
00280 FOR M=J+1 TO N
00290 LET S6=0
00300 FOR H=1 TO P
00310 LET S6=S6+(X(H,J)-X(H,M))^2
00320 NEXT H
00330 WRITE #2,S6
00340 LET S1=S1+S6
00350 LET S2=S2+S6^2
00360 NEXT M,J
00370 REM---COMPUTE DISTANCES FROM TRUNCATED COUNTS
00380 LET K=5
00390 LET S3=S4=S5=0
00400 RESTORE #2
00410 FOR J=1 TO N-1
```

```
00420 FOR M=J+1 TO N
00430 LET S6=0
00440 FOR H=1 TO P
00450 IF X(H,J)<=K THEN 480
00460 LET A=K
00470 GO TO 490
00480 LET A=X(H,J)
00490 IF X(H,M)<=K THEN 520
00500 LET B=K
00510 GO TO 530
00520 LET B=X(H,M)
00530 LET S6=S6+(A-B)^2
00540 NEXT H
00550 LET S3=S3+S6
00560 LET S4=S4+S6^2
00570 READ #2,A
00580 LET S5=S5+S6*A
00590 NEXT M,J
00600 LET A=S2-S1^2/N1
00610 LET B=S4-S3^2/N1
00620 LET C=S5-S1*S3/N1
00630 LET D=C^2/(A*B)
00640 PRINT K,SQR(D),1-D
00650 LET K=K+5
00660 IF 1-D>.01 THEN 390
00670 END

RAWD            14:11           01-MAR-77

01000 45,2,9,26,3,5,2,90,31,16,18,92,32,48,73,80,95,13,92,78
01010 3,40,5,83,68,27,2,17,1,23,10,61,11,3,32,2,39,2,8,6
01020 9,53,99,21,49,81,72,6,90,62

.R BASIC

READY
OLD STRESC
READY
RUN

STRESC          14:14           01-MAR-77

PROGRAM NAME --- STRESC
==================================================
NUMBER OF SPECIES P ?5
NUMBER OF QUADRATS N ?10

 K              CORRELATION    STRESS
 5              9.24958E-2     0.991445
 10             0.277735       0.922863
 15             0.379077       0.856301
 20             0.488919       0.760959
 25             0.569557       0.675604
 30             0.640074       0.590306
 35             0.715967       0.487391
```

40	0.768257	0.409781
45	0.817181	0.332215
50	0.857721	0.264314
55	0.889989	0.207919
60	0.913295	0.165893
65	0.936184	0.123559
70	0.955287	8.74266E-2
75	0.972166	5.48941E-2
80	0.982968	3.37737E-2
85	0.992263	1.54139E-2
90	0.997155	5.68229E-3

TIME: 9.92 SECS.

```
EUCD          16:19          04-JAN-77

00010 PRINT "PROGRAM NAME --- EUCD"
00020 REM--- THIS PROGRAM COMPUTES EUCLIDEAN DISTANCE BETWEEN
00030 REM QUADRATS FROM RAW OR NORMALIZED DATA.  THE DATA ARE READ
00040 REM FROM DISK FILE RAWD9 ARRANGED AS P SETS OF N NUMBERS.  P
00050 REM SIGNIFIES THE NUMBER OF SPECIES AND N THE NUMBER OF
00060 REM QUADRATS. THE COMPUTED DISTANCES ARE WRITTEN INTO
00065 REM DISK FILE DIS.
00070 PRINT "========================================================"
00080 FILES RAWD9, DIS
00090 SCRATCH #2
00100 DIM X(3,5), Y(5)
00120 REM---EXPLANATION OF ARRAY SYMBOLS:
00140 REM          X - A P*N ARRAY
00150 REM          Y - AN N-VALUED VECTOR
00180 REM---READ DATA, NORMALIZE QUADRAT VECTORS
00190 PRINT "NUMBER OF SPECIES P";
00200 INPUT P
00210 PRINT "NUMBER OF QUADRATS N";
00220 INPUT N
00230 PRINT "TYPE 1 IF NORMALIZATION REQUIRED ELSE TYPE 0";
00240 INPUT Z
00250 MAT Y=ZER
00260 FOR I=1 TO P
00270 FOR J=1 TO N
00280 READ #1,A
00290 LET X(I,J)=A
00300 LET Y(J)=Y(J)+X(I,J)^2
00310 NEXT J,I
00320 IF Z=1 THEN 340
00330 MAT Y=CON
00340 FOR I=1 TO N
00350 LET Y(I)=SQR(Y(I))
00360 NEXT I
00370 REM---COMPUTE DISTANCES
00380 PRINT "TYPE 1 IF PRINTING OF DISTANCES REQUIRED ELSE TYPE 0";
00390 INPUT V
00400 PRINT
00410 FOR J=1 TO N-1
00420 FOR K=J+1 TO N
00430 LET S=0
00440 FOR I=1 TO P
00450 LET S=S+(X(I,J)/Y(J)-X(I,K)/Y(K))^2
00460 NEXT I
00470 WRITE #2,SQR(S)
00480 NEXT K,J
00490 IF V=0 THEN 590
00500 RESTORE #2
00510 PRINT "UPPER HALF OF DISTANCE MATRIX"
00520 FOR J=1 TO N-1
```

```
00530 FOR K=J+1 TO N
00540 READ #2,A
00550 PRINT A;
00560 NEXT K
00570 PRINT
00580 NEXT J
00590 END

RAWD9          16:28          04-JAN-77

01000 2,5,2,1,0,0,1,4,3,1,3,4,1,0,0

.R BASIC

READY
OLD EUCD
READY
RUN

EUCD           16:29          04-JAN-77

PROGRAM NAME --- EUCD
=======================================================
NUMBER OF SPECIES P ?3
NUMBER OF QUADRATS N ?5
TYPE 1 IF NORMALIZATION REQUIRED ELSE TYPE 0 ?0
TYPE 1 IF PRINTING OF DISTANCES REQUIRED ELSE TYPE 0 ?1

UPPER HALF OF DISTANCE MATRIX
 3.31662  4.47214  4.3589  3.74166
 5.19615  6        6.40312
 1.73205  3.74166
 2.23607

TIME:  1.13 SECS.

READY


OBLIK          12:28          12-JAN-77

00010 PRINT "PROGRAM NAME --- OBLIK"
00020 REM--- THIS PROGRAM COMPUTES EUCLIDEAN DISTANCES BASED ON
00030 REM OBLIQUE CO-ORDINATES.  THE DATA ARE READ FROM DISK
00040 REM FILE RAWD9 ARRANGED AS P SETS OF N NUMBERS. P INDICATES
00050 REM THE NUMBER OF SPECIES AND N THE NUMBER OF QUADRATS.
00060 REM THE COMPUTED QUADRAT DISTANCES ARE WRITTEN INTO DISK FILE
00070 REM DIS.  RUN PROGRAM EUCD WITH OPTION Z=0 AND CORRS WITH
00080 REM OPTIONS Z=1 AND V=1 BEFORE RUNNING PROGRAM OBLIK.
00090 PRINT "======================================================="
00100 FILES RAWD9, OBD, DIS, COR
00110 SCRATCH #2
00120 DIM X(3,5), R(3,3)
00140 REM---EXPLANATION OF ARRAY SYMBOLS:
```

```
ØØ16Ø REM         X - A P*N ARRAY
ØØ17Ø REM         R - A P*P ARRAY
ØØ19Ø REM---READ DATA, CENTER DATA WITHIN SPECIES
ØØ2ØØ PRINT "NUMBER OF SPECIES P";
ØØ21Ø INPUT P
ØØ22Ø PRINT "NUMBER OF QUADRATS N";
ØØ23Ø INPUT N
ØØ24Ø PRINT "TYPE 1 IF PRINTING OF DISTANCES REQUIRED ELSE TYPE Ø";
ØØ25Ø INPUT Z
ØØ26Ø PRINT
ØØ27Ø FOR I=1 TO P
ØØ28Ø FOR J=1 TO N
ØØ29Ø READ #1,X(I,J)
ØØ3ØØ NEXT J,I
ØØ31Ø REM---READ FILES DIS AND COR
ØØ32Ø FOR H=1 TO P
ØØ33Ø FOR I=H TO P
ØØ34Ø READ #4,R(H,I)
ØØ35Ø NEXT I,H
ØØ36Ø REM---COMPUTE DISTANCE MATRIX
ØØ37Ø FOR J=1 TO N-1
ØØ38Ø FOR K=J+1 TO N
ØØ39Ø LET S=Ø
ØØ4ØØ FOR H=1 TO P-1
ØØ41Ø FOR I=H+1 TO P
ØØ42Ø LET S=S+(X(H,J)-X(H,K))*(X(I,J)-X(I,K))*R(H,I)
ØØ43Ø NEXT I,H
ØØ44Ø READ #3,D
ØØ45Ø LET A=SQR(D^2+2*S)
ØØ46Ø WRITE #2,A
ØØ47Ø NEXT K,J
ØØ48Ø REM---PRINT DISTANCE MATRIX
ØØ49Ø IF Z=Ø THEN 57Ø
ØØ5ØØ RESTORE #2
ØØ51Ø PRINT "UPPER HALF OF DISTANCE MATRIX"
ØØ52Ø FOR J=1 TO N-1
ØØ53Ø FOR K=J+1 TO N
ØØ54Ø READ #2,A
ØØ55Ø PRINT A;
ØØ56Ø NEXT K
ØØ57Ø PRINT
ØØ58Ø NEXT J
ØØ59Ø END

RAWD9          12:3Ø          12-JAN-77

Ø1ØØØ 2,5,2,1,Ø,Ø,1,4,3,1,3,4,1,Ø,Ø

.R BASIC

READY
OLD EUCD
READY
RUN
```

```
EUCD          12:33          12-JAN-77

PROGRAM NAME --- EUCD
==========================================================
NUMBER OF SPECIES P ?3
NUMBER OF QUADRATS N ?5
TYPE 1 IF NORMALIZATION REQUIRED ELSE TYPE Ø ?Ø
TYPE 1 IF PRINTING OF DISTANCES REQUIRED ELSE TYPE Ø ?Ø

TIME:  Ø.8Ø SECS.

READY
OLD CORRS
READY
RUN

CORRS         12:36          12-JAN-77

PROGRAM NAME --- CORRS
======================================================
NUMBER OF SPECIES P ?3
NUMBER OF QUADRATS N ?5
TYPE 1 TO NORMALIZE SPECIES VECTORS ELSE TYPE Ø ?1
TYPE 1 TO COMPUTE SPECIES SCALAR PRODUCTS ELSE TYPE Ø ?1
TYPE 1 IF PRINTING OF SCALAR PRODUCTS
REQUIRED ELSE TYPE Ø ?Ø

TIME:  Ø.81 SECS.

READY
OLD OBLIK
READY
RUN

OBLIK         12:38          12-JAN-77

PROGRAM NAME --- OBLIK
======================================================
NUMBER OF SPECIES P ?3
NUMBER OF QUADRATS N ?5
TYPE 1 IF PRINTING OF DISTANCES REQUIRED ELSE TYPE Ø ?1

UPPER HALF OF DISTANCE MATRIX
 3.77471  5.34569  5.9Ø937  5.33475
 7.4475   8.68483  8.73552
 1.83525  3.51611
 2.Ø8552

TIME:  Ø.93 SECS.

READY

GEND          16:35          12-JAN-77

ØØØ1Ø PRINT "PROGRAM NAME --- GEND"
```

```
00020 REM--- THIS PROGRAM COMPUTES THE GENERALIZED DISTANCE OF TWO
00030 REM QUADRAT VECTORS. THE DATA ARE STORED AS 2 SETS OF P NUMBERS
00040 REM IN DISK FILE EXI. P INDICATES THE NUMBER OF SPECIES. RUN
00050 REM PROGRAM CORRS WITH RAWD9 AND OPTIONS Z=0, V=1 BEFORE
00060 REM RUNNING PROGRAM GEND. (RAWD9 HOLDS THE DATA FROM WHICH
00070 REM THE P*P COVARIANCE MATRIX IS COMPUTED IN CORRS.)
00110 PRINT "=============================================================="
00120 FILES EXI, COR
00140 DIM M(3), X(3), R(3,3), I(3,3), Q(3), G(3)
00160 REM---EXPLANATION OF ARRAY SYMBOLS:
00180 REM          R, I - P*P ARRAYS
00190 REM          M, X, Q, G - P-VALUED VECTORS
00220 PRINT "NUMBER OF SPECIES P";
00230 INPUT P
00300 PRINT
00310 MAT M=ZER
00330 FOR J=1 TO P
00340 READ #1,G(J)
00360 NEXT J
00362 FOR J=1 TO P
00364 READ #1,X(J)
00380 NEXT J
00390 FOR H=1 TO P
00400 FOR I=H TO P
00410 READ #2,A
00420 LET R(H,I)=A
00430 LET R(I,H)=A
00440 NEXT I,H
00490 REM---COMPUTE DISTANCE
00500 MAT I=INV(R)
00510 MAT M=X-G
00520 MAT M=TRN(M)
00530 MAT Q=M*I
00540 LET A=0
00550 FOR I=1 TO P
00560 LET A=A+Q(I)*M(I)
00570 NEXT I
00580 LET A=SQR(A)
00600 PRINT A
00640 END

EXI            16:39          12-JAN-77

01000 2,0,3,5,1,4

.R BASIC

READY
OLD CORRS
READY
RUN

CORRS          16:40          12-JAN-77

PROGRAM NAME --- CORRS
```

```
=======================================================
NUMBER OF SPECIES P ?3
NUMBER OF QUADRATS N ?5
TYPE 1 TO NORMALIZE SPECIES VECTORS ELSE TYPE 0 ?0
TYPE 1 TO COMPUTE SPECIES SCALAR PRODUCTS ELSE TYPE 0 ?1
TYPE 1 IF PRINTING OF SCALAR PRODUCTS
REQUIRED ELSE TYPE 0 ?0

TIME:  0.88 SECS.

READY
OLD GEND
READY
RUN

GEND            16:40           12-JAN-77

PROGRAM NAME --- GEND
=======================================================
NUMBER OF SPECIES P ?3

 2.75299

TIME:  0.92 SECS.

READY
```

```
CORRS           17:37           06-JAN-77

00010 PRINT "PROGRAM NAME --- CORRS"
00020 REM--- THIS PROGRAM COMPUTES SCALAR PRODUCTS BETWEEN P SPECIES
00030 REM VECTORS (OPTION V=1) OR N QUADRAT VECTORS (OPTION V=0).
00040 REM THE DATA ARE CENTERED WITHIN SPECIES. AN OPTION FOR
00050 REM NORMALIZATION (Z=1) WITHIN SPECIES IS PROVIDED.  THE DATA
00060 REM ARE STORED IN DISK FILE RAWD9 AS P SETS OF N NUMBERS.  THE
00070 REM CORRELATIONS ARE WRITTEN INTO DISK FILE COR.
00090 PRINT "======================================================="
00100 FILES RAWD9, COR
00110 SCRATCH #2
00120 DIM X(3,5), I(5,3), R(3,3), Q(5,5)
00140 REM---EXPLANATION OF ARRAY SYMBOLS:
00160 REM          X - A P*N ARRAY
00170 REM          I - A TRANSPOSE OF X
00180 REM          R - A P*P ARRAY
00190 REM          Q - AN N*N ARRAY
00220 REM---READ DATA, CENTER AND NORMALIZE DATA WITHIN SPECIES
00240 PRINT "NUMBER OF SPECIES P";
00250 INPUT P
00260 PRINT "NUMBER OF QUADRATS N";
00270 INPUT N
00280 PRINT "TYPE 1 TO NORMALIZE SPECIES VECTORS ELSE TYPE 0";
00290 INPUT Z
00300 PRINT "TYPE 1 TO COMPUTE SPECIES SCALAR PRODUCTS ELSE TYPE 0";
```

```
00320 INPUT V
00330 PRINT "TYPE 1 IF PRINTING OF SCALAR PRODUCTS  "
00340 PRINT "REQUIRED ELSE TYPE 0";
00350 INPUT Y
00351 PRINT
00352 IF Y=O THEN 360
00354 PRINT "VECTOR SCALAR PRODUCTS"
00360 FOR I=1 TO P
00370 LET M=S=0
00380 FOR J=1 TO N
00390 READ #1,A
00400 LET A=A/SQR(N-1)
00410 LET X(I,J)=A
00420 LET M=M+A
00430 LET S=S+A^2
00440 NEXT J
00450 LET S=SQR(S-M^2/N)
00460 LET M=M/N
00470 IF Z=1 THEN 490
00480 LET S=1
00490 FOR J=1 TO N
00500 LET X(I,J)=(X(I,J)-M)/S
00510 LET I(J,I)=X(I,J)
00520 NEXT J,I
00530 IF V=0 THEN 650
00540 MAT R=X*I
00550 FOR H=1 TO P
00560 FOR I=H TO P
00570 WRITE #2,R(H,I)
00580 IF Y=0 THEN 600
00590 PRINT R(H,I);
00600 NEXT I
00610 IF Y=0 THEN 630
00620 PRINT
00630 NEXT H
00640 GO TO 720
00650 MAT Q=I*X
00660 FOR J=1 TO N
00670 FOR K=J TO N
00680 WRITE #2,Q(J,K)
00690 IF Y=0 THEN 710
00700 PRINT Q(J,K);
00710 NEXT K
00720 IF Y=0 THEN 740
00730 PRINT
00740 NEXT J
00750 END

RAWD9          17:40          06-JAN-77

01000 2,5,2,1,0,0,1,4,3,1,3,4,1,0,0

.R BASIC

READY
```

```
OLD CORRS
READY
RUN

CORRS           17:47           Ø6-JAN-77

PROGRAM NAME --- CORRS
=======================================================
NUMBER OF SPECIES P ?3
NUMBER OF QUADRATS N ?5
TYPE 1 TO NORMALIZE SPECIES VECTORS ELSE TYPE Ø ?Ø
TYPE 1 TO COMPUTE SPECIES SCALAR PRODUCTS ELSE TYPE Ø ?1
TYPE 1 IF PRINTING OF SCALAR PRODUCTS
REQUIRED ELSE TYPE Ø ?1

VECTOR SCALAR PRODUCTS
 3.5 -Ø.5  3.
 2.7 -1.6
 3.3

TIME:  Ø.89 SECS.

READY
```

```
PINDEX          13:17           23-DEC-76

ØØØ1Ø PRINT "PROGRAM NAME --- PINDEX"
ØØØ2Ø REM---   THIS PROGRAM COMPUTES ONE VERSION OF THE PROBABILISTI
ØØØ3Ø REM SIMILARITY INDEX FOR THE COLUMNS OF THE DATA MATRIX.  THE
ØØØ4Ø REM INPUT DATA ARE READ FROM DISK FILE RAWD ARRANGED AS P
ØØØ5Ø REM SETS OF N NUMBERS.    THE COMPUTED VALUES OF CHI SQUARE AR
ØØØ6Ø REM WRITTEN INTO DISK FILE CHI.
ØØØ7Ø PRINT "======================================================="
ØØØ8Ø FILES RAWD, CHI
ØØØ9Ø SCRATCH #2
ØØ1ØØ DIM X(5,1Ø), D(5,45), P(5)
ØØ11Ø REM---EXPLANATION OF ARRAY SYMBOLS:
ØØ13Ø REM          X - A P*N ARRAY
ØØ14Ø REM          D - A P*N(N-1)/2 ARRAY
ØØ15Ø REM          P - A P-VALUED VECTOR
ØØ18Ø REM---READ DATA
ØØ19Ø PRINT
ØØ2ØØ PRINT "NUMBER OF ROWS P";
ØØ21Ø INPUT P
ØØ22Ø PRINT "NUMBER OF COLUMNS N";
ØØ23Ø INPUT N
ØØ24Ø FOR H=1 TO P
ØØ25Ø FOR J=1 TO N
ØØ26Ø READ #1,X(H,J)
ØØ27Ø NEXT J,H
ØØ28Ø REM---COMPUTE DIFFERENCES
ØØ29Ø FOR H=1 TO P
ØØ3ØØ LET A=Ø
```

```
00310 FOR J=1 TO N-1
00320 FOR K=J+1 TO N
00330 LET A=A+1
00340 LET D(H,A)=ABS(X(H,J)-X(H,K))
00350 NEXT K,J,H
00360 REM---DETERMINE PROPORTIONS AND COMPUTE CHI SQUARES
00370 PRINT "CHI SQUARE VALUES"
00380 FOR J=1 TO N-1
00390 FOR K=J+1 TO N
00400 LET Q=0
00410 FOR H=1 TO P
00420 LET C=0
00430 LET D=ABS(X(H,J)-X(H,K))
00440 FOR I=1 TO A
00450 IF D(H,I)>D THEN 470
00460 LET C=C+1
00470 NEXT I
00480 IF C>0 THEN 500
00490 LET C=1
00500 LET B=C/A
00510 LET Q=Q-2*LOG(B)
00520 NEXT H
00530 WRITE #2,Q
00540 PRINT Q;
00550 NEXT K
00560 PRINT
00570 NEXT J
00580 REM---WHEN C=0 THE LOGARITHM IS INDETERMINATE. TO RESOLVE
00590 REM    INDETERMINACY THE VALUE OF C IS SET EQUAL TO 1
00600 END

RAWD            13:18           23-DEC-76

01000 45,2,9,26,3,5,2,90,31,16,18,92,32,48,73,80,95,13,92,78
01010 3,40,5,83,68,27,2,17,1,23,10,61,11,3,32,2,39,2,8,6
01020 9,53,99,21,49,81,72,6,90,62

.R BASIC

READY
OLD PINDEX
READY
RUN

PINDEX          13:19           23-DEC-76

PROGRAM NAME --- PINDEX
=================================================================
NUMBER OF ROWS P ?5
NUMBER OF COLUMNS N ?10

CHI SQUARE VALUES

 2.76587  13.0518  8.44779  4.0769  4.317  8.00294  16.3  11.0273
 6.51841
```

```
5.33273  4.41584  15.3182  11.7907  16.8436  2.28712  10.6981
10.712
5.9031  6.87363  10.1315  10.4465  6.45816  13.0507  9.36509
7.68409  8.61957  3.14895  9.62788  7.29767  8.91253
11.4522  11.5202  2.79612  5.16194  11.0879
12.1384  11.492  11.6431  17.4365
3.06485  15.0172  9.97839
4.86847  7.4063
11.6465

TIME:  3.12 SECS.

READY



HYPD          10:00          28-FEB-77

00010 PRINT "PROGRAM NAME --- HYPD"
00020 REM--- THIS PROGRAM COMPUTES CALHOUN DISTANCES AMONG QUADRATS.
00030 REM THE INPUT DATA ARE READ FROM DISK FILE CAL. IN THIS THE
00040 REM DATA ARE ARRANGED AS N SETS OF P NUMBERS. N REPRESENTS
00050 REM THE NUMBER OF QUADRATS AND P THE NUMBER OF SPECIES.
00060 PRINT "===================================="
00070 FILES CAL
00080 DIM X(10,2)
00090 REM---EXPLANATION OF ARRAY SYMBOL:
00100 REM          X - AN N*P ARRAY OF DATA
00110 REM--- READ DATA
00120 PRINT "NUMBER OF QUADRATS N";
00130 INPUT N
00140 PRINT "NUMBER OF SPECIES P";
00150 INPUT P
00160 PRINT
00170 PRINT "UPPER HALF OF DISTANCE MATRIX"
00180 FOR I=1 TO N
00190 FOR J=1 TO P
00200 READ #1,X(I,J)
00210 NEXT J,I
00220 REM CHOOSE TARGET VECTORS
00230 FOR I=1 TO N-1
00240 FOR J=I+1 TO N
00250 REM CHOOSE FEATURE VECTOR
00260 FOR K=1 TO N
00270 IF K<>I THEN 290
00280 LET K=K+1
00290 IF K<>J THEN 310
00300 LET K=K+1
00310 IF K<=N THEN 360
00320 LET K=N
00330 GO TO 800
00340 REM TEST IF THE KTH VECTOR HAS AT LEAST ONE INTERVENING
00350 REM COMPONENT VALUE
00360 FOR L=1 TO P
00370 LET T1=X(I,L)
```

```
00380 LET T2=X(J,L)
00390 LET T3=X(K,L)
00400 IF T1<T2 THEN 450
00410 LET A=T1
00420 LET B=T2
00430 LET T1=B
00440 LET T2=A
00450 IF T2<=T3 THEN 510
00460 IF T3<=T1 THEN 510
00470 LET N1=N1+1
00480 LET L=P
00490 LET F=2
00500 GO TO 520
00510 LET F=1
00520 NEXT L
00530 IF F=2 THEN 800
00540 REM TEST IF THE KTH VECTOR HAS ONE TIE IN AT LEAST ONE
00550 REM COMPONENT VALUE
00560 FOR L=1 TO P
00570 LET T1=X(I,L)
00580 LET T2=X(J,L)
00590 LET T3=X(K,L)
00600 IF T1=T2 THEN 670
00610 IF T3=T1 THEN 630
00620 IF T3<>T2 THEN 670
00630 LET N2=N2+1
00640 LET L=P
00650 LET F=2
00660 GO TO 680
00670 LET F=1
00680 NEXT L
00690 IF F=2 THEN 800
00700 REM TEST IF THE KTH VECTOR HAS TWO TIES IN A GIVEN COMPONENT
00710 FOR L=1 TO P
00720 LET T1=X(I,L)
00730 LET T2=X(J,L)
00740 LET T3=X(K,L)
00750 IF T1<>T2 THEN 790
00760 IF T3<>T1 THEN 790
00770 LET N3=N3+1
00780 LET L=P
00790 NEXT L
00800 NEXT K
00810 PRINT 6*N1+3*N2+2*N3;
00820 LET N1=N2=N3=0
00830 NEXT J
00840 PRINT
00850 NEXT I
00860 END

CAL            10:01           28-FEB-77

01000 4,3,7,6,5,4,7,2,5,2,1,7,4,8,8,3,2,2,7,4

.R BASIC
```

```
READY
OLD HYPD
READY
RUN

HYPD            10:02           28-FEB-77

PROGRAM NAME --- HYPD
==================================
NUMBER OF QUADRATS N ?10
NUMBER OF SPECIES P ?2

UPPER HALF OF DISTANCE MATRIX
 27  12  27  15  30  27  33  12  24
 9  30  30  36  27  18  39  5
 24  21  30  21  24  27  9
 11  48  45  15  30  23
 45  42  27  17  24
 9  48  36  39
 39  42  30
 42  12
 36

TIME:  1.46 SECS.

READY
```

```
INFC            10:37           07-JAN-77

00010 PRINT "PROGRAM NAME --- INFC"
00020 REM--- THIS PROGRAM FORMS TWO-DIMENSIONAL TABLES OF JOINT
00030 REM FREQUENCIES BETWEEN THE ROWS OF THE DATA MATRIX.
00040 REM THE RAW DATA ARE INPUT FROM DISK FILE RAWD8 ARRANGED
00050 REM AS P SETS OF N NUMBERS.  P SIGNIFIES THE NUMBER OF ROWS
00060 REM AND N THE NUMBER OF COLUMNS.  THE FOLLOWING QUANTITIES
00070 REM ARE COMPUTED:
00080 REM          1.  MULTIPLE OF ENTROPY OF ORDER ONE
00090 REM          2.  JOINT INFORMATION
00100 REM          3.  MUTUAL INFORMATION
00110 REM          4.  EQUIVOCATION INFORMATION
00120 REM          5.  RAJSKI'S METRIC
00130 REM          6.  COHERENCE COEFFICIENT
00140 REM THE VALUES OF RAJSKI'S METRIC ARE WRITTEN INTO DISK FILE
00150 REM RAJ.  IT IS ASSUMED THAT THE DATA REPRESENT DISCRETE CLASS
00160 REM SYMBOLS CODED AS INTEGER NUMBERS.
00170 PRINT "=============================================="
00180 FILES RAWD8, RAJ
00190 SCRATCH #2
00200 DIM X(2,25),I(2),J(2,2),C(25),K(25),F(25)
00220 REM---EXPLANATION OF ARRAY SYMBOLS:
00240 REM         X - A P*N ARRAY
00250 REM         I - A P-VALUED VECTOR
00260 REM         C, K, F - N-VALUED VECTORS
```

```
00270 REM          J - A P*P ARRAY
00290 REM---READ DATA
00300 PRINT "NUMBER OF ROWS P";
00310 INPUT P
00320 PRINT "NUMBER OF COLUMNS N";
00330 INPUT N
00340 PRINT "SPECIFY THE INFORMATION FUNCTION TO BE COMPUTED BY"
00350 PRINT "TYPING A NUMBER FROM 1 TO 6";
00360 INPUT Z
00370 PRINT
00380 FOR H=1 TO P
00390 FOR J=1 TO N
00400 READ #1,A
00410 LET X(H,J)=A
00420 NEXT J,H
00430 REM---COUNT FREQUENCIES, COMPUTE INFORMATION
00440 FOR H=1 TO P
00450 FOR I=H TO P
00460 LET C(1)=X(H,1)
00470 LET K(1)=X(I,1)
00480 MAT F=CON
00490 LET Q=1
00500 FOR J=2 TO N
00510 LET A=X(H,J)
00520 LET B=X(I,J)
00530 FOR L=1 TO Q
00540 IF C(L)<>A THEN 580
00550 IF K(L)<>B THEN 580
00560 LET F(L)=F(L)+1
00570 GO TO 620
00580 NEXT L
00590 LET Q=Q+1
00600 LET C(Q)=A
00610 LET K(Q)=B
00620 NEXT J
00630 REM---AT THIS POINT, VECTORS C AND K HOLD THE REALIZED
00640 REM   JOINT CLASS SYMBOLS AND VECTOR F THE JOINT
00650 REM   FREQUENCIES FOR ROWS H AND I
00660 LET S=0
00670 FOR J=1 TO Q
00680 LET A=F(J)
00690 LET S=S+A*LOG(A)
00700 NEXT J
00710 LET J(H,I)=N*LOG(N)-S
00720 NEXT I
00730 REM---THE H,I ELEMENT IN MATRIX J HOLDS THE JOINT
00740 REM   INFORMATION OF ROW H AND I
00750 MAT F=CON
00760 LET C(1)=X(H,1)
00770 LET Q=1
00780 FOR J=2 TO N
00790 LET A=X(H,J)
00800 FOR L=1 TO Q
00810 IF C(L)<>A THEN 840
00820 LET F(L)=F(L)+1
```

```
00830 GO TO 870
00840 NEXT L
00850 LET Q=Q+1
00860 LET C(Q)=A
00870 NEXT J
00880 REM---AT THIS POINT, VECTOR C HOLDS THE CLASS SYMBOLS AND
00890 REM   VECTOR F THE CLASS FREQUENCIES FOR ROW H
00900 LET S=0
00910 FOR J=1 TO Q
00920 LET A=F(J)
00930 LET S=S+A*LOG(A)
00940 NEXT J
00950 LET I(H)=N*LOG(N)-S
00960 NEXT H
00970 REM---AT THIS POINT, ELEMENT H IN VECTOR I HOLDS THE
00980 REM   N-MULTIPLE OF ENTROPY IN ROW H
00990 IF Z=1 THEN 1050
01000 IF Z=2 THEN 1080
01010 IF Z=3 THEN 1160
01020 IF Z=4 THEN 1240
01030 IF Z=5 THEN 1310
01040 IF Z=6 THEN 1410
01050 PRINT "MULTIPLE OF ENTROPY OF ORDER ONE"
01060 MAT PRINT I;
01070 GO TO 1310
01080 PRINT "JOINT INFORMATION"
01090 FOR H=1 TO P-1
01100 FOR I=H+1 TO P
01110 PRINT J(H,I);
01120 NEXT I
01130 PRINT
01140 NEXT H
01150 GO TO 1310
01160 PRINT "MUTUAL INFORMATION"
01170 FOR H=1 TO P-1
01180 FOR I=H+1 TO P
01190 PRINT I(H)+I(I)-J(H,I);
01200 NEXT I
01210 PRINT
01220 NEXT H
01230 GO TO 1310
01240 PRINT "EQUIVOCATION INFORMATION"
01250 FOR H=1 TO P
01260 FOR I=H TO P
01270 PRINT 2*J(H,I)-I(H)-I(I);
01280 NEXT I
01290 PRINT
01300 NEXT H
01310 PRINT "RAJSKI'S METRIC"
01320 FOR H=1 TO P
01330 FOR I=H TO P
01340 LET M=2-(I(H)+I(I))/J(H,I)
01350 WRITE #2,M
01360 PRINT M;
01370 NEXT I
```

```
01380 PRINT
01390 NEXT H
01400 GO TO 1500
01410 PRINT 'COHERENCE COEFFICIENT'
01420 FOR H=1 TO P
01430 FOR I=H TO P
01440 LET M=2-(I(H)+I(I))/J(H,I)
01450 PRINT SQR(1-M^2);
01460 NEXT I
01470 PRINT
01480 NEXT H
01490 GO TO 1310
01500 END

RAWD8          10:47          07-JAN-77

01000 1,1,5,5,1,5,5,5,0,6,5,5,1,5,1,4,1,4,3,2,1,1,3,1,3,3,4
01010 3,3,1,3,5,4,6,2,5,5,5,1,5,1,5,0,0,1,1,2,0,1,5,

.R BASIC

READY
OLD INFC
READY
RUN

INFC           10:57          07-JAN-77

PROGRAM NAME --- INFC
================================================
NUMBER OF ROWS P ?2
NUMBER OF COLUMNS N ?25
SPECIFY THE INFORMATION TO BE COMPUTED BY
TYPING A NUMBER FROM 1 TO 6 ?1

MULTIPLE OF ENTROPY OF ORDER ONE
 39.3792  44.4864
RAJSKI'S METRIC
 0  0.727425
 0

TIME:  1.36 SECS.

READY



RANKIN         21:57          21-DEC-76

00010 PRINT 'PROGRAM NAME --- RANKIN'
00020 REM---  THIS PROGRAM RANKS SPECIES ACCORDING
00030 REM TO WEIGHTS REPRESENTING MUTUAL
00040 REM INFORMATION.  THE DATA ARE READ FROM
00050 REM DISK FILE DATA WHERE THE INDIVIDUAL
00060 REM VALUES SIGNIFY DISCRETE STATES IN
```

```
00070 REM A NOMINAL OR ORDERED VARIABLE.
00080 REM DISK FILE DATA CONSISTS OF P SETS
00090 REM OF N NUMBERS.  SYMBOL P SPECIFIES THE
00100 REM NUMBER OF SPECIES AND N THE NUMBER OF
00110 REM QUADRATS.
00120 PRINT "====================================="
00130 FILES DATA
00140 DIM X(10,50),D(10,50),I(10),G(10),R(10,50),K(10,50),F(50)
00150 REM--- EXPLANATION OF ARRAY SYMBOLS:
00170 REM          X,D,R,K - P*N ARRAYS
00180 REM          I G - P-VALUED VECTORS
00190 REM          F - N-VALUED VECTOR
00210 PRINT "NUMBER OF SPECIES P";
00220 INPUT P
00230 PRINT "NUMBER OF QUADRATS N";
00240 INPUT N
00250 FOR H=1 TO P
00260 FOR J=1 TO N
00270 READ #1,D(H,J)
00280 NEXT J,H
00290 MAT X=D
00300 LET P1=P
00310 LET R=1
00320 GOSUB 840
00330 GOSUB 1060
00340 LET T=0
00350 FOR H=1 TO P
00360 IF I(H)=-111111 THEN 380
00370 LET T=T+I(H)
00380 NEXT H
00390 LET T1=T-I
00400 PRINT "MUTUAL INF OF REMAINING"P1"SPECIES"T1
00410 LET L1=0
00420 LET P1=P1-1
00430 LET L1=L1+1
00440 LET D=A=T=0
00450 FOR H=1 TO P
00460 IF I(H)=-111111 THEN 540
00470 LET D=D+1
00480 IF L1=D THEN 540
00490 LET A=A+1
00500 FOR J=1 TO N
00510 LET X(A,J)=D(H,J)
00520 NEXT J
00530 LET T=T+I(H)
00540 NEXT H
00550 GOSUB 1060
00560 LET G(L1)=T1-(T-I)
00570 IF L1=P1+1 THEN 590
00580 GO TO 430
00590 LET A=M=0
00600 FOR H=1 TO P
00610 IF I(H)=-111111 THEN 670
00620 LET A=A+1
00630 IF M>G(A) THEN 670
```

```
00640 LET M=G(A)
00650 LET S=H
00660 LET S1=A
00670 NEXT H
00680 LET A=0
00690 IF R=P THEN 720
00700 PRINT "SPECIES"S"HAS RANK"R"AND MUTUAL INF"G(S1)
00710 GO TO 740
00720 PRINT "SPECIES"S"HAS RANK"R
00730 STOP
00740 LET R=R+1
00750 LET I(S)=-111111
00760 FOR H=1 TO P
00770 IF I(H)=-111111 THEN 820
00780 LET A=A+1
00790 FOR J=1 TO N
00800 LET X(A,J)=D(H,J)
00810 NEXT J
00820 NEXT H
00830 GO TO 330
00840 MAT R=CON
00850 FOR H=1 TO P
00860 LET K(H,1)=X(H,1)
00870 LET Q=1
00880 FOR J=2 TO N
00890 LET A=X(H,J)
00900 FOR L=1 TO Q
00910 IF K(H,L)<>A THEN 940
00920 LET R(H,L)=R(H,L)+1
00930 GO TO 970
00940 NEXT L
00950 LET Q=Q+1
00960 LET K(H,Q)=A
00970 NEXT J
00980 LET S=0
00990 FOR J=1 TO Q
01000 LET A=R(H,J)
01010 LET S=S+A*LOG(A)
01020 NEXT J
01030 LET I(H)=N*LOG(N)-S
01040 NEXT H
01050 RETURN
01060 FOR H=1 TO P1
01070 LET K(H,1)=X(H,1)
01080 NEXT H
01090 MAT F=CON
01100 LET Q=1
01110 FOR J=2 TO N
01120 FOR L=1 TO Q
01130 FOR H=1 TO P1
01140 LET A=X(H,J)
01150 IF K(H,L)<>A THEN 1190
01160 NEXT H
01170 LET F(L)=F(L)+1
01180 GO TO 1240
```

```
01190 NEXT L
01200 LET Q=Q+1
01210 FOR H=1 TO P1
01220 LET K(H,Q)=X(H,J)
01230 NEXT H
01240 NEXT J
01250 LET S=0
01260 FOR L=1 TO Q
01270 LET A=F(L)
01280 LET S=S+A*LOG(A)
01290 NEXT L
01300 LET I=N*LOG(N)-S
01310 RETURN
01320 END
```

```
DATA          11:21          22-DEC-76

01010     0,0,3,2,0,0,0,3,3,2,0,0,0
01020     2,0,3,2,2,3,2,3,2,3,0,3
01030     5,5,5,3,2,3,0,3,3,0,3,3,2
01040     0,3,0,0,2,5,2,0,0,2,0,0
01050     0,0,0,0,0,0,0,0,0,0,0,0,0
01060     0,0,0,0,0,0,0,0,0,0,0,0
01070     0,0,0,0,0,0,0,0,0,0,0,0,0
01080     0,0,0,0,0,0,0,9,9,9,9,9
01090     0,0,0,0,0,0,0,0,0,0,0,0,0
01100     0,0,0,0,0,0,0,0,0,0,0,0
01110     0,0,0,0,0,0,0,0,0,0,8,7,9
01120     9,7,0,0,0,0,0,0,0,0,0,0
01130     0,0,0,0,2,5,5,2,0,0,2,0,2
01140     0,0,2,3,2,2,2,3,2,2,2,8
01150     8,5,5,5,7,8,9,5,5,2,0,0,2
01160     0,5,0,0,0,0,0,0,0,0,0,0
01170     0,0,0,0,0,7,7,8,9,8,0,0,0
01180     0,0,0,0,0,0,0,0,0,0,0,0
01190     0,0,0,0,0,0,0,0,0,5,0,0,0
01200     0,0,0,0,0,0,0,0,0,0,0,0
01210     0,0,0,0,0,0,0,0,0,0,0,0,0
01220     0,0,0,0,0,0,0,0,0,0,0,0
01230     0,0,0,0,0,0,0,0,0.0,0,0,0
01240     0,0,9,8,9,9,9,0.0,0,0,0
01250     0,0,0,0,2,5,0,2,0.2,0,0,0
01260     0,0,5,5,5,3,7,8,8,7.7,0
01270     2,3,3,3,5,2,3,3,5,3,0,0,0
01280     0,5,0,0,0,0,0,0,0,0,0,0
01290     0,0,0,0,0,8,7,2,3,5,0,0,0
01300     0,0,8,7,7,5,5,0,0,3,2,2
01310     0,0,0,0,2,3,0,0,0,8,3,0,0
01320     0,5,0,0,2,3,0,0,0,0,0,0
01330     7,9,8,9,8,5,3,3,3,3,0,0,0
01340     0,0,0,0,0,0,0,0,0,0,0,0
01350     0,0,0,0,0,0,0,0,0,0,0,0,0
01360     0,0,0,0,0,0,0,0,0,0,0,0
01370     0,0,0,0,0,0,0,0,0,0,7,8,8
01380     9,9,5,3,3,3,3,3,3,3,3,3
```

```
01390      5,2,5,5,3,3,3,5,5,5,5,5,7
01400      5,5,0,0,0,0,0,0,0,0,0,0

.R BASIC

READY
OLD RANKIN
READY
RUN

RANKIN          11:22           22-DEC-76

PROGRAM NAME --- RANKIN
======================================
NUMBER OF SPECIES P ?10
NUMBER OF QUADRATS N ?50

MUTUAL INF OF REMAINING 10 SPECIES 282.114

SPECIES 10 HAS RANK 1 AND MUTUAL INF 71.173
MUTUAL INF OF REMAINING 9 SPECIES 210.941

SPECIES 7 HAS RANK 2 AND MUTUAL INF 69.009
MUTUAL INF OF REMAINING 8 SPECIES 141.932

SPECIES 4 HAS RANK 3 AND MUTUAL INF 59.3438
MUTUAL INF OF REMAINING 7 SPECIES 82.5886

SPECIES 8 HAS RANK 4 AND MUTUAL INF 39.0365
MUTUAL INF OF REMAINING 6 SPECIES 43.5521

SPECIES 9 HAS RANK 5 AND MUTUAL INF 23.1958
MUTUAL INF OF REMAINING 5 SPECIES 20.3563

SPECIES 1 HAS RANK 6 AND MUTUAL INF 16.305
MUTUAL INF OF REMAINING 4 SPECIES 4.05125

SPECIES 5 HAS RANK 7 AND MUTUAL INF 2.3112
MUTUAL INF OF REMAINING 3 SPECIES 1.74005

SPECIES 6 HAS RANK 8 AND MUTUAL INF 1.18334
MUTUAL INF OF REMAINING 2 SPECIES 0.556704

SPECIES 3 HAS RANK 9 AND MUTUAL INF 0.556704
MUTUAL INF OF REMAINING 1 SPECIES 0

SPECIES 2 HAS RANK 10

TIME:  58.35 SECS.
```

```
PCAR            15:41           23-DEC-76

ØØØ1Ø PRINT "PROGRAM NAME --- PCAR"
ØØØ2Ø REM--- COMPONENT COEFFICIENTS ARE COMPUTED FOR SPECIES
ØØØ3Ø REM AND COMPONENT SCORES FOR QUADRATS. THE INPUT DATA
ØØØ4Ø REM ARE READ FROM DISK FILE RAWD2 WITH THE DATA
ØØØ5Ø REM ARRANGED AS P SETS OF N NUMBERS. P REPRESENTS
ØØØ6Ø REM THE NUMBER OF SPECIES AND N THE NUMBER OF
ØØØ7Ø REM QUADRATS.  COMPONENT SCORES ARE WRITTEN INTO DISK
ØØØ8Ø REM FILE DOMS.
ØØØ9Ø PRINT "============================================="
ØØ1ØØ FILES RAWD2, DOMS
ØØ11Ø SCRATCH #2
ØØ12Ø DIM X(2,5),R(2,2),B(2,2),T(5,2),Q(2),C(5,2)
ØØ13Ø REM---EXPLANATION OF ARRAY SYMBOLS:
ØØ15Ø REM          X - A P*N ARRAY
ØØ16Ø REM          T, C - N*P ARRAYS
ØØ17Ø REM          R, B - P*P ARRAYS
ØØ18Ø REM          Q - A P-VALUED VECTOR
ØØ21Ø REM---READ DATA.
ØØ23Ø PRINT "NUMBER OF SPECIES P";
ØØ24Ø INPUT P
ØØ25Ø PRINT "NUMBER OF QUADRATS N";
ØØ26Ø INPUT N
ØØ27Ø PRINT "TYPE 1 FOR COVARIANCE, Ø FOR CORRELATION";
ØØ28Ø INPUT Z1
ØØ29Ø PRINT "TYPE 1 IF PRINTING OF SCALAR PRODUCTS REQUIRED"
ØØ3ØØ PRINT "ELSE TYPE Ø";
ØØ31Ø INPUT Z2
ØØ32Ø LET C=L=N
ØØ33Ø LET R=N=P
ØØ34Ø MAT B=IDN
ØØ35Ø FOR I=1 TO R
ØØ36Ø LET A=B=Ø
ØØ37Ø FOR J=1 TO C
ØØ38Ø READ #1,Q
ØØ39Ø LET X(I,J)=Q
ØØ4ØØ LET A=A+Q
ØØ41Ø LET B=B+Q^2
ØØ42Ø NEXT J
ØØ43Ø LET B=SQR(ABS(B-A^2/C))
ØØ44Ø LET A=A/C
ØØ45Ø FOR J=1 TO C
ØØ46Ø IF Z1=Ø THEN 48Ø
ØØ47Ø LET B=SQR(C-1)
ØØ48Ø LET X(I,J)=(X(I,J)-A)/B
ØØ49Ø NEXT J,I
ØØ5ØØ MAT T=TRN(X)
ØØ51Ø MAT R=X*T
ØØ52Ø IF Z2=Ø THEN 56Ø
ØØ53Ø PRINT
```

```
00540 PRINT "SCALAR PRODUCTS"
00550 MAT PRINT R;
00560 REM---EIGENVALUE AND VECTOR PROCEDURE
00570 LET A=0.00000001
00580 LET C=0
00590 FOR I=2 TO N
00600 FOR J=1 TO I-1
00610 LET C=C+2*(R(I,J)^2)
00620 NEXT J,I
00630 LET Y=SQR(C)
00640 LET O=(A/N)*Y
00650 LET T=Y
00660 LET D=0
00670 LET T=T/N
00680 FOR Q=2 TO N
00690 FOR P=1 TO Q-1
00700 IF ABS(R(P,Q))< T THEN 1010
00710 LET D=1
00720 LET V=R(P,P)
00730 LET Z=R(P,Q)
00740 LET E=R(Q,Q)
00750 LET F=.5*(V-E)
00760 IF F=0 THEN 790
00770 LET G=-(SGN(F))
00780 GO TO 800
00790 LET G=-1
00800 LET G=G*Z/(SQR(Z^2+F^2))
00810 LET H=G/(SQR(2*(1+SQR(1-G^2))))
00820 LET K=SQR(1-H^2)
00830 FOR I=1 TO N
00840 IF I=P THEN 920
00850 IF I=Q THEN 920
00860 LET C=R(I,P)
00870 LET F=R(I,Q)
00880 LET R(Q,I)=C*H+F*K
00890 LET R(I,Q)=R(Q,I)
00900 LET R(P,I)=C*K-F*H
00910 LET R(I,P)=R(P,I)
00920 LET C=B(I,P)
00930 LET F=B(I,Q)
00940 LET B(I,Q)=C*H+F*K
00950 LET B(I,P)=C*K-F*H
00960 NEXT I
00970 LET R(P,P)=V*K^2+E*H^2-2*Z*H*K
00980 LET R(Q,Q)=V*H^2+E*K^2+2*Z*H*K
00990 LET R(P,Q)=(V-E)*H*K+Z*(K^2-H^2)
01000 LET R(Q,P)=R(P,Q)
01010 NEXT P
01020 NEXT Q
01030 IF D<>1 THEN 1060
01040 LET D=0
01050 GO TO 680
01060 IF T>O THEN 670
01070 FOR I=1 TO N
01080 LET Q(I)=I
```

```
Ø1Ø9Ø NEXT I
Ø11ØØ LET J=Ø
Ø111Ø LET V1=Ø
Ø112Ø LET J=J+1
Ø113Ø FOR I=1 TO N-J
Ø114Ø IF R(I,I)>=R(I+1,I+1) THEN 122Ø
Ø115Ø LET V1=1
Ø116Ø LET V2=R(I,I)
Ø117Ø LET R(I,I)=R(I+1,I+1)
Ø118Ø LET R(I+1,I+1)=V2
Ø119Ø LET P=Q(I)
Ø12ØØ LET Q(I)=Q(I+1)
Ø121Ø LET Q(I+1)=P
Ø122Ø NEXT I
Ø123Ø IF V1<>Ø THEN 111Ø
Ø124Ø FOR J=1 TO N
Ø125Ø PRINT
Ø126Ø LET K=Q(J)
Ø127Ø PRINT "ROOT"; J; "="; R(J,J)
Ø128Ø PRINT "VECTOR";J
Ø129Ø LET V=Ø
Ø13ØØ FOR I=1 TO N
Ø131Ø LET V=V+B(I,K)^2
Ø132Ø NEXT I
Ø133Ø FOR I=1 TO N
Ø134Ø LET B(I,K)=B(I,K)*SQR(1/V)
Ø135Ø PRINT B(I,K);
Ø136Ø NEXT I
Ø137Ø PRINT
Ø138Ø PRINT
Ø139Ø NEXT J
Ø14ØØ LET C=L
Ø141Ø MAT C=T*B
Ø142Ø FOR J=1 TO R
Ø143Ø PRINT "COMPONENT";J
Ø144Ø PRINT "COMPONENT SCORES"
Ø145Ø LET K=Q(J)
Ø146Ø FOR I=1 TO C
Ø147Ø LET Q=C(I,K)
Ø148Ø WRITE #2,Q
Ø149Ø PRINT Q;
Ø15ØØ NEXT I
Ø151Ø PRINT
Ø152Ø PRINT
Ø153Ø NEXT J
Ø154Ø END

RAWD2          16:Ø1          23-DEC-76

Ø1ØØØ 2,5,2,1,Ø,Ø,1,4,3,1

.R BASIC

READY
OLD PCAR
```

```
READY
RUN

PCAR            16:44           23-DEC-76

PROGRAM NAME --- PCAR
=========================================================
NUMBER OF SPECIES P ?2
NUMBER OF QUADRATS N ?5
TYPE 1 FOR COVARIANCE, Ø FOR CORRELATION ?1
TYPE 1 IF PRINTING OF SCALAR PRODUCTS REQUIRED
ELSE TYPE Ø ?Ø

ROOT 1 = 3.74Ø31
VECTOR 1
 Ø.9Ø13Ø3 -Ø.433189

ROOT 2 = 2.45969
VECTOR 2
 Ø.433189  Ø.9Ø13Ø3

COMPONENT 1
COMPONENT SCORES
 Ø.38987  1.52523 -Ø.4765Ø8 -Ø.71Ø565 -Ø.728Ø28

COMPONENT 2
COMPONENT SCORES
-Ø.811173  Ø.289262  Ø.991434  Ø.324188 -Ø.79371

TIME:  1.42 SECS.

READY



PCAD            2Ø:12           23-DEC-76

ØØØ1Ø PRINT "PROGRAM NAME --- PCAD"
ØØØ2Ø REM--- COMPONENT SCORES ARE COMPUTED FOR N INDIVIDUALS
ØØØ3Ø REM BASED ON THE D-ALGORITHM OF COMPONENT ANALYSIS.  THE
ØØØ4Ø REM INPUT DATA CONTAIN THE UPPER HALF OF AN EUCLIDEAN DISTANCE
ØØØ5Ø REM MATRIX, INCLUDING ZEROS IN THE PRINCIPAL DIAGONAL, STORED
ØØØ6Ø REM IN DISK FILE DIST3. COMPUTED COMPONENT SCORES ARE WRITTEN
ØØØ7Ø REM INTO DISK FILE COMS. TO COMPUTE THE Q-ALGORITHM OF COMPONENT
ØØØ8Ø REM ANALYSIS REPLACE STATEMENTS 13Ø, 38Ø, 39Ø BY THE FOLLOWING:
ØØØ9Ø REM                13Ø FILES COR, COMS
ØØ1ØØ REM                38Ø LET R(I,J)=A
ØØ11Ø REM                39Ø LET R(J,I)=A
ØØ12Ø PRINT "========================================="
ØØ13Ø FILES DIST3, COMS
ØØ14Ø SCRATCH #2
ØØ15Ø DIM R(5,5),B(5,5),Q(5)
ØØ17Ø REM---EXPLANATION OF ARRAY SYMBOLS:
ØØ19Ø REM            R, B - N*N ARRAYS
ØØ2ØØ REM            Q - AN N-VALUED VECTOR
```

```
00230 REM---READ DATA
00250 PRINT "NUMBER OF INDIVIDUALS N";
00260 INPUT N
00270 PRINT "TO DIVIDE DATA BY SQR(N-1) TYPE 1"
00280 PRINT "ELSE TYPE A NUMBER OTHER THAN 1";
00290 INPUT I
00300 IF I<>1 THEN 330
00310 LET P=N-1
00320 GO TO 340
00330 LET P=1
00340 FOR I=1 TO N
00350 FOR J=I TO N
00360 READ #1,A
00370 LET A=A/SQR(P)
00380 LET R(I,J)= -0.5*A^2
00390 LET R(J,I)= -0.5*A^2
00400 NEXT J,I
00410 MAT B=IDN
00420 REM---EIGENVALUE AND VECTOR PROCEDURE
00430 LET A=0.00000001
00440 LET C=0
00450 FOR I=2 TO N
00460 FOR J=1 TO I-1
00470 LET C=C+2*(R(I,J)^2)
00480 NEXT J,I
00490 LET Y=SQR(C)
00500 LET O=(A/N)*Y
00510 LET T=Y
00520 LET D=0
00530 LET T=T/N
00540 FOR Q=2 TO N
00550 FOR P=1 TO Q-1
00560 IF ABS(R(P,Q))< T THEN 870
00570 LET D=1
00580 LET V=R(P,P)
00590 LET Z=R(P,Q)
00600 LET E=R(Q,Q)
00610 LET F=.5*(V-E)
00620 IF F=0 THEN 650
00630 LET G=-(SGN(F))
00640 GO TO 660
00650 LET G=-1
00660 LET G=G*Z/(SQR(Z^2+F^2))
00670 LET H=G/(SQR(2*(1+SQR(1-G^2))))
00680 LET K=SQR(1-H^2)
00690 FOR I=1 TO N
00700 IF I=P THEN 780
00710 IF I=Q THEN 780
00720 LET C=R(I,P)
00730 LET F=R(I,Q)
00740 LET R(Q,I)=C*H+F*K
00750 LET R(I,Q)=R(Q,I)
00760 LET R(P,I)=C*K-F*H
00770 LET R(I,P)=R(P,I)
00780 LET C=B(I,P)
```

```
00790 LET F=B(I,Q)
00800 LET B(I,Q)=C*H+F*K
00810 LET B(I,P)=C*K-F*H
00820 NEXT I
00830 LET R(P,P)=V*K^2+E*H^2-2*Z*H*K
00840 LET R(Q,Q)=V*H^2+E*K^2+2*Z*H*K
00850 LET R(P,Q)=(V-E)*H*K+Z*(K^2-H^2)
00860 LET R(Q,P)=R(P,Q)
00870 NEXT P
00880 NEXT Q
00890 IF D<>1 THEN 920
00900 LET D=0
00910 GO TO 540
00920 IF T>O THEN 530
00930 FOR I=1 TO N
00940 LET Q(I)=I
00950 NEXT I
00960 LET J=0
00970 LET V1=0
00980 LET J=J+1
00990 FOR I=1 TO N-J
01000 IF R(I,I)>=R(I+1,I+1) THEN 1080
01010 LET V1=1
01020 LET V2=R(I,I)
01030 LET R(I,I)=R(I+1,I+1)
01040 LET R(I+1,I+1)=V2
01050 LET P=Q(I)
01060 LET Q(I)=Q(I+1)
01070 LET Q(I+1)=P
01080 NEXT I
01090 IF V1<>0 THEN 970
01100 FOR J=1 TO N
01110 IF R(J,J)<0 THEN 1270
01120 PRINT
01130 LET K=Q(J)
01140 PRINT "ROOT"; J; "="; R(J,J)
01150 PRINT "COMPONENT SCORES"
01160 LET V=0
01170 FOR I=1 TO N
01180 LET V=V+B(I,K)^2
01190 NEXT I
01200 FOR I=1 TO N
01210 LET A=B(I,K)*SQR(R(J,J)/V)
01220 PRINT A;
01230 WRITE #2,A
01240 NEXT I
01250 PRINT
01260 PRINT
01270 NEXT J
01280 END

DIST3          20:13          23-DEC-76

01000 0,6,0
```

```
.R BASIC

READY
OLD PCAD
READY
RUN

PCAD          20:14          23-DEC-76

PROGRAM NAME --- PCAD
=========================================
NUMBER OF INDIVIDUALS N ?2
TO DIVIDE DATA BY SQR(N-1) TYPE 1
ELSE TYPE A NUMBER OTHER THAN 1 ?0

ROOT 1 = 18
COMPONENT SCORES
 3. -3

TIME:  1.15 SECS.

READY
```

```
OFORD          12:36          13-JAN-77

00010 PRINT "TITLE -- OFORD"
00020 REM--- THIS PROGRAM COMPUTES THE COEFFICIENTS IN
00030 REM ORTHOGONAL FUNCTIONS OF SPECIES TO DETERMINE
00040 REM ORDINATION CO-ORDINATES FOR QUADRATS.  THE
00050 REM DATA ARE INPUT FROM DISK FILE RAWD9 WHICH
00060 REM CONTAINS P SETS OF N NUMBERS. P REPRESENTS
00070 REM THE NUMBER OF SPECIES AND N THE NUMBER
00080 REM OF QUADRATS. THE CO-ORDINATES ARE
00100 REM WRITTEN INTO DISK FILE COOR.
00110 PRINT "========================================"
00120 FILES RAWD9,COOR
00130 SCRATCH #2
00140 DIM S(3,3), R(3,3), X(3,5), T(5,3), A(3,3)
00150 DIM B(3,3), V(3), Q(3), Y(3,5), Z(3)
00170 REM---EXPLANATION OF ARRAY SYMBOLS:
00190 REM          S, R, A, B - P*P ARRAYS
00200 REM          X, Y - P*N ARRAYS
00210 REM          T - AN N*P ARRAY
00220 REM          V, Q, Z - P-VALUED VECTORS
00240 PRINT "NUMBER OF SPECIES P";
00250 INPUT P
00260 PRINT "NUMBER OF QUADRATS N";
00270 INPUT N
00280 PRINT "TO ORDER SPECIES BY SPECIFIC"
00290 PRINT "VARIANCE TYPE 1 ELSE TYPE 0";
00300 INPUT G1
00310 GOSUB 910
00320 IF G1=1 THEN 350
```

```
00330 GOSUB 1820
00340 GO TO 360
00350 GOSUB 1300
00360 GOSUB 390
00370 GOSUB 1110
00380 STOP
00390 FOR H=1 TO P
00400 LET K=Q(H)
00410 LET Z(H)=V(K)
00420 FOR I=1 TO P
00430 LET S(H,I)=R(K,I)
00440 NEXT I,H
00450 FOR H=1 TO P
00460 LET K=Q(H)
00470 FOR I=1 TO P
00480 LET R(I,H)=S(I,K)
00490 NEXT I
00500 FOR J=1 TO N
00510 LET T(J,H)=X(K,J)
00520 NEXT J,H
00530 MAT S=R
00540 MAT X=TRN(T)
00550 FOR I=1 TO P
00560 LET V(I)=Z(I)
00570 NEXT I
00580 FOR H=1 TO P
00590 LET B(H,1)=S(H,1)
00600 NEXT H
00610 FOR I=2 TO P
00620 FOR J=1 TO I-1
00630 LET S=0
00640 FOR T=I-1 TO 1 STEP -1
00650 LET S=S+A(J,T)*B(I,T)
00660 NEXT T
00670 LET B(I,J)=S(I,J)-S
00680 LET A(I,J)=B(I,J)/V(J)
00690 NEXT J
00700 IF G1=1 THEN 760
00710 LET S=0
00720 FOR J=1 TO I-1
00730 LET S=S+A(I,J)*B(I,J)
00740 NEXT J
00750 LET V(I)=R(I,I)-S
00760 NEXT I
00770 PRINT "ORTHOGONAL FUNCTIONS"
00780 LET K=Q(1)
00790 PRINT "FUNCTION 1 SPV"V(1)
00800 PRINT "Y1=X"K
00810 FOR I=2 TO P
00820 PRINT "FUNCTION"I"SPV"V(I)
00830 LET K=Q(I)
00840 PRINT "Y"I"=X"K;
00850 FOR J=I-1 TO 1 STEP -1
00860 PRINT -A(I,J)"Y"J;
00870 NEXT J
```

```
ØØ88Ø PRINT
ØØ89Ø NEXT I
ØØ9ØØ RETURN
ØØ91Ø PRINT "TYPE 1 IF STANDARDIZATION OF SPECIES"
ØØ92Ø PRINT "REQUIRED ELSE TYPE Ø";
ØØ93Ø INPUT G
ØØ94Ø FOR I=1 TO P
ØØ95Ø LET M=S=Ø
ØØ96Ø FOR J=1 TO N
ØØ97Ø READ #1,A
ØØ98Ø LET X(I,J)=A
ØØ99Ø LET M=M+A
Ø1ØØØ LET S=S+A*A
Ø1Ø1Ø NEXT J
Ø1Ø2Ø FOR J=1 TO N
Ø1Ø3Ø LET X(I,J)=X(I,J)-M/N
Ø1Ø4Ø IF G=Ø THEN 1Ø6Ø
Ø1Ø5Ø LET X(I,J)=X(I,J)/SQR(S-M*M/N)
Ø1Ø6Ø NEXT J,I
Ø1Ø7Ø MAT T=TRN(X)
Ø1Ø8Ø MAT R=X*T
Ø1Ø9Ø MAT R=(1/(N-1))*R
Ø11ØØ RETURN
Ø111Ø PRINT
Ø1115 PRINT "QUADRAT CO-ORDINATES"
Ø112Ø FOR J=1 TO N
Ø113Ø LET Y(1,J)=X(1,J)
Ø114Ø FOR I=2 TO P
Ø115Ø LET S=Ø
Ø116Ø FOR T=I-1 TO 1 STEP -1
Ø117Ø LET S=S+A(I,T)*Y(T,J)
Ø118Ø NEXT T
Ø119Ø LET Y(I,J)=X(I,J)-S
Ø12ØØ NEXT I,J
Ø121Ø FOR I=1 TO P
Ø122Ø PRINT "FUNCTION"I"(SPECIES"Q(I)")"
Ø123Ø FOR J=1 TO N
Ø124Ø PRINT Y(I,J);
Ø125Ø WRITE #2,Y(I,J)
Ø126Ø NEXT J
Ø127Ø PRINT
Ø128Ø NEXT I
Ø129Ø RETURN
Ø13ØØ LET M1=Ø
Ø131Ø LET M2=P+1
Ø132Ø MAT Z=CON
Ø133Ø FOR H=1 TO P
Ø134Ø LET M1=M1+R(H,H)
Ø135Ø NEXT H
Ø136Ø PRINT
Ø137Ø PRINT "SPECIFIC VARIANCES IN P-SPECIES SAMPLE"
Ø138Ø FOR C=1 TO P
Ø139Ø IF Z(C)=Ø THEN 16ØØ
Ø14ØØ MAT S=R
Ø141Ø LET O=Ø
```

```
01420 FOR M=1 TO P
01430 IF Z(M)=0 THEN 1570
01440 IF M=C THEN 1570
01450 LET O=O+1
01460 FOR H=1 TO P
01470 IF S(M,M)<=0 THEN 1485
01480 LET Y(H,1)=S(H,M)/SQRT(S(M,M))
01482 GO TO 1490
01485 LET Y(H,1)=0
01490 NEXT H
01500 FOR H=1 TO P
01510 FOR I=1 TO P
01520 LET S(H,I)=S(H,I)-Y(H,1)*Y(I,1)
01530 NEXT I,H
01540 FOR H=1 TO P
01550 LET S(M,H)=S(H,M)=0
01560 NEXT H
01570 NEXT M
01580 LET V(C)=S(C,C)
01590 PRINT "SPECIES"C;V(C)
01600 NEXT C
01610 LET M=M1
01620 FOR I=1 TO P
01630 IF Z(I)=0 THEN 1670
01640 IF M<=V(I) THEN 1670
01650 LET M=V(I)
01660 LET C=I
01670 NEXT I
01680 PRINT "MIN SPV"V(C)"SPEC"C
01690 IF M2=2 THEN 1720
01700 PRINT " SPECIFIC VARIANCES IN RESIDUAL SAMPLE"
01710 PRINT "AFTER SPECIES"C"HAS BEEN REMOVED"
01720 LET Z(C)=0
01730 LET M2=M2-1
01740 LET Q(M2)=C
01750 IF M2=1 THEN 1770
01760 GO TO 1380
01770 PRINT " SPV FILE"
01780 FOR I=1 TO P
01790 PRINT "SPECIES"I;V(I)
01800 NEXT I
01810 RETURN
01820 PRINT "TYPE SPECIES LABELS IN DESIRED SEQUENCE"
01830 FOR I=1 TO P
01840 INPUT Q(I)
01850 LET V(I)=R(I,I)
01860 NEXT I
01870 PRINT
01880 RETURN
01890 END

RAWD9          12:44          13-JAN-77

01000 2,5,2,1,0,0,1,4,3,1,3,4,1,0,0
```

```
.R BASIC

READY
OLD OFORD
READY
RUN

OFORD          12:46          13-JAN-77

TITLE -- OFORD
========================================
NUMBER OF SPECIES P ?3
NUMBER OF QUADRATS N ?5
TO ORDER SPECIES BY SPECIFIC
VARIANCE TYPE 1 ELSE TYPE Ø ?Ø
TYPE 1 IF STANDARDIZATION OF SPECIES
REQUIRED ELSE TYPE Ø ?Ø
TYPE SPECIES LABELS IN DESIRED SEQUENCE
 ?1
 ?2
 ?3

ORTHOGONAL FUNCTIONS
FUNCTION 1 SPV 3.5
Y1=X1
FUNCTION 2 SPV 2.62857
Y2=X2   Ø.142857Y1
FUNCTION 3 SPV Ø.2Ø6522
Y3=X3   Ø.445652Y2 -Ø.857143Y1

QUADRAT CO-ORDINATES
FUNCTION 1 (SPECIES 1 )
 Ø  3  Ø -1 -2
FUNCTION 2 (SPECIES 2 )
-1.8 -Ø.371429  2.2  1.Ø5714 -1.Ø8571
FUNCTION 3 (SPECIES 3 )
 Ø.597826 -Ø.336957  Ø.38Ø435 -Ø.271739 -Ø.369565

TIME:  2.91 SECS.

READY



PVO            11:21          Ø4-JAN-77

ØØØ1Ø PRINT "PROGRAM NAME --- PVO"
ØØØ2Ø REM--- METRIC CO-ORDINATES ARE COMPUTED FOR N
ØØØ3Ø REM QUADRATS BASED ON POSITION VECTORS ORDINATION.  THE
ØØØ4Ø REM INPUT DATA CONTAIN THE UPPER HALF OF A SCALAR PRODUCT
ØØØ5Ø REM MATRIX, INCLUDING ELEMENTS IN THE PRINCIPAL DIAGONAL CELLS,
ØØØ6Ø REM STORED IN DISK FILE COR3.  THE METRIC CO-ORDINATES ARE
ØØØ7Ø REM WRITTEN INTO DISK FILE COMS.
ØØØ8Ø PRINT "=========================================================
ØØØ9Ø FILES COR3, COMS
```

```
00100 SCRATCH #2
00110 DIM X(6,6), K(6)
00130 REM---EXPLANATION OF ARRAY SYMBOLS:
00150 REM          X - AN N*N ARRAY
00160 REM          K - A K-VALUED VECTOR
00190 REM---READ DATA
00210 PRINT "NUMBER OF QUADRATS N";
00220 INPUT N
00225 PRINT
00230 LET Q=0
00240 FOR H=1 TO N
00250 FOR J=H TO N
00260 READ #1,A
00270 LET X(H,J)=A
00280 LET X(J,H)=A
00290 NEXT J
00300 LET Q=Q+X(H,H)
00310 NEXT H
00320 LET W=0
00330 LET S=0
00340 FOR H=1 TO N
00350 FOR J=1 TO N
00360 LET S=S+X(H,J)^2
00370 NEXT J
00380 IF X(H,H)<=0 THEN 410
00390 LET C=S/X(H,H)
00400 GO TO 420
00410 LET C=0
00420 IF C<W THEN 450
00430 LET W=C
00440 LET Z=H
00450 LET S=0
00460 NEXT H
00470 IF W=0 THEN 760
00500 PRINT "VECTOR "Z
00520 PRINT "SUM OF SQUARES "W;Q;W/Q
00540 PRINT "QUADRAT CO-ORDINATES"
00550 FOR J=1 TO N
00560 IF X(Z,Z)<=0 THEN 580
00570 GO TO 600
00580 LET W=0
00590 GO TO 610
00600 LET W=X(Z,J)/SQR(X(Z,Z))
00610 LET K(J)=W
00620 PRINT W;
00630 WRITE #2,W
00640 NEXT J
00650 PRINT
00660 FOR H=1 TO N
00670 FOR J=H TO N
00680 LET W=X(H,J)-K(H)*K(J)
00690 LET X(H,J)=W
00700 LET X(J,H)=W
00710 NEXT J
00720 LET X(Z,H)=0
```

```
00730 LET X(H,Z)=0
00740 NEXT H
00750 GO TO 320
00760 END
```

```
COR3            11:25           04-JAN-77

01000 .722222,-.444444,-.444444,-.111111,.722222,-.444444,.388889
01010 .388889,-.277778,-.444444,.388889,.388889,-.277778,-.444444
01020 .388889,1.05556,-.111111,-.277778,.722222,-.444444,.388889

.R BASIC

READY
OLD PVO
READY
RUN

PVO             11:31           04-JAN-77

PROGRAM NAME --- PVO
==========================================================
NUMBER OF QUADRATS N ?6

VECTOR  6
SUM OF SQUARES  2.38095  3.66667  0.649349
QUADRAT CO-ORDINATES
-0.712696  0.62361  0.62361 -0.445436 -0.712696  0.62361
VECTOR  4
SUM OF SQUARES  1.28572  3.66667  0.350649
QUADRAT CO-ORDINATES
-0.462909  0  0  0.925822 -0.462909  0
VECTOR  5
SUM OF SQUARES  4.50388E-6  3.66667  1.22833E-6
QUADRAT CO-ORDINATES
 1.50065E-3  0  0  0  1.50065E-3  0

TIME:  1.31 SECS.

READY
```

```
BCOAX           10:21           19-JUL-77

00010 PRINT "TITLE -- BCOAX"
00020 REM--- THIS PROGRAM COMPUTES MULTIDIMENSIONAL SCALING BASED ON
00030 REM A MODIFIED BRAY AND CURTIS METHOD.    THE MODIFICATIONS
00040 REM IN THE METHOD INVOLVE CORRECTIONS FOR OBLIQUE ORDINATION
00050 REM AXES.  THE DATA ARE INPUT FROM DISK FILE DIS WHICH
00060 REM INCLUDES THE UPPER HALF OF AN EUCLIDEAN DISTANCE MATRIX
00070 REM WITH ZEROS IN THE PRINCIPAL DIAGONAL CELLS.  THE OUTPUT
00080 REM INCLUDES SETS OF CO-ORDINATES ON TWO AXES.    THE ENTITIES
00090 REM ORDINATED ARE THE N QUADRATS BETWEEN WHICH THE DISTANCES
00100 REM ARE COMPUTED.  IN THIS PROGRAM THE ENTITIES ARE REFERRED
```

```
00110 REM TO AS QUADRATS. THE METHODOLOGY, HOWEVER, IS VALIDLY
00120 REM APPLIED TO OBJECTS OF OTHER KINDS.    RECTANGULAR
00130 REM CO-ORDINATES ARE WRITTEN INTO DISK FILE COOR.
00140 PRINT "=========================================================="
00150 FILES DIS, COOR
00160 SCRATCH #2
00170 DIM C(5,5),Y(2,5),Z(5)
00180 REM---EXPLANATION OF ARRAY SYMBOLS:
00190 REM           C - AN N*N ARRAY
00200 REM           Y - A 2*N ARRAY
00210 REM           Z - AN N-VALUED VECTOR
00220 PRINT "NUMBER OF QUADRATS N";
00230 INPUT N
00240 PRINT "SPECIFY QUADRATS TO SERVE AS ORDINATION POLES"
00250 PRINT "FOR THE FIRST AXIS";
00260 INPUT A,B
00270 PRINT "SPECIFY QUADRATS TO SERVE AS ORDINATION POLES"
00280 PRINT "FOR THE SECOND AXIS";
00290 INPUT C,D
00300 FOR J=1 TO N
00310 FOR K=J TO N
00320 READ #1,C(J,K)
00330 LET C(K,J)=C(J,K)
00340 NEXT K,J
00350 REM---COMPUTE CO-ORDINATES ON OBLIQUE AXES
00360 FOR J=1 TO N
00370 LET Y(1,J)=(C(A,J)^2+C(A,B)^2-C(B,J)^2)/(2*C(A,B))
00380 LET Y(2,J)=(C(C,J)^2+C(C,D)^2-C(D,J)^2)/(2*C(C,D))
00390 NEXT J
00400 REM---TRANSFER TO RECTANGULAR CO-ORDINATES
00410 LET D1=ABS(C(A,D)^2+C(B,C)^2-C(A,C)^2-C(D,B)^2)/(2*C(A,B))
00420 LET D2=D1/C(D,C)
00430 LET S=SQR(1-D2^2)
00440 LET D3=S/D2
00450 FOR J=1 TO N
00460 LET D=ABS(ABS(Y(1,J)-Y(1,C))-Y(2,J)*D2)/D3
00470 IF Y(2,J)*S>ABS(Y(1,J)-Y(1,C))*D3 THEN 490
00480 LET D=-D
00490 LET Z(J)=Y(2,J)*S+D
00500 NEXT J
00510 PRINT
00520 PRINT "CO-ORDINATES 0N OBLIQUE AXES"
00530 FOR J=1 TO 2
00540 PRINT "AXIS "J
00550 FOR K=1 TO N
00560 PRINT Y(J,K)
00570 NEXT K
00580 PRINT
00590 NEXT J
00600 PRINT "RECTANGULAR CO-ORDINATES"
00610 PRINT "AXIS 1"
00620 FOR J=1 TO N
00630 PRINT Y(1,J),
00640 WRITE #2,Y(1,J)
00650 NEXT J
```

```
00660 PRINT
00670 PRINT "AXIS 2"
00680 FOR J=1 TO N
00690 PRINT Z(J),
00700 WRITE #2,Z(J)
00710 NEXT J
00720 END

DIS             10:22           19-JUL-77

01000 0,1.2,0.84,0.96,0.93
01010 0,1.0,0.63,0.54
01020 0,0.9,0.84
01030 0,0.12
01040 0

.R BASIC

READY
OLD BCOAX
READY
RUN

BCOAX           10:23           19-JUL-77

TITLE -- BCOAX
==============================================================
NUMBER OF QUADRATS N ?5
SPECIFY QUADRATS TO SERVE AS ORDINATION POLES
FOR THE FIRST AXIS ?1,2
SPECIFY QUADRATS TO SERVE AS ORDINATION POLES
FOR THE SECOND AXIS ?3,4

CO-ORDINATES ON OBLIQUE AXES
AXIS  1
 0              1.2             0.477333        0.818625        0.838875
AXIS  2
 0.33           0.785056        0               0.9             0.834

RECTANGULAR CO-ORDINATES
AXIS 1
 0              1.2             0.477333        0.818625        0.838875
AXIS 2
 0.161015       0.55226         0               0.832778        0.753152

TIME:  1.32 SECS.

READY



RQT             19:32           13-JAN-77

00010 PRINT "PROGRAM NAME --- RQT"
00020 REM--- THIS PROGRAM COMPUTES THE EIGENVALUE AND VECTOR
```

```
00030 REM ALGORITHM OF RECIPROCAL ORDERING WITH OPTIONS FOR
00040 REM ADJUSTMENTS TO GIVE WEIGHT TO SPECIES
00050 REM AND FOR CONVERSION TO SCORES IN RECIPROCAL AVERAGING.
00060 REM THE DATA ARE READ FROM DISK FILE RAWD WITH ELEMENTS
00070 REM ARRANGED AS R SETS OF C NUMBERS.  R INDICATES THE NUMBER
00080 REM OF SPECIES AND C THE NUMBER OF QUADRATS.  SPECIES SCORES
00090 REM ARE WRITTEN INTO DISK FILE SOMS AND THE QUADRAT SCORES INTO
00100 REM DISK FILE COMS.
00110 PRINT "========================================================="
00120 FILES RAWD, SOMS, COMS
00130 SCRATCH #2,3
00140 DIM X(5,10),R(5,5),B(5,5),T(10,5),Q(5)
00150 DIM C(10,5),S(5),O(10),D(5,10),U(5,5)
00160 MAT S=ZER
00170 MAT O=ZER
00190 REM---EXPLANATION OF ARRAY SYMBOLS:
00210 REM         X, D - R*C ARRAYS
00220 REM         T, C - C*R ARRAYS
00230 REM         R, B, U - R*R ARRAYS
00240 REM         Q, S - R-VALUED VECTORS
00250 REM         O - A C-VALUED VECTOR
00270 REM---READ DATA
00280 PRINT "NUMBER OF SPECIES R";
00290 INPUT R
00300 PRINT "NUMBER OF QUADRATS C";
00310 INPUT C
00320 PRINT "TYPE 1 IF PRINTING OF SCALAR PRODUCTS REQUIRED"
00330 PRINT "ELSE TYPE 0";
00340 INPUT Z2
00350 PRINT "TYPE 1 IF TRANSFORMATION TO RECIPROCAL"
00360 PRINT "AVERAGING SCORES REQUIRED ELSE TYPE 0";
00370 INPUT Z3
00380 IF Z3=1 THEN 450
00390 PRINT "IF COMMON OR RARE SPECIES ARE GIVEN HIGH WEIGHT"
00400 PRINT "TYPE 0, TYPE 1 IF NO WEIGHTING REQUIRED";
00410 INPUT Z4
00420 PRINT "IF RARE SPECIES ARE GIVEN HIGH WEIGHT"
00430 PRINT "TYPE 1 ELSE TYPE 0";
00440 INPUT Z5
00450 PRINT
00460 LET L=C
00470 LET N=R
00480 MAT B=IDN
00490 LET A=0
00500 FOR I=1 TO R
00510 FOR J=1 TO C
00520 READ #1,Q
00530 LET X(I,J)=Q
00540 LET D(I,J)=Q
00550 LET S(I)=S(I)+Q
00560 LET O(J)=O(J)+Q
00570 NEXT J
00580 LET A=A+S(I)
00590 NEXT I
00600 LET Q2=A
```

```
00610 FOR I=1 TO R
00620 FOR J=1 TO C
00630 LET X(I,J)=X(I,J)/SQR(S(I)*O(J))
00640 LET X(I,J)=X(I,J)-SQR(S(I)*O(J))/A
00650 NEXT J
00660 NEXT I
00670 MAT T=TRN(X)
00680 MAT R=X*T
00690 IF Z2=0 THEN 720
00700 PRINT "SCALAR PRODUCTS"
00710 MAT PRINT R;
00720 REM---EIGENVALUE AND VECTOR PROCEDURE
00730 LET A=0.00000001
00740 LET C=0
00750 FOR I=2 TO N
00760 FOR J=1 TO I-1
00770 LET C=C+2*(R(I,J)^2)
00780 NEXT J,I
00790 LET Y=SQR(C)
00800 LET O=(A/N)*Y
00810 LET T=Y
00820 LET D=0
00830 LET T=T/N
00840 FOR Q=2 TO N
00850 FOR P=1 TO Q-1
00860 IF ABS(R(P,Q))<T THEN 1170
00870 LET D=1
00880 LET V=R(P,P)
00890 LET Z=R(P,Q)
00900 LET E=R(Q,Q)
00910 LET F=.5*(V-E)
00920 IF F=0 THEN 950
00930 LET G=-(SGN(F))
00940 GO TO 960
00950 LET G=-1
00960 LET G=G*Z/(SQR(Z^2+F^2))
00970 LET H=G/(SQR(2*(1+SQR(1-G^2))))
00980 LET K=SQR(1-H^2)
00990 FOR I=1 TO N
01000 IF I=P THEN 1080
01010 IF I=Q THEN 1080
01020 LET C=R(I,P)
01030 LET F=R(I,Q)
01040 LET R(Q,I)=C*H+F*K
01050 LET R(I,Q)=R(Q,I)
01060 LET R(P,I)=C*K-F*H
01070 LET R(I,P)=R(P,I)
01080 LET C=B(I,P)
01090 LET F=B(I,Q)
01100 LET B(I,Q)=C*H+F*K
01110 LET B(I,P)=C*K-F*H
01120 NEXT I
01130 LET R(P,P)=V*K^2+E*H^2-2*Z*H*K
01140 LET R(Q,Q)=V*H^2+E*K^2+2*Z*H*K
01150 LET R(P,Q)=(V-E)*H*K+Z*(K^2-H^2)
```

```
01160 LET R(Q,P)=R(P,Q)
01170 NEXT P,Q
01180 IF D<>1 THEN 1210
01190 LET D=0
01200 GO TO 840
01210 IF T>O THEN 830
01220 FOR I=1 TO N
01230 LET Q(I)=I
01240 NEXT I
01250 LET J=0
01260 LET V1=0
01270 LET J=J+1
01280 FOR I=1 TO N-J
01290 IF R(I,I)>=R(I+1,I+1) THEN 1370
01300 LET V1=1
01310 LET V2=R(I,I)
01320 LET R(I,I)=R(I+1,I+1)
01330 LET R(I+1,I+1)=V2
01340 LET P=Q(I)
01350 LET Q(I)=Q(I+1)
01360 LET Q(I+1)=P
01370 NEXT I
01380 IF V1<>0 THEN 1260
01390 IF Z3=0 THEN 1420
01400 GOSUB 1730
01410 STOP
01420 FOR J=1 TO N
01430 IF R(J,J)<0 THEN 1560
01440 LET K=Q(J)
01450 PRINT "  EIGENVALUE";R(J,J)
01460 PRINT "SET"J"OF SPECIES SCORES"
01470 FOR I=1 TO N
01480 IF Z5=0 THEN 1500
01490 LET B(I,K)=B(I,K)*(Q2/S(I))
01500 IF Z4=0 THEN 1520
01510 LET B(I,K)=B(I,K)/SQR(S(I)/Q2)
01520 PRINT B(I,K);
01530 WRITE #2,B(I,K)
01540 NEXT I
01550 PRINT
01560 NEXT J
01570 LET C=L
01580 MAT T=TRN(D)
01590 MAT C=T*B
01600 FOR J=1 TO R
01610 IF R(J,J)<0 THEN 1710
01620 PRINT "  SET "J" OF QUADRAT SCORES"
01630 LET K=Q(J)
01640 FOR I=1 TO C
01650 LET Q=C(I,K)/(O(I)*SQR(R(J,J)))
01660 WRITE #3,Q
01670 PRINT Q;
01680 NEXT I
01690 PRINT
01700 NEXT J
```

```
01710 GOSUB 2210
01720 STOP
01730 FOR J=1 TO N
01740 IF R(J,J)<0 THEN 2200
01750 LET K=Q(J)
01760 FOR I=1 TO N
01770 LET B(I,K)=B(I,K)/SQR(S(I))
01780 NEXT I
01790 PRINT "  RANGE"R(J,J)
01800 PRINT "SET"J"OF SPECIES SCORES"
01810 LET M1=10^20
01820 LET M2=-10^20
01830 FOR I=1 TO N
01840 IF M1<=B(I,K) THEN 1860
01850 LET M1=B(I,K)
01860 IF M2>=B(I,K) THEN 1880
01870 LET M2=B(I,K)
01880 NEXT I
01890 LET M3=M2-M1
01910 LET M5=1/M3
01920 FOR I=1 TO N
01930 LET U(I,K)=(B(I,K)-M1)*M5
01940 PRINT U(I,K);
01950 NEXT I
01960 PRINT
01970 GOSUB 2090
01980 PRINT "--OR--"
01990 PRINT "  SET"J"OF SPECIES SCORES"
02000 FOR I=1 TO N
02010 LET U(I,K)=U(I,K)*R(J,J)
02020 PRINT U(I,K);
02030 WRITE #2,U(I,K)
02040 NEXT I
02050 PRINT
02060 GOSUB 2090
02070 NEXT J
02080 RETURN
02090 PRINT "  SET"J"OF QUADRAT SCORES"
02100 FOR G=1 TO L
02110 LET Q=0
02120 FOR I=1 TO N
02130 LET Q=Q+D(I,G)*U(I,K)
02140 NEXT I
02150 LET Q=Q/O(G)
02160 PRINT Q;
02170 WRITE #3,Q
02180 NEXT G
02190 PRINT
02200 RETURN
02210 IF Z4=0 THEN 2280
02220 PRINT "  CANONICAL CORRELATIONS"
02230 FOR J=1 TO N
02240 IF R(J,J)<0 THEN 2270
02250 PRINT "SET"J"R(X,Y)="SQR(R(J,J))
02260 NEXT J
```

```
02270 PRINT "GRAND TOTAL IN MATRIX D"Q2
02280 END

RAWD            19:38           13-JAN-77

01000 45,2,9,26,3,5,2,90,31,16,18,92,32,48,73,80,95,13,92,78
01010 3,40,5,83,68,27,2,17,1,23,10,61,11,3,32,2,39,2,8,6
01020 9,53,99,21,49,81,72,6,90,62

.R BASIC

READY
OLD RQT
READY
RUN

RQT             19:42           13-JAN-77

PROGRAM NAME --- RQT
=======================================================
NUMBER OF SPECIES R ?5
NUMBER OF QUADRATS C ?10
TYPE 1 IF PRINTING OF SCALAR PRODUCTS REQUIRED
ELSE TYPE 0 ?1
TYPE 1 IF TRANSFORMATION TO RECIPROCAL
AVERAGING SCORES REQUIRED ELSE TYPE 0 ?0
IF COMMON OR RARE SPECIES ARE GIVEN HIGH WEIGHT
TYPE 0, TYPE 1 IF NO WEIGHTING REQUIRED ?1
IF RARE SPECIES ARE GIVEN HIGH WEIGHT
TYPE 1 ELSE TYPE 0 ?0

SCALAR PRODUCTS
 0.299985    -6.84761E-2 -1.51015E-2 -5.49618E-2 -7.99155E-2
-6.84761E-2  3.04182E-2 -1.62509E-2  1.15827E-2  1.68363E-2
-1.51015E-2 -1.62509E-2  0.129276    -1.02631E-2 -5.80477E-2
-5.49618E-2  1.15827E-2 -1.02631E-2  7.37661E-2 -1.12381E-2
-7.99155E-2  1.68363E-2 -5.80477E-2 -1.12381E-2  8.11857E-2
  EIGENVALUE 0.351403
SET 1 OF SPECIES SCORES
 2.59054 -0.375602  9.52410E-2 -0.584892 -0.523676
  EIGENVALUE 0.163559
SET 2 OF SPECIES SCORES
-0.510831 -0.112589  2.26283  0.195453 -0.840983
  EIGENVALUE 7.87536E-2
SET 3 OF SPECIES SCORES
 0.188011  9.87786E-2 -0.679524  2.81944 -0.760493
  EIGENVALUE 2.09145E-2
SET 4 OF SPECIES SCORES
 7.70121E-2 -1.33843  0.479915  1.10292  0.908719
  SET  1  OF QUADRAT SCORES
 1.97543 -0.605376 -0.5029  0.414539 -0.431461 -0.502718 -0.729606
 2.97285 -4.53156E-2 -0.197281
  SET  2  OF QUADRAT SCORES
-0.6935  0.463462 -1.23622  2.07723  1.1997 -0.230698 -0.707885
-0.263207 -1.09215 -0.212224
```

```
  SET  3  OF QUADRAT SCORES
 1.2388  1.63748 -0.978097 -1.06868  0.230041 -1.19631  1.07926
 0.215173 -0.508063 -0.677047
  SET  4  OF QUADRAT SCORES
 1.72124E-3  0.324964  2.76404 -6.73696E-4  0.460377 -0.635438
-0.579378  0.288916 -0.923847 -1.09026
  CANONICAL CORRELATIONS
SET 1 R(X,Y)= 0.592793
SET 2 R(X,Y)= 0.404425
SET 3 R(X,Y)= 0.280631
SET 4 R(X,Y)= 0.144619
GRAND TOTAL IN MATRIX D 1835

TIME:  3.52 SECS.

READY



STEREO          12:06          15-FEB-77

00010 PRINT "PROGRAM NAME --- STEREO"
00020 REM--- THIS PROGRAM FINDS STEREO CO-ORDINATES FOR A SAMPLE
00030 REM OF N POINTS IN THREE SPACE. DATA ARE READ FROM DISK FILES
00040 REM PTS1, PTS2 AND PTS3, EACH CONSISTING OF N NUMBERS. IT
00050 REM IS ADVISED THAT THE FILE WITH MAXIMUM DISPERSION
00060 REM BE ASSIGNED TO THE X1-AXIS AND THE FILE WITH
00070 REM MINIMUM DISPERSION BE ASSIGNED TO THE Y-AXIS.  BY
00080 REM PERMUTING THE ASSIGNMENT OF FILES TO CO-ORDINATE AXES,
00090 REM SIX DIFFERENT VIEWS CAN BE OBTAINED. THE OUTPUT IS IN
00100 REM THE FORM OF A LEFT AND RIGHT SET OF CO-ORDINATES
00101 REM WHICH ARE READ INTO DISK FILE STC. THIS FILE CAN BE USED
00102 REM IN PLOTTING.
00110 PRINT "=========================================================="
00120 REM  MAKE PERMUTATION OF FILES IN FOLLOWING LINE
00130 FILES PTS1,PTS2,PTS3,STC
00135 SCRATCH #4
00140 DIM Y(150),X1(150),X2(150),M1(150),M2(150),N1(150),N2(150)
00150 DIM Z(150)
00160 REM---EXPLANATION OF ARRAY SYMBOLS:
00180 REM           X1,X2,Y,M1,M2,N1,N2,Z -  N-VALUED VECTORS
00210 REM---READ DATA
00230 PRINT "SAMPLE SIZE N";
00240 INPUT N
00250 REM  A,B AND C ARE THE DIMENSIONS OF A BOX CORRESPONDING TO
00260 REM  THE X1,X2 AND Y AXES
00270 PRINT "LENGTH A, WIDTH B, HEIGHT C";
00280 INPUT A,B,C
00290 REM  (R1.R2,H) AND (L1,L2,H) ARE THE VIEWING POINTS IN
00300 REM  THREE SPACE
00310 PRINT "L1,L2,R1,R2,H";
00320 INPUT L1,L2,R1,R2,H
00330 FOR I=1 TO N
00340 READ #1,X1(I)
00350 READ #2,X2(I)
```

```
00360 READ #3,Y(I)
00370 NEXT I
00380 REM  MAXIMUM AND MINIMUM VALUES OF Y.X1 AND X2 ARE FOUND VIA
00390 REM  SUBROUTINES. THEY RETURN WITH A MAXIMUM F AND A MINIMUM D.
00395 REM  F AND D ARE VALUES IN THE DATA FILES FOR Y,X1 AND X2
00400 FOR I=1 TO N
00410 LET Z(I)=Y(I)
00420 NEXT I
00430 GOSUB 1060
00440 LET Y6=F
00450 GOSUB 1150
00460 LET Y7=D
00470 FOR I=1 TO N
00480 LET Z(I)=X1(I)
00490 NEXT I
00500 GOSUB 1060
00510 LET X6=F
00520 GOSUB 1150
00530 LET X7=D
00540 FOR I=1 TO N
00550 LET Z(I)=X2(I)
00560 NEXT I
00570 GOSUB 1060
00580 LET X8=F
00590 GOSUB 1150
00600 LET X9=D
00610 REM  TRANSLATE DATA INTO A BOX WITH DIMENSIONS A,B,C,
00620 REM  SUCH THAT ONE VERTEX IS AT THE ORIGIN WITH THE SIDES
00630 REM  EXTENDING IN THE POSITIVE DIRECTIONS
00640 FOR I=1 TO N
00650 LET Y(I)=C*(Y(I)-Y7)/(Y6-Y7)
00660 LET X1(I)=A*(X1(I)-X7)/(X6-X7)
00670 LET X2(I)=B*(X2(I)-X9)/(X8-X9)
00680 NEXT I
00690 REM  PROJECT IMAGES ONTO THE PLANE X1,X2.
00700 REM  (M1,M2) ARE THE POINTS SIGHTED FROM (L1,L2,H)
00710 REM  (N1,N2) ARE THE POINTS SIGHTED FROM (R1,R2,H)
00720 FOR I=1 TO N
00730 LET M1(I)=(H*X1(I)-L1*Y(I))/(H-Y(I))
00740 LET M2(I)=(H*X2(I)-L2*Y(I))/(H-Y(I))
00750 LET N1(I)=(H*X1(I)-R1*Y(I))/(H-Y(I))
00760 LET N2(I)=(H*X2(I)-R2*Y(I))/(H-Y(I))
00770 NEXT I
00780 PRINT
00790 PRINT " ","LEFT IMAGE",,"RIGHT IMAGE"
00800 PRINT " ","---- -----",,"----- -----"
00810 PRINT
00820 PRINT " ","    X1","    X2","    X1","    X2"
00830 FOR I=1 TO N
00840 PRINT I,M1(I),M2(I),N1(I),N2(I)
00845 WRITE #4,M1(I),M2(I),N1(I),N2(I)
00850 NEXT I
00860 REM  BOX CO-ORDINATES ARE SIGHTED AND RESULTS CAN
00870 REM  BE HAND PLOTTED. BASIC ASSUMES PARAMETERS
00880 REM  TO BE ZERO UNLESS GIVEN A VALUE
```

```
ØØ89Ø LET R(5)=R(6)=R(7)=R(8)=A
ØØ9ØØ LET S(1)=S(2)=S(5)=S(6)=B
ØØ91Ø LET T(2)=T(3)=T(6)=T(7)=C
ØØ92Ø PRINT
ØØ93Ø PRINT
ØØ94Ø PRINT "BOX CO-ORDINATES:"
ØØ95Ø PRINT
ØØ96Ø PRINT "X1","X2",,"X1","X2"
ØØ97Ø FOR I=1 TO 8
ØØ98Ø LET P1(I)=(H*S(I)-L1*R(I))/(H-R(I))
ØØ99Ø LET P2(I)=(H*T(I)-L2*R(I))/(H-R(I))
Ø1ØØØ LET Q1(I)=(H*S(I)-R1*R(I))/(H-R(I))
Ø1Ø1Ø LET Q2(I)=(H*T(I)-R2*R(I))/(H-R(I))
Ø1Ø2Ø PRINT P1(I),P2(I),,Q1(I),Q2(I)
Ø1Ø3Ø NEXT I
Ø1Ø4Ø GO TO 123Ø
Ø1Ø5Ø REM  MAXIMUM SUBROUTINE FOLLOWS
Ø1Ø6Ø LET F=Z(1)
Ø1Ø7Ø FOR I=2 TO N
Ø1Ø8Ø LET G=Z(I)
Ø1Ø9Ø IF F<=G THEN 111Ø
Ø11ØØ GO TO 112Ø
Ø111Ø LET F=G
Ø112Ø NEXT I
Ø113Ø RETURN
Ø114Ø REM MINIMUM SUBROUTINE FOLLOWS
Ø115Ø LET D=Z(1)
Ø116Ø FOR I=2 TO N
Ø117Ø LET E=Z(I)
Ø118Ø IF D>=E THEN 12ØØ
Ø119Ø GO TO 121Ø
Ø12ØØ LET D=E
Ø121Ø NEXT I
Ø122Ø RETURN
Ø123Ø END

PTS1            12:Ø7           15-FEB-77

Ø1ØØØ -Ø.635979,-Ø.692536,-Ø.598581,-Ø.477787,-Ø.648752,Ø.411381
Ø1Ø1Ø Ø.171221,-2.83154E-2,-8.95978E-2,-Ø.255626,-Ø.765841,-Ø.729417
Ø1Ø2Ø -Ø.746579,-Ø.683717,-Ø.545Ø74,-1.Ø1783,-1.15413,-1.23175
Ø1Ø3Ø -1.12819,-1.Ø5361,Ø.153463,8.53Ø82E-2,Ø.135327,9.15428E-2
Ø1Ø4Ø Ø.159698,Ø.485431,Ø.6949Ø3,Ø.58Ø857,Ø.464751,Ø.347436
Ø1Ø5Ø Ø.51Ø772,Ø.372775,Ø.45Ø639,Ø.51969,Ø.34699,Ø.388Ø96
Ø1Ø6Ø Ø.6421Ø5,Ø.645,Ø.673458,Ø.5627Ø1,Ø.118751,-5.53191E-3
Ø1Ø7Ø Ø.316913,Ø.799955,-7.4984ØE-2,Ø.359327,Ø.544674,Ø.66Ø549
Ø1Ø8Ø Ø.456941,Ø.413157

PTS2            12:Ø8           15-FEB-77

Ø1ØØØ -Ø.354Ø16,-Ø.472Ø6,-Ø.4Ø8873,-Ø.46531,-Ø.469283,-1.4Ø5Ø8
Ø1Ø1Ø -1.13242,-Ø.9ØØ627,-1.Ø6235,-Ø.963755,6.946ØØE-2,6.83131E-2
Ø1Ø2Ø Ø.1Ø9535,-4.77853E-2,-Ø.111727,Ø.252231,Ø.285967,Ø.3Ø43Ø2
Ø1Ø3Ø Ø.281314,Ø.274189,Ø.513252,Ø.599746,Ø.694593,Ø.691816
Ø1Ø4Ø Ø.6Ø5322,-Ø.534189,-Ø.231618,-Ø.319227,-Ø.289357,-Ø.141332
```

```
01060 5.42801E-2,8.27617E-2,-8.15380E-2,-0.224721,-2.32304E-2
01070 8.44658E-2,0.28755,-3.92530E-2,-7.59002E-2,9.92967E-2
01080 0.566492,0.711521,0.886654,0.245939,0.829434,6.83335E-2
01090 0.254677,0.211215,0.311884,0.309107

PTS3          12:09          15-FEB-77

01000 -0.272153,-0.32045,-0.321688,-9.46153E-2,-0.337374,0.167644
01010 5.03254E-2,-3.63526E-2,-8.41426E-2,-0.174874,-0.78363,-0.708017
01020 -0.876336,-0.741786,-0.814243,0.770054,0.964595,1.07297
01030 0.927754,0.864644,-8.84063E-2,-0.162801,-0.21158,-0.194656
01040 -0.120261,5.56885E-2,0.127029,0.181674,0.12788,6.04346E-2
01050 0.231502,0.18617,0.162458,0.251371,0.181573,0.110422
01060 0.15042,0.278766,0.239025,0.290089,-0.410611,-0.398166
01070 -0.375816,-3.61338E-2,-0.511079,0.129504,0.137597,0.205721
01080 6.64646E-2,8.33887E-2

.R BASIC

READY
OLD STEREO
READY
RUN

STEREO          12:10          15-FEB-77

PROGRAM NAME --- STEREO
=======================================================
SAMPLE SIZE N ?50
LENGTH A, WIDTH B, HEIGHT C ?2.92557,3.3,2.80692
L1,L2,R1,R2,H ?1.287,1.287,2.112,1.287,9.9

              LEFT IMAGE                      RIGHT IMAGE
              ---- -----                      ----- -----

               X1            X2               X1            X2
 1            0.816542      1.53531          0.737057      1.53531
 2            0.731534      1.34848          0.658962      1.34848
 3            0.878806      1.44744          0.80641       1.44744
 4            1.05985       1.36173          0.95401       1.36173
 5            0.801428      1.35265          0.731253      1.35265
 6            2.55922      -0.230415         2.41152      -0.230415
 7            2.13444       0.253292         2.00592       0.253292
 8            1.79496       0.648365         1.68013       0.648365
 9            1.69123       0.390181         1.58379       0.390181
10            1.41905       0.561464         1.32531       0.561464
11            0.662468      2.13471          0.651192      2.13471
12            0.709193      2.14257          0.688488      2.14257
13            0.698626      2.18098          0.698626      2.18098
14            0.779207      1.96777          0.762739      1.96777
15            0.986067      1.86762          0.978548      1.86762
16           -2.09765E-4    2.73264         -0.259977      2.73264
17           -0.317989      2.85485         -0.619675      2.85485
18           -0.5093        2.92619         -0.835774      2.92619
19           -0.255688      2.83438         -0.549189      2.83438
```

20	-0.092917	2.80162	-0.372669	2.80162
21	2.08625	2.95328	1.97946	2.95328
22	1.96709	3.07213	1.87155	3.07213
23	2.04148	3.20931	1.95318	3.20931
24	1.97356	3.21012	1.88276	3.21012
25	2.09217	3.09356	1.99024	3.09356
26	2.65861	1.24888	2.52923	1.24888
27	3.02845	1.75855	2.88747	1.75855
28	2.85071	1.61389	2.70067	1.61389
29	2.6406	1.66126	2.49949	1.66126
30	2.42965	1.90378	2.29951	1.90378
31	2.74392	2.25786	2.58545	2.25786
32	2.49753	2.29879	2.34672	2.29879
33	2.62468	2.01599	2.47785	2.01599
34	2.76432	1.78064	2.60245	1.78064
35	2.45269	2.11761	2.30266	2.11761
36	2.50771	2.28864	2.36946	2.28864
37	2.94604	2.63927	2.8012	2.63927
38	2.98827	2.104	2.82167	2.104
39	3.02545	2.03537	2.86569	2.03537
40	2.84893	2.34589	2.68036	2.34589
41	1.99245	2.95175	1.93251	2.95175
42	1.80149	3.17943	1.73982	3.17943
43	2.30405	3.45805	2.23928	3.45805
44	3.15369	2.5292	3.03884	2.5292
45	1.68694	3.32593	1.64065	3.32593
46	2.46316	2.26469	2.32177	2.26469
47	2.77784	2.58078	2.63512	2.58078
48	2.99339	2.52172	2.8393	2.52172
49	2.61356	2.66074	2.48244	2.66074
50	2.54408	2.66002	2.41023	2.66002

BOX CO-ORDINATES:

X1	X2	X1	X2
3.3	0	3.3	0
3.3	2.80692	3.3	2.80692
0	2.80692	0	2.80692
0	0	0	0
4.14439	-0.539859	3.79833	-0.539859
4.14439	3.44448	3.79833	3.44448
-0.539859	3.44448	-0.885922	3.44448
-0.539859	-0.539859	-0.885922	-0.539859

TIME: 4.90 SECS.

8. Programs From Chapter 4

```
ALC            13:53           04-JAN-77

00010 PRINT "PROGRAM NAME --- ALC"
00020 REM--- A HIERARCHICAL CLASSIFICATION IS COMPUTED OF
00030 REM N INDIVIDUALS. THE METHOD USED IS AVERAGE LINKAGE
00040 REM CLUSTERING. THE INPUT DATA ARE READ FROM
00050 REM DISK FILE COR2 CONTAINING THE UPPER HALF OF AN
00060 REM N*N SIMILARITY MATRIX EXCLUDING THE ELEMENTS
00070 REM IN THE PRINCIPAL DIAGONAL POSITIONS.  THE
00075 REM SIMILARITY VALUES ARE ASSUMED TO BE IN THE
00077 REM ZERO TO ONE RANGE.
00080 PRINT "=========================="
00090 FILES COR2
00100 DIM S(5,5),R(5,5),N(5)
00120 REM---EXPLANATION OF ARRAY SYMBOLS:
00140 REM            S,R - N*N ARRAYS
00160 REM            N - AN N-VALUED VECTOR
00190 REM---READ DATA
00200 PRINT "NUMBER OF INDIVIDUALS N";
00210 INPUT N
00220 FOR L=1 TO N-1
00230 FOR M=L+1 TO N
00240 READ  1,A
00250 LET S(L,M)=A
00260 LET S(M,L)=A
00270 NEXT M,L
00280 MAT N=CON
00290 MAT R=ZER
00300 FOR L=1 TO N
00310 LET R(L,1)=L
00320 NEXT L
00330 REM---FIND VALID FUSIONS
00340 LET Q=0
00350 FOR I=1 TO N-1
00360 FOR J=I+1 TO N
00370 IF S(I,J)<=Q THEN 410
00380 LET Q=S(I,J)
00390 LET L=I
00400 LET M=J
00410 NEXT J,I
00420 IF Q=0 THEN 870
00430 LET C=0
00440 REM---UPDATE GROUP REGISTERS
00450 FOR I=N(L)+1 TO N(L)+N(M)
00460 LET C=C+1
00470 LET R(L,I)=R(M,C)
00480 LET R(M,C)=0
00490 NEXT I
00500 LET N1=N(L)
00510 LET N2=N(M)
00520 LET N3=N1+N2
```

```
00540 LET N(L)=N(L)+N(M)
00560 LET N(M)=0
00570 REM---PRINT FUSION STATISTICS
00600 LET C6=C6+1
00610 PRINT
00620 PRINT "  CLUSTERING PASS"C6
00630 PRINT "NUMBER OF INDIVIDUALS"N(L)
00640 PRINT "AVERAGE SIMILARITY"Q
00650 PRINT "INDIVIDUALS:"
00660 FOR J=1 TO N
00670 IF R(L,J)=0 THEN 700
00680 PRINT R(L,J);
00690 NEXT J
00700 PRINT
00720 REM---COMPUTE AVERAGE SIMILARITIES
00730 FOR J=1 TO N
00740 IF S(L,J)=-1000 THEN 810
00750 IF M=J THEN 810
00755 IF J=L THEN 810
00760 LET A=(N1/N3)*S(L,J)
00770 LET B=(N2/N3)*S(M,J)
00780 LET D=((N1*N2)/(N32))*(1-S(L,M))
00790 LET S(L,J)=A+B+D
00800 LET S(J,L)=A+B+D
00810 NEXT J
00820 FOR J=1 TO N
00830 LET S(M,J)=-1000
00840 LET S(J,M)=-1000
00850 NEXT J
00860 GO TO 330
00870 END

COR2           14:06          04-JAN-77

01000 0.94,0.42,0.18,0,0.61,0.39,0.15,0.97,0.87,0.95

.R BASIC

READY
OLD ALC
READY
RUN

ALC            14:07          04-JAN-77

PROGRAM NAME --- ALC
===========================
NUMBER OF INDIVIDUALS N ?5

  CLUSTERING PASS 1
NUMBER OF INDIVIDUALS 2
AVERAGE SIMILARITY 0.97
INDIVIDUALS:
 3  4
```

```
  CLUSTERING PASS 2
NUMBER OF INDIVIDUALS 2
AVERAGE SIMILARITY 0.94
INDIVIDUALS:
 1  2

  CLUSTERING PASS 3
NUMBER OF INDIVIDUALS 3
AVERAGE SIMILARITY 0.9175
INDIVIDUALS:
 3  4  5

  CLUSTERING PASS 4
NUMBER OF INDIVIDUALS 5
AVERAGE SIMILARITY 0.33
INDIVIDUALS:
 1  2  3  4  5

TIME:  0.69 SECS.

READY



SSA            15:45          04-JAN-77

00010 PRINT "PROGRAM NAME --- SSA"
00020 REM--- THIS PROGRAM COMPUTES A HIERARCHICAL CLASSIFICATION
00030 REM FOR N INDIVIDUALS BASED ON EUCLIDEAN DISTANCES.
00040 REM THE DISTANCES ARE READ FROM DISK FILE DIS2 CONTAINING
00050 REM THE UPPER HALF OF THE DISTANCE MATRIX EXCLUDING
00060 REM ZEROS IN THE PRINCIPAL DIAGONAL POSITIONS.
00070 PRINT "================================"
00080 FILES DIS2
00090 DIM D(5,5),R(5,5),N(5),Q(5),A(5,5),X(5)
00110 REM---EXPLANATION OF ARRAY SYMBOLS:
00130 REM         D, R, A - N*N ARRAYS
00140 REM         N, Q, X - N-VALUED VECTORS
00170 REM---READ DATA
00180 PRINT "NUMBER OF INDIVIDUALS N";
00190 INPUT N
00200 FOR J=1 TO N-1
00210 FOR K=J+1 TO N
00220 READ #1,A
00230 LET D(J,K)=A*A/2
00240 LET D(K,J)=A*A/2
00250 LET A(J,K)=A*A
00260 LET A(K,J)=A*A
00270 NEXT K,J
00280 REM---AT THIS POINT MATRIX D CONTAINS THE WITHIN GROUP SUM OF
00290 REM    SQUARES FOR ALL POTENTIAL FUSIONS
00300 LET D=0
00310 MAT N=CON
00320 MAT R=ZER
00330 MAT Q=ZER
```

```
00340 FOR I=1 TO N
00350 LET R(I,1)=I
00360 NEXT I
00370 REM---SEARCH FOR VALID FUSIONS
00380 LET D=D+1
00390 LET Q=10^10
00400 FOR J=1 TO N-1
00410 IF R(J,1)=0 THEN 500
00420 FOR K=J+1 TO N
00430 IF R(K,1)=0 THEN 490
00440 LET W=D(J,K)-Q(J)-Q(K)
00450 IF W>=Q THEN 490
00460 LET Q=W
00470 LET L=J
00480 LET M=K
00490 NEXT K
00500 NEXT J
00510 LET C=0
00520 FOR I=N(L)+1 TO N(L)+N(M)
00530 LET C=C+1
00540 LET R(L,I)=R(M,C)
00550 LET R(M,C)=0
00560 NEXT I
00570 PRINT
00580 PRINT "   CLUSTERING PASS"D
00590 LET N(L)=N(L)+N(M)
00600 LET N(M)=0
00610 LET Q(L)=D(L,M)
00620 LET Q(M)=0
00630 FOR I=1 TO N
00640 LET D(M,I)=0
00650 LET D(I,M)=0
00660 NEXT I
00670 PRINT "GROUPS IN FUSION("L"+"M")"
00680 PRINT "NUMBER OF INDIVIDUALS IN GROUP"N(L)
00690 PRINT "SUM OF SQUARES"Q(L)
00700 PRINT "INDIVIDUALS:"
00710 FOR J=1 TO N(L)
00720 PRINT R(L,J);
00730 NEXT J
00735 PRINT
00740 REM---GENERATE NEW D MATRIX
00750 FOR I=1 TO N
00760 LET S=0
00770 IF R(I,1)=0 THEN 940
00780 IF I=L THEN 940
00790 FOR J=1 TO N(I)
00800 LET X(J)=R(I,J)
00810 NEXT J
00820 FOR H=1 TO N(L)
00830 LET J=J+1
00840 LET X(J)=R(L,H)
00850 NEXT H
00860 FOR J=1 TO N(I)+N(L)-1
00870 LET A=X(J)
```

```
00880 FOR H=J+1 TO N(I)+N(L)
00890 LET B=X(H)
00900 LET S=S+A(A,B)
00910 NEXT H,J
00920 LET D(I,L)=S/(N(I)+N(L))
00930 LET D(L,I)=D(I,L)
00940 NEXT I
00950 IF D=N-1 THEN 970
00960 GO TO 380
00970 END

DIS2            15:59           04-JAN-77

01000 0.34641,1.07703,1.28062,1.41421,0.883176,1.10454,1.30384
01010 0.244949,0.509902,0.316228

.R BASIC

READY
OLD SSA
READY
RUN

SSA             15:60           04-JAN-77

PROGRAM NAME --- SSA
==============================
NUMBER OF INDIVIDUALS N ?5

  CLUSTERING PASS 1
GROUPS IN FUSION( 3 + 4 )
NUMBER OF INDIVIDUALS IN GROUP 2
SUM OF SQUARES 0.03
INDIVIDUALS:
 3  4

  CLUSTERING PASS 2
GROUPS IN FUSION( 1 + 2 )
NUMBER OF INDIVIDUALS IN GROUP 2
SUM OF SQUARES 5.99999E-2
INDIVIDUALS:
 1  2

  CLUSTERING PASS 3
GROUPS IN FUSION( 3 + 5 )
NUMBER OF INDIVIDUALS IN GROUP 3
SUM OF SQUARES 0.14
INDIVIDUALS:
 3  4  5

  CLUSTERING PASS 4
GROUPS IN FUSION( 1 + 3 )
NUMBER OF INDIVIDUALS IN GROUP 5
SUM OF SQUARES 1.808
INDIVIDUALS:
```

```
 1  2  3  4  5

TIME:  1.28 SECS.

READY


TRGRPS          14:30          24-FEB-77

00010 PRINT "PROGRAM NAME -- TRGRPS"
00020 REM--- THIS PROGRAM PERFORMS A DETERMINISTIC
00030 REM TEST ON THE HYPOTHESIS THAT THE
00040 REM SAMPLE DIVIDES INTO R GROUPS ALONG
00050 REM NATURAL DISCONTINUITIES.  IF NO SOLUTION
00060 REM IS FOUND UNDER THE HYPOTHESIS OF R GROUPS, THE
00070 REM TEST CONTINUES WITH A MODIFIED HYPOTHESIS OF R-1
00080 REM GROUPS.  THE DATA ARE INPUT FROM DISK FILE
00090 REM RAWD4 ARRANGED AS P SETS OF N NUMBERS. THE
00100 REM SYMBOLS P AND N INDICATE RESPECTIVELY THE NUMBER
00110 REM OF SPECIES AND QUADRATS.
00120 PRINT"================================="
00130 FILES RAWD4
00140 DIM D(2,85),A(85,85),N(85),M(2,85),P(85),O(85)
00150 DIM S(85,85),B(85,85),X(85),Y(85,85),Z(85,85),W(85)
00160 REM---EXPLANATION OF ARRAY SYMBOLS:
00170 REM        D, M - P*N ARRAYS
00180 REM        A, S, B, Z, Y - N*N ARRAYS
00190 REM        N, P. O, X, W - N-VALUED VECTORS
00200 PRINT "NUMBER OF SPECIES P";
00210 INPUT P
00220 PRINT "NUMBER OF QUADRATS N";
00230 INPUT N
00240 PRINT "MINIMUM GROUP SIZE REQUIRED S";
00250 INPUT S
00260 PRINT "NUMBER OF GROUPS REQUIRED R";
00270 INPUT R
00280 REM---READ DATA, GENERATE DISTANCE MATRIX
00290 MAT A=ZER
00300 LET C9=1
00310 FOR I=1 TO P
00320 FOR K=1 TO N
00330 READ #1,D(I,K)
00340 NEXT K,I
00350 FOR K=1 TO N-1
00360 FOR J=K+1 TO N
00370 FOR I=1 TO P
00380 LET A(J,K)=A(J,K)+(D(I,K)-D(I,J))^2
00390 NEXT I
00400 LET A(K,J)=A(J,K)
00410 NEXT J,K
00420 FOR H=1 TO 51
00430 IF H>N THEN 550
00440 LET M=10^30
00450 FOR I=1 TO N-1
```

```
00460 FOR J=I+1 TO N
00470 IF A(I,J)<=W(C9) THEN 500
00480 IF M<=A(I,J) THEN 500
00490 LET M=A(I,J)
00500 NEXT J,I
00510 IF M=10^30 THEN 550
00520 LET W(H)=M
00530 LET C9=H
00540 NEXT H
00550 LET C=W(1)
00560 PRINT "INITIAL VALUE OF NEIGHBOURHOOD RADIUS ="C
00570 PRINT "CONFIRM BY TYPING "C"ELSE TYPE DESIRED VALUE:";
00580 INPUT C
00590 PRINT "SELECT A VALUE FROM THE LIST BELOW"
00600 PRINT "TO SERVE AS THE INCREMENT IN NEIGHBOURHOOD RADIUS:"
00610 FOR I=1 TO C9-1
00620 LET W(I)=W(I+1)-W(I)
00630 NEXT I
00640 GOSUB 1640
00650 FOR I=1 TO C9-1
00660 IF I=1 THEN 680
00670 IF W(I)=W(I-1) THEN 690
00680 PRINT W(I);
00690 NEXT I
00700 PRINT
00710 PRINT "TYPE SELECTED VALUE:"
00720 INPUT D
00730 LET O1=C
00740 MAT B=ZER
00750 MAT N=CON
00760 LET N2=N3=0
00770 FOR J=1 TO N
00780 LET B(J,1)=J
00790 NEXT J
00800 MAT W=N
00810 LET U=N3
00820 MAT Z=B
00830 LET O2=N2
00840 FOR J=1 TO N-1
00850 FOR K=J+1 TO N
00860 IF A(J,K)>C THEN 1240
00870 REM   SEARCH ARRAY B FOR LABEL J
00880 FOR H=1 TO N
00890 FOR I=1 TO N
00900 IF B(H,I)=0 THEN 950
00910 IF B(H,I)<>J THEN 940
00920 LET R1=H
00930 GO TO 970
00940 NEXT I
00950 NEXT H
00960 REM   SEARCH ARRAY B FOR LABEL K
00970 FOR H=1 TO N
00980 FOR I=1 TO N
00990 IF B(H,I)=0 THEN 1040
01000 IF B(H,I)<>K THEN 1030
```

```
01010 LET R2=H
01020 GO TO 1050
01030 NEXT I
01040 NEXT H
01050 IF R1=R2 THEN 1240
01060 IF R1>R2 THEN 1160
01070 LET N2=0
01080 FOR I=N(R1)+1 TO N(R1)+N(R2)
01090 LET N2=N2+1
01100 LET B(R1,I)=B(R2,N2)
01110 LET B(R2,N2)=0
01120 NEXT I
01130 LET N(R1)=N(R1)+N(R2)
01140 LET N(R2)=0
01150 GO TO 1240
01160 LET N2=0
01170 FOR I=N(R2)+1 TO N(R1)+N(R2)
01180 LET N2=N2+1
01190 LET B(R2,I)=B(R1,N2)
01200 LET B(R1,N2)=0
01210 NEXT I
01220 LET N(R2)=N(R1)+N(R2)
01230 LET N(R1)=0
01240 NEXT K,J
01250 LET N2=N3=0
01260 FOR I=1 TO N
01270 IF N(I)=0 THEN 1310
01280 LET N3=N3+1
01290 IF N(I)<S THEN 1310
01300 LET N2=N2+1
01310 NEXT I
01320 LET N9=N8
01330 LET N8=C
01340 LET C=C+D
01350 IF N2<>R THEN 1380
01360 IF N3<>R THEN 1380
01370 GO TO 1480
01380 IF N3>R THEN 800
01390 PRINT "NO SOLUTION AT R="R"-- R REDUCED TO "R-1
01400 LET R=R-1
01410 LET C=O1
01420 IF R>1 THEN 740
01430 PRINT "ONLY ONE GROUP FOUND, ANALYSIS TERMINATED"
01440 STOP
01450 REM   AT THIS POINT ARRAY B HOLDS
01460 REM   INFORMATION ABOUT GROUP CONTENTS.
01470 REM   VECTOR N CONTAINS GROUP SIZES
01480 PRINT N2"--GROUPS"
01490 PRINT "MAXIMUM NEIGHBOURHOOD RADIUS"N8
01500 MAT X=N
01510 LET U=0
01520 FOR I=1 TO R
01530 PRINT
01540 PRINT "GROUP" I
01550 PRINT
```

```
01560 LET U=U+1
01570 IF X(U)=0 THEN 1560
01580 FOR K=1 TO X(U)
01590 PRINT B(U,K);
01600 NEXT K
01610 PRINT
01620 NEXT I
01630 STOP
01640 FOR H=1 TO C9
01650 FOR I=1 TO C9-1
01660 IF W(I)<W(I+1) THEN 1700
01670 LET C1=W(I)
01680 LET W(I)=W(I+1)
01690 LET W(I+1)=C1
01700 NEXT I,H
01710 RETURN
01720 END

RAWD4           14:33           24-FEB-77

01000       19.5,25.5,17.5,23.2,12.3,19.2,11.4,17.0
01010       22.6,16.5,21.2,22.1,27.8,25.7,31.9,33.4
01020       37.3,33.2,36.8,38.0,40.3,42.1,46.8,43.1
01030       40.6,46.5,47.9,47.8,52.6,54.0,53.1,63.2
01040       68.2,62.6,61.5,68.1,65.2,65.6,69.5,73.3
01050       72.8,66.8,62.7,68.9,61.0,60.0,64.1,57.6
01060       55.2,50.5,58.8,48.8,46.9,42.9,36.1,29.5
01070       35.5,57.5,38.1,41.2,36.5,40.4,43.0,46.0
01080       45.1,46.2,57.1,61.5,62.5,67.5,61.1,68.3
01090       72.9,80.5,83.0,79.1,85.5,84.4,89.1,93.1
01100       91.1,93.2,85.9,86.2,92.6
01110       12.5,17.8,19.5,22.2,23.8,25.8,31.7,31.0
01120       33.0,38.0,39.2,44.8,42.1,49.1,48.1,53.4
01130       50.7,58.3,55.5,59.5,57.9,54.1,58.4,62.1
01140       63.8,68.2,64.1,61.4,60.9,62.8,67.9,66.8
01150       65.5,61.2,59.5,59.7,56.9,52.1,52.9,57.4
01160       48.7,49.1,44.3,41.8,41.0,37.2,34.7,32.0
01170       32.8,29.9,26.9,25.8,18.9,20.9,14.2,12.9
01180       20.8,64.1,34.2,37.9,37.8,41.4,40.0,39.0
01190       43.0,44.9, 9.6, 7.4,12.0,15.6,17.9,21.3
01200       26.1,26.0,27.0,33.1,34.5,40.1,38.0,44.0
01210       49.1,15.2,10.3,16.4,17.6

.R BASIC

READY
OLD TRGRPS
READY
RUN

TRGRPS          14:38           24-FEB-77

PROGRAM NAME -- TRGRPS
================================
NUMBER OF SPECIES P ?2
```

```
NUMBER OF QUADRATS N ?85
MINIMUM GROUP SIZE REQUIRED S ?4
NUMBER OF GROUPS REQUIRED R ?4
INITIAL VALUE OF NEIGHBOURHOOD RADIUS = 4.1
CONFIRM BY TYPING  4.1 ELSE TYPE DESIRED VALUE: ?4.1
SELECT A VALUE FROM THE LIST BELOW
TO SERVE AS THE INCREMENT IN NEIGHBOURHOOD RADIUS:
 4.05312E-6  9.99975E-3  4.00009E-2  4.99989E-2  5.00007E-2
 6.00021E-2  7.99975E-2  8.00025E-2  9.00049E-2  0.119995  0.140002
 0.15  0.160002  0.169997  0.2  0.229996  0.230001  0.230002  0.23999
 0.28  0.28  0.3  0.33  0.330005  0.339996  0.349997  0.350001
 0.369998  0.370001  0.419999  0.440002  0.450001  0.479997  0.520003
 0.520003  0.55  0.559992  0.560001  0.630001  0.719999  0.750002
 0.760004  0.85  0.860001  0.870001  0.929998  0.929999  1.5  2.45
TYPE SELECTED VALUE:
 ?2.45
 4 --GROUPS
MAXIMUM NEIGHBOURHOOD RADIUS 58.

GROUP 1
 1  2  4  6  8  7  9  56  3  10  11  12  14  13  15  16  19  20  21
 22  18  17  23  27  28  26  24  25  29  30  58  31  32  33  34  35
 36  37  38  42  39  41  40  43  45  46  47  48  49  51  50  52  5
 44  55  57  53  54
GROUP 2
 59  61  60  63  62  64  65  66
GROUP 3
 67  68  69  70  72  71  73  74  75  76  77  79  78  80  81
GROUP 4
 82  85  83  84

TIME:  253.33 SECS.

READY
```

```
TSTCOV          09:15          10-JAN-77

00010 PRINT "PROGRAM NAME --- TSTCOV"
00020 REM---  THE EQUALITY OF K COVARIANCE MATRICES
00030 REM IS TESTED BASED ON USING A CHI SQUARE CRITERION.
00040 REM THE DATA ARE INPUT FROM DISK FILE RAWD3. IN THIS FILE
00050 REM THE DATA ARE ARRANGED AS N1 SETS OF P NUMBERS.
00060 PRINT "=================================================="
00070 FILES RAWD3
00080 DIM X(10,3),U(10,3),M(3),N(3),R(3,3),B(3,3),E(3,3)
00090 DIM A(3,3),Y(3,3),Q(3),T(3,10)
00110 REM---EXPLANATION OF ARRAY SYMBOLS:
00120 REM    X, U - MAX(N)*P ARRAYS WHERE MAX(N) INDICATES
00130 REM              THE SIZE OF THE LARGEST GROUP AND P THE NUMBER
00140 REM              OF VARIATES
00150 REM    N - A K-VALUED VECTOR WHERE K INDICATES
00160 REM              THE NUMBER OF GROUPS
00170 REM    R, B, E, A, Y - P*P ARRAYS
```

```
00180 REM    M, Q - P-VALUED VECTORS
00190 REM    T - A P*MAX(N) ARRAY
00210 REM---READ DATA
00220 PRINT "TOTAL NUMBER OF INDIVIDUALS N1";
00230 INPUT N1
00240 PRINT "NUMBER OF VARIATES P";
00250 INPUT P
00260 LET N=P
00270 PRINT "NUMBER OF GROUPS K";
00280 INPUT K
00290 LET G=K
00300 PRINT "K-VALUED VECTOR OF GROUP SIZES";
00310 MAT INPUT N
00320 LET O=0
00330 LET S1=Z1=0
00340 MAT A=ZER
00350 LET F4=N1-G
00360 LET F5=1/F4
00370 LET O=O+1
00380 MAT M=ZER
00390 FOR H=1 TO N(O)
00400 FOR I=1 TO N
00410 READ #1,X(H,I)
00420 LET M(I)=M(I)+X(H,I)
00430 NEXT I,H
00440 FOR H=1 TO N(O)
00450 FOR I=1 TO N
00460 LET X(H,I)=X(H,I)-M(I)/N(O)
00470 NEXT I,H
00480 MAT T=TRN(X)
00490 MAT R=ZER
00500 REM---COMPUTE WITHIN GROUP SUMS OF SQUARES AND CROSS PRODUCTS
00510 FOR H=1 TO N
00520 FOR I=H TO N
00530 FOR J=1 TO N(O)
00540 LET R(H,I)=R(H,I)+X(J,H)*X(J,I)
00550 NEXT J
00560 LET R(I,H)=R(H,I)
00570 NEXT I,H
00580 MAT A=A+R
00590 PRINT
00600 PRINT "   ANALYSIS OF GROUP";O
00610 PRINT "      COVARIANCE MATRIX"
00620 LET Z3=N(O)-1
00630 LET Z=1/Z3
00640 REM---COMPUTE WITHIN GROUP COVARIANCE MATRIX AND ITS INVERSE
00650 MAT R=(Z)*R
00660 MAT PRINT R;
00670 MAT Y=INV(R)
00680 PRINT "      INV OF COV MAT"
00690 MAT PRINT Y;
00700 LET D1=DET
00710 PRINT "   DETERMINANT =";D1
00720 LET S1=S1+(Z3*LOG(D1))
00730 LET Z1=Z1+Z
```

```
00740 GOSUB 970
00750 IF O<G THEN 370
00760 LET Z=F5
00770 MAT R=(Z)*A
00780 MAT Y=INV(R)
00790 PRINT
00800 PRINT "POOLED COV MAT"
00810 MAT PRINT R;
00820 PRINT "INV OF POOLED COV MAT"
00830 MAT PRINT Y;
00840 LET D2=DET
00850 PRINT "  DETERMINANT OF POOLED COV MAT =";D2
00860 GOSUB 970
00870 REM---COMPUTE STATISTIC M AND 1/C
00880 LET M=F4*LOG(D2)-S1
00890 LET C1=1-(((2*N^2+3*N-1)/(6*(N+1)*(G-1)))*(Z1-F5))
00900 LET C=M*C1
00910 PRINT "M=";M
00920 PRINT "1/C ="; C1
00930 PRINT "CHI SQUARE ="; C
00940 LET D7=.5*(G-1)*N*(N+1)
00950 PRINT "THIS IS APPROXIMATELY DISTRIBUTED AS A CHI SQUARE";
00955 PRINT "VARIATE WITH";D7;"DF"
00960 STOP
00970 LET Z3=SQR(Z)
00980 LET O1=O
00990 G9=G
01000 REM---EIGENVECTORS PROCEDURE
01010 MAT B=IDN
01020 LET A=0.00000001
01030 LET C=0
01040 FOR I=2 TO N
01050 FOR J=1 TO I-1
01060 LET C=C+2*(R(I,J)^2)
01070 NEXT J,I
01080 LET Y=SQR(C)
01090 LET O=(A/N)*Y
01100 LET T=Y
01110 LET D=0
01120 LET T=T/N
01130 FOR Q=2 TO N
01140 FOR P=1 TO Q-1
01150 IF ABS(R(P,Q))<T THEN 1460
01160 LET D=1
01170 LET V=R(P,P)
01180 LET Z=R(P,Q)
01190 LET E=R(Q,Q)
01200 LET F=.5*(V-E)
01210 IF F=0 THEN 1250
01220 LET G=-(SGN(F))
01230 LET G=G*Z/(SQR(Z^2+F^2))
01240 GO TO 1260
01250 LET G=-1
01260 LET H=G/(SQR(2*(1+SQR(1-G^2))))
01270 LET K=SQR(1-H^2)
```

```
01280 FOR I=1 TO N
01290 IF I=P THEN 1370
01300 IF I=Q THEN 1370
01310 LET C=R(I,P)
01320 LET F=R(I,Q)
01330 LET R(Q,I)=C*H+F*K
01340 LET R(I,Q)=R(Q,I)
01350 LET R(P,I)=C*K-F*H
01360 LET R(I,P)=R(P,I)
01370 LET C=B(I,P)
01380 LET F=B(I,Q)
01390 LET B(I,Q)=C*H+F*K
01400 LET B(I,P)=C*K-F*H
01410 NEXT I
01420 LET R(P,P)=V*K^2+E*H^2-2*Z*H*K
01430 LET R(Q,Q)=V*H^2+E*K^2+2*Z*H*K
01440 LET R(P,Q)=(V-E)*H*K+Z*(K^2-H^2)
01450 LET R(Q,P)=R(P,Q)
01460 NEXT P,Q
01470 IF D<>1 THEN 1500
01480 LET D=0
01490 GO TO 1130
01500 IF T>O THEN 1120
01510 FOR I=1 TO N
01520 LET Q(I)=I
01530 NEXT I
01540 LET J=0
01550 LET V1=0
01560 LET J=J+1
01570 FOR I=1 TO N-J
01580 IF R(I,I)>=R(I+1,I+1) THEN 1660
01590 LET V1=1
01600 LET V2=R(I,I)
01610 LET R(I,I)=R(I+1,I+1)
01620 LET R(I+1,I+1)=V2
01630 LET P=Q(I)
01640 LET Q(I)=Q(I+1)
01650 LET Q(I+1)=P
01660 NEXT I
01670 IF V1<>0 THEN 1550
01680 LET O=O1
01690 LET G=G9
01700 RETURN
01710 END
```

```
RAWD3           09:21           10-JAN-77
```

```
01000 7.5,20.1,42,10.9,20.4,43.2,10.8,17.9,40.4,12.2,18,84.3,5.1,15.4
01010 69.5,6.2,12.1,99.8,3,21,110.4,6.6,16.1,101.4,12.9,13.8,87.7,4
01020 18,115.2,15.2,11,94.7,11.7,13.5,70.8,9.1,7.8,87.5,7,10.9,125.6
01030 8.9,15.1,71,11.7,15.3,133.5,10.1,13.8,77.7,6.2,16,70.3,20.6,9.6
01040 121.9,20.8,4.1,101,17.1,5.8,83.3,15.1,12.1,121.4,16.3,8.9,105.4
01050 14.9,11.9,128.8,17.3,5.8,103.6,13.7,8.7,163.3,15.1,10,126.5

.R BASIC
```

```
READY
OLD TSTCOV
READY
RUN

TSTCOV         Ø9:25          1Ø-JAN-77

PROGRAM NAME --- TSTCOV
=====================================================
TOTAL NUMBER OF INDIVIDUALS N1 ?27
NUMBER OF VARIATES P ?3
NUMBER OF GROUPS K ?3
K-VALUED VECTOR OF GROUP SIZES ?1Ø,8,9

  ANALYSIS OF GROUP 1
    COVARIANCE MATRIX
 12.5218 -1.Ø7844 -52.36Ø9
-1.Ø7844  8.53511 -26.6524
-52.36Ø9 -26.6524  84Ø.434
    INV OF COV MAT
 Ø.12Ø714     4.29954E-2  8.88425E-3
 4.29954E-2  Ø.145355     7.2883ØE-3
 8.88425E-3  7.2883ØE-3  1.9745ØE-3
  DETERMINANT = 53538.4

  ANALYSIS OF GROUP 2
    COVARIANCE MATRIX
 8.32696 -1.18536  7.98125
-1.18536  7.85643 -12.6739
 7.98125 -12.6739  634.51
    INV OF COV MAT
 Ø.123638     1.66829E-2 -1.22196E-3
 1.66829E-2  Ø.133773     2.46219E-3
-1.22196E-3  2.46219E-3  1.64Ø57E-3
  DETERMINANT = 39Ø2Ø.

  ANALYSIS OF GROUP 3
    COVARIANCE MATRIX
 6.2525   -3.93958 -3Ø.7371
-3.93958  7.76278  34.7115
-3Ø.7371  34.7115  515.228
    INV OF COV MAT
 Ø.2649Ø3     9.12663E-2  9.65469E-3
 9.12663E-2  Ø.2158Ø2   -9.Ø9415E-3
 9.65469E-3 -9.Ø9415E-3  3.12955E-3
  DETERMINANT = 1Ø549.9

POOLED COV MAT
 9.2Ø853 -2.Ø6334 -27.5532
-2.Ø6334  8.Ø7972 -2.12Ø72
-27.5532 -2.12Ø72  671.971
INV OF POOLED COV MAT
 Ø.133227     3.54857E-2  5.57475E-3
 3.54857E-2  Ø.133321     1.87579E-3
 5.57475E-3  1.87579E-3  1.72266E-3
```

```
  DETERMINANT OF POOLED COV MAT = 40718.9
M= 8.63953
1/C = 0.817295
CHI SQUARE = 7.06105
THIS IS APPROXIMATELY DISTRIBUTED AS A CHI SQUARE VARIATE WITH 12 DF

TIME:  2.66 SECS.

READY
```

```
EMV            15:08          07-JAN-77

00010 PRINT "PROGRAM NAME --- EMV"
00020 REM--- THIS PROGRAM COMPUTES QUANTITIES BASED ON WHICH
00030 REM THE EQUALITY OF K GROUP MEAN VECTORS CAN BE TESTED.  THE
00040 REM INPUT DATA ARE STORED IN DISK FILE RAWD3 AS N SETS OF P
00050 REM NUMBERS. N IS THE TOTAL NUMBER OF INDIVIDUALS
00060 REM AND P THE NUMBER OF VARIATES.  THE FOLLOWING
00070 REM ASSUMPTIONS MUST BE MET:
00080 REM        (1)  INDIVIDUALS ARE ASSIGNED TO GROUPS
00090 REM             BASED ON CRITERIA INDEPENDENT FROM THE VARIATES
00100 REM             USED IN THE ANALYSIS
00110 REM        (2)  THE UNDERLYING DISTRIBUTION IS MULTIVARIATE
00120 REM             NORMAL
00130 REM        (3)  THE OBSERVATIONS ARE COMMENSURABLE
00140 REM        (4)  THE POPULATION COVARIANCE MATRICES ARE EQUAL
00150 PRINT "=================================================="
00160 FILES RAWD3
00170 DIM X(27,3),T(3,3),Y(27),C(3),G(3),N(3),Z(27)
00180 DIM H(3,3), E(3,3), R(3,3), B(3,3), Q(3)
00200 REM---EXPLANATION OF ARRAY SYMBOLS:
00220 REM       X - AN N*P ARRAY
00230 REM       T - A K*P ARRAY
00240 REM       Y,Z - N-VALUED VECTORS
00250 REM       C - A K-VALUED VECTOR
00260 REM       G - A P-VALUED VECTOR
00270 REM       N - A K-VALUED VECTOR
00280 REM       H,E,R,B - P*P ARRAYS
00290 REM       Q - A P-VALUED VECTOR
00310 REM---READ DATA, PERFORM ARCSIN TRANSFORMATION IF REQUIRED,
00320 REM   COMPUTE MARGINAL TOTALS
00330 PRINT "TOTAL NUMBER OF INDIVIDUALS N";
00340 INPUT N
00350 PRINT "NUMBER OF VARIATES P";
00360 INPUT P
00370 PRINT "NUMBER OF GROUPS K";
00380 INPUT K
00390 PRINT "K-VALUED VECTOR OF GROUP SIZES";
00400 MAT INPUT N
00410 PRINT "SET Z1 EQUAL TO 1 IF ARCSIN TRANSFORMATION REQUIRED";
00420 INPUT Z1
00430 PRINT "IF Z1=1 THEN SPECIFY NUMBER OF RANDOM TRIALS ELSE"
00440 PRINT "TYPE ANY NUMBER";
```

```
00450 INPUT Z2
00460 MAT Y= ZER
00470 FOR I=1 TO N
00480 FOR H=1 TO P
00490 READ #1,W
00500 IF Z1<>1 THEN 570
00510 IF W<Z2 THEN 530
00520 LET W=W-0.00001
00530 LET W=(W/Z2)^.5
00540 LET W=W/((1-W^2)^.5)
00550 LET X(I,H)=ATN(W)*(180/3.14159)
00560 GO TO 580
00570 LET X(I,H)=W
00580 LET Y(I)=Y(I)+X(I,H)
00590 NEXT H,I
00600 MAT Z=Y
00610 MAT C=ZER
00620 MAT G=ZER
00630 LET G=0
00640 FOR H=1 TO P
00650 LET U=0
00660 FOR J=1 TO K
00670 LET T(J,H)=0
00680 FOR I=1 TO N(J)
00690 LET U=U+1
00700 LET T(J,H)=T(J,H)+X(U,H)
00710 NEXT I
00720 LET C(J)=C(J)+T(J,H)
00730 LET G(H)=G(H)+T(J,H)
00740 NEXT J
00750 LET G=G+G(H)
00760 NEXT H
00770 REM---COMPUTE MEAN VECTORS
00780 PRINT
00790 PRINT "  PROFILE DATA"
00800 LET U=0
00810 FOR J=1 TO K
00820 PRINT "MEAN VECTOR OF GROUP";J
00830 FOR H=1 TO P
00840 PRINT T(J,H)/N(J);
00850 NEXT H
00860 PRINT
00870 NEXT J
00880 PRINT "SAMPLE MEAN VECTOR"
00890 FOR H=1 TO P
00900 PRINT G(H)/N;
00910 NEXT H
00920 PRINT
00930 PRINT "SAMPLE MEAN ";G/N
00940 REM---COMPUTE BETWEEN GROUPS SUMS OF SQUARES AND CROSS
00950 REM   PRODUCTS (MATRIX H)
00960 MAT H=ZER
00970 FOR R=1 TO P
00980 FOR S=R TO P
00990 FOR J=1 TO K
```

```
01000 LET H(R,S)=H(R,S)+(T(J,R)*T(J,S))/N(J)
01010 NEXT J
01020 LET H(R,S)=H(R,S)-(G(R)*G(S))/N
01030 LET H(S,R)=H(R,S)
01040 NEXT S,R
01050 PRINT "     H-MATRIX"
01060 MAT PRINT H;
01070 MAT E=ZER
01080 REM---COMPUTE WITHIN GROUP SUMS OF SQUARES AND CROSS
01090 REM    PRODUCTS (MATRIX E)
01100 FOR R=1 TO P
01110 FOR S=R TO P
01120 FOR I=1 TO N
01130 LET E(R,S)=E(R,S)+X(I,R)*X(I,S)
01140 NEXT I
01150 LET Y=0
01160 FOR J=1 TO K
01170 LET Y=Y+T(J,R)*T(J,S)/N(J)
01180 NEXT J
01190 LET E(R,S)=E(R,S)-Y
01200 LET E(S,R)=E(R,S)
01210 NEXT S,R
01220 PRINT "     E-MATRIX"
01230 MAT PRINT E;
01240 MAT R=H
01250 MAT H=INV(E)
01260 MAT E=R*H
01270 MAT R=E
01280 PRINT "     INV(E)"
01290 MAT PRINT H;
01300 PRINT "     H*INV(E)"
01310 MAT PRINT R;
01320 LET N1=N
01330 LET N=P
01340 LET K1=K
01350 REM---EIGENVECTORS PROCEDURE
01360 MAT B=IDN
01370 LET A=0.00000001
01380 LET C=0
01390 FOR I=2 TO N
01400 FOR J=1 TO I-1
01410 LET C=C+2*(R(I,J)^2)
01420 NEXT J,I
01430 LET Y=SQR(C)
01440 LET O=(A/N)*Y
01450 LET T=Y
01460 LET D=0
01470 LET T=T/N
01480 FOR Q=2 TO N
01490 FOR P=1 TO Q-1
01500 IF ABS(R(P,Q))<T THEN 1810
01510 LET D=1
01520 LET V=R(P,P)
01530 LET Z=R(P,Q)
01540 LET E=R(Q,Q)
```

```
01550 LET F=.5*(V-E)
01560 IF F=0 THEN 1590
01570 LET G=-(SGN(F))
01580 GO TO 1600
01590 LET G=-1
01600 LET G=G*Z/(SQR(Z^2+F^2))
01610 LET H=G/(SQR(2*(1+SQR(1-G^2))))
01620 LET K=SQR(1-H^2)
01630 FOR I=1 TO N
01640 IF I=P THEN 1720
01650 IF I=Q THEN 1720
01660 LET C=R(I,P)
01670 LET F=R(I,Q)
01680 LET R(Q,I)=C*H+F*K
01690 LET R(I,Q)=R(Q,I)
01700 LET R(P,I)=C*K-F*H
01710 LET R(I,P)=R(P,I)
01720 LET C=B(I,P)
01730 LET F=B(I,Q)
01740 LET B(I,Q)=C*H+F*K
01750 LET B(I,P)=C*K-F*H
01760 NEXT I
01770 LET R(P,P)=V*K^2+E*H^2-2*Z*H*K
01780 LET R(Q,Q)=V*H^2+E*K^2+2*Z*H*K
01790 LET R(P,Q)=(V-E)*H*K+Z*(K^2-H^2)
01800 LET R(Q,P)=R(P,Q)
01810 NEXT P,Q
01820 IF D<>1 THEN 1850
01830 LET D=0
01840 GO TO 1480
01850 IF T>O THEN 1470
01860 FOR I=1 TO N
01870 LET Q(I)=I
01880 NEXT I
01890 LET J=0
01900 LET V1=0
01910 LET J=J+1
01920 FOR I=1 TO N-J
01930 IF R(I,I)>=R(I+1,I+1) THEN 2010
01940 LET V1=1
01950 LET V2=R(I,I)
01960 LET R(I,I)=R(I+1,I+1)
01970 LET R(I+1,I+1)=V2
01980 LET P=Q(I)
01990 LET Q(I)=Q(I+1)
02000 LET Q(I+1)=P
02010 NEXT I
02020 IF V1<>0 THEN 1900
02030 FOR J=1 TO N
02040 LET K=Q(J)
02050 PRINT "ROOT"; J; "="; R(J,J)
02060 NEXT J
02070 LET K=K1
02080 LET P=N+1
02090 LET N=N1
```

```
02100 PRINT "  TEST ON THE EQUALITY OF GROUP MEAN VECTORS"
02110 PRINT "S=";(K-1);"OR";(P-1);"WHICHEVER IS SMALLEST"
02120 PRINT "M=";(ABS(K-P)-1)/2
02130 PRINT "N=";(N-K-P)/2
02140 PRINT "THETA=";R(1,1)/(1+R(1,1))
02150 END

RAWD3          15:10          07-JAN-77

01000 7.5,20.1,42,10.9,20.4,43.2,10.8,17.9,40.4,12.2,18,84.3,5.1,15.4
01010 69.5,6.2,12.1,99.8,3,21,110.4,6.6,16.1,101.4,12.9,13.8,87.7,4
01020 18,115.2,15.2,11,94.7,11.7,13.5,70.8,9.1,7.8,87.5,7,10.9,125.6
01030 8.9,15.1,71,11.7,15.3,133.5,10.1,13.8,77.7,6.2,16,70.3,20.6,9.6
01040 121.9,20.8,4.1,101,17.1,5.8,83.3,15.1,12.1,121.4,16.3,8.9,105.4
01050 14.9,11.9,128.8,17.3,5.8,103.6,13.7,8.7,163.3,15.1,10,126.5

.R BASIC

READY
OLD EMV
READY
RUN

EMV            15:18          07-JAN-77

PROGRAM NAME --- EMV
==================================================
TOTAL NUMBER OF INDIVIDUALS N ?27
NUMBER OF VARIATES P ?3
NUMBER OF GROUPS K ?3
K-VALUED VECTOR OF GROUP SIZES ?10,8,9
SET Z1 EQUAL TO 1 IF ARCSIN TRANSFORMATION REQUIRED ?0
IF Z1=1 THEN SPECIFY NUMBER OF RANDOM TRIALS ELSE
TYPE ANY NUMBER ?0

  PROFILE DATA
MEAN VECTOR OF GROUP 1
 7.92  17.28  79.39
MEAN VECTOR OF GROUP 2
 9.9875  12.925  91.3875
MEAN VECTOR OF GROUP 3
 16.7667  8.54444  117.244
SAMPLE MEAN VECTOR
 11.4815  13.0778  95.563
SAMPLE MEAN  120.122
    H-MATRIX
 396.096 -363.471  1657.22
-363.471  361.733 -1559.13
 1657.22 -1559.13  6985.9
    E-MATRIX
 221.005 -49.5202 -661.277
-49.5202  193.913 -50.8975
-661.277 -50.8975  16127.3
    INV(E)
 5.55111E-3  1.47857E-3  2.32282E-4
```

```
 1.47857E-3  5.55505E-3  7.81584E-5
 2.32282E-4  7.81584E-5  7.17777E-5
    H*INV(E)
 2.0463 -1.30392  0.182549
-1.84497  1.35017 -0.168066
 8.51681 -5.6647  0.764514

ROOT 1 = 12.1929
ROOT 2 = 0.361267
ROOT 3 =-8.39324
  TEST ON THE EQUALITY OF GROUP MEAN VECTORS
S= 2 OR 3 WHICHEVER IS SMALLEST
M= 0
N= 10
THETA= 0.924202

TIME:  4.83 SECS.

READY
```

```
MTFDT           13:45          08-MAR-77

00010 PRINT "PROGRAM NAME --- MOUNTFORD'S TESTS."
00020 REM--- THIS PROGRAM COMPUTES A TEST ON THE HYPOTHESIS THAT
00030 REM THERE IS NO DIFFERENCE IN AVERAGE SIMILARITY BETWEEN TWO
00040 REM GROUPS. THE INPUT DATA CONSIST OF THE UPPER HALF OF A
00050 REM SIMILARITY MATRIX (EXCLUDING VALUES IN THE PRINCIPAL DIAG-
00060 REM ONAL POSITIONS). THE DATA ARE READ FROM DISK FILE MOTFOD.
00070 PRINT "================================================"
00080 FILES MOTFOD
00090 DIM S(5,5),G(5)
00100 REM--- EXPLANATION OF ARRAY SYMBOLS:
00110 REM            S - AN N*N ARRAY. IF N1 IS THE SIZE OF THE FIRST
00120 REM                GROUP, N2 THE SIZE OF THE SECOND THEN N=N1+N2
00130 REM            G - AN N-VALUED VECTOR
00140 MAT S=ZER
00150 PRINT "GROUP SIZES N1,N2";
00160 INPUT N1,N2
00170 PRINT
00180 LET N=N1+N2
00190 LET U=N*(N-1)/2
00200 FOR J=1 TO N-1
00210 FOR K=J+1 TO N
00220 READ #1,S(J,K)
00230 LET S(K,J)=S(J,K)
00240 NEXT K,J
00250 LET S1=S2=S3=0
00260 FOR J=1 TO N1-1
00270 FOR K=J+1 TO N1
00280 LET S1=S1+S(J,K)
00290 NEXT K,J
00300 FOR J=N1+1 TO N-1
00310 FOR K=J+1 TO N
```

```
00320 LET S2=S2+S(J,K)
00330 NEXT K,J
00340 FOR J=1 TO N1
00350 FOR K=N1+1 TO N
00360 LET S3=S3+S(J,K)
00370 NEXT K,J
00380 LET A=S1/(N1*(N1-1)/2)
00390 LET B=S2/(N2*(N2-1)/2)
00400 LET C=S3/(N1*N2)
00410 LET Q=A+B-2*C
00420 LET S=(S1+S2+S3)/U
00430 PRINT "AVERAGE SIMILARITY:"
00440 PRINT "GROUP 1="A;" GROUP 2="B;" BETWEEN GROUPS 1 AND 2="C
00450 PRINT "UNION GROUP 1+2="S
00460 PRINT "Q="Q
00470 FOR J=1 TO N
00480 FOR K=1 TO N
00490 IF J=K THEN 510
00500 LET G(J)=G(J)+S(J,K)
00510 NEXT K
00520 LET G(J)=(G(J)-(N-1)*S)/(N-2)
00530 NEXT J
00540 PRINT "G VALUES"
00550 MAT PRINT G;
00560 LET S4=0
00570 FOR J=1 TO N-1
00580 FOR K=J+1 TO N
00590 LET S4=S4+(S(J,K)-S-G(J)-G(K))^2
00600 NEXT K,J
00610 PRINT "VARIANCE S^2="S4
00620 LET V2=(2*(N-1)*(N-2))/(N1*(N1-1)*N2*(N2-1))
00630 PRINT "V^2="V2
00640 PRINT "T="(((U-N-1)*Q^2/V2)/(S4-Q^2/V2))^0.5
00650 PRINT "DF="U-N-1
00660 PRINT "B="Q/((V2*S4)^0.5)
00670 END

MOTFOD          13:46           08-MAR-77

01000 39.4341,47.0039,22.0879,19.333,38.8828
01010 18.2515,15.9789,21.7693,19.0553,8.96148

.R BASIC

READY
OLD MTFDT
READY
RUN

MTFDT           13:48           08-MAR-77

PROGRAM NAME --- MOUNTFORD'S TESTS.
================================================
GROUP SIZES N1,N2 ?3,2
```

```
AVERAGE SIMILARITY:
GROUP 1= 41.7736  GROUP 2= 8.96148  BETWEEN GROUPS 1 AND 2= 19.4127
UNION GROUP 1+2= 25.0758
Q= 11.9098
G VALUES
 9.18521  4.08134  8.80268 -9.74436 -12.3249
VARIANCE S^2= 79.6159
V^2= 2
T= 5.71211
DF= 4
B= 0.943819

TIME:  0.89 SECS.

READY
```

```
SSSIM2          13:26          03-JAN-77

00010 PRINT "PROGRAM NAME --- SSSIM2"
00020 REM--- DENDROGRAMS ARE COMPUTED FOR P ENTITIES,
00030 REM THEIR TOPOLOGY MATRICES DETERMINED AND THESE MATRICES
00040 REM WRITTEN INTO DISK FILE TOPOL. THE DATA ARE GENERATED
00050 REM INTERNALLY IN THE PROGRAM BY A RANDOM NUMBER GENERATOR.
00060 REM THE FOLLOWING PARAMETERS HAVE TO BE SPECIFIED:
00070 REM          1. NUMBER OF QUADRATS,
00080 REM          2. NUMBER OF SPECIES, AND
00090 REM          3. ESTIMATED TOTAL NUMBER OF INDIVIDUALS PER SPECIES
00100 REM SSSIM2 IS FOLLOWED BY SSSIM3 TO DETERMINE AN
00110 REM EMPIRICAL DISTRIBUTION FOR I(M;O).
00120 PRINT "=================================================="
00130 RANDOMIZE
00140 FILES TOPOL
00150 SCRATCH #1
00160 DIM X(25,7), V(7), R(7), L(7), T(7,2), D(7,7)
00170 DIM Y(7,7), F(7,7), N(7,2), Q(7,7), S(7)
00190 REM---EXPLANATION OF ARRAY SYMBOLS:
00200 REM          X - AN N*P ARRAY
00210 REM          D, Y, F, Q - P*P ARRAYS
00220 REM          T, N - P*2 ARRAYS
00230 REM          V, R, L, S - P-VALUED VECTORS
00240 PRINT "NUMBER OF QUADRATS N";
00250 INPUT N
00260 LET K0=N
00270 PRINT "NUMBER OF SPECIES P";
00280 INPUT P
00290 LET N=P
00300 PRINT "SPECIES TOTALS (ELEMENTS IN VECTOR S)"
00310 MAT INPUT S
00320 PRINT "NUMBER OF DENDROGRAMS REQUIRED";
00330 INPUT Y2
00340 PRINT "SPECIFY SAMPLING RATIO";
00350 INPUT F3
00360 PRINT "TYPE 1 IF NORMALIZATION OF QUADRAT VECTORS "
```

```
00370 PRINT "REQUIRED ELSE TYPE ANY NUMBER OTHER THAN 1";
00380 INPUT Y8
00390 PRINT
00400 MAT R=CON
00410 MAT D=ZER
00420 MAT L=CON
00430 MAT Y=ZER
00440 MAT X=ZER
00450 FOR L=1 TO N
00460 LET Y(L,1)=L
00470 NEXT L
00480 MAT F=ZER
00490 LET M=K0
00500 GOSUB 2080
00510 REM---COMPUTE DISTANCES
00520 IF Y8=0 THEN 610
00530 FOR K=1 TO N
00540 LET S=0
00550 FOR Z=1 TO M
00560 LET S=S+X(Z,K)^2
00570 NEXT Z
00580 FOR Z=1 TO M
00590 LET X(Z,K)=X(Z,K)/SQR(S)
00600 NEXT Z,K
00610 FOR K=1 TO N-1
00620 FOR L=K+1 TO N
00630 LET S=0
00640 FOR Z=1 TO M
00650 LET S=S+(X(Z,K)-X(Z,L))^2
00660 NEXT Z
00670 LET D(K,L)=S
00680 LET D(L,K)=S
00690 NEXT L,K
00700 GOSUB 760
00710 GOSUB 1780
00720 LET C3=C3+1
00730 PRINT "FINISHED PART"C3
00740 IF C3<Y2 THEN 400
00750 STOP
00760 LET A=N9=0
00770 LET K=N
00780 REM---CLUSTERING ROUTINE
00790 FOR M=1 TO K
00800 LET Y=U=10000000
00810 FOR L=1 TO K
00820 IF M<K THEN 840
00830 IF L=K THEN 930
00840 IF M<>L THEN 860
00850 LET L=L+1
00860 LET F=R(M)+R(L)
00870 LET Z=(D(M,L)+D(M,M)+D(L,L))/F
00880 LET Z=Z-D(M,M)/R(M)-D(L,L)/R(L)
00890 IF Z>=U THEN 920
00900 LET U=Z
00910 LET H=L
```

```
00920 NEXT L
00930 FOR L=1 TO K
00940 IF H<K THEN 960
00950 IF L=K THEN 1050
00960 IF L<>H THEN 980
00970 LET L=L+1
00980 LET F=R(H)+R(L)
00990 LET Z=(D(H,L)+D(H,H)+D(L,L))/F
01000 LET Z=Z-D(H,H)/R(H)-D(L,L)/R(L)
01010 IF Z>=Y THEN 1040
01020 LET Y=Z
01030 LET B=L
01040 NEXT L
01050 IF B<>M THEN 1270
01060 LET T(M,1)=M
01070 LET T(M,2)=H
01080 LET V(M)=R(M)+R(H)
01090 LET A1=R(M)
01100 LET A2=R(H)
01110 FOR L=1 TO R(M)
01120 LET A2=A2+1
01130 LET Y(H,A2)=Y(M,L)
01140 NEXT L
01150 FOR L=1 TO R(H)
01160 LET A1=A1+1
01170 LET Y(M,A1)=Y(H,L)
01180 NEXT L
01190 IF M>H THEN 1260
01200 LET N9=N9+1
01210 FOR L=1 TO V(M)
01220 LET F(N9,L)=Y(M,L)
01230 NEXT L
01240 LET N(N9,1)=V(M)
01250 LET N(N9,2)=(D(M,M)+D(H,H)+D(M,H))/V(M)
01260 GO TO 1300
01270 LET T(M,1)=M
01280 LET T(M,2)=0
01290 LET V(M)=R(M)
01300 NEXT M
01310 LET W=0
01320 FOR M=1 TO K
01330 IF T(M,2)>=T(M,1) THEN 1350
01340 IF T(M,2)>0 THEN 1430
01350 LET W=W+1
01360 LET T(W,1)=T(M,1)
01370 LET T(W,2)=T(M,2)
01380 LET R(W)=V(M)
01390 FOR L=1 TO K
01400 IF Y(M,L)=0 THEN 1420
01410 LET Y(W,L)=Y(M,L)
01420 NEXT L
01430 NEXT M
01440 LET K=W
01450 LET A=A+1
01460 FOR M=1 TO K
```

```
01470 LET J=T(M,1)
01480 LET E=T(M,2)
01490 IF E=0 THEN 1520
01500 LET D(M,M)=D(J,J)+D(E,E)+D(J,E)
01510 GO TO 1530
01520 LET D(M,M)=D(J,J)
01530 LET Y=D(M,M)/R(M)
01540 LET Z=Y/R(M)
01550 NEXT M
01560 FOR M=1 TO K-1
01570 LET J=T(M,1)
01580 LET E=T(M,2)
01590 FOR L=M+1 TO K
01600 LET G=T(L,1)
01610 LET C=T(L,2)
01620 IF E+C=0 THEN 1660
01630 IF E+C=C THEN 1680
01640 IF E+C=E THEN 1700
01650 GO TO 1720
01660 LET Z=D(J,G)
01670 GO TO 1730
01680 LET Z=D(J,G)+D(J,C)
01690 GO TO 1730
01700 LET Z=D(J,G)+D(E,G)
01710 GO TO 1730
01720 LET Z=D(J,G)+D(J,C)+D(E,G)+D(E,C)
01730 LET D(M,L)=Z
01740 LET D(L,M)=Z
01750 NEXT L,M
01760 IF K>1 THEN 780
01770 RETURN
01790 PRINT "TOPOLOGY MATRIX"
01800 FOR H=1 TO N-1
01810 FOR L=H+1 TO N
01820 LET A1=K1=A2=K2=0
01830 FOR E=1 TO N
01840 LET K4=K5=0
01850 FOR J=1 TO N
01860 IF F(E,J)=0 THEN 1960
01870 IF F(E,J)<>H THEN 1910
01880 LET A1=A1+1
01890 LET K1=1
01900 LET K4=1
01910 IF F(E,J)<>L THEN 1950
01920 LET A2=A2+1
01930 LET K2=1
01940 LET K5=1
01950 NEXT J
01960 IF K1+K2<2 THEN 1990
01970 LET Q(H,L)=A1+A2-1
01980 GO TO 2000
01990 NEXT E
02000 NEXT L
02010 NEXT H
02020 MAT PRINT Q;
```

```
Ø2Ø3Ø FOR H=1 TO N-1
Ø2Ø4Ø FOR J=H+1 TO N
Ø2Ø5Ø WRITE #1,Q(H,J)
Ø2Ø6Ø NEXT J,H
Ø2Ø7Ø RETURN
Ø2Ø8Ø REM---RANDOM DATA GENERATOR
Ø2Ø9Ø LET F=F3
Ø21ØØ LET P=N
Ø211Ø LET Q=M
Ø212Ø FOR H=1 TO P
Ø213Ø LET N7=Ø
Ø214Ø LET R7=INT(RND*1ØØ)+1
Ø215Ø IF R7>Q THEN 214Ø
Ø216Ø LET N7=N7+1
Ø217Ø IF N7>F*S(H) THEN 22ØØ
Ø218Ø LET X(R7,H)=X(R7,H)+1
Ø219Ø GO TO 214Ø
Ø22ØØ NEXT H
Ø221Ø RETURN
Ø222Ø END

.R BASIC

READY
OLD SSSIM2
READY
RUN

SSSIM2          13:27          Ø3-JAN-77

PROGRAM NAME --- SSSIM2
=================================================
NUMBER OF QUADRATS N ?25
NUMBER OF SPECIES P ?7
SPECIES TOTALS (ELEMENTS IN VECTOR S)
 ?1994,473,48,37,21,12,8
NUMBER OF DENDROGRAMS REQUIRED ?5
SPECIFY SAMPLING RATIO ?1
TYPE 1 IF NORMALIZATION OF QUADRAT VECTORS
REQUIRED ELSE TYPE ANY NUMBER OTHER THAN 1 ?1
TOPOLOGY MATRIX

 Ø   1   2   3   3   4   3

 Ø   Ø   2   3   3   4   3

 Ø   Ø   Ø   2   2   3   2

 Ø   Ø   Ø   Ø   1   2   1

 Ø   Ø   Ø   Ø   Ø   1   1

 Ø   Ø   Ø   Ø   Ø   Ø   1

 Ø   Ø   Ø   Ø   Ø   Ø   Ø
```

FINISHED PART 1
TOPOLOGY MATRIX

Ø	1	3	2	4	5	6
Ø	Ø	3	2	4	5	6
Ø	Ø	Ø	2	2	3	4
Ø	Ø	Ø	Ø	3	4	5
Ø	Ø	Ø	Ø	Ø	2	3
Ø	Ø	Ø	Ø	Ø	Ø	2
Ø	Ø	Ø	Ø	Ø	Ø	Ø

FINISHED PART 2
TOPOLOGY MATRIX

Ø	1	2	4	3	4	4
Ø	Ø	2	4	3	4	4
Ø	Ø	Ø	3	2	3	3
Ø	Ø	Ø	Ø	2	1	1
Ø	Ø	Ø	Ø	Ø	2	2
Ø	Ø	Ø	Ø	Ø	Ø	1
Ø	Ø	Ø	Ø	Ø	Ø	Ø

FINISHED PART 3
TOPOLOGY MATRIX

Ø	1	2	3	4	5	6
Ø	Ø	2	3	4	5	6
Ø	Ø	Ø	2	3	4	5
Ø	Ø	Ø	Ø	2	3	4
Ø	Ø	Ø	Ø	Ø	2	3
Ø	Ø	Ø	Ø	Ø	Ø	2
Ø	Ø	Ø	Ø	Ø	Ø	Ø

FINISHED PART 4
TOPOLOGY MATRIX

Ø	1	2	3	4	1	1

```
 0  0  2  3  4  1  1

 0  0  0  2  3  1  1

 0  0  0  0  2  1  1

 0  0  0  0  0  1  1

 0  0  0  0  0  0  1

 0  0  0  0  0  0  0

FINISHED PART 5

TIME:  39.06 SECS.

READY
```

```
SSSIM3          15:43          03-JAN-77

00010 PRINT "PROGRAM NAME --- SSSIM3"
00020 REM--- IF RUN AFTER SSSIM2 THIS PROGRAM COMPUTES AN
00030 REM EMPIRICAL DISTRIBUTION FOR I(M;O).
00040 PRINT "=================================================================="
00050 FILES TOPOL
00060 DIM T(5,21), I(10), O(10)
00070 REM---EXPLANATION OF ARRAY SYMBOLS:
00090 REM          T - A K*N ARRAY WHERE N=Q*(Q-1)/2, Q IS
00110 REM              THE ORDER AND K THE NUMBER OF TOPOLOGY MATRICES
00120 REM          I, O - K*(K-1)/2-VALUED VECTOR
00150 REM---READ DATA
00160 PRINT "ORDER OF A TOPOLOGY MATRIX Q";
00170 INPUT Q
00180 PRINT "NUMBER OF TOPOLOGY MATRICES K";
00190 INPUT K
00200 LET N=Q*(Q-1)/2
00210 FOR H=1 TO K
00220 FOR J=1 TO N
00230 READ #1,T(H,J)
00240 NEXT J,H
00250 LET C=0
00260 FOR H=1 TO K-1
00270 FOR J=H+1 TO K
00280 LET I=0
00290 FOR E=1 TO N
00300 LET M=(T(H,E)+T(J,E))/2
00310 LET I=I+T(H,E)*LOG(T(H,E)/M)
00320 LET I=I+T(J,E)*LOG(T(J,E)/M)
00330 NEXT E
00340 LET C=C+1
00350 LET I(C)=I
00360 NEXT J,H
00370 LET C=K*(K-1)/2
```

```
ØØ38Ø LET R=C+1
ØØ39Ø FOR H=1 TO C
ØØ40Ø LET M=Ø
ØØ41Ø FOR J=1 TO C
ØØ42Ø IF M>I(J) THEN 45Ø
ØØ43Ø LET M=I(J)
ØØ44Ø LET Q=(J)
ØØ45Ø NEXT J
ØØ46Ø LET R=R-1
ØØ47Ø LET O(R)=M
ØØ48Ø LET I(Q)=-I(Q)
ØØ49Ø NEXT H
ØØ5ØØ PRINT "ORDERED I(M,O) VALUES"
ØØ51Ø MAT PRINT O;
ØØ52Ø END

TOPOL          15:44          Ø3-JAN-77

Ø1ØØØ 1,2,3,3,4,3,2,3,3,4,3,2,2,3,2,1,2,1,1,1,1,1,3,2,4,5
Ø1Ø1Ø 6,2,2,3,4,3,4,5,2,3,2,1,2,4,3,4,4,2,4,3,4,4,3,2,3,3
Ø1Ø2Ø 2,1,1,2,2,1,1,2,3,4,5,6,2,3,4,5,6,2,3,4,5,6,2,3,4.5
Ø1Ø3Ø 2,3,4,2,3,2,1,2,3,4,1,1,2,3,4.1,1,2,3,1,1,2,1,1,1,1,1

.R BASIC

READY
OLD SSSIM3
READY
RUN

SSSIM3         15:45          Ø3-JAN-77

PROGRAM NAME --- SSSIM3
==========================================================
ORDER OF A TOPOLOGY MATRIX Q ?7
NUMBER OF TOPOLOGY MATRICES K ?5

ORDERED I(M,O) VALUES

 Ø.858753  1.16765  3.Ø8399  4.2Ø792  4.25Ø91  4.5Ø119  5.1982
 5.72924  11.6427  12.2468

TIME:  1.24 SECS.
```

GLOSSARY

Absolute value function: A metric with functional form described in Chapter 2. See **Metric.**

Additivity: A property of commensurable and independent variables. See **Commensurability.**

Agglomerative: A classification which combines individuals into groups.

a posteriori: Based on observation.

a priori: Deduced from basic principles or derived from external sources prior to the survey or experiment.

Association: The tendency toward common occurrence (*positive association*) or mutual exclusion (*negative association*).

Attribute: In this book, a character. See **Character.**

Average linkage: A classification in which average similarity represents the clustering criterion.

Binomial: A probability distribution completely described by the expansion of $(p + q)^n$ where $p + q = 1$ and n indicates the number of random trails. Often assumed when the data represent frequencies. See **Distribution.**

Bray & Curtis distance: A potential semimetric, with changing scale of measure, derived from the absolute value function. Its functional form is given in Chapter 2.

Categorical: Data consisting of zeros or positive integer numbers.

Centroid: The tip of a population or sample mean vector. See **Mean vector.**

Centroid clustering: A classification which combines groups based on the nearness of centroids.

Character: In this book, a distinctive feature which may take on different values or states, or may be constant.

Chord distance: The length of the chord connecting two points on the circle or some sphere. Its functional form is given in Chapter 2.

Classification: The process which partitions a sample into groups.

Clumping: Formation of overlapping groups.

Cluster: Any arbitrary group of points,

Coherence coefficient: The square root of the one complement of the square of **Rajski's metric.** Its values are between zero and one. Its functional form is given in Chapter 2.

Complete enumeration: A survey in which all elements of a population are located and measured. See **Population.**

Component analysis: The method of summarizing linear continuous data structures on orthogonal axes. See **Orthogonal.**

Commensurability: A property of qualitatively comparable variables whose unit of measure is the same.

Consistent procedure: A logical sequence of statements indicating operations.

Continuity: A property of data structures which can be represented by a single point cluster of even density.

Continuity analysis: A method for summarizing non-linear continuous data structures. See **Continuity, Data structure.**

Continuous variable: One which can assume any value on the real number axis within natural limits. See **Variable.**

Correlation: (a) Any expression of relatedness. (b) A scalar (inner) product of normalized vectors with origin in the centroid. See **Scalar product, Normalization, Centroid.**

Covariance: The scalar (inner) product of two vectors with origin in the centroid. The vectors are not normalized. See **Scalar product, Centroid, Normalization.**

***D*-technique**: The *D*-algorithm of component analysis described in Chapter 3.

Data structure: The manner in which sample points are spatially arranged, e.g., linear, curved, continuous, disjoint, etc.

Dendrogram: A classification tree indicating hierarchical relationships.

Density: Number of individuals per unit area or data points per unit volume.

Density function: Co-ordinate function of the normal distribution. See **Normal.**

Dependence: A property of variables responding to the influence of others.

Deterministic: (a) A mathematical model without random components. (b) The approach concerned with an exact determination of population parameters.

Direct gradient analysis: A method for displaying ecological information by arranging species or vegetation stands within an environmentally defined reference system. See **Ordination.**

Discrete variable: Whose values are positive integer numbers or zero.

Discriminant analysis: A family of statistical methods which use the generalized distance or other measures of affinity to distinguish between groups, or to find the most likely parent population for an individual or a group. See **Identification.**

Dissection: Partition of a continuous sample into groups. See **Continuity.**

Distortion: Scrambling in the data structure due to transformations or other adjustments in the resemblance function or method of analysis.

Distribution: A statistical concept for describing the frequencies with which different values or states occur in a variable.

Distribution function: The area function (integral) of a distribution. See **Normal.**

Divergence: A functional expression of difference between distributions. See **Distribution.**

Efficiency: In this book, the extent to which an ordination or classification can account for variation in the data.

Equivocation: An information quantity representing the difference between joint and mutual information. Its functional form is given in Chapter 2. See **Joint information, Mutual information.**

Empirical distribution: Frequency distribution derived on the basis of an experiment.

Entropy: The average information per observation. Its functional form is given in Chapter 2.

Estimate. A sample value which replaces the unknown population value in statistical analysis.

Euclidean: The geometry or space in which vector scalar (inner) products, such as the covariance, represent a meaningful concept. See **Euclidean distance, Covariance**.

Euclidean distance: The common notion of distance as we know it from everyday experience. Functional forms are given in Chapter 2.

Euclidean space: See **Euclidean.**

Formal: Objective methods relying on mathematics and probability.

Function: (a) The changing structure. (b) An algebraic formulation. See **Structure.**

Gaussian: Bell-shaped. See **Normal.**

Generalized distance: A distance in standard units. Functional forms are given in Chapters 2 and 5.

Gengerelli's distance: Euclidean distance computed from oblique coordinates. Its functional form is given in Chapter 2.

Geodesic metric: The length of the shorter arc between two points. The functional form is given in Chapter 2.

Hierarchy: A nested clustering in which increasingly larger (or smaller) groups are formed by combining (or breaking up) groups in successive steps.

Identification: The process of finding the most likely parent population for an individual.

***I*-divergence**: A one-way information divergence such as $\mathbf{F}^0 \rightarrow \mathbf{F}$ where **F** is an observed distribution and $\mathbf{F}^0$ is a specified standard. See Chapter 2 for functional forms.

Importance value: In this book, the amount of sum of squares or other quantity specifically accounted for by a species in the sample.

Independence: Lack of correlation, often meaning a zero covariance.

Indirect gradient analysis: Ordinations which use vegetation data to infer about trends in environmental variation. See **Ordination.**

Individual: The basic unit subjected to analysis.

Informal: In this book, methods not based on mathematical formulations.

Information: Physical property of data; surprisal value; a multiple of entropy; or a divergence. See **Divergence, Entropy**.

Information clustering: Grouping together objects based on information content. See **Information**.

Inverse: A matrix derived from another matrix by certain algebraic manipulations. See Chapter 4 for example.

***J*-divergence**: A two-way information divergence such as $\mathbf{F}_1 \leftrightharpoons \mathbf{F}_2$ where $\mathbf{F}_1$ and $\mathbf{F}_2$ are two observed distributions. Functional form is given in Chapter 2. See **Information, Distribution**.

Joint frequency: A property of joint observations.

Joint information: A multiple of the entropy in joint frequencies. Functional form is given in Chapter 2. See **Information, Joint frequency**.

Latent structure analysis: See Chapter 4 for reference.

Linear: In this book, (a) a data structure characterized by points falling within an ellipse or some ellipsoid, (b) a relationship between variables.

Mahalanobis distance: See **Generalized distance.**

Matching coefficient: The one complement of the squared Euclidean distance computed from presence data expressed as proportions. See Chapter 2 for functional form.

Matrix: A two-dimensional array of numbers.

Mean vector: Whose elements represent the mean values of variates.

Metric: Which satisfies the metric space axioms; a distance. See Chapter 2 for functional forms.

Metric space: A set of points and a metric which defines relative spatial placement.

Metric space axioms: Described in Chapter 2.

Minimum discrimination information statistic: An *I*-divergence with functional forms given in Chapter 2.

Model: A set of statements in a formal algebraic language.

Monothetic: A clustering method in which groups are recognized on the basis of the presence of one or several common characters; decisions based on a single species (variable). See **Character.**

Multidimensional scaling: A method of ordination which derives co-ordinates from a matrix of similarity (dissimilarity) values. See **Ordination.**

Multivariate: Incorporating several correlated variables.

Mutual informations: Shared information, common between two distributions. Functional forms given in Chapter 2. See **Information.**

Normal: A probability distribution characterized by a bell-shaped curve. Often assumed when the data represent measurements. Functional forms are given in Chapter 1. See **Distribution.**

Normalization: Adjustment of elements in a vector in such a way that their sum of squares is unity.

Object: See **Individual.**

Ochiai coefficient: The scalar (inner) product of normalized vectors of binary elements. Functional form is given in Chapter 2. See **Normalization, Scalar Product.**

Ordination: Ordering points on axes to achieve different objectives such as summarization of variation, multidimensional scaling, trend seeking, or reciprocal ordering. Descriptions are given in Chapter 3.

Ordination efficiency: See **Efficiency**.

Orthogonal: In this book, a relationship of variables characterized by zero covariance.

Poisson: A probability distribution assumed for counts under specific circumstances. Functional form is given in Chapter 1. See **Distribution.**

Poisson-logarithmic: A probability distribution for counts in certain aggregated populations. See Chapter 1 for reference, and **Distribution.**

Poisson-Poisson: A probability distribution for counts in certain aggregated populations. See Chapter 1 for reference, and **Distribution.**

Polythetic: A classification based on the degree of resemblance between objects; decision based on two or more species (variables). See **Monothetic.**

Population: (a) The totality of individuals sufficiently alike to be classified as members of the same group. (b) All possible states or values of a variable characterized by some probability distribution. See **Variable.**

Population unit: The basic unit distinguished in sampling. See **Individual.**

Position vector: A directed line from the origin of the co-ordinate system to a given point in space.

Position vectors ordination: An ordination which manipulates vectors. See Chapter 2 for description, and **Ordination.**

Prediction: A statistical statement about an unknown state or value of a variable.

Preferential sampling: Subjective selection of units for description. Not a statistical method.

Probability: The likelihood of a value or state.

Probabilistic: (a) A mathematical model with random components. (b) Methods of analysis concerned with estimation and statistical inference.

Probability theory: All the notions about probability.

Q-technique: In this book the Q-algorithm of component analysis; Q-clustering described in Chapters 3, 4.

Rajski's metric: A distance based on the equivocation information. Functional form is given in Chapter 2. See **Equivocation.**

Random: According to chance.

Random point: Determined by random co-ordinates.

Random sampling: In which every individual is given an equal chance to get into the sample. See **Sample.**

Randomly sited systematic sample: A systematic sample in which the pivot point is randomly sited.

Ranking: In this book, ordering species in terms of their specific share of the total sum of squares, specific variance or mutual information. See Chapters 1, 2 for descriptions.

Reciprocal ordering: Ordering quadrats based on species scores, or species based on quadrat scores. See descriptions in Chapter 3.

Regression analysis: Fitting a line or a surface to points in space.

Resemblance: The likeness or unlikeness of objects measured in terms of the characters which they possess. See **Resemblance function.**

Resemblance function: A mathematical formulation to measure the similarity of objects. See descriptions in Chapter 2.

Resemblance structure: A property of samples or populations manifested in a matrix of similarity or dissimilarity values.

Restricted random: Random sampling within compartments. See Chapter 1 for reference.

RQ-technique: See **Reciprocal ordering.**

Sample: A subset of the population.

Sample space: A set of points and a resemblance function which defines their relative spatial placement. See **Resemblance function.**

Sampling unit: See population unit.

Scalar product of vectors: An inner product of vectors, i.e. the product of their lengths and the cosine of the enclosed angle.

Shannon's entropy: See **Entropy.**

Stochastic: Which accords with some probability law.

Stress: The departure of resemblance structure from a standard.
Structure: See resemblance structure.
Subdivisive: Clustering by subdivisions.
Successive clustering: Clustering in which a group is completely formed be-re the formation of another begins.
Sum of squares: Sum of squared deviations.
Symbols:

→ or ←	One-way divergence.
⇌	Two-way divergence.
:=	Becomes.

Sytematic: Sampling in which individuals are taken at regular intervals.
Topology: In this book, the pattern of fusions in a dendrogram. See **Dendrogram**.
Topology matrix: In this book, a matrix of fusion counts. See example in Chapter 4, and **Topology**.
Transpose: A matrix derived by interchanging the rows and columns of another matrix.
Type: An abstraction; equivalent to population. See **Population** (a).
Univariate: Incorporating a single variable.
Variable: A property, character, that varies. See **Character**.
Variate: In this book, a variable. See **Variable**.
Vegetation type: See **Type**.

AUTHOR INDEX

(When several variants of a name exist in the reference list due to inconsistent use of initials by authors, the one believed to be correct is retained in the index.)

Page numbers

SUBJECT INDEX